Other Homes and Garbage

Other Homes and Garbage

Designs for Self-sufficient Living

by Jim Leckie, Gil Masters, Harry Whitehouse and Lily Young

SIERRA CLUB BOOKS ~ SAN FRANCISCO

1975

Library of Congress Cataloging in Publication Data
Main entry under title:

Other homes and garbage.

Includes bibliographies and index.
1. Environmental engineering. I. Leckie, James O., 1939-
TA170.B47 620.8 75-8913
ISBN 0-87156-141-7

Design by Joe Fay

Cover design by Anita Scott

Pen and ink illustrations by Bonnie Russell

Production by Charlsen + Johansen & Others

Printed in the United States of America

The Sierra Club, founded in 1892 by John Muir, has devoted itself to the study and protection of the nation's scenic and ecological resources—mountains, wetlands, woodlands, wild shores and rivers. All Club publications are part of the nonprofit effort the Club carries on as a public trust. There are some 50 chapters coast to coast, in Canada, Hawaii, and Alaska. Participation is invited in the Club's program to enjoy and preserve wilderness everywhere.
Address: 530 Bush Street, San Francisco, California 94108.

Acknowledgments:

The assistance of many fine people was essential in putting this effort together. For figures and diagrams, we are grateful to Mui Ho for those in Chapter 2 (Architecture), Jay Watkins for his imaginative help in Chapter 4 (Solar heating) and Irene Findikakis for her monumental efforts in all the chapters. We thank Bonnie Russell for the fine illustrations, Linda Gunnarson and Jon Goodchild for their help with the final stages of production, and Wendy Goldwyn for the coordination of all phases. Lastly, our thanks go to Larry Miller for his herculean achievement in copy editing this book which he carried out with patience, perseverance and good humor.

J.O. Leckie
G. Masters
H. Whitehouse
L.Y. Young

Grateful acknowledgment is also made to the following:

The American Society of Civil Engineers, for permission to reprint material from *Transactions of the American Society of Civil Engineers*, Volume 122, Paper No. 2849.

The American Society of Heating, Refrigerating and Air-conditioning Engineers, Inc., for permission to reprint material from *ASHRAE Guide and Data Book*, 1967; *ASHRAE Guide: Design and Evaluation Criteria for Energy Conservation in New Buildings (Proposed Standard 90-P)*, 1974; *ASHRAE Transactions*, Vol. 80, Part II, 1974; *ASHRAE Guide: Systems*, 1970; *Handbook of Fundamentals*, 1967.

The Chemical Rubber Co., for permission to reprint material from *Standard Mathematical Tables*, Twenty-first edition. © 1973 The Chemical Rubber Co.

Dunham-Bush, Inc., West Hartford, Conn., for permission to reprint material from Dunham-Bush Form No. 6001-2.

Industrial Press, Inc., N.Y., for permission to reprint material from *Handbook of Air Conditioning, Heating and Ventilating* by Strock & Koral, 1965.

McGraw-Hill Book Co., for permission to reprint material from *Introduction* to the *Utilization of Solar Energy* by Zarem & Erway. © 1963 by McGraw-Hill Book Company.

The New Alchemy Institute and Richard Merrill and John Fry, for permission to reprint material from *Methane Digesters for Fuel Gas and Fertilizer* by Richard Merrill and John Fry. © Richard Merrill and John Fry.

Portola Institute, Menlo Park, Ca., and Kim Mitchell for permission to reprint material by Kim Mitchell from *The Energy Primer*. © 1974 by Portola Institute.

The Rodale Press, Emmaus, Pa., for permission to adapt and reprint material from *The Complete Book of Composting* by J.I. Rodale. © 1966 by J.I. Rodale; *The Encyclopedia of Organic Gardening* by J.I. Rodale. © 1973 by J.I. Rodale; *How to Grow Fruits and Vegetables by the Organic Method* by J.I. Rodale. © 1966 by J.I. Rodale; *Organic Gardening and Farming*, February 1972. © 1972 by the Rodale Press.

Slant/Fin Corporation, Greenville, N.Y., for permission to reprint material from a series of data sheets published by Slant/Fin Corporation. © 1966 by Slant/Fin Corporation.

Contents

Small-Scale Generation of Electricity From Renewable Energy Sources by Gil Masters, with Angelos Findikakis 32

Solar Heating by Harry Whitehouse 74

Contents

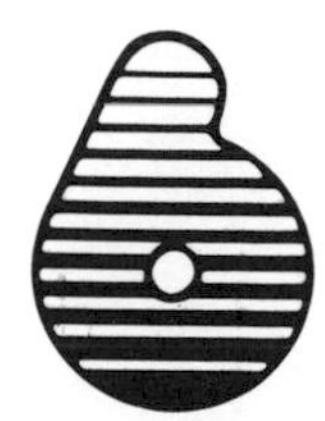

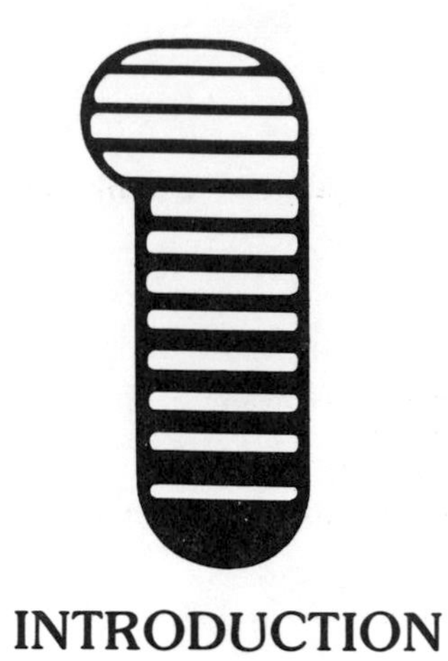

INTRODUCTION

Human needs and attitudes and the earth's potential

Self-sufficiency and self-reliance: changing current trends

Creating your own living environment

Designing to fit your specific requirements

Fitting engineering to small-scale individual needs

Where We Are Going

WHAT THIS BOOK IS ABOUT

The book adventure on which you are about to embark owes its existence to a rather unusual set of circumstances. It may be of some interest (and perhaps provide some encouragement) for us to spend a few moments here describing the birth process. During the fall quarter, 1973, a group of us from the School of Engineering at Stanford began talking about organizing a small experimental workshop on self-sufficient living systems. We offered the course in the winter quarter through Stanford's SWOPSI (Stanford Workshop on Social and Political Issues) but, instead of the twenty or twenty-five students that we expected, over one hundred interested students and community people showed up on the first night. Consequently, we had to redesign our small workshop framework into sub-courses centered around the themes of the various chapters found in this book, with the general dissemination of information to occur through the format of a final consolidated report. It was on the basis of the initial rough sub-course reports that this book had its beginning. The original material has been reworked considerably, and a substantial amount of new material has been added by the coauthors. We want to acknowledge the student energy and enthusiasm which carried the original idea to fruition.

This book represents an attempt by engineers and other technically trained people to communicate practical, useful technical information in an interesting format and in terms that are comprehensible to nonspecialized people—you! We want to help you gain the ability to design—for and by yourself—technologies which will allow you to establish a lifestyle which is energetically and materially more conservative than those most of us now lead. We hope to help people realize that such simplified lifestyles are possible without sacrificing the things that give quality to our lives.

> *If I am content with little, then enough is as good as a feast.*—Dean Swift

It is inherent in human nature to want to be self-sufficient and self-reliant. Modern society has removed many of the opportunities for self-reliance by burying in technological jargon and terminology much of the information needed by nontechnical people for development of intelligent choices. We are here trying to remove many of the artificial barriers which can deter you from designing your own methane digester, solar heater, or whatever. What we *cannot* remove is your expense of time and energy to acquire the necessary information for alternative choices. It may be of some help if we develop some of our guiding philosophy to set the general context of our overall effort.

> *The distance is nothing; it is only the first step that is difficult.*—Marquise du Deffaud

WHERE ARE WE COMING FROM?

In modern times we have treated the earth as if its reserves of usable matter and energy were in never-ending supply, and as if time would quickly heal all wounds caused by our exploitation. As long as the human population was small and our activities constrained by limited access to energy, the damage we inflicted upon the environment was limited and usually correctable through the earth's natural cycles. Now, however, the demands made on the environment are often beyond nature's regenerative capacities. Humankind's narrow understanding of conservation and our shortsighted technological approach to satisfying only our immediately perceived needs have begun to seriously deplete stored reserves. Fossil fuels are being used at increasing rates, and the steady dwindling of accessible supplies is becoming apparent. Vast quantities of water are being used injudiciously and contaminated before their return to large water systems; pollution has nearly destroyed such rivers as the Hudson and such lakes as Erie. Even more far-reaching effects are being realized as the damage makes its way through food chains and into groundwaters. The Aswan Dam, built as a necessary instrument for Egypt's economic growth, prevents the Nile from depositing its rich silt on the surrounding agricultural valley during seasonal floods. As a result the Aswan reservoir is filling with sediment, human parasites are increasing in the stored water, crops downstream are suffering, and the Mediterranean Sea is lacking a major nutrient source. What is the real gain in situations like these?

To be self-sufficing is the greatest of all wealth. —Porphyry

The earth's ecosystems are balanced and its resources are finite. With our present consumptive attitudes, fossil fuels *will* run out. Water systems *will* fail from pollution. Fertile soil *will* deteriorate and be eroded into sterile desert. These effects *will* alter catastrophically every dependent organic system, including our own.

It is true that more fossil fuels can form over the next few millions of years, and many damaged forests and lakes can heal in less than a thousand. But on a time-scale of human dimensions we must look toward shorter regeneration cycles to supply a larger part of our energy needs. All regenerative cycles, including those for water and fossil fuels, are dependent upon steady radiation from the sun. Each system thrives only when this energy is used efficiently. Specialization and biosimplification—monoculturing, for example—decrease efficiency since a single species cannot utilize all available energy as effectively as a diversified community. So, while "excellent rangeland" in the U.S. may yield 5000 kilograms of cattle per square kilometer per year, an African savannah community may yield closer to 35,000 kilograms of large edible animals in the same area, and the American bison herds of two hundred years ago had an even larger yield. Yet the savannah and prairie communities, including the thriving indigenous flora, were self-sustaining and did not require extensive artificial energy investments to continue to be productive.

If there is sufficient energy for a large natural community to maintain a stable existence, then we also can have enough energy if only we enter into a rational, ecological relationship with the earth. We must recognize the complexity of the natural world and acknowledge the limits of our understanding. The natural environment in its diversity can be viewed as a unique library of genetic information. From this library can be drawn new food crops, new drugs and vaccines, and new biological pest controls. The loss of a species is the loss forever of an opportunity to improve human welfare. The "public-service" possibilities of the global environment cannot be replaced by technology either now or in the foreseeable future, in some cases because the process by which the service is provided is not understood scientifically, in other cases because no technological equivalent for the natural process has yet been devised. In almost all cases there would be no need to create technological substitutes if only we would learn to live in harmony with our natural environment.

Nature . . . invites us to lay our eye level with her smallest leaf, and take an insect view of its plain. —Thoreau

Choosing among a multiplicity of trivial options has been a constant burden to mankind. We can all generate a long list of choices which have caused us to waste much time and emotional energy. Modern technologies and distribution practices have increased the range of different deodorants, but have had little impact on the choices involving the real and unchangeable values of life—the attitudes, needs, and desires that determine happiness or suffering, hope or despair. We must all still struggle with the same appetites, passions, and hopes that motivated Homer's or Shakespeare's heroes. The genetic code acquired by the human species more than 50,000 years ago is so stable that it still determines the conditions necessary for human health, comfort, and happiness, regardless of ephemeral changes in technological and political systems. Even though modern technology provides us with synthetic fabrics and electrical heaters, we still try to achieve the same body temperature as the Eskimos with their fur parkas and igloos, and as Stone Age people sought to achieve with animal skins in their caves. Regardless of whether we live in isolated free-standing houses or on top of skyscrapers, we still seek to relate in a personal way to the number of people once present in primitive hunting tribes or neolithic villages. All the social and technological futures we invent turn out to be mere reformulations, in a contemporary context, of the ancient ways of life; when Old Stone Age people, in the semitropical savannah, had achieved fitness to their biological and social environment—the natural harmony with nature we have lost.

Beginning with the great migrations of the Stone Age

from his semitropical Arcadia, man has suffered from various levels of Future Shock. In the half century between 1850 and 1900, we have seen the introduction of railroads, steamships, and electricity; of the telephone, telegraph, and photography; of antisepsis, vaccinations, anesthesia, radiography, and most of the innovations which have revolutionized the practices of public health and medicine. All of these advances have penetrated deeply and rapidly into the Western world. More recently we have seen the assimilation of aviation, television, the transistor radio, antibiotics, hormones, tranquilizers, contraceptive pills, and pesticides. It is not the development of these technologies which we regret but rather their misuse. We have allowed synthetic barriers to come between us and nature. Five thousand years ago, the Sumerians recorded on clay tablets their anguish about the generation gap, the breakdown of the social order, the corruption of public and domestic servants. They asked themselves a question that has been asked by every age: *Where are we going?* There seems to be a strong resistance to accepting the very simple answer implied in T.S. Eliot's words:

> *The end of all our exploring*
> *Will be to arrive where we started*
> *And know the place for the first time.*

To those of us involved in the effort represented by this book, Eliot's words mean that we shall continue to question our values, extend our knowledge, develop new technological forms, rediscover old technological answers, and experiment with new ways of life. Change there will be and change there must be, because this is an essential condition of life. But we are now discovering that spectacular innovations are not the best approach to the improvement of life and indeed commonly create more problems than they solve. We believe that the emphasis in the future will be less on the development of esoteric technologies than on the development of a conservatively decent world, designed to satisfy those needs of human nature that were woven into our genetic fabric during our evolutionary past. To state the obvious, in terms of mankind, evolution is not the solution to environmental disruption.

It is not pessimism to believe that there is no lasting security—it is simple realism. Fortunately, man has displayed a remarkable ability to change the course of social trends and start new ventures, often taking advantage of apparently hopeless situations to develop entirely new formats for living. Trends are not necessarily destiny. Whatever the circumstances, we must use our minds to select among the conditions and materials available in a given environment and organize them into new, humanized forms. With this in mind, we have brought together here some of the tools and information which will help individuals such as yourselves to begin the process of change.

> *the brilliant young intellects of our age*
> *make their homes*
> *in geometrically decorated apartments*
> *and conduct their lives algebraically*
> *in aristotelian fashion*
> *they pursue their precise pleasures*
> *eating only*
> *in medically approved restaurants*
> *and in an objective scientific manner*
> *do everything to their lives*
> *except live them*
>
> Roy Hamilton

PREPARING FOR THE TRIP

The journey of a thousand miles begins with one step. —Lao-Tsze

If you are a single man or woman with a demanding job in the city, you probably require living quarters quite different than those of a farmer in Colorado or a fisherman living in Maine. If you have a family of six children, each one demanding accommodation for his or her dog, cat, rabbit, or gerbil, you will want a place with amenities different from those you would find just right if you were a bachelor whose hobby is playing the piano. The ideal size, location, and design of a home are different in almost every case. Yet whoever you are and whatever your circumstances, you can be sure that if you have the understanding and willingness to take the time, you can design to fit your requirements.

Design is the making of plans which we know—or think we know—how to carry out. In this it differs from prophecy, speculation, and fantasy, though all of these may enter into design. Most often, we think of design as being concerned with physical objects, with a chair, a house, a city, or an energy system. But the broader the scope of our plans, the less feasible it is to separate the tangible object from the less tangible system of values or way of life which it is intended to support or complement. Politics and education enter into such broader plans not only as *means*, as part of the essential "how," but also as *ends*, as factors which modify the character of these plans. We do not, after all, make plans for the fun of it (at least most of us!). Fantasy is fun; design is hard work. We do it because we want to increase or encourage the things we like and remove or reduce the features of things—of our environment, of our lives—which we dislike. In this sense everyone is a designer, though only a few are professionals. We all seek not only to understand the world around us but to change it, to bring it nearer to an ideal and, in this way, create our own living environment. Design is thus always fundamentally both ethical and aesthetic.

Don't part with your illusions. When they are gone, you may still exist, but you have ceased to live. —Mark Twain

It may not be so obvious that the design process is also bound up with knowledge, and thus with science and engineering. Design is not fantasy alone; we must *know how* to achieve what we imagine and desire. Engineering and scientific information and tools are the means by which we change the situation we have to the one we would like to have. The history of science, technology, and design, at least in the sense we are discussing it, is a story of extraordinary success; to a great extent it is, in fact, the story of the human race. Yet this is not necessarily a fashionable view. It is more usual these days—indeed amongst the counterculture youth it is almost an article of faith—to say that science, technology, and design have failed. In part, this view stems from a weakness of the historical imagination; the full horror of the general condition of life, even but a hundred years ago, is simply not grasped by most of us who rail against the present. However, the criticism is also true—because design always fails to some degree.

From a philosophical point of view, it is not a paradox to say that design can succeed grandly and yet must fail. That is, the success of design does not relieve anxiety, because the satisfaction of a primary need nearly always allows secondary drives to come into play. So, relieved of the constant threat of starvation and other perils, we worry about economic crises or, more humanely, about the starvation and suffering of people remote from us. The relief of anxiety is accomplished not by the practical life, of which design is an important part, but by the contemplative life, which is often but quite mistakenly seen as opposed to design, whereas in fact it is design's necessary complement.

Thus happiness depends, as Nature shows,
Less on exterior things than most suppose.
Cowper

On a more immediate level, it is important to recognize that, even on its own terms, design is likely to fail. Designers are human, as are we all, and there are very severe limitations on our capacity to imagine many factors simultaneously. The "side effects" of a design—those unintentional results which follow upon the intended; the difficulty of imagining the needs and feelings of many different people; the effects of technical, social, and political change—these are common and indeed notorious causes of failure. Computers cannot solve these problems. Computers are marvelous tools, but, like other tools, they can magnify mistakes and errors of judgment just as much as they can assist well-conceived plans. And each of these potential pitfalls of design is enlarged and deepened as the scale on which we attempt to design is increased. An individual designing a solar unit for his or her specific needs is more likely to succeed than the designer of an individual building; town planning runs more risks than planning an individual building. We want to encourage you to become your own designer. The basic purpose of this book is to place design tools and information into the hands of the nonspecialist. Only the individual can set the context and limits of his specific design project. Individual design is not only possible but usually successful, if only we choose our problems according to our capacities, abstain from putting all our eggs in one basket, and generally adopt the principle that engineering can be suited to small-scale individual needs, always keeping in mind the overall objective of compatibility with nature. Our designs and, ultimately, our living and working environments will approach a more harmonious level if we design within the framework of the principle of interdependency, remembering that everything we do affects something or someone else.

The information and design examples presented in this book should help you add a new dimension to your daily existence; to do intensive gardening in your urban backyard, to design and construct a solar preheater for your suburban hot-water system, or to design a totally independent electrical system for your country home. Whatever your living context and lifestyle, you will find ideas and information to allow you to gain a little more control over your own existence.

Out of the light that blinds my eyes
White as a laboratory coat,
I thank the test-tube gods so wise,
For now my psychic soul can't gloat.

In the fell clutch of modern hands
I have been twisted by the crowd.
Under environment's strong commands
My head is perfumed but not proud.

Beyond this place of gas and gears
Looms a horizon yet unknown.
And yet if war comes in my years
I'll find I'll not go there alone.

It matters not how others wait,
How charged with fees I get my bill,
I am your patient, Doctor Fate:
Let me tell you my troubles still.

Luis Miguel Valdez, "Victus: A Parody"

THE ROAD MAP

From the many types of low-impact technologies described in the following chapters, you must select the ones appropriate for your situation, for existing and/or proposed buildings or structures into which the alternative technology must be integrated in a compatible manner.

In Chapter 2 (Alternative Architecture), we have provided a technical and visual foundation for the incorporation of alternative technology within a single structure. However, we also feel that we should explore the sociological implications behind an alternative lifestyle movement. The origin of this movement, we believe, is the instinct to survive—survive the environmental and psychological degradations created by a technology of consumer convenience and infinite industrial growth. In the beginning, those making use of alternative technologies will be a small minority, the vanguard of a group which may lead us away from world destruction. In creating the forms to put alternative technology to use, we must allow for the variety of these people who are willing to change. We feel the idea of "optimum" solutions is too inflexible in the sense that a transitional movement is essentially experimental; and so we attempt where possible to provide criteria which will guide the user to his own best solution.

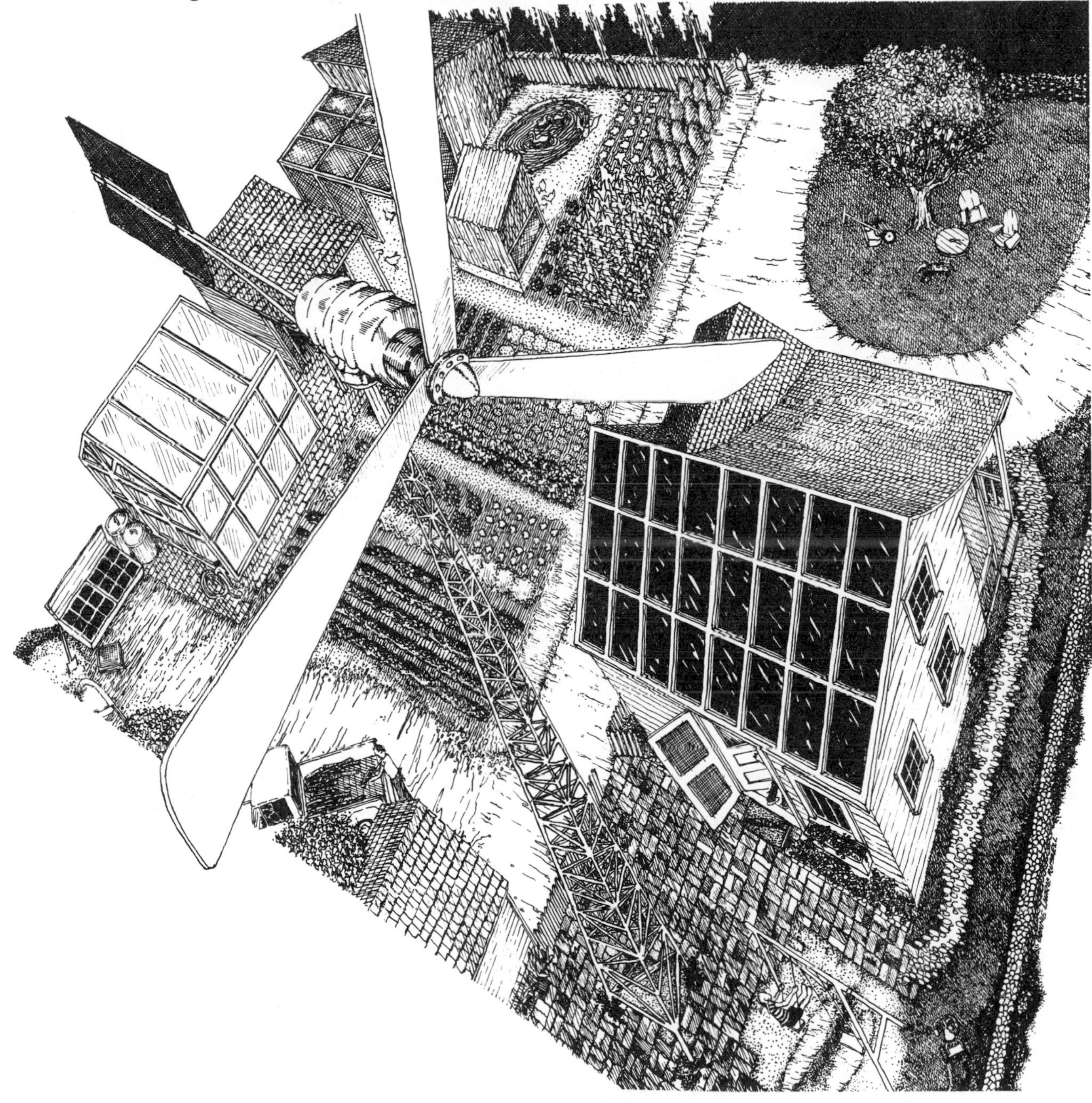

Low-impact technology must include ways and means by which natural, renewable energy sources may be used on a small scale to provide electricity for a single home or a small cluster of homes. In Chapter 3 we first discuss the fundamental concepts of electricity which are essential to the design of a home electrical system, including methods for estimating your own electrical energy demand. There are presently only two economically realistic technologies using renewable sources of energy which can be used to meet that electrical demand—wind power and water power—and these are discussed next.

In the section on wind, techniques for measuring wind speeds and estimating the energy available are given. The functions of each piece of electrical equipment necessary in a complete wind-electric system are described and techniques for calculating the necessary component specifications are given. Similar design calculations are then given for water-power systems. The intent is to lead you through sufficient amounts of material to allow you to design a generating system that is suited to the conditions of your own locale. In essence we would like to help you "get it together."

Utilization of solar energy is a very large element of any successful attempt at alternative-lifestyle living. In Chapter 4 we are concerned with eco–house heating. The material is presented in four sections. The status of our indoors environment is treated in the bioclimatic section where the discussion includes biological thermal control, the subjectivity of human comfort, and health considerations. Included is a comprehensive evaluation intended to help you determine qualities of comfort for your own living environment. Climatology is treated next as we deal with the determination of outside design criteria for a given specific house site. Known heating requirements are compared to solar energy availability in an effort to estimate the practicality of a given design and construction. Average winter temperature maps for the continental United States are included, as well as tables and maps showing the quantity and intensity of solar energy received at any particular place in the United States. Data from these maps are utilized in heat-loss calculations and solar-collector sizing. The third section discusses general heat-transfer theory, categories of heat loss for dwellings, and minimization of heat loss, accompanied by heat-loss and heat-load calculation examples. Transmission-coefficient tables for standard types of construction are included; calculation examples for various nonstandard "ecodesigns" are treated in some detail. With appropriate figures from this heat-loss section, we are ready to consider the design of particular solar heating systems, presented in the last section. Here we consider waterbed, southwall, and greenhouse designs. Component systems, consisting of solar collectors, circulation and control systems, and storage media are next treated. An original low-cost collector design is proposed and its operation described. A short section on auxiliary (nonsolar) heating systems concludes the chapter. Hopefully you will acquire sufficient insight into heating problems to be able to design and operate your own low-cost, ecologically sound living system.

Chapter 5 deals with methods and aspects of recycling organic waste. Since the average individual in the United States produces five pounds of solid municipal waste per day (agricultural wastes are still greater), there is considerable energy available here from conversion, in addition to

the possible return of nutrients to the food cycle. Four conversion systems are described and analyzed: the methane digester, which produces bio-gas and high-nutrient sludge through anaerobic bacterial action; the Clivus Multrum and outhouse arrangements; anaerobic decomposition in septic tanks; and bacterial-algal symbiosis in oxidation ponds.

With the growing consumption of freshwater in the United States approaching 50 percent of the supply flowing in rivers daily, cleansing and recycling of water is necessary if severe depletion and pollution are to be avoided. In Chapter 6 several areas of the problem are studied for small-demand users: the sources (open-body, ground, recycled, and artificially collected) and transportation of water; impurity types and levels, disease sources, and available treatment options; and finally storage, in terms of water quality, intended usage, and the recharge rate of a given system. In each area the systems are considered for three- and fifteen-person units living an alternative lifestyle, but the methods and analyses are applicable to specific situations through extrapolation. Immediate expenditures in time and energy may be higher than for conventional systems, but the long-range cost should prove to be far less.

A self-sufficient living unit must be able to feed itself. In the chapter on agriculture and aquaculture (Chapter 7), we explain how this goal may be achieved, in some instances using waste material. Agricultural crops and techniques are covered in some detail, though not exhaustively. There are hundreds of other crops and techniques for their culture, but we have chosen to describe a few which we feel can be applied successfully to creating a self-sufficient home. We also consider whether aquaculture can be used to supply significant amounts of food. Aquaculture is aquatic agriculture, in which the crops are fish, shellfish, or algae. Aquaculture has existed for centuries and, although it is not practiced extensively in the United States, it should not be ignored because its food-producing potential is generally greater for a given area than that of agriculture. As in the other chapters, our study has aimed toward supplying the needs of a single family or small cluster of families on limited acreage. Attempts have been made to allow for various climatic and soil conditions, and much of the information can be applied to both urban and suburban settings, as well as rural.

We wish you well on your journey through these pages and trust that you will enjoy learning and creating as much as we did in the preparation of the material. Remember: things worthwhile seldom come easy. We only hope that we have made the job somewhat easier and more enjoyable.

He who would arrive at the appointed end must follow a single road and not wander through many ways. —Seneca

ARCHITECTURE

Changing energy practices:
the philosophy of transition

Land, wind, water, sun:
determining the best site

Light, natural and man-made

Your house and what goes into it

Homes of yesterday;
consciousness of tomorrow

Alternative Architecture

OUR NEED FOR CHANGE

American building technologies and practices have developed under a natural blessing: abundant land on which to scatter our homes; abundant materials with which to build and rebuild almost at will; abundant energy to process, transport, fabricate, and demolish these materials; and yet more energy to heat, cool, light, and power our buildings at heretofore unattainable levels of comfort and dependability. This abundance has shaped the whole of our culture. Furthermore, the notion that this abundance is unlimited has given legitimacy to the unrestrained depletion of our natural resources: we move to the suburbs rather than maintain or improve existing urban environments; we build houses to last twenty years instead of two hundred; to condition our structures, we adjust a thermostat rather than open a window or pull a drape. And so disappear the land, the materials, and the energy.

It is now possible to detect a growing disillusionment in the United States with the shortsighted, self-interested technology of the past. We have only begun to feel the effects of the depletion of once-abundant resources, and to see and smell the aftermath of their misuse, but it has been enough to raise to consciousness the desirability of changing our technologies and attendant lifestyles: from the *energy-intensive* practices which created modern society, to the *energy-conservative* technology which will be needed, not only to repair the damage (where possible), but merely to maintain our viability.

The goal of alternative architecture is an end to resource depletion. But its successful adoption involves far more than facts and figures: a new relationship to our world is called for, characterized by both respect and reverence for the place we inhabit. We cannot "rule the world" without paying a heavy price; far better to enter into a symbiotic relationship and preserve both the world and ourselves. An alternative lifestyle may well be the first step in an evolutionary process which might eventually bring about a more reasonable approach to the environment by society in general. Indeed, we now face an alternative technology which grows in a new context, stemming less from a concern with comfort and efficiency than from a need to reduce the undesirable impacts of our old technology. For the first time, our new tools did not develop in response to a lifestyle; rather a new lifestyle evolved and is evolving as a response to old tools.

Architectural design is traditionally a response to a modeled use. Essential to the design of any living unit is a program, an orientation to the task, which supplies both objectives and constraints. For example, the five-day work week constrains all systems which require maintenance: no homeowner existing in that current lifestyle model will easily adapt to the increased investment of time called for by most aquaculture schemes. Certain constraints seem to be almost universal (dependence on a car, the work week, the need for sleep), while others are less easily generalized (the size and composition of the "family," the extent of personal flexibility, and so on). Given the difficulties regarding goals and constraints, it is not surprising to find differences of opinion among designers. Some, like Ken Kern, Art Boericke, and Barry Shapiro, are convinced that alternatives can only (or best) grow in the context of the forest and its lifestyles. Others, including Soleri and Kenzo Tange, interpret "alternative" as a lifestyle based on self-sufficiency in an urban context. What seems clear is that, in an era of rapidly altering roles and models, alternative architecture must place a premium on design flexibility.

New technologies will demand some significant departures from accepted cultural norms. Some may require a degree of social reorganization. For example, to conserve the most energy, the heat source should be located in the center of a room. Americans, however, have traditionally relied upon perimeter heating which keeps walls and corners warm. Consider the extent to which our room arrangements utilize the walls and corners. Unlike the Japanese, who focus their attention on the center of a room, Americans tend to line the walls of rooms with furniture and reserve the central areas as a circulation buffer. A single heat source placed at the center of a room or house will require that this longstanding, almost subconscious cultural norm be modified. Or if, for instance, the economics of heat loss demand a reduction of natural light in the house (fewer windows), we may see the rebirth, in appropriate climates, of the outdoor porch, with all its possibilities for more communication between neighbors.

To some degree our current atmosphere of experimentation calls for structural flexibility as well. Errors and failures sometimes demand a change in room organization or the capacity to install new or modify old mechanical systems. For this reason, designers should aim to develop a flexible arrangement which would allow some physical modification of the building, perhaps clustering the mechanical systems in a "core" area with easy access.

Generally we can view the philosophy of an alternative lifestyle as a philosophy of transition: when faced with a new technology, we must be capable of creatively adapting both the forms of our buildings and the forms of our lives. The kinds of architecture, the physical forms that will be generated in creating a resource-conservative relationship between the building and its environment, are still ill defined and experimental. But in other cultures and at previous times in our culture, architectures have existed and still exist which make both conservative use of consumable resources and maximum use of renewable resources.

As inspiration for an alternative architecture, we can look to what Bernard Rudofsky has termed "architecture without architects," the primitive architectures of the world that have, more often than not, developed in response to shortages of material and energy resources. In the absence of mechanical technologies, indigenous societies have depended on the ability of their building techniques to make the most effective use of landscape, climate, and architectural form, and the most permanent and economic use of materials in providing a habitable environment. We marvel at the ability of these builders who take a few materials—earth, stone, wood, grasses—to shape and assemble by hand, and yet who produce an architecture that adapts incredibly well to daily and seasonal climatic variations and that maintains a delicate balance of consumption and regeneration with its environment.

We cannot turn our backs on the fact that we live in a highly industrialized society. Where sophisticated technologies serve our goal and can be used effectively to reduce the depletion of resources, they should become a part of our alternative architecture. But we advocate an architecture that relates to its environment *in the same way* that primitive architecture does: symbiotically. To reduce energy and material waste, we must design simpler, smaller structures that use less highly processed building materials, that reuse discarded materials, and that make more permanent use of both. We must use fewer machines in the construction of our dwellings. We must learn to use fewer mechanical appliances to service our houses. And we must, above all, be willing to adjust our habits, physiology, and psychology, opening ourselves to the wonders and limits of the earth where we abide our little while.

Although our focus here is on the design of a new structure, much of the discussion and many of the ideas are directly applicable to the modification and improvement of existing structures, in both rural and urban settings. The scope of this chapter is limited to general relationships, concepts, and ideas; only occasionally do we touch on specific designs or solutions. The material is rather more integrative than differential—specific details on various technologies can be found in chapters 3 through 7. Here we try to provide a general framework for the more detailed information to come.

SITE DETERMINANTS IN HOUSE DESIGN

Elements of the natural environment were of vital concern in the design of early dwellings, placing major restrictions on the form of these structures. But upon the advent of cheap energy sources and high-density housing developments, the only environmental inputs attended to were maximum and minimum temperatures, for determination of the amount of insulation and the size of any necessary heating or cooling plants. It was difficult to design in harmony with nature simply because there was very little that was natural in the typical neighborhood.

That situation is changing. No longer is the abundance of concentrated energy resources so apparent, and there is a generally increasing awareness of ways in which the human sphere of activity interacts with the natural realm, as well as a growing desire for more direct interaction. In the field of domestic architecture, we are encouraged to take advantage of the topographical and climatic features of a given site to the fullest possible extent, minimizing the energy requirements of our houses and maximizing human comfort and pleasure within the home environment.

The site on which a house is located commits the house and its occupants to a physical and climatic environment. In the design of the alternative house, we are seeking two objectives in the relationship between the physical and climatic site characteristics and the house itself. The first, and by far the more important, is to use these characteristics to provide natural conditioning—non-resource depleting—cooling, heating, lighting, and powering. Also, if possible, we can try to use the site as a source of materials with which to build and maintain our structure.

Subsurface Characteristics

The composition of soils may influence the design of the house in several respects. Soil composition affects the design of building foundations. In general, this is an engineering question related to foundation size and reinforcing. But in some cases, adverse soil compositions, or water tables near the ground surface (as in marshy areas), may prohibit building or limit buildable area (Figure 2.1).

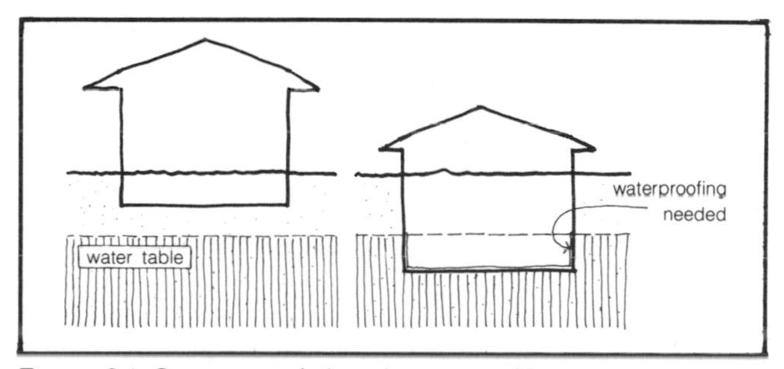

Figure 2.1 Construction below the water table.

Both conditions also determine the ease and practicality of subsurface excavation, either for the construction of buried or semi-buried structures, or for the benching of a sloped site to provide a flat building area. This is particularly important if hand excavation is anticipated. Excavation may also be limited or eliminated entirely because of the presence of subsurface rock, to the dismay of an ambitious planner. A high water table considerably complicates construction work and poses problems in waterproofing buried walls and in providing proper subsurface drainage around the building. And, if your soil happens to lend itself to adobe construction, this is a useful fact to note. For soil and water-table testing techniques, see *Site Planning* by Kevin Lynch.

As we will see in later chapters, soil composition and groundwater conditions may indirectly affect the location of a house on a site when the positioning of wells or on-site sewage systems is considered.

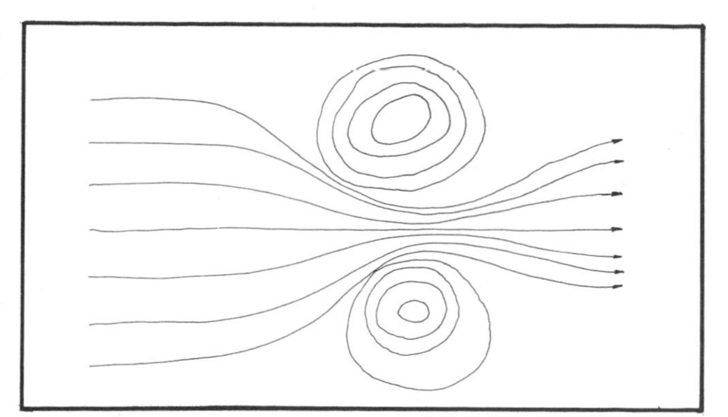
Figure 2.2 Wind channeled by hills.

Topography

Landforms determine the natural water runoff patterns and, wherever possible, these drainage patterns should be avoided in siting the house, to reduce the need to divert water runoff around the building and to lower the risk of a flooded house in a particularly wet season. Extremely flat areas or slightly depressed areas, where ponding is likely to occur, should also be avoided. Ideally, building sites should have slopes from 2 to 4 percent; that is, enough slope to provide good drainage but not so steep that the building process is complicated.

The angle of the slope determines where building is or isn't practical. Building on steep slopes, above 10 percent, is considerably more difficult, consuming far more material, time, and money than building on flatter slopes. The inclination and orientation of the slope also affect both the amount of sunlight received and the resulting air temperature on the site. As you might suppose, southern slopes are generally warmer in winter.

Topography affects wind patterns by constricting the wind (Figure 2.2), increasing its velocities in certain areas, while sheltering others protectively. Slopes to the leeward side of winter winds are preferable for building sites (Figure 2.3). Hill crests, where wind velocities are increased, should be avoided in all but hot, humid climates. The bottoms of valleys and ravines, as well as topographic depressions, are likely to channel and trap cold air masses during the night and winter, and so should be avoided as building sites.

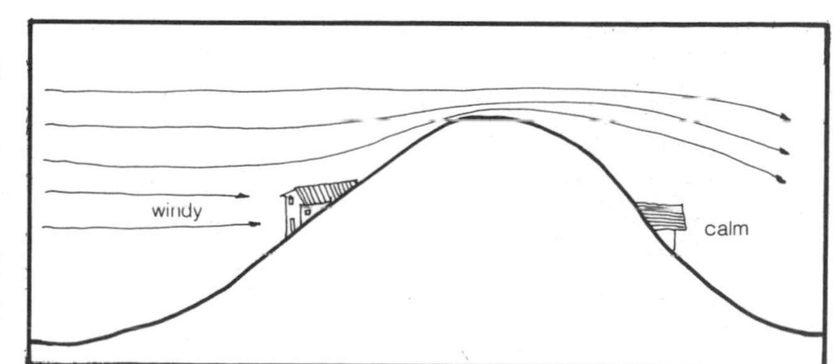

Figure 2.3 Building on slopes.

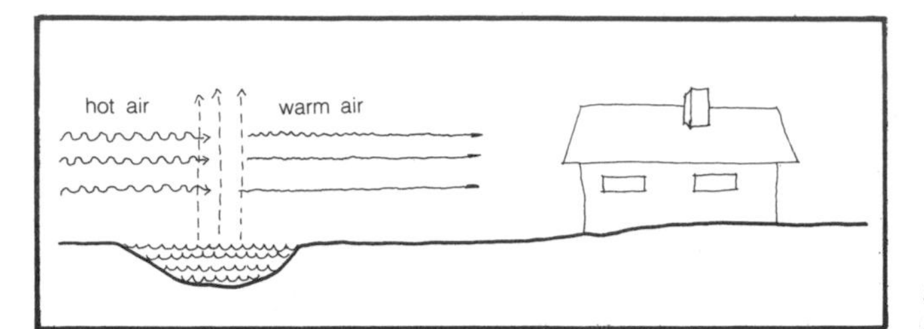

Figure 2.4 Cooling by a water-body.

Surface Water

Because of their ability to absorb heat during warm periods and release it during cold periods, large water-bodies such as lakes or the ocean exert a moderating influence on air temperatures, throughout the day as well as the year. Even small streams and ponds, in the process of evaporation, cool air temperatures in summer. In addition, there are usually breezes sweeping across the surface of a lake or down a river valley. Where possible, site water courses and water-bodies, such as aquaculture ponds, should be used in conjunction with prevailing summer winds as conditioning elements for the house (Figure 2.4). A further effect from unshaded bodies of water is added heat from reflected light; thoughtful siting and use of strategic shading devices can offset this gain when it is undesirable. It is always best to build on high ground when close to large water sources, for both drainage purposes and protection against flooding.

Vegetation

Natural vegetation has numerous influences on the microclimate near the house (Figure 2.5). Healthy trees and shrubs provide shade, reduce glare, and in dense configurations are effective windbreaks. They also have a cooling effect in the summer, using heat for evapotranspirational processes in their foliage. They reduce the albedo (reflectivity) of the land surface, thereby reducing the possible heat gain of a house from outdoor reflection. Dense growths effectively absorb sound, enhance privacy, and are remarkably efficient air filters for dust and other particles. Vegetation generally looks nice and often smells better. It can support an animal population, which may or may not be desirable. However, in order to permit access to summer breezes, aid drainage, and reduce problems from pests, it is advisable to discourage forests or extensive vegetation too near the house.

One of the major contributions of trees on a site is the shade that they provide. Deciduous trees are ideal in this regard because shading is needed in summer, but is generally undesirable in winter when solar radiation is a benefit; and the leaf-bearing period of deciduous trees native to any particular latitude spans exactly the period when shading is required for that latitude. These trees have the added advantage that, in an unusually cold spring or warm autumn, the leaves appear late and remain later, correspondingly, which is highly desirable. Trees are primarily effective at low sun angles, which makes them useful for intercepting early morning and late afternoon sun. Hence, for these purposes, their positioning on the east-southeast and west-southwest sides of the house is most efficient in northern latitudes. At higher sun angles, structural features of the house, such as louvers or screens, can provide needed shade.

The other classic function of vegetation is to provide a windbreak. In this respect, the ideal arrangement allows in summer breezes and blocks out cold winter winds. Obviously, this is not always possible. In many localities, however, the prevailing winds in summer and winter come from different directions, which makes the placement problem for hedges or rows of trees relatively simple. For this situation, evergreens are highly desirable, as they maintain dense growth in the winter. In other situations, it is possible to use deciduous hedges or deciduous trees with a low, open branch structure to direct summer air flow into your house, while these same plantings would present little obstruction to air flow in the winter.

In some cases, it is possible to place strategic plantings to combine shading and wind-directing benefits for optimum effect. In most cases, however, compromise is necessary. The full spectrum of plant use in house design is presented in *Plants, People, and Environmental Quality* by Gary Robinette. See also "Wind Protection" in this chapter for diagrams.

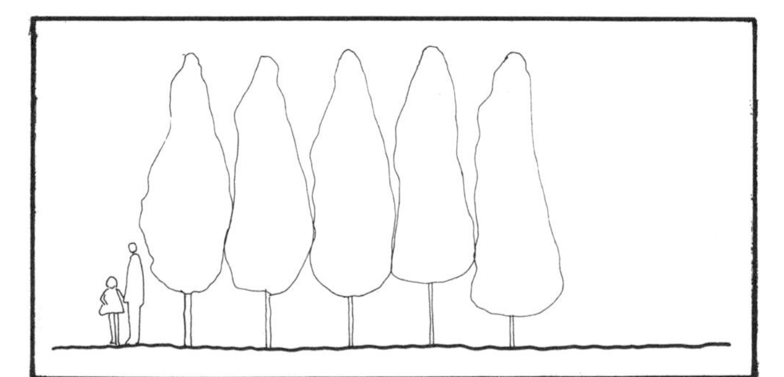

Figure 2.5a Vegetation as fenestration.

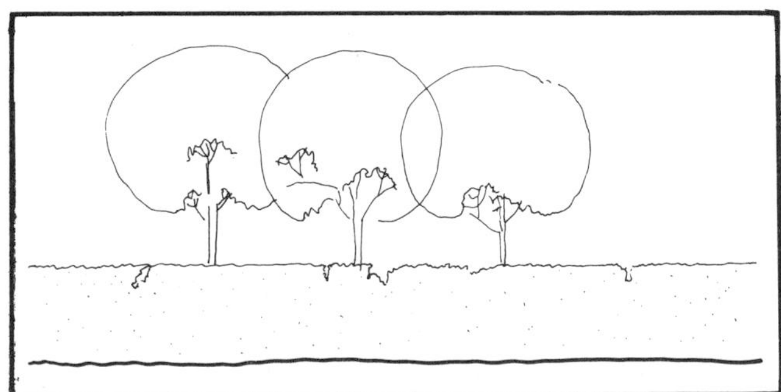

Figure 2.5b Vegetation as fenestration.

Man-made Characteristics

The man-made environment on and surrounding your site also influences the air temperature, sun, and wind. Air temperatures are raised by an abundance of hard, reflective surfaces like those of buildings or paving (Figure 2.6). Buildings, fences, and high walls also channel air movements and change wind patterns.

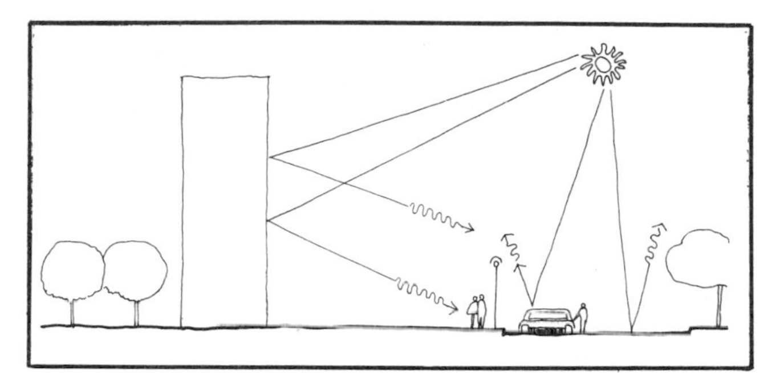

Figure 2.6 Temperature raised by reflective surfaces.

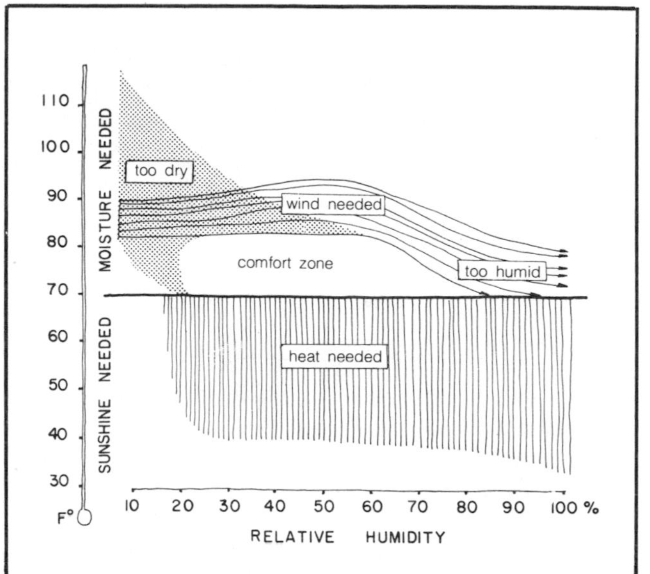

Figure 2.7 A bioclimatic chart.

CLIMATIC FACTORS IN HOUSE DESIGN

Climatic factors are obviously very important and, of these, temperature has the greatest impact. The extremes and averages of temperature as well as the duration of the various temperature ranges, both in terms of daily and yearly cycles, influence the size of the temperature-controlling facilities a house requires. They also dictate whether a house must be designed for optimum efficiency of cold exclusion or of heat dissipation, or whether some compromise of the two will prevail.

The other climatic factor which most affects human comfort is humidity. The amount of humidity has significance when it is tied to air temperature. Within the range of variation of these two factors, a human "comfort zone" can be described. Figure 2.7, a schematic diagram of Victor Olgyay's Bioclimatic Chart, depicts this zone and also shows corrective measures to be taken when conditions fall outside its limits. This diagram is naturally subject to variation on an individual basis, but gives workable guidelines. The shading line represents a limit above which shading is necessary and below which solar radiation becomes useful for heating.

The major effect contributed by solar radiation is heat, but light and consequent glare also have to be considered. Solar angles during different hours and seasons, and the distribution of sunny and cloudy periods are important, particularly in terms of the feasibility of solar heating for the house. Wind is significant as a potential ventilation aid and cooling agent, as well as a potentially destructive force. Maximum velocities, average velocities, directions, and daily and yearly variations of all these factors are important, especially if you are considering a windmill as a power source. Finally, seasonal variations in rates of precipitation, total amounts, predominant direction (if any), and maximum snow loads (if any) influence the structural design.

In design, we encounter these climatic variables in three distinct relationships. The first is *general climate*, which determines the overall climatic character of a structure. While the differences in general climate are myriad, we usually refer to four distinct types in discussing specific design ideas: cold climates, in which our emphasis is principally on heat conservation, sun utiliza-

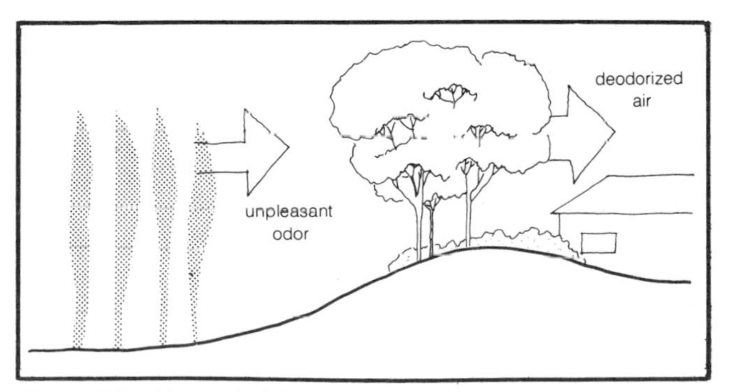

Figure 2.5c Air purified by vegetation.

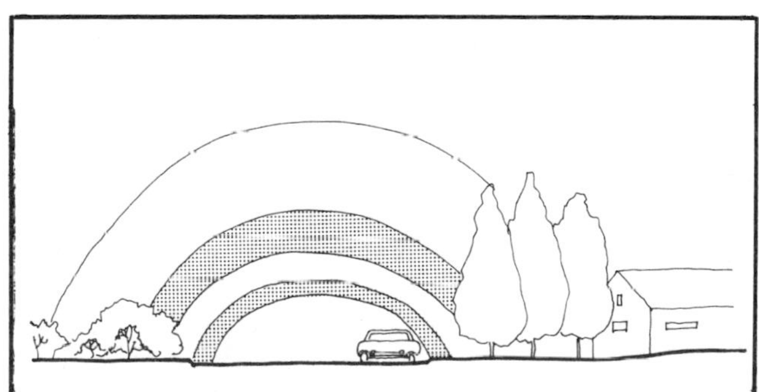

Figure 2.5d Vegetation as a sound barrier.

tion, wind protection, and rain and snow protection; temperate climates, in which the emphasis is on striking a balance of all conditions; hot, arid climates, where our concern is with heat and sun control, wind utilization, rain utilization, and increased humidity; and lastly, hot, humid climates, where we try to promote heat, sun, and rain protection, wind utilization, and humidity reduction. Olgyay (*Design with Climate*) gives detailed methods for designing in response to climatic conditions, as well as specific design responses based on these four regional types.

The second relationship consists of the climatic variables as they interact with particular site characteristics to produce the *microclimate* of each individual site. It is these microclimatic conditions that are used in the conditioning of the resource-conservative house. The classic study of microclimate seems to be Geiger's *The Climate Near the Ground*, from which Olgyay and Lynch draw much of their information

The third relationship consists of the microclimate as it affects our feelings of environmental comfort. We are reacting to this *bioclimate* when we feel hot and sticky on a summer day. (Additional discussion on bioclimate and climatology can be found in Chapter 4.) One purpose of design is to provide a comfortable bioclimate for our bodies, "comfortable" being somewhere between 70° and 80°F and 20 percent to 80 percent relative humidity (Figure 2.7). Where we refer to "warm" and "cool" periods of the year, we mean those times when climatic conditions fall either above or below this comfort zone.

Temperature and the Reduction of Heat Transfer

Temperature averages and extremes throughout the year dictate important design considerations of your house. Regional information is generally sufficient and can be found in the *Climatic Atlas of the United States*, as well as other sources. During at least a portion of each day, a difference in temperature will exist between the bioclimate you wish to maintain and the actual microclimate. This temperature differential encourages a process of heat transfer (convection) through the building materials of the house, heat traveling from higher to lower temperature areas. Building materials and forms may be used to prevent unwanted heat transfer, since they have different transfer resistances. Where we wish to reduce heat transfer, those materials and techniques offering high resistance should be used. Resistances of various building materials and methods of calculating heat transfer are presented in Chapter 4. Climatic factors surrounding the house can also be controlled to reduce temperature differentials. Examples of no-cost/low-cost techniques for conserving energy in the typical home are given in *Technical Note 789*, issued by the National Bureau of Standards.

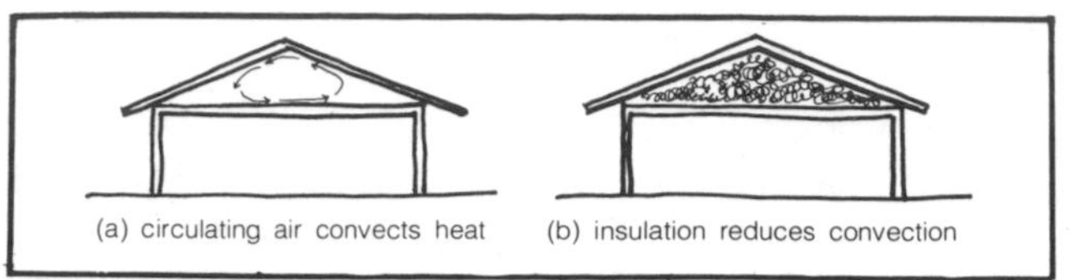

Figure 2.8 Heat convection in an attic.

Insulation

Building insulation generally refers to materials of high resistance—glass fiber or mineral wool blankets and fill, polystyrene and polyurethane foam boards, vermiculite and perlite fill—that are effective because of small cellular voids in their structure. They are generally highly processed materials designed to be used with standard construction techniques, particularly to fill large structural cavities in stud-frame and concrete-block construction.

Good insulation conserves energy otherwise spent on heating and cooling, not only by reducing heat transfer, but by making available energy more useful. In winter, a person feels colder in an uninsulated house than in an insulated one, *even if* the internal air temperatures are the same. This is due to the cold-wall phenomenon. The cold interior surfaces of uninsulated walls absorb radiant heat rather than reflect it as the warm surfaces of insulated walls do, consequently causing a greater loss of body heat. Most people compensate for this chilling effect by raising the heat. This adjustment is unnecessary with good insulation. In summer, cold wall surfaces are desirable; insulation accomplishes this effect and also reduces, yearlong, drafts produced by temperature differences between walls and air. Finally, it reduces room-to-room and floor-to-ceiling temperature contrasts.

An insulating air space within a wall or roof is not as effective as that same space filled with an insulating material (Figure 2.8): temperature differences across the space cause the air to circulate and transfer heat. In standard wood-frame construction, the most common, cheapest manufactured insulation is mineral wool or fiberglass. It is furnished in widths suited to 16-inch and 24-inch stud and joist spacing, in blanket rolls three to seven inches thick (available with reflective foil or vapor barrier on one side). Fiberglass has the advantage of being highly fire resistant and the disadvantage of being itchingly uncomfortable to work with.

Inch for inch, the best insulating material is polyurethane foam. One inch of urethane is equivalent to about two inches of fiberglass. An extremely versatile construction material, it is available in rigid sheets one-half to two inches thick or in liquid form with a catalyst for on-the-job foaming. The liquid may be poured into forms or sprayed on with special equipment. In the hands of a skilled applicator, this material can be rendered into almost any sculptural form and will provide considerable structural support. In the hands of an unskilled applicator, it can easily turn into an utter mess and its high price makes experimentation a serious venture. If ignited, urethane burns explosively and emits noxious gases; it should be covered on the inside with a fireproof wall of plaster or sheet rock.

In house construction, a common type of rigid insulation is sheathing board made of processed wood or vegetable products, impregnated with an asphalt compound to provide water resistance. It is usually affixed to the exterior of a stud frame, often in addition to flexible insulation applied between the studs. Sheathing board also serves as a main structural component of the wall.

Loose-fill insulations like vermiculite, sawdust, shavings, shredded redwood bark, or blown-in fiberglass can be used in walls of existing houses that were not insulated during construction. They are also commonly added between ceiling joists in unheated attics. Vermiculite and perlite are often mixed with concrete aggregates to reduce heat loss.

Insulating was a common practice long before the advent of modern manufactured materials, and we might do well to consider the use of less energy-intensive materials than those commonly used in modern construction. Any shelter sunk into the earth is cheaply insulated. About thirty inches of packed dry earth is equivalent to one inch of urethane, but a lot cheaper. Although the earth may absorb a considerable amount of heat from the room during the day, this energy is not entirely lost, since the warmed earth will keep the room at a more uniform temperature throughout the cold night. Blocks of sod and moss are commonly used in northern countries where insulation is particularly important. Not to be overlooked for its insulation value is wood: as siding it is significantly better than plaster, and, as we mentioned earlier, sawdust is almost as good as fiberglass although it tends to settle in vertical wall cavities. Dried grass and straw matting work well as roof insulation when sandwiched between layers of plastic, although their fire hazard is great. Other useful examples are given in Kahn's *Shelter*.

The air inside a house normally contains much more water vapor than the outside air, due to cooking, laundering, bathing, and other domestic activities. In cold weather the vapor may pass through wall and ceiling materials and condense inside the wall or attic space, damaging finish or even causing decay of structural members. To prevent this penetration, a vapor barrier should be applied on the interior side of insulation, the most effective being asphalt-laminated paper, aluminum foil, and plastic film. Some building materials, such as fiberglass insulation and gypsum board, are available with vapor barriers factory affixed.

Surface Area

Heat transfer that occurs between the interior and exterior of your house is principally dependent on the surface area of materials separating inside from outside spaces. In extreme climates, we try to reduce these

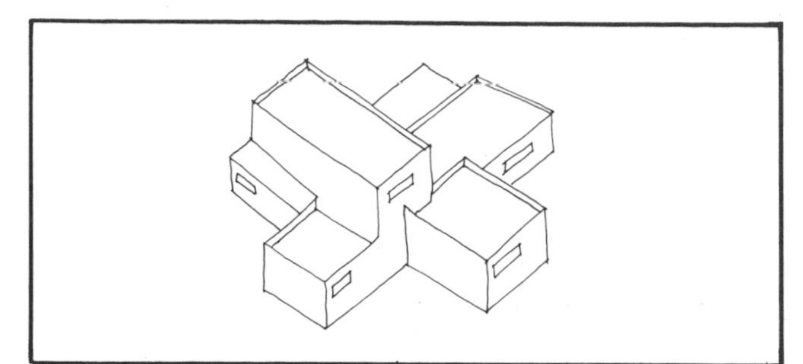

Figure 2.9 The reduction of wall surface by clustering.

exposed surface areas. Simple, compact forms—domes, cubes, and other regular polygonal structures—offer less surface area for a given floor area than elongated or complex forms. Clustering of housing units, both horizontally and vertically, also reduces exposed surface areas of walls and roofs (Figure 2.9). In reasonably dry soils, buried, semi-buried, or excavated houses are feasible and have obvious insulation advantages.

Window Areas

Window glass has very little resistance to heat transfer; it transmits about twelve times as much heat per square foot as a fiberglass-insulated stud wall. When designing to prevent heat transfer, reduce window areas to the limit of your psychological need for visual contact with the outdoors. Minimum window-area standards (based on floor area) for dwellings are found in section 1405 of the Uniform Building Code. New recommendations which incorporate floor area and window types are described in *Technical Note 789*, issued by the National Bureau of Standards. Techniques for making the best use of a limited amount of window area, both for natural lighting and for visual contact, are presented later in this chapter.

When your psychological needs won't allow for a great deal of reduction in window area, a half-inch air space between two sheets of glass can cut heat transfer in half. Also, two separate windows, one to four inches apart, produces the same reduction (Figure 2.10). When only a rough sense of visual contact is needed, glass blocks perform the identical service. But the most common and least expensive method is the use of curtains,

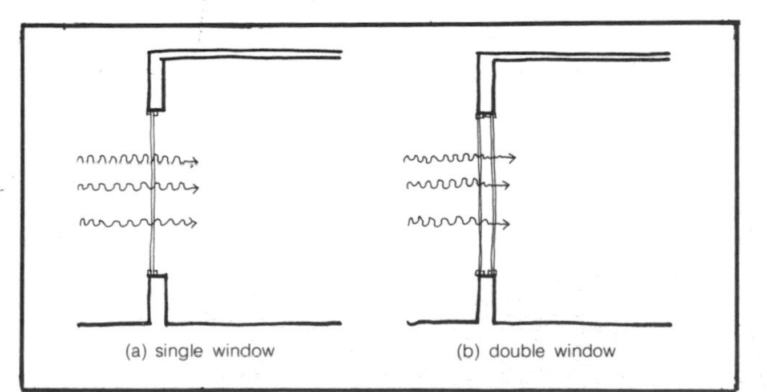

Figure 2.10 Heat transfer through windows.

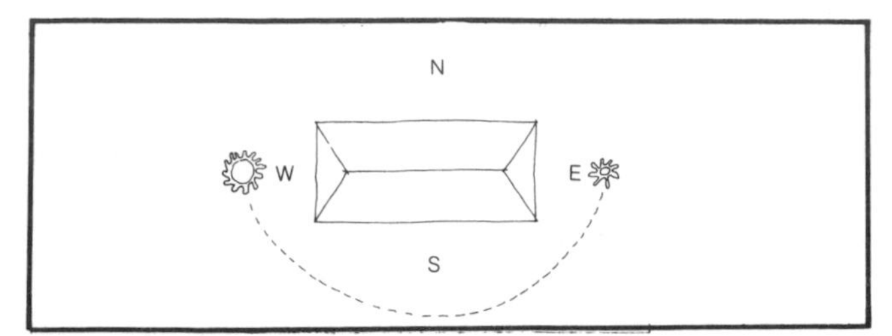

Figure 2.11 Orient the short side of the house toward the west.

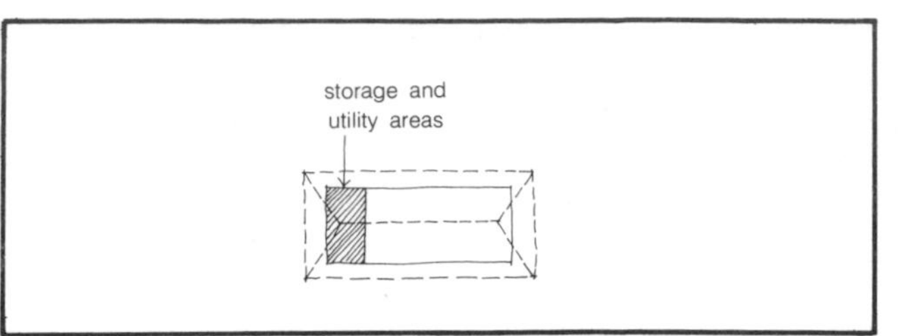

Figure 2.12 Locate nonhabitable space on the western side.

blinds, or shutters on all window areas. Similar to double glazing, they reduce heat transfer by creating a dead air space between you and the outside environment. They are also useful for controlling light penetration into your house.

Window frames are also points of heat transfer, through cracks between window and frame and, if metal frames are used, through the frame itself. Weatherstripping and wood frames should be used throughout.

Solar Protection

All sunlight, whether direct or reflected, whether striking directly the outer surfaces of your house or streaming through windows onto the floors or walls, is converted to heat. Light striking nearby ground surfaces and objects heats these materials, which also warms the air immediately surrounding the house. However, such heat loads can be minimized using several techniques.

House Orientation

The critical period for solar heat gain is usually a late summer afternoon when the sun is low. Orient the shorter side of your house toward the west (Figure 2.11), put rooms that require small windows on the western side, or locate here your storage, garage, and toilet facilities as thermal buffers (Figure 2.12).

In rooms with large windows, the windows should open to the south, where summer sun penetration can be controlled by overhangs or other shading devices; in climates where heat loss is not critical, they can open both to the north and south.

Shade

South walls can be shaded with roof overhangs but, due to early morning and late afternoon sun angles, overhangs are less effective on east and west walls (Figure 2.13). In hot climates, you can use porches, verandas, arcades, or double roofs to shade your walls and provide cooler outdoor spaces. However, their use in temperate and cool climates obstructs direct solar radiation, a point to consider in winter.

Deciduous trees and bushes provide shade in the summer and allow sun penetration in winter (Figure 2.14). Trees located to the southeast and southwest of the house will provide shade for both the roof and walls (Figure 2.15). East and west walls may be screened with low trees or bushes. A horizontal trellis with deciduous

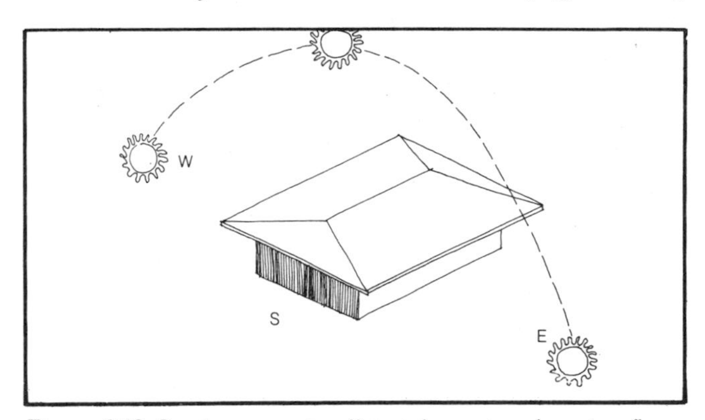

Figure 2.13 Overhangs are insufficient for east and west walls.

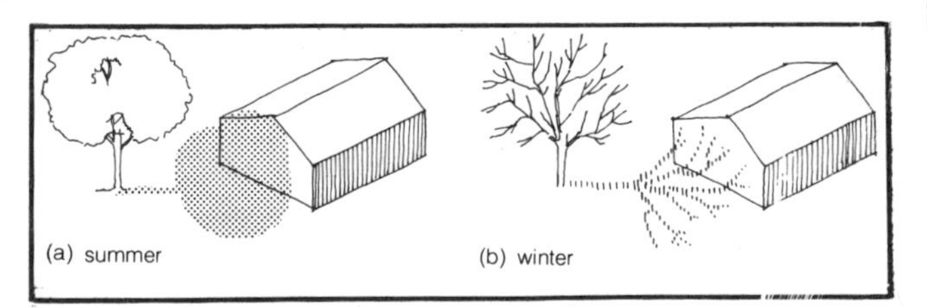

Figure 2.14 Shading by deciduous trees.

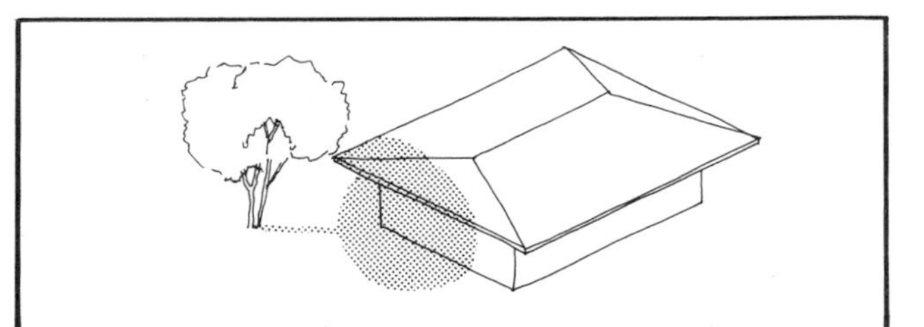

Figure 2.15 Shading SE or SW walls of the house.

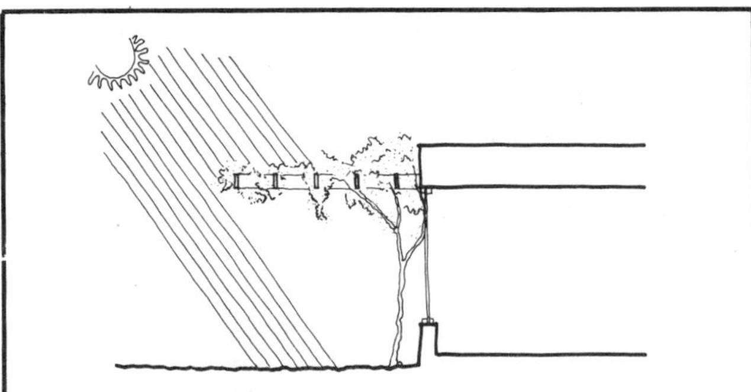

Figure 2.16 A deciduous vine arbor.

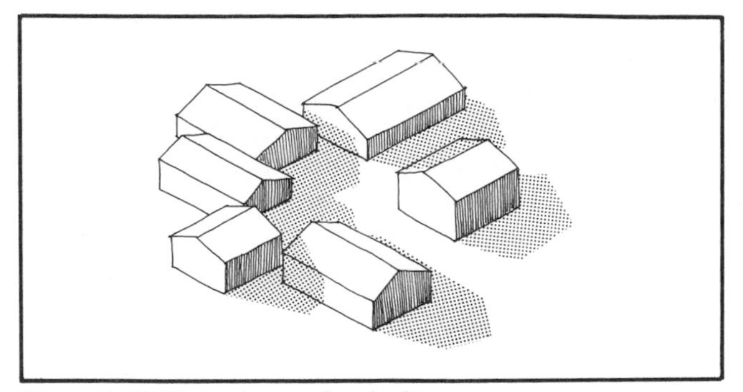

Figure 2.17 Cluster houses for shading.

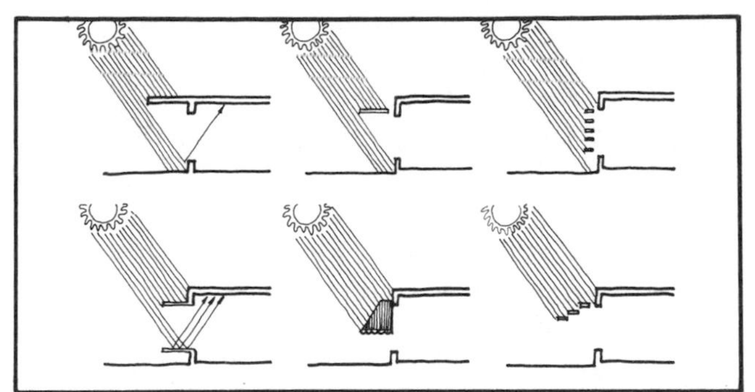

Figure 2.19 Horizontal shading devices.

vines (such as grapes) can provide the same seasonal protective variability (Figure 2.16). Tree and bush sizes and shapes are shown in Ramsey's *Architectural Graphic Standards* and Lynch's *Site Planning*.

The heat gain from sunlight, as that caused by air temperature, is directly dependent upon the exposed surface area of your house; hence the same control techniques are applicable. In hot, arid climates, it is possible to cluster housing units (Figure 2.17) and to use minimum-surface structures. Buried or semi-buried (Figure 2.18) structures are also feasible.

Techniques for shading exterior surfaces are also obviously applicable for shading windows. In addition, specific window devices can be used. On a southern exposure, where incident sunlight is predominantly vertical, horizontal shading devices are most effective (Figure 2.19). On eastern and western exposures, vertical or parallel devices are required for the low sun angles. For maximum protection, use combinations of various devices. If these are a fixed part of your house, they should be sized to exclude sun only during the summer. Removable or retractable awnings, movable exterior louvers or shutters can be adjusted to the season. See *Design with Climate* (Olgyay), *Sun Protection* (Danz), and *Architectural Graphic Standards* (Ramsey) for detailed discussions and examples of technology.

Exterior shading devices are seldom completely effective; curtains, blinds, shades, and shutters are recommended. Curtains uniformly reduce sunlight while blinds allow both a reduction in intensity and a redistribution of light. Heat penetration may be reduced by up to 50 percent with either blinds or curtains, and by up to 75 percent with roller shades.

The effective use of shading devices as a means of solar control will depend on the designer's ability to calculate the sun's position. Sun paths and altitudes for various latitudes in the United States can be determined by using the Sun Angle Calculator produced by Libby-Owens-Ford and provided with *Climate and House Design*, published by the United Nations. They may also be calculated by means shown in *Architectural Graphic Standards*.

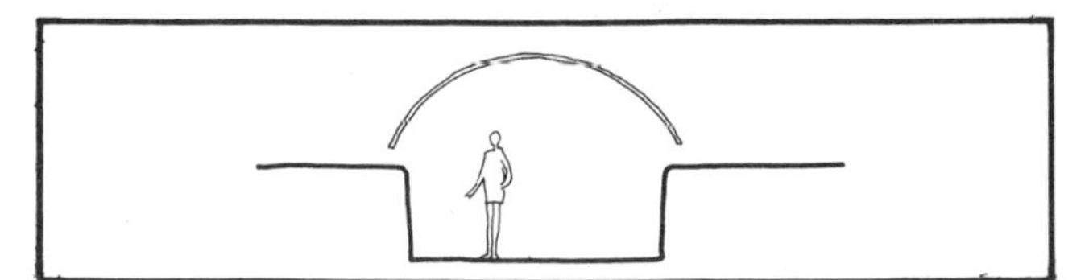

Figure 2.18 A semi-buried structure.

Reflectivity

Glare from nearby water-bodies or a sea of parked cars can be controlled both by proper siting of your house and by strategic use of bushes, berms, and fences (Figure 2.20). You can minimize excess glare from unshaded ground surfaces near the house by the use of ground cover (grass, ivy) that absorbs a fair amount of light. Reflectivities of various ground surfaces are shown in *Plants, People, and Environmental Quality* (Robinette).

The more sunlight reflected off the surface of your house, the less will be absorbed into building materials. In hot climates, use white or metallic surfaces; even in temperate areas, light colors should be used for protection during warm spells.

Figure 2.20 Reducing glare from objects and water-bodies.

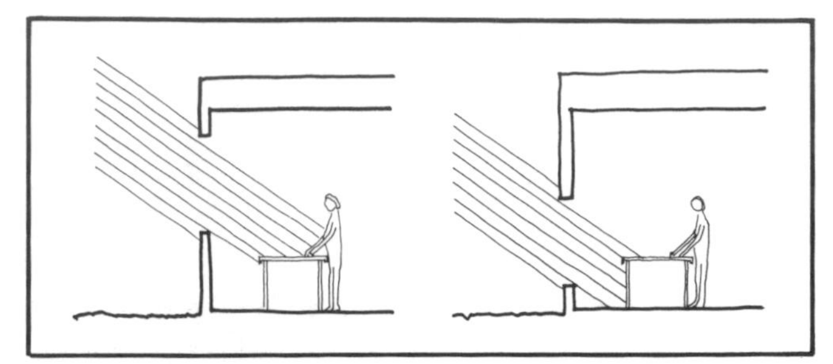

Figure 2.21 Light from high and low windows.

Solar Lighting

A well-integrated lighting system in any structure must make the best use of both natural and artificial light sources. To use daylight for inside lighting, the first obvious requirement is that the sky be bright enough to provide some lighting potential. Before going further, it would be wise to define the difference between sunlight and daylight. *Sunlight* comes in a straight line while *daylight* is reflected or refracted (also called *glare*). There are several factors that affect sky brightness for each particular location: latitude, altitude, time of year, time of day, amount of air pollution, and relative humidity. For any particular lot, the effect of the local terrain, landscaping, and nearby buildings must also be considered. It is handy to research the average number of clear days per year in your part of the country, or else the percentage of time that skies are clear during sunlight hours. Derek Phillips (*Lighting in Architectural Design*) devotes one chapter to daylighting, giving a method to calculate daylight levels in interior spaces and also general suggestions for the design and placement of windows.

Window Location

Because the reflectivity of the sky is generally much higher than that of the landscape, the quantity of light entering a window is directly related to the amount of sky visible through the window. Low windows transmit little light, and this light is at a poor angle for most activities (Figure 2.21). Raising most of your window area, but still keeping it low enough for a standing adult to see out, best compromises conflicting needs for visual contact and maximum light (Figure 2.22). Small, lower windows should then be provided only at specific sitting areas, such as in dining or living rooms.

Overhead skylights, clerestories, and monitors,

Figure 2.22 Specific uses of high and low windows.

which are less apt to be affected by obstructions surrounding the house, efficiently admit light into interior spaces. They can be used in such areas as bathrooms or bedrooms, where light is necessary but visual contact is not (Figure 2.23). Diffusing glass can be used to bathe the room with light and reduce over-contrast.

Kitchens, reading and writing areas, and work spaces all require intensive lighting and should be located near large windows. Areas requiring less intensive lighting—living or dining areas, toilets, bathrooms and bedrooms—should have correspondingly smaller windows (see also "Artificial Lighting" in this chapter).

Figure 2.23 A skylight.

Window Efficiency

There are three general classifications of glasses; each one has different advantages and disadvantages, but any choice must be made with heat-loss characteristics somewhere in mind. High-transmittance materials pass light easily and allow clear vision in either direction. Low-transmittance materials have brightness control, which increases as transmittance decreases. Finally, diffusing materials (suggested for skylight use) include opal and surface-coated or patterned glass and plastic. They are directionally nonselective: brightness is nearly constant from any viewing angle. This property is especially pronounced in highly diffused materials, but transmittance and brightness decrease as the level of diffusion increases. In other words, to get a uniform diffusion of light through a skylight with diffusion glass, you must sacrifice the level of brightness in the room and a sharp vision of the sky overhead.

The light transmission of windows is also impaired by external or internal shading devices. But compensations can be made. Permanent external shades should be sized to exclude sunlight only during warm periods. Even then, they can be designed to exclude direct sunlight, but to include some reflected and diffused light (Figure 2.24). Light reflected from the ground, from reflective window sills, and from adjustable or removable reflectors can also be used to increase diffused light entering interior spaces.

Contrast

Contrast between the outside environment viewed through a window and the darkness of an interior space can cause the discomforts of glare and eye fatigue as your

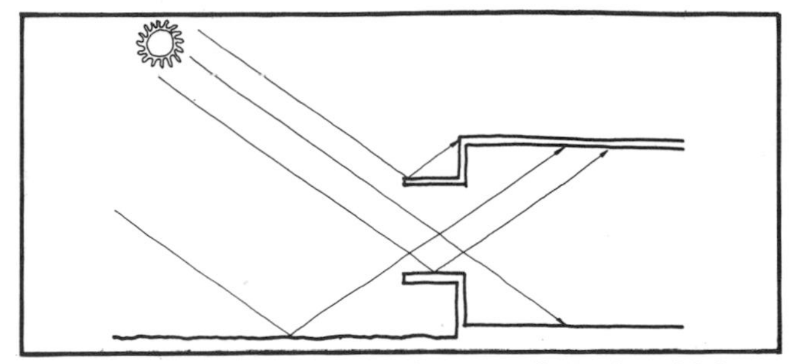

Figure 2.24 Sizing of a shading device to allow reflected light.

eyes constantly adjust from one lighting condition to another. This contrast can be reduced by using light-colored window frames, walls, ceilings, and floors throughout the interior of the house, because these reflective objects increase overall brightness in any space (see also "Interior Coloring" in this chapter).

Windows may be placed adjacent to perpendicular interior walls, so that the light reflected from those walls produces a contrast-reducing transition in light intensities (Figure 2.25) rather than a brilliant hole of light surrounded by an unlit wall. Where walls have considerable thickness, as in adobe construction, windows should be located on the wall's outer surface. The window opening should then be beveled, to form a transition surface between window and interior wall surface and also to allow more light penetration (Figure 2.26). This technique is also useful for skylights. Curtains and blinds, by reducing and redirecting sunlight, can also be used to reduce contrast and illuminate dark areas.

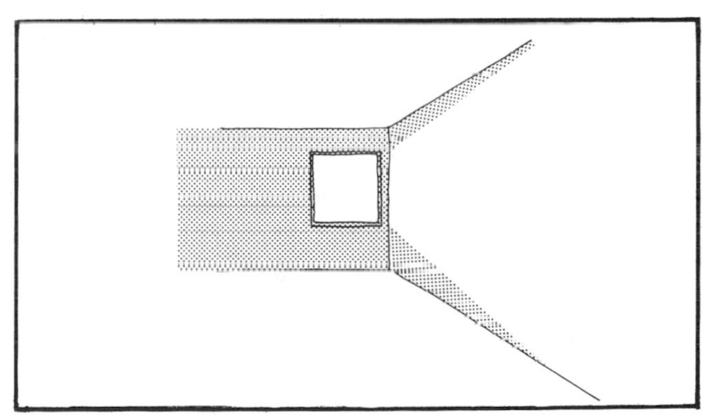

Figure 2.25 The transition of light provided by a wall.

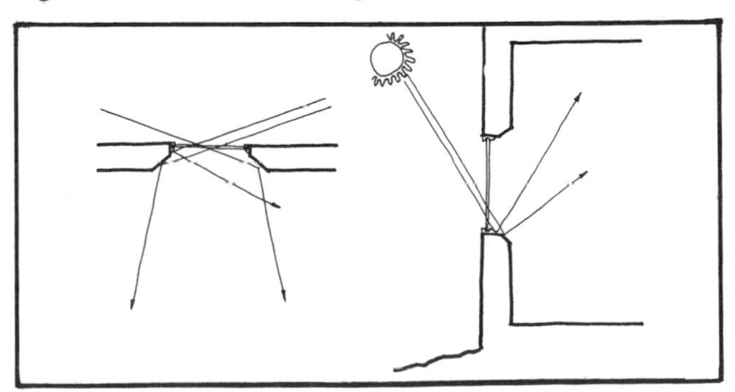

Figure 2.26 Reflected light from beveled openings.

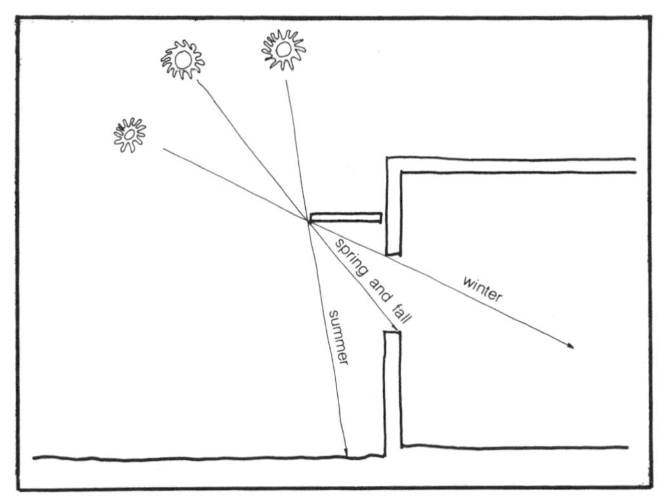

Figure 2.27 The sizing of a fixed shading device.

Solar Heating

Around the country there are many homes and businesses that use solar power in some significant way, whether to heat a swimming pool or power an entire structure. As an abundant and nonpolluting source of energy, sunlight should be used as a principal method of heating a resource-conservative house. Chapter 4 offers an extended discussion of this topic, but we mention here a few general points in passing.

Site Orientation

The slope of your site affects its solar potential at different times of the year. During the summer, total daily solar radiation will be approximately the same for northern, southern, eastern, or western grades of up to 10 percent. But in winter, southern slopes receive more sunlight because of low sun angle; this also raises microclimatic air temperatures. Southern slopes are preferred as building sites for regions with cold winters.

Solar Penetration

During cool periods, direct sunlight can be used as a method of heating. If fixed sunshades are used for protection during warm periods, they should be sized to allow partial light penetration during spring and fall and complete light penetration during winter (Figure 2.27).

Heat Transfer

The capacity of certain materials to absorb heat has been used for centuries in certain climates to both heat and cool buildings. The method depends first on the use of such "massive" materials as earth, stone, brick, or concrete to enclose space; and second on a climate of predominantly warm or sunny days and cool nights. The massive material absorbs the heat of outside air temperature and incident sunlight during the day, stores the heat within its mass, and then reradiates the heat to the cool

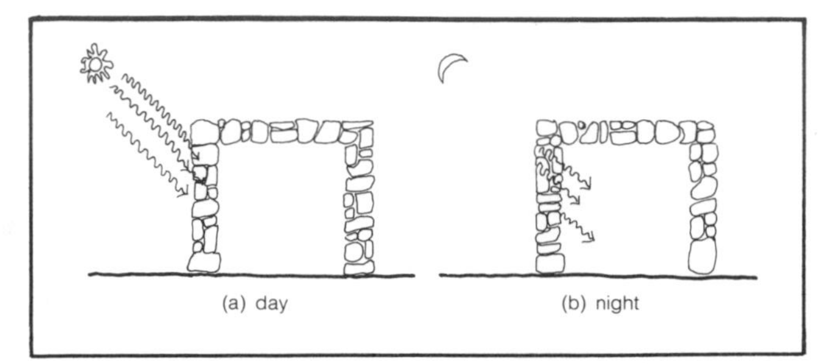

Figure 2.28 Heat transfer in massive construction materials.

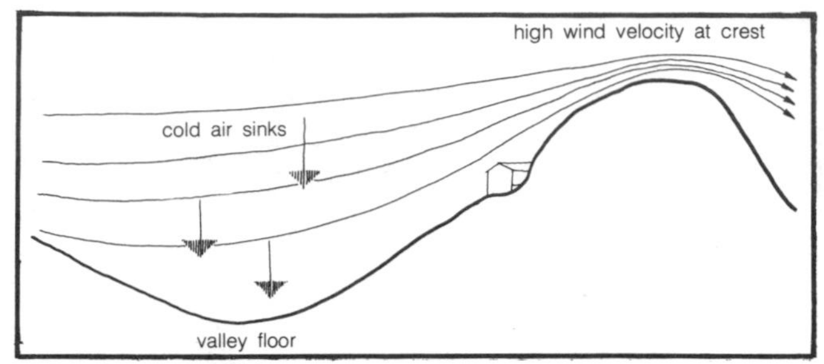

Figure 2.29 The location of a house on a hilly site.

night air (Figure 2.28). As the quantity of a thermal mass surrounding any space is increased, and hence, as its ability to store heat is increased, the temperature variation within the enclosed space will diminish, coming closer to daily and seasonal temperature averages. For this reason, caves and other underground dwellings maintain almost constant temperatures.

Sunny-day/cool-night climatic conditions are most prevalent in hot, arid regions where, unsurprisingly, we find, even today, extensive use of stone and earth as building materials and the indigenous use of underground housing. But, by incorporating thermal masses with other forms of heating and ventilating, and with the proper use of insulating materials, the range of climatic conditions in which they are applicable as space conditioners can be broadened. Chapter 4 presents several examples of contemporary houses that rely on this method of heat transfer for conditioning.

Solar Heaters

Ultimately, we seek a more predictable use of sunlight than primitive techniques may be able to supply. One solution was the development of a solar "heat-collection and storage" machine: a solar heater. The principle behind the heater is simple—a sheet of black, heat-absorbent material backed with heavy insulation to prevent thermal leakage is positioned to face the sun. Glass covers this collector to keep heat from reflecting back into space. Air or water pumped through the enclosed space is warmed and flows down into a well-insulated storage tank. The stored heat is then recirculated into interior spaces as needed.

Wind Protection

In contrast to the sun, wind should be utilized during warm periods and blocked during cool periods to aid in the natural conditioning of the house. In designing for wind protection and wind use, directions and velocities of the wind should be known in relation to cool and warm periods of the day and year. Of all climatic variables, wind is the most affected by your individual site conditions; general climatic data will probably be insufficient. Air movement along the outer surfaces of your house convects heat away from those surfaces and increases the heat transfer through building materials. Knowing the prevailing wind direction during the cool period, you can take steps to provide protection. Techniques for determining wind velocities and directions are discussed in Chapter 3.

Wind Paths

Both natural and man-made landforms and structures channel different climatic air movements into particular patterns. In all but hot, humid climates, these natural wind paths should be avoided when locating your house. On small sites there may not be much choice. If the site is hilly, the mid-portions of slopes are best, away from both high winds at the crests and cold air movements along the valley floors (Figure 2.29).

House Orientation

In areas of cold or constant winds, the house should offer as little exposed surface area to the wind as possible. Designs should be compact and clustering of houses may be considered to reduce exposed wall area. In suitable soils, semi-buried structures are a possibility. Also, where cold winds are severe, your house plan should be organized so that it turns its back on the wind. The wall facing the wind should be windowless and well insulated. Closets, storage areas, toilets, laundry, and garage can also be used as buffers on the windward side. House entrances, large windows, and outdoor areas should then be located on the protected side.

Heat losses through building materials and through door and window cracks are directly dependent on exposure to and velocity of wind. As you can see, all the various techniques discussed under solar protection and heat transfer are appropriate for wind protection.

Tornadoes, hurricanes, and other destructive winds can demolish your house. In hurricane regions (generally the same hot, humid climatic regions that require light, open structures), enclosures can be designed that allow solar protection when open and wind protection when closed. In tornado and severe wind regions, use shutters to protect your glass.

Windbreaks

Fences, bushes, trees, and other site objects acting as wind barriers create areas of relative calm on their

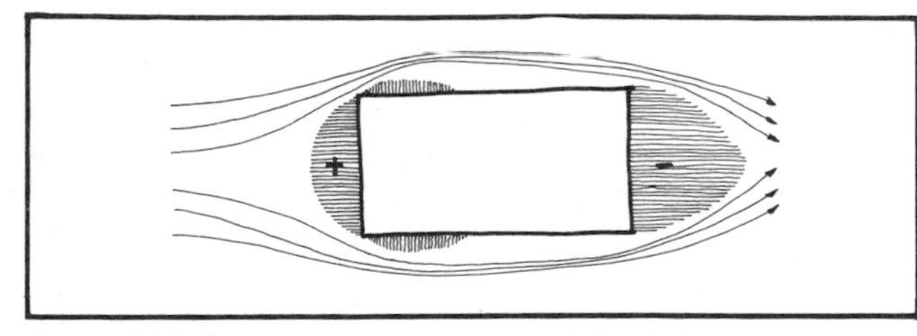

Figure 2.30 Air movement pattern around a house.

leeward side (Figure 2.30). Wind acting perpendicular to a more solid windbreak (a wall, earth berm, or building) is reduced in velocity from 100 percent at the break to about 50 percent at distances equivalent to about 10 or 15 heights from the break. Open windbreaks, such as trees and bushes, offer a maximum reduction in wind velocity of about 50 percent at a distance equivalent to about 5 heights. Evergreens, which retain their foliage throughout the year, are best for winter protection.

Windbreaks may also be an integral part of the house structure. Such protrusions as parapets or fin walls on the windward side of a house divert air movement away from other wall and roof surfaces (Figure 2.31).

Wind Use

Channeling Wind

Wind convects heat away from roof and wall surfaces. Consequently, windbreaks should be used to channel wind toward your house during warm periods (Figure 2.32). Where prevailing summer and winter winds come from the same direction, deciduous trees and bushes may be used to direct summer winds toward the house (when foliage acts as a barrier to wind movement); their winter bare branches allow cold winds to pass by, undeflected (Figure 2.33). Where summer and winter winds come from different directions, ventilation openings and windows should be placed in the direction of summer winds (Figure 2.34). If, however, such a window arrangement conflicts with proper solar protection, summer winds can be directed into the house by additional windbreaks (Figure 2.35).

In hot, humid climates, where maximum ventilation is required, the velocity of the wind, and hence its effectiveness as a cooling agent, can be increased by using windbreaks to constrict and accelerate wind flow in the vicinity of the house. Robert White shows many configurations of trees and hedges and their use in modifying air movements into and around the house (see Bibliography).

Ventilation

The use of natural ventilation in cooling the alternative home is advisable because this is one of the few existing techniques which can replace the modern air conditioner. Air conditioning is not only expensive and a

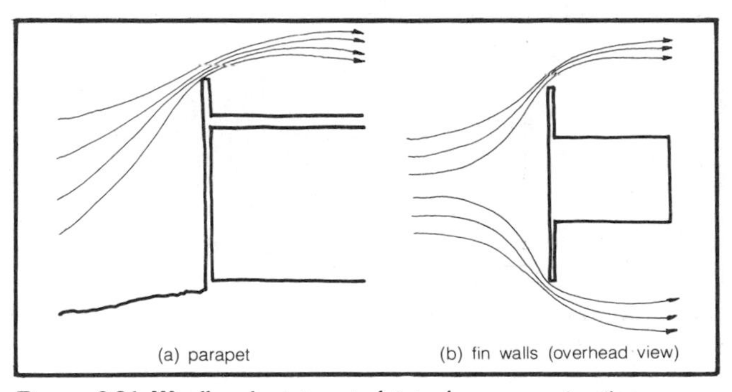

Figure 2.31 Windbreaks integrated into house construction.

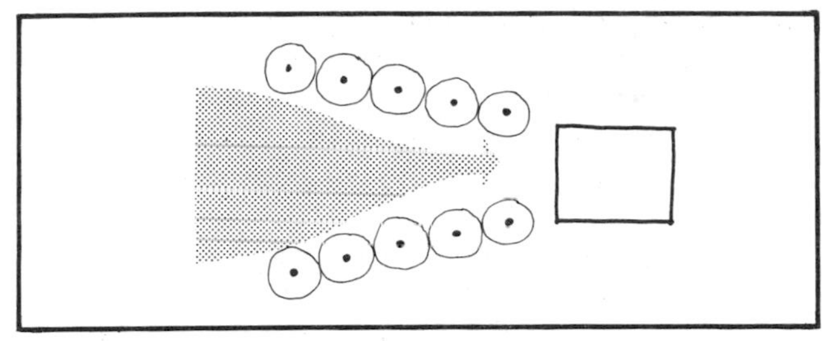

Figure 2.32 Channeling wind by trees.

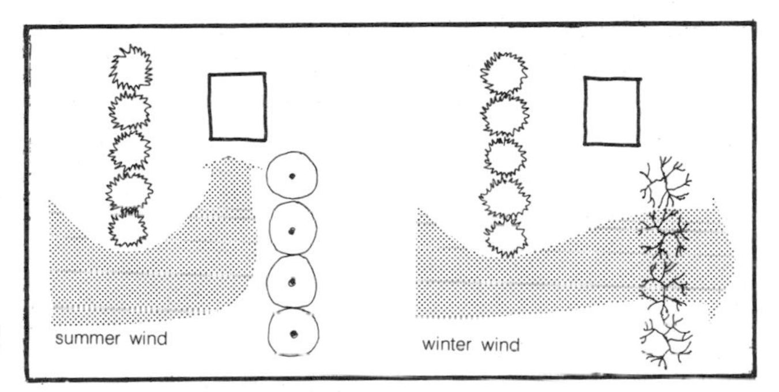

Figure 2.33 Channeling summer and winter winds from the same direction.

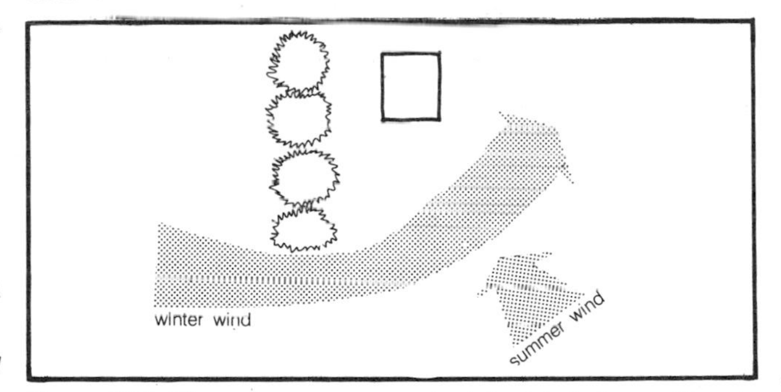

Figure 2.34 Channeling summer and winter winds from different directions.

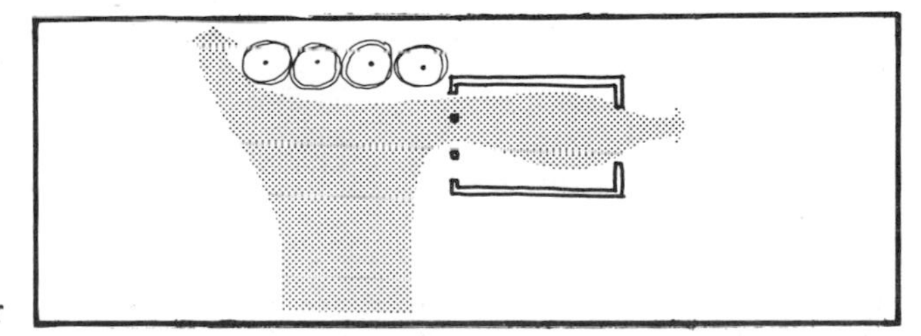

Figure 2.35 The deflection of summer winds into a house.

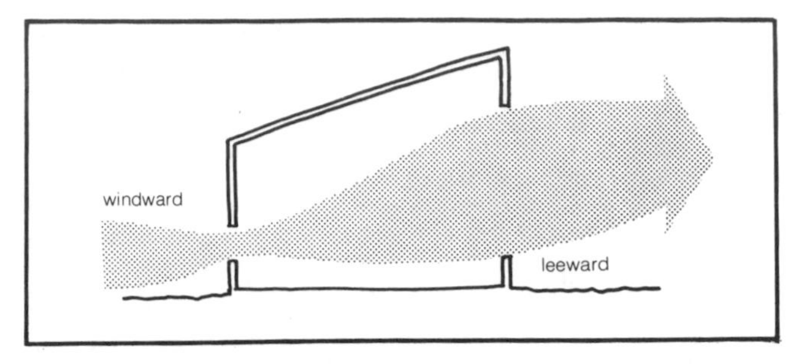

Figure 2.36 Effective ventilation.

terrible energy drain, but can also damage your health since unnatural temperature differences are created. Proper ventilation both evacuates warm or stale air from your house and cools your body by encouraging evaporation of moisture from the skin. Two natural circulation techniques can be used: wind-induced *cross-ventilation* and *gravity ventilation*.

Two openings are necessary for proper cross-ventilation: one as an inlet, preferably on the windward side of the house where air pressures are high; and the other as an outlet on the leeward side where air pressures are low. Because of these pressure differences, rooms are most effectively ventilated by using a small inlet located near the bottom or middle of the windward wall and a large outlet in any position on the leeward wall (Figure 2.36).

The pattern and velocity of cross-ventilation within a room is modified by the location of vegetation, roof overhangs, and sunshades placed near the air inlets, and also by the type of inlets used. In general, use exterior elements to increase the wind pressure in the vicinity of your inlets. For example, proper location of trees near the house increases the velocity of air movement at ground level, and, hence, air pressure. Roof overhangs and sun

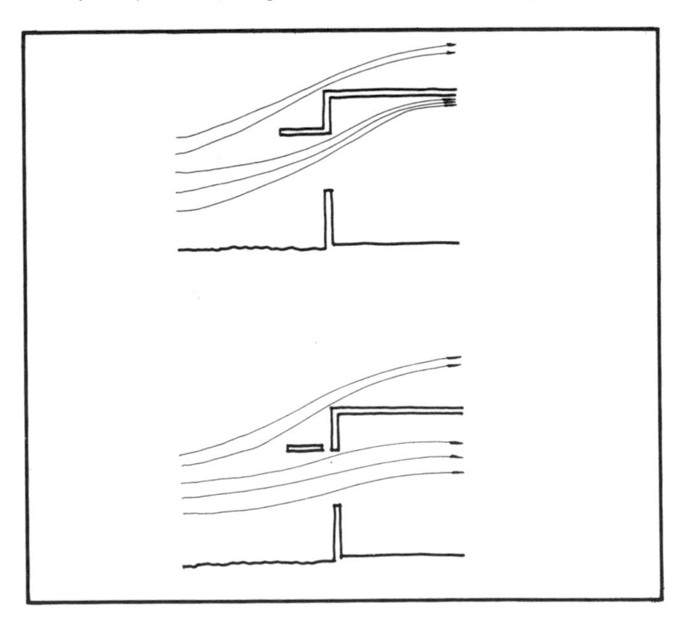

Figure 2.37 Air movement with different types of inlets.

protectors tend to trap the wind, similarly creating higher air pressures. As shown in Figure 2.37, different types of window inlets can be used to direct air either into or above living areas. And if there are no winds to use, or if the house, even with large vents, still feels too stuffy, you can resort to the use of electric fans, to be mounted in either the exhaust vent or the ceiling.

Gravity ventilation of spaces is dependent on the fact that cool, dense air displaces warm, less dense air, forcing the warm air to rise. By placing vents at different levels in interior spaces, cool air is drawn in through the lower inlet, while warm air is forced out through the higher outlet. The rate at which air circulates is directly dependent on the difference in air temperature, the height difference between the two vents, and the sizes of the two apertures. Air inlets should be as low as possible in areas likely to have the lowest air temperature, such as the north wall of the house. Keep these vents clear of shrubbery. Outlets should be located as high as possible and preferably in areas where wind movement can be used to create a suction or "stack" effect to aid in the ventilating process.

Gravity ventilation can be used for summer cooling, or in specific locations, such as bathrooms and kitchens, where air exchange rather than cooling is required. In all cases, the inlets and outlets should be closable to prevent heat transfer during cool periods.

We can also make direct use of the wind with airscoop ventilation. In this method, various types of scoops are placed on the roof or at the wind-blown sides of the house, with piping to circulate air into the home. Simple, low-power ventilating fans can successfully replace the use of such a system, usually eliminating the necessity for intricate piping arrangements. By using electric fans in conjunction with gravity air flow and intelligent shading and orientation techniques, the air temperature within a house can be significantly lowered.

The key to this claim lies in effective attic insulation and ventilation. During the course of the day the roof becomes very hot, radiating heat into the attic. An oven effect is created if this hot air is not allowed to escape; as night comes, it continues to radiate heat into the rooms below. This phenomenon is known as *heat lag*. With proper ventilation, the attic acts as an insulator by day and heat lag is eliminated at night. Roofs with a high pitch (and so a tall attic) can be ventilated adequately by placing vents at the upper and lower extremes of the attic. This method of gravity ventilation is employed when the attic height exceeds five feet, since height differences are important in determining air flow rates; on flatter roofs electric fans must be used. Although electric fans can be mounted on the roof in the same manner as vents, to simply pump air out of the attic, in yet another installation air is pumped from the *interior* of the house into the attic, where the air escapes through vents. Thus the whole house is cooled. Here the fan should be centrally located in your attic so that uniform air circulation is promoted. It is also advisable that the underside of the roof be insulated, preferably by reflective insulation. A successful approach to roof ventilation without attics is the use of a double roof.

The fans mentioned earlier should be the slow type. These are quieter, more efficient, and well worth the extra price. Never use cheap fans, since they often need frequent repairs. Large attic fans should be wired to a thermostat, to prevent unnecessary operation. Do not forget that screens and louvers cut down on fan efficiency, so vent openings must be increased.

In warm climates it is imperative that the walls of the house remain cool, due to the fact that people cool themselves by radiating heat to the walls. When wall temperatures exceed 83°F (skin temperature), people become very uncomfortable even if breezes are introduced into the room. The most efficient means of cooling walls is to insulate them properly. A relatively new and untried way of cooling walls is to ventilate them through their core. Since hot air rises and the walls have considerable height, openings at the bottom and the top of the wall generate a strong air current. It is still important that these ventilated walls be heavily insulated, and all openings should be screened to prevent insects and small animals from entering. Ventilated walls should be used only in warm climates, since heat loss can be quite severe in cold weather.

Crawl space ventilation is of only minor importance. Minimum requirements are intended to drive off moisture accumulated under the house. During hot weather, the air in crawl spaces is generally cooler than the outdoor air, so thorough venting is actually undesirable.

In all climatic regions some ventilation is needed, but in cold, temperate and hot, arid climates this requirement should have little effect on the plan of your house. However, in hot, humid climates, plan your house around the need to maximize ventilation. Rooms should be open and elongated perpendicular to the wind. Areas that generate humidity, heat, and odors—kitchens and bathrooms—should be separated from other areas and also be well ventilated (Figure 2.38). Floors should be raised above the warm ground to allow circulation under the house (Figure 2.39).

EQUIPMENT

Artificial Lighting

The two most widely used artificial light sources today are incandescent and fluorescent lamps. In homes, incandescent lights are by far the more popular. The principle behind incandescent lighting is to heat metal to such a degree that it gives off white light, a wasteful process at best.

Fluorescent Light

Fluorescent lights work on a more complicated basis: light is produced by fluorescent powders which are excited by ultraviolet energy. But it is less important to understand how fluorescent lights work than it is to understand their efficiency, an efficiency on the order of 20 percent, compared to about 5 percent for incandescent lamps. Hence they use only a quarter the power for the same amount of light. Prove to yourself which source creates more waste heat by touching both types (be careful with the incandescent one).

In economic terms, the controlling factor in choosing a light source is not the initial cost or even the lifespan of the lamp, but rather the cost of operation. Since the initial cost and lifespans of incandescent and fluorescent lights are comparable, it is a wise choice to use fluorescent lights for general lighting needs. In the winter do not be fooled into thinking you can use the waste heat for warmth; solar power does the same job far more efficiently.

There are three general categories of hot cathode fluorescent lamps: preheat, instant-start, and rapid-start. In the preheat type, there is a short delay between the time the circuit is turned on and the time the lamp produces light. The current is allowed through the cathodes to heat them; when this is accomplished the starter automatically opens the circuit so the light can start. In the instant-start variety, enough voltage is immediately supplied between the cathodes to operate the light as your fingers hit the switch. The rapid-start type, now the most common, uses low-resistance cathodes which are continually heated with but small energy losses. This variation requires lower starting voltages than

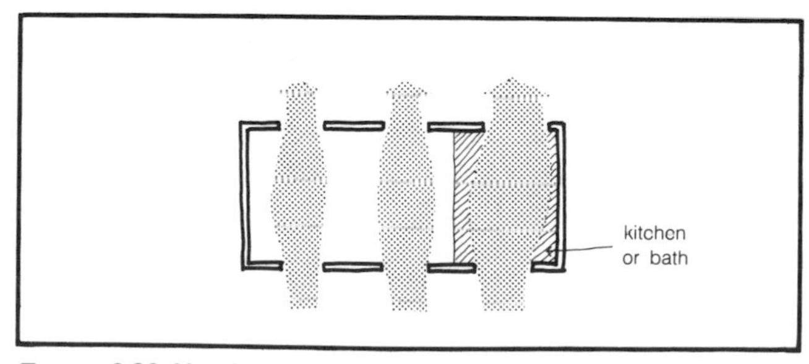

Figure 2.38 Ventilate odorous areas separately.

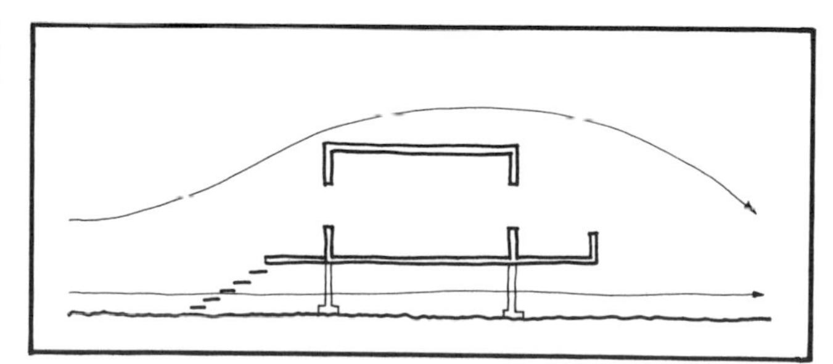

Figure 2.39 Air circulation with a raised floor.

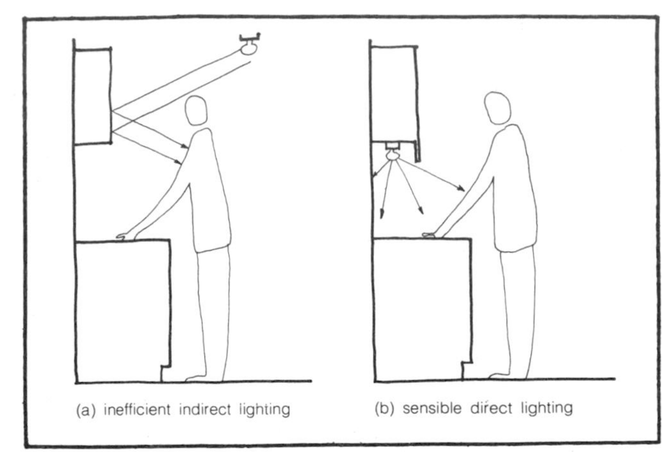

Figure 2.40 The location of kitchen lighting.

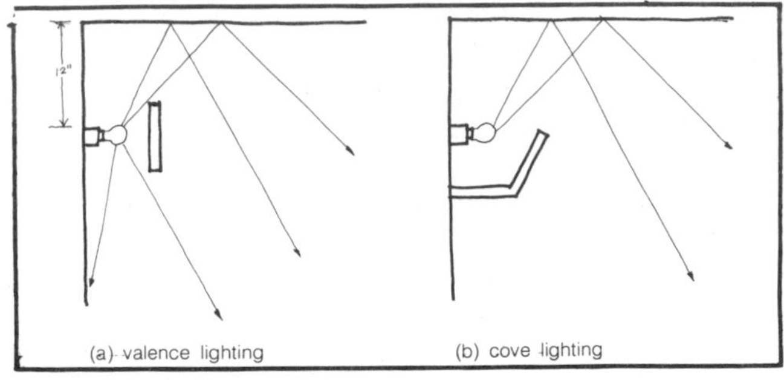

Figure 2.41 Lighting techniques.

instant-start lamps, and produces light in one to two seconds.

In a resource-conservative home, the sacrifice of one or two seconds certainly does not warrant the use of instant-start lamps. The rapid-start lamp appears to be your best choice of the three. Keep in mind that the frequency of starting a lamp is a major factor governing its lifespan. Lamp sizes range from 6 to 96 inches in length and from ⅝ to 2⅛ inches in diameter. By the way, if a windmill provides your energy, both incandescent and fluorescent lamps are available which run off direct current.

Lighting Techniques

In lighting your house, first consider the activities for which any space is used. Second, choose a means of lighting which is most practical. Direct lighting (or task lighting) should be used wherever possible. For example, one sixty-watt bulb provides plenty of light to read by—even if the rest of the room is fairly dark. This procedure is far more sensible than putting lights all over the ceiling, lighting every corner of a room to the intensity necessary for reading in one specific spot. The same concept applies to living areas and bedrooms. There will be times, however, when small islands of light do not suffice; here indirect lighting can be used.

In kitchens, most activity is concentrated on the counter tops. Common sense suggests that lighting should be concentrated there, but kitchens traditionally have been lit with one central light of staggering intensity. Instead, smaller fluorescent lights should be placed underneath the upper cabinets as shown in Figure 2.40, an arrangement less wasteful and which avoids any problem with shadows over your work area.

The rule for artificial lighting is *use direct light wherever possible*, especially in all stationary activities—reading, kitchen work, sewing, etc. Another good idea is to place low-intensity light bulbs in closets to eliminate the need to turn on all the lights in your bedroom to choose a wardrobe. (It might also please any others who are trying to sleep at the time.)

For other activities where more general lighting is needed, our tendency has been to overlight. One of the most pleasant commodities we may discover in an alternative lifestyle is atmosphere. It takes only enough light to be able to read expressions to carry on a conversation. There are many satisfying ways to achieve indirect lighting, a few of which are diagramed in Figure 2.41. And, if we can enjoy any "inconveniences" which may occur as a result of using naturally supplied commodities, it will all be worthwhile. If the windmill should be calm for too long and there is no power for lighting, candles give ample light, burn cleanly, and provide wonderful atmosphere. Use common sense and take advantage of nature's surprises to produce a well-lit home.

Interior Coloring

The cheapest and most pleasant interior lighting during daytime hours is through natural sources. But in considering a self-sufficient system, we must also consider heat loss; windows, which account for approximately 15 percent of the overall heat loss, should be kept to a minimum. This is why interior coloring is so important. Light colors, because of their reflective capacities, spread the light around, giving you the best "mileage" from any particular light source (see Table 2.1). White, naturally, is the best color for this purpose, but other reflective colors can be used to relieve monotony.

Colors can also control your "psychological" room temperature to some extent. That is, colors with shorter wave lengths (green, blue, violet) create an impression of

Table 2.1 Reflectivity of Colors[a]

Color	Light Reflected
White	80–90%
Pale pastel (yellow, rose)	80%
Pale pastel (beige, lilac)	70%
Cool colors (blue, green pastels)	70–75%
Full yellow hue (mustard)	35%
Medium brown	35%
Blue and green	20–30%
Black	10%

Notes: a. From K. Kern.

being cold, while colors with longer wave lengths (yellow, orange, red) appear warm. Use warm colors on the north side of your house, or areas where there is minimal sunlight; on the south side, cool colors should be used.

Appearance is markedly affected by contrast. Thus a central chromatic area appears brighter if surrounded by a sufficiently large and relatively dark area, but dimmer if surrounded by a relatively light one. Interior glare content, for this reason, becomes an important category to be evaluated, since 1 percent contrast lost by glare requires a 15 percent increase in illumination. Comfortable reading on a glossy table requires considerably greater illumination than on a dull surface.

Lighting Standards

There has been a tendency over the years to continually raise the recommended light values for various tasks. Largely as a result of an intensive campaign financed and conducted by the electrical industry, these high recommendations have been followed closely. We can examine the standards in Table 2.2 as a representative example. Ken Kern, author of *The Owner Built Home*, writes: "Illumination experts specify an artificial light intensity of from 50 to 100 footcandles for most visual tasks. But experts in the field of light and color conditioning warn against the use of more than 30 to 35 footcandles. . . . Further light intensity is apt to cause visual distraction."

So what exactly is the minimum amount of light a person needs to be comfortable? William Lam, a lighting consultant in Cambridge, Massachusetts, has a very simple definition: "Good lighting is lighting which creates a visual environment appropriate and comfortable for the purpose. This is a criterion that is measurable only by people using their own eyes and brains." He summarized his insights concerning lighting in six points:

> First: we see well over a tremendous range of light levels.
>
> Second: *we see by the balance of light more than by the quantity.*
>
> Third: once 10–15 footcandles has been achieved, task visibility can be improved far more easily through quality changes rather than adding quantity.
>
> Fourth: *apparent brightness is determined by brightness relationships, not absolute values.*
>
> Fifth: we look at tasks only a small part of the time, but react to the environment all of the time.
>
> Sixth: whether our response to the environment is to be favorable or unfavorable *cannot be forecasted or explained by numbers—but by the exact design of everything in relation to what we want to see.*

To set absolute personal lighting standards is impossible; therefore test for yourself the minimum amount of lighting you need for different activities.

Table 2.2 Lighting Standards[a]

Task	Footcandles Required
Kitchen Activities:	
Sink	70
Working surfaces	50
Table games	30
Reading and writing:	
Books, magazines, newspapers	30
Handwriting, reproduction, poor copies	70
Desks (study purposes)	70
Music:	
Reading simple scores	30
Advanced scores	70
Grooming	30
General lighting:	
Passageways	10
Relaxation and recreation	10
Areas involving visual tasks	30

Notes: a. From I.E.S. *Standards for Various Home Activities.*

Fireplace Design

Many of the changes made in creating the alternative home concern combining older, simpler, less energy-intensive designs with the technological advances of the present. A typical modern home makes scant use of its fireplace as a heat source, a waste implicit in the rather crude fireplaces now being installed. We are principally interested in heat-ventilating and free-standing fireplaces, both exceptional finds today. Few people are even familiar with some of the basic principles of fireplace operation.

Perhaps one of the most critical aspects of efficient use is the correct tending of the fire. A fire should always burn brightly and briskly. Whenever starting a fire, use a small amount of wood. Add more timber only after this initial wood is burning vigorously. Never force a fire to eat its way through a pile of wood. Obviously, other fuels should be burned with the same diligence and economy.

A healthy fire uses a tremendous amount of air both to burn its fuel and to dispose of waste gases and smoke. In order to accommodate this demand, the fire is raised above the hearth by means of a grate. Studies have shown that this step increases burning efficiency by 15 percent. Vents or windows must also bring air into the room from the outside; lacking air in a house weather-stripped and insulated properly, a fire does not burn well and smoke ebbs into the room.

Equally important is the exhaust of hot gases and smoke. Here we are concerned with the design of the chimney breast, throat, and flue. The required depth of the chimney throat for all fireplaces is between four and five inches. The small size of the throat, which can be adjusted by a damper, prevents drafts of cold air from blowing down your flue; yet it permits smoke and waste

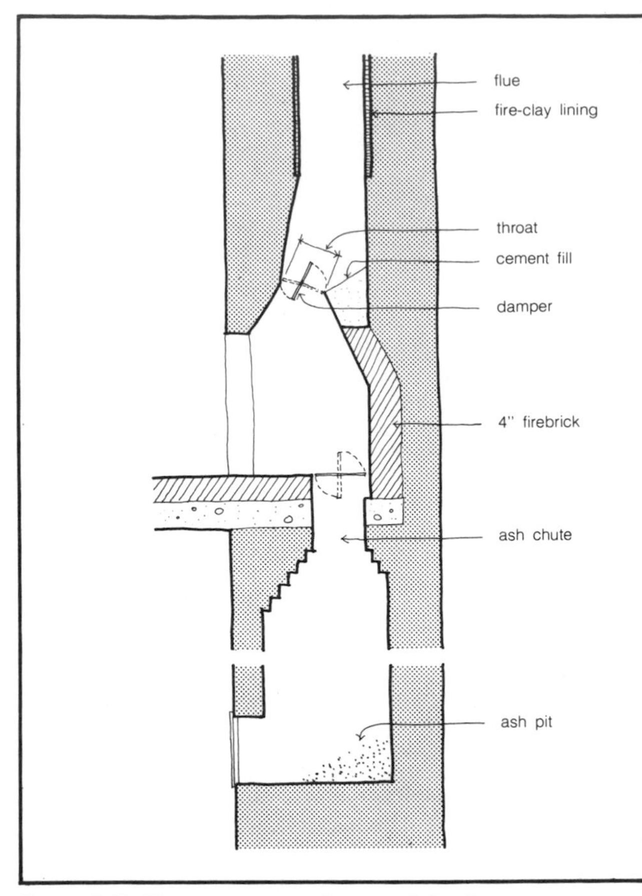

Figure 2.42 The construction of a typical fireplace.

gases to rise. (Figure 2.42). Because smoke spirals as it rises, round flues are more effective than square or cornered ones. The strength of the draft is also related to the temperature of the flue lining, so your flue should be well insulated. *Architectural Graphic Standards* (Ramsey) features data by which flue diameters can be calculated.

The only usable warmth escaping from the ordinary fireplace is radiant heat. Roughly three-quarters of the heat produced by the fire is used to heat up exhaust gases, smoke, and the wall of the fireplace. In addition, a fire needs air to burn properly, so cold air is necessarily sucked into the room. Through the extensive use of ducting, heat-ventilating fireplaces (Figure 2.43) take advantage of this situation. They not only heat the fresh air entering the room, but also utilize 40 percent of the heat given off directly by the fire. The basic arrangement involves an intake vent connected to ducting which leads air across the back of the firebox. This hot air is then ducted to vents above the mantel. A natural air current is created by the hot air rising in the ducts as well as by suction created through the burning fire. A number of commercial models are available. They are provided with only a minimum of fireproof ducting, so if you wish to heat other rooms, cost for this rather expensive item must be considered. Fans can also be installed to accelerate the air flow.

Although the construction of fireplaces appears to be fairly rigid, fireplaces have been built chiefly out of salvaged materials, using, for example, a discarded water boiler as a fire box. For details see *The Owner Built Home* (Kern). Another effective heating device is the free-standing fireplace with a metal hood. Heat radiates in all directions from the fire, and once the hood is warm, it too gives off tremendous heat.

MATERIALS

Due to our once-abundant supply of energy, there developed a tendency to use building materials—aluminum, steel, concrete, and plastics—which require intensive processing. Until this last century, the process of converting raw materials into building materials was accomplished principally by human labor. Because of this

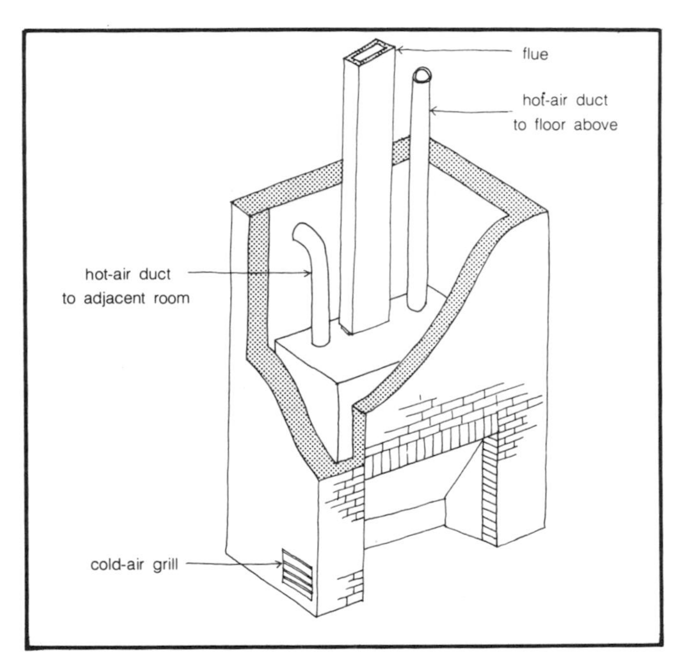

Figure 2.43a The construction of a heat-ventilating fireplace.

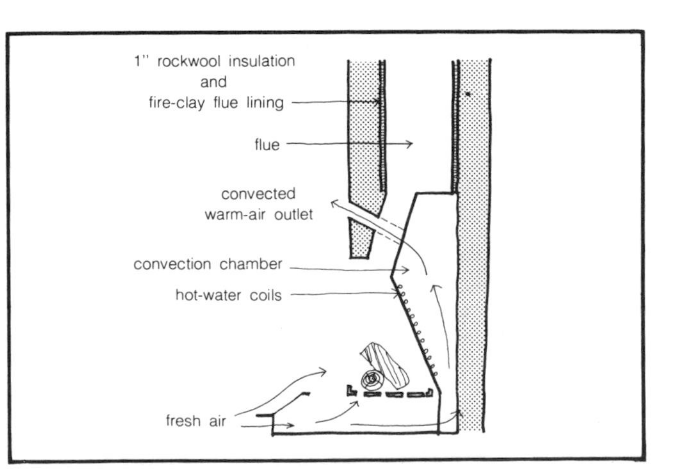

Figure 2.43b Air flows in a heat-ventilating fireplace.

direct relationship between human effort and its product, the materials were used economically and constructed to last. We can still find wooden houses erected over two centuries ago, or stone houses dating back two millennia. Despite the fact that highly processed materials are inherently more durable, their high initial cost (attributable to expensive processing) requires that they be used in the smallest quantities possible. Today, most structures are designed and built under the guideline of *lowest initial cost*; consequently, they have a very short useful life.

Where building ordinances are not a problem, natural or alternative materials can be incorporated into design and used, with a little know-how, to produce sound construction. There are various sources for these low-cost and usually low-energy materials. Scrounge secondhand building materials—bricks, timber, steel, and glass—from demolition sites and dumps. One enterprising couple obtained enough timber to build their own home by offering to demolish dilapidated structures for other people, in exchange for the timber they could salvage. Or use processed materials that are considered waste products. Scrap bins in lumber yards are a good source. And bottles make excellent translucent bricks; you get not only a wall but light penetration as well. Tin cans have produced shelter. Sulphur, a waste product from oil and coal production, can be used for cement, and with sand aggregate, produces sulphur concrete (see Rybczynski's article, "From Pollution to Housing," listed in the Bibliography). One architecture student in London constructed a four-room dwelling from scrap materials and garbage collected within one mile of where the dwelling was built.

Natural materials that are readily available—stone, earth, driftwood, trees—require only manual energy. For centuries, people made stone cottages, log cabins, and adobe homes. These techniques, although a lost art among urban dwellers, can still provide structures that are as sound as conventional homes and often have superior insulating properties. However, some knowledge of their limitations is necessary, and some skill is required if they are to be used effectively.

The energy required to produce and transport the materials of a typical house is equal to about ten years of household energy consumption. To reduce the need for transporting materials, wood should be the dominant building material in forested areas, earth in dry climates, stone where it is available, grasses where they grow rapidly and easily, and so forth. Where possible, materials from the building site itself should be incorporated. Useful information on design and construction techniques is contained in *The Owner Built Home* (Kern), *Your Engineered House* (Roberts), and *Adobe—Build it Yourself* (McHenry). The National Bureau of Standards has also undertaken extensive experiments to establish the structure, sound reduction, heat transfer, water permeability, and other relevant properties of house construction, for a wide range of materials. Their work is especially valuable because of an emphasis on materials for low-cost construction.

Primitive Architecture

There are four general traits which characterize primitive architecture: unsupplemented use of natural, locally available building materials and local construction skills; planning and massing as a result of specific functional requirements and site conditions, regardless of symmetry and generally accepted taste; an absence of ornamentation which is not part of the structure; and the identity of enclosing form and enclosed space. This architecture is a simple and original response, the most economic shaping of space and form for the maximum benefit of body and soul. Several interesting examples are found in *Native Genius in Anonymous Architecture* (Moholy-Nagy), *Shelter* (Kahn), *Architecture Without Architects* (Rudofsky), and in an article by Suzanne Stephens, "Before the Virgin Met the Dynamo." We mention a few general types in the next section.

Types and Techniques

Troglodytic Shelters: In areas where soil and climate conditions are suitable, men have lived in caves throughout history. Simple caves can be enlarged and altered if the local rock or earth is sufficiently soft and porous. Because these shelters are below ground and thus insulated by the earth, they are usually cool in summer and warm in winter.

Nomadic Architecture: Nomadic structures must be readily portable as the lifestyle of the nomad is above all one of transience and motion, moving and changing with trade, livestock, and the seasons. Some examples include the tipi of the American Indian, the tents of the Bedouin, the yurts of the tribes of Asia and Asia Minor, and the igloo of the Eskimo.

Aquatic Architecture: The proximity of a body of water has always been an important consideration in siting a community. For thousands of years people have been living in houseboats, in pile dwellings, and in more conventional structures in canal cities linked by waterways rather than streets. An expanse of water serves as a cooling plant during the hot season; a bath or drink is never far away. On a houseboat your home is also your transportation.

Tree Houses: People of many cultures have lived in and amongst the tree tops in tropical and near-tropical zones. In cooler zones one is shaded from the summer sun by a leafy parasol while the bare branches of winter let the sun shine through. Air circulation is excellent.

Movable Houses: In some parts of the world, when a family or community moves, the house is either completely disassembled or picked up in large sections and then moved to a new location, provided the distance is not too far. These structures are usually lightweight and are often made from grasses or reed.

Towers: The usual function of towers has been either symbolic or defensive. They often express religious sentiments: grief, faith, hope, or prayer. In some agricultural cultures they are used for grain storage or as pigeon roosts.

Arcades and Covered Streets: Arcades and covered streets are an example of private property given over to an entire community. Their presence reminds us that city streets exist for people. They provide shelter from the elements, offering shade from the summer sun and protection from winter wind and rains.

Building Materials

Materials and climate determine what is possible in terms of shelter within a particular context. The fact that over three hundred house forms have emerged from nine basic climate zones suggests that these specific forms derive more directly from socio-cultural factors than from restraints of either climate or materials. In spite of this diversity, almost all examples of vernacular architecture emphasize simple yet profound solutions to the problems of human comfort, using elementary building materials which require little or no basic transformation to be utilized.

The cost of a building material is roughly proportional to the ecological damage caused by its removal and refinement. In aesthetic terms, the less molecular rearrangement a material has undergone, the better it feels to be around and the more gracefully it will age. More often than not, natural, locally available building materials are the least refined and least polluting; they may also be self-regenerating.

Adobe: Adobe is the world's most abundant building material since its primary ingredient is earth. It is also the oldest and most popular form of earth construction. Ingredients vary but usually consist of clay soil mixed with sand, shredded grass, roots, straw, pottery shards, or gravel. These are formed into bricks and set out to bake in the sun. Adobe houses are most practical in areas with little rainfall; wet climate may erode an adobe structure into a pile of mud. It is best used on a well-drained site. Walls are short and also thick for strength and insulation; doors and windows are placed away from corners and the roof is supported by rafters.

Rammed Earth: To make rammed-earth blocks, one uses a dampened mixture of humus-free soil and a stabilizing agent (usually portland cement) which is formed into blocks in a hand-operated press. (The Cinva-Ram press is commercially available.) Walls of rammed earth must be reinforced by a corner or pillar every ten feet. As with adobe, it is best to build on dry, flat, solid ground or rock as the compression of block upon block can cause sinking or settling.

Stone: There is little information available on laying up stone, partly because of the traditional secrecy of stone masons and partly because stone differs so much from area to area. Stonework is heavy and time-consuming but the materials are a gift of the earth and the structure often blends unobtrusively with the landscape. One can use mortar or simply pile stone upon stone as did the ancient Egyptians in their pyramids.

Sod: Sod is a good building material for treeless grasslands. It is usually cut directly from the earth in long solid ribbons which are then sectioned into blocks measuring about 4-by-24-by-36 inches with a sharp spade. The sods are laid without mortar, grass down like huge bricks. Walls are two or three sods thick, with staggered joints. Every third or fourth layer is set crosswise to the vertical pilings for stability. The walls, which are excellent insulators, settle about six to eight inches in the first year. Sod roofs are also excellent insulators, but miserable water repellers; a wooden roof is best.

Snow: Although snow can be used only in the polar regions, there it has proved to be an instantly available building material with superior insulating qualities. After a bit of experience, one need only cut blocks of suitably dense snow and lay them in an upward spiraling fashion to form a dome. With a little body heat and a small heat source, the interior walls form an ice glaze which solidifies the structure. In subzero exterior conditions, Eskimos are often stripped to the waist in the warm interior of their igloo.

Wood: Wood is an ideal building material for many climates and cultures as it is relatively light, strong, durable, pleasant smelling, easy to work, readily available, and can be regenerated. Forms vary from simple twig and brush shelters and log cabins to the most elaborate, ornate structures conceivable. It is also a material which lends itself to recycling if one salvages from old buildings and other discards of the twentieth century.

Grasses: Various grasses have been employed as building materials throughout history. Reed has been used for frameworks, walls, and roofs; it insulates well and is light and easy to work with. Drawbacks include easy flammability and a relatively short life. Bamboo is flexible yet tough, light but very strong. This versatile plant splits easily in only one direction; it is pliant or rigid as the situation demands; it can be compressed sufficiently to remain sturdily in place in holes; after heating, it will bend and retain its new shape; one of the world's fastest growing plants, it also grows straight. As with reed, typical construction must include lashing of structural members.

In tropical climates woven grasses are often formed into houses. Here woven matting is supported by a simple sapling superstructure to form walls and a roof. Thatch is perhaps the world's most commonly used roofing material. Although it is a time-consuming process to thatch a roof, the end product is a waterproof, insulated, biodegradable roof of reeds, straw, or fronds. Truly waterproof thatch requires a steep pitch and the overlap of all its elements.

Baled hay was used as a building material in arid grassland regions of the United States where the soil is too sandy for sod houses. Most of these structures now have concrete foundations or wooden floors. The best hay to use is harvested in the fall: it is tough and woody. The bales are two to four feet long and about two feet square. They are stacked like bricks, one bale deep with the joints staggered. Mortar can be used; man-high wooden poles are driven through the bales to hold them firmly together. The roof is usually a wooden frame with shingles. The insulating properties of baled hay are obvious. Equally obvious drawbacks include its fire hazard and the fact that hay is a choice breeding ground for many insects.

Skin and Fabric: Skins or woven fabrics stretched over a light portable framework have provided shelter for nomadic people throughout the ages. In the summer fabric tents are cooler than skin tents, although in the rainy season skin tents are far more waterproof. Typical materials include buffalo hide, woven sheep and goat hair, felt and canvas, and sheep and goat skin. Commonly available in the United States, canvas is easy to

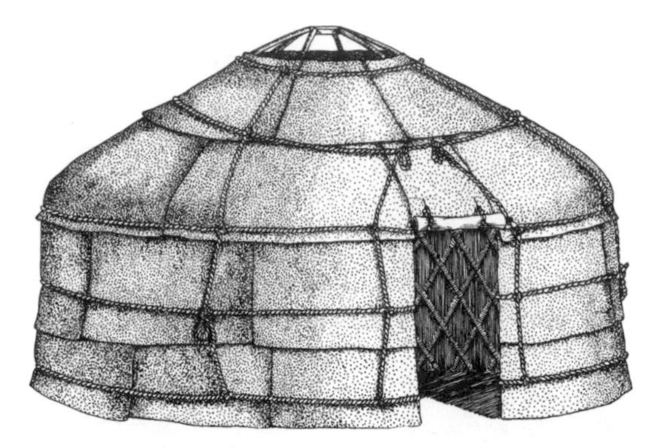

work with, cheap, covers space quickly, and lights with a translucent glow. It is relatively short-lived (five to ten years), biodegradable, and can be rendered water- and fire-resistant. Cool in the summer, canvas can also be adapted to winter conditions with proper insulation and a heat source.

Implications

The more we do for ourselves, the greater our individual freedom and independence. But rather than turn our backs completely on present-day technology, we should seek a responsible and sensitive balance between the skills and wisdom of the past and the sustainable products of the twentieth century. To strike this balance, we have much to learn from the past. These so-called "primitive" or "vernacular" structures, built by self-taught people on the basis of experience acquired through generations, were in one sense ideal: they usually fulfilled the needs and aspirations of a community in a design which was both serviceable and timelessly beautiful.

All architecture testifies to the essential nature of its creators, revealing through physical expression the private history of a culture, its ongoing struggle for material and spiritual survival. If the maintenance of human life is based on economy, the first premise of community life is the organization and upkeep of our resources, taking into account both human needs and environmental factors. Most vernacular architecture reflects an organic relationship to its setting; the builders and inhabitants felt no need either to dominate or to submit to their surroundings. They achieved a sort of mutual coexistence in accord with the basic functions of both man and nature, a spontaneous and continuing experience of peoplehood within a community of experience quite a bit larger than merely human. We of the twentieth century have a long way to go in reordering our priorities. We have yet to learn that any significant departure from that greater community of experience is, in fact, pollution.

Bibliography

American Society of Heating, Refrigerating and Air-Conditioning Engineers. 1972. *Handbook of Fundamentals*. New York.

Anderson, L.O. 1971. *Wood Frame House Construction*. Los Angeles: Craftsman Book Company of America.

______. 1973. *How to Build a Wood Frame House*. New York: Dover Publications.

Aronin, J. 1953. *Climate and Architecture*. New York: Reinhold Publishing.

Banham, R. 1969. *The Architecture of the Well-Tempered Environment*. Chicago: University of Chicago Press.

Blackburn, G. 1974. *Illustrated House Building*. Woodstock, New York: Overlook Press.

Brand, S. 1974. *Whole Earth Epilog*. Baltimore: Penguin.

Caudill, W.W.; Crites, S.E.; and Smith, E.G. 1951. *Some General Considerations in the Natural Ventilation of Buildings*. Texas Engineering Experiment Station, Research Report no. 22.

Caudill, W.W.; Lawyer, F.D.; and Bullock, T.A. 1974. *A Bucket of Oil: The Humanistic Approach to Building Design for Energy Consideration*. Boston: Cahners Books.

Chermayeff, S., and Alexander, C. 1963. *Community and Privacy: Toward a New Architecture of Humanism*. Garden City, New York: Doubleday & Co.

Conklin, G. 1958. *The Weather Conditioned House*. New York: Reinhold Publishing.

Daniels, F. 1964. *Direct Use of the Sun's Energy*. New Haven, Connecticut: Yale University Press.

Danz, E. 1967. *Sun Protection: An International Architectural Survey (Architecture and the Sun)*. New York: Praeger.

Dietz, A.G.H. 1971. *Dwelling House Construction*. Cambridge, Mass.: Massachusetts Institute of Technology Press.

Drew, J. and Fry, M. 1964. *Tropical Architecture in the Dry and Humid Zones*. New York: Reinhold Publishing.

Freestone Collective. 1974. *Dwelling*. Albion, Calif.: Freestone Publishing.

Fry, M., and Drew, J. 1956. *Tropical Architecture in the Humid Zone*. New York: Reinhold Publishing.

Geiger, R. 1957. *The Climate Near the Ground*. Cambridge, Mass.: Harvard University Press.

Gray, R.L. 1974. *The Impact of Solar Energy on Architecture*. Master's thesis. University of Oregon.

International Conference of Building Officials. 1973. *Uniform Building Code*. 1973 ed. Whittier, Calif.

Kahn, L., ed. 1974. *Shelter*. Bolinas, Calif.: Shelter Publications.

Kaufman, J.E. 1966. *IES Lighting Handbook*. 4th ed. New York: Illuminating Engineering Society.

Kern, K. 1972. *The Owner Built House*. Homestead Press.

Libby-Owens-Ford. 1951. *Sun Angle Calculator*. Toledo, Ohio: Libby-Owens-Ford Glass Company.

Lowry, W.P. 1967. "The Climate of Cities." *Scientific American*, vol. 217, no. 2.

Lynch, K. 1971. *Site Planning*. 2nd ed. Cambridge, Mass.: Massachusetts Institute of Technology Press.

McHarg, I. 1969. *Design With Nature*. Garden City, New York: Doubleday & Co.

Moholy-Nagy, S. 1957. *Native Genius in Anonymous Architecture*. New York: Horizon Press.

Moorcraft, C. 1973. "Solar Energy in Housing." *Architectural Design* 43:656.

National Association of Home Builders Research Foundation. 1971. *Insulation Manual*.

National Bureau of Standards. 1973. *Technical Options for Energy Conservation in Buildings*. Technical Note no. 789. Washington, D.C.

National Oceanic and Atmospheric Administration. 1968. *Climatic Atlas of the United States*. Washington, D.C.: NOAA, U.S. Department of Commerce.

Olgyay, V. 1963. *Design with Climate: A Bioclimatic Approach to Architectural Regionalism*. Princeton, N.J.: Princeton University Press.

Pawley, M. 1973. "Garbage Housing." *Architectural Design* 43:764–76.

Phillips, D. 1964. *Lighting in Architectural Design*. New York: McGraw-Hill.

Portola Institute. 1971. *Whole Earth Catalog*. Menlo Park, Calif.: Portola Institute.

Ramsey, C.G., and Sleeper, H.R. 1972. *Architectural Graphic Standards*. New York: John Wiley & Sons.

Robinette, G.O. 1972. *Plants, People, and Environmental Quality*. Washington, D.C.: National Park Service, U.S. Department of Interior.

Rubenstein, H.M. 1969. *A Guide to Site and Environmental Planning*. New York: John Wiley & Sons.

Rudofsky, B. 1965. *Architecture Without Architects*. New York: Museum of Modern Art.

Rybczynski, W. 1973. "From Pollution to Housing." *Architectural Design* 43:785–89.

______. 1974. "People in Glass Houses. . . ." *On Site Magazine*.

Stein, R.G. 1973. "Architecture and Energy." *Architectural Forum* 139:38–58.

Stephens, S. 1974. "Before the Virgin Met the Dynamo." *Architectural Forum* 139:76–78.

"The Climate Controlled House": a series of monthly articles. *House Beautiful*, October 1949–January 1951.

United Nations. 1950. *Adobe and Rammed Earth*. New York: United Nations, Department of Economic and Social Affairs.

______. 1971. *Climate and House Design*. New York: United Nations, Department of Economic and Social Affairs.

United States Geological Survey. 1970. *The National Atlas of the United States of America*. Washington, D.C.: U.S. Geological Survey, U.S. Department of Interior.

White, R.F. 1954. *Effects of Landscape Development on the Natural Ventilation of Buildings and Their Adjacent Areas*. Texas Engineering Experiment Station, Research Report no. 45.

Wolfskiill, L.A. (n.d.). *Handbook for Building Homes of Earth*. Washington, D.C.: Division of International Affairs, U.S. Department of Housing and Urban Development.

ELECTRICITY

Windmills and dams

Demythologizing volts and watts

Figuring your electrical needs

Measuring wind and water:
what you need to know and how to know it

Choosing equipment:
what you need to get and where to get it

Small-Scale Generation Of Electricity From Renewable Energy Sources

GENERATING YOUR OWN ELECTRICITY

Most of us living in the United States have become so accustomed to cheap, abundant electrical power, available at the flick of a switch, that it is usually considered an absolute necessity. While it would probably be quite easy for some people to give up air conditioners, dishwashers, trash compactors, and that array of silly electric gadgets that are supposed to make our lives easier—electric toothbrushes, carving knives, can-openers, tie racks, and the like—it is more painful to contemplate doing without the hi-fi, electric lights, and refrigerator. People, of course, have done without all these things during most of our country's history, and indeed the majority of the world's people still does without them today. So they aren't necessary—but they are nice.

The question is, do we have to contribute to the long and growing list of environmental insults associated with the centralized production of electricity just to enjoy a few relatively simple pleasures? Can't we somehow "tune out" the big utility and "tune into" more natural, renewable, nonpolluting sources to generate our own modest supply of electricity?

This chapter explores in some detail two techniques for doing just that. Small-scale generation of electricity from the wind or from a stream can be a realistic alternative right now if you live in the country. Unfortunately, we're going to have to wait a few more years, until solar cells become quite a bit cheaper, before self-sufficiency in electricity will be feasible in the city.

At this point it is useful to point out that electricity is a very *high* form of energy. By this we mean that it is extremely versatile; it can be used, for example, to power a motor, process information in a computer, heat and illuminate our rooms, electroplate the chrome onto our bumpers, and create sound and images in our entertainment equipment. Try to imagine doing all that with any other form of energy (low-temperature heat, for example) and you'll get an idea of what a marvelous tool electricity is.

As another indicator of its status in the hierarchy of energy forms, note that it takes about three units of high-temperature heat energy in a steam powerplant to generate one unit of electric energy; this is why it is inherently wasteful to use electricity for most simple heating applications. For cooking and space heating, you should turn to alternatives such as solar power and methane, saving the electricity for tasks where its unique properties are required. In other words, plan on using as little electricity as possible.

Obviously, something as useful as electricity is not without its price, even if you generate it yourself. The price combines dollars with physical and mental effort, and the three can be traded off, one against the other. To produce it cheaply, you must become involved with your energy. Some of you will now be receiving your first exposure to electrical theory; it may not be the easiest of all exercises, but hang in there. The concepts are relatively simple and the calculations will be kept on a level that you should be able to handle with a minimum of pain.

Some Basic Electricity

The amount of theory required to design a home electrical system is fortunately small. One need only understand the relationships between five basic concepts: current,

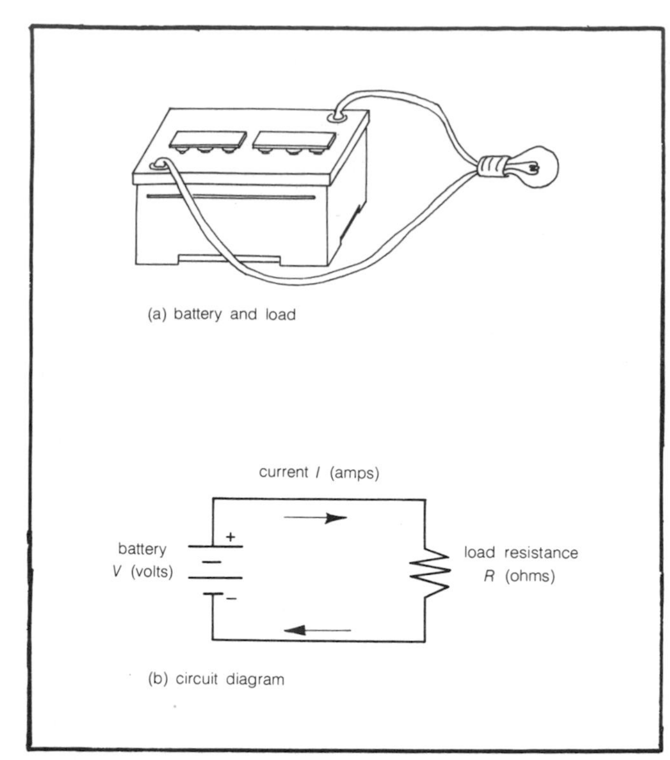

Figure 3.1 A simple electrical circuit.

voltage, resistance, power, and energy.

Consider the very simple electrical circuit of Figure 3.1, which consists of a battery and a resistance load (e.g., a lightbulb) connected together by some wire conductors. The battery forces electrons to move through the circuit and, as the electrons move through the load, work is done (in this case the bulb emits heat and light). The number of electrons passing a given point in the circuit per unit time is called the *current*. Current is measured in amperes (amps, for short) where one amp corresponds to the flow of 6.24×10^{18} electrons per second.

In this battery-driven circuit, the electrons are always moving in the same direction down the wire; this is called *direct* current (dc). The current that you get out of the wall plug at home is *alternating* current (ac), in which the electrons head in one direction for a short time, then turn around and head back the other way for a while. This back-and-forth movement is characterized by its *frequency*, how many times it changes direction; in the United States that frequency is 60 cycles per second (now called 60 Hertz). Power companies prefer to work with ac because it is easier to generate and because it is easier to change the voltage from one level to another, which they must do to avoid heavy losses in their transmission lines.

Back to our circuit. The battery supplies a certain amount of *voltage*. Voltage is a measure of the "pressure" that is trying to force electrons down the wire. Increasing the voltage across a given load will increase the current through the load. Voltages are measured, handily enough, in volts and these may also be either ac or dc. A car battery typically puts out 12 volts dc and the wall socket voltage is somewhere between 110 and 120 volts ac.

The load in our circuit is characterized by its *resistance* to the flow of electrons. The amount of this resistance depends upon the material, its composition, thickness, density, temperature, etc. Insulators have very high resistances while conductors have very much lower resistances. The resistance is measured in ohms (and given the symbol Ω).

So now we have voltages, currents, and resistances. The equation relating the three is known as Ohm's law and states that

E. 3.1 $$V = IR$$

V is voltage in volts, I is current in amps, and R is resistance in ohms. For example, a 12-volt battery will suppply 6 amps to a 2-ohm load.

As electrons move through a load, they do work: lighting a lightbulb, turning a motor, heating a toaster, etc. The rate at which work is done is known as *power* and it is measured in watts or kilowatts (KW), where 1 KW equals 1000 watts (some useful conversions: 1 KW equals 1.34 horsepower equals 3412 Btu/hr). It is interesting to note that the power equivalent of consuming 3000 food calories per day is about 150 watts—so you might say the human body is roughly equivalent to a 150-watt machine.

The electrical power (P) consumed in a load is equal to the product of the current through the load multiplied by the voltage across the load:

E. 3.2 $$P = VI$$

Alternative forms, $P = I^2R$ and $P = V^2/R$, can be derived from Ohm's Law.

The final parameter of importance is electrical *energy* Energy is simply the product of power and the length of time that the power is being consumed. Electrical energy is measured in watt-hours or kilowatt-hours (1000 watt-hours equals 1 kilowatt-hour equals 1 KWH). For example, a 100-watt bulb burning for 5 hours uses 500 watt-hours or 0.5 KWH of energy. A typical house in the United States consumes about 550 KWH per month; hopefully you will be designing your system to supply much less than this.

And that's about it! With this small amount of theory we can go a long way, so let's get started.

Resistance Losses in the Wire

Let us illustrate the relationships from the last section by considering an often overlooked, but important, aspect of electrical design—the losses in the wires which connect a power source (e.g., a battery) to the load. We will see the advantages of using high-voltage sources, heavy connecting wires, and short distances between source and load.

Example: Calculate the current in the circuit of Figure 3.1 when the battery voltage is 12 volts and the load is receiving 120 watts of power. Compare the current to that which would result if the same amount of power, 120 watts, is being delivered from a 120-volt battery.

Solution: For a 12-volt battery:

$$I = \frac{P}{V} = \frac{120 \text{ watts}}{12 \text{ volts}} = 10 \text{ amps}$$

In the case of a 120-volt battery:

$$\frac{120 \text{ watts}}{120 \text{ volts}} = 1 \text{ amp}$$

Now, compare the losses which would be encountered in the connecting wires in each of the two above cases. The wires have some resistance to them, call it R_w. Then the power lost in the wires, which is given by $P_w = I^2R_w$, will be 100 times as much in a circuit with I equaling 10 amps as in one carrying only 1 amp! By raising the system voltage by a factor of 10, from 12 volts to 120 volts, losses in the connecting wires have been decreased by a factor of 100. In other words, the higher the system voltage, the lower the line losses. This is why electric companies, which must transmit power over hundreds or thousands of miles, raise their voltages on the transmission lines as high as they possible can—some lines today carry 756,000 volts and lines are under development that will carry two million volts! And this is why, in your home electrical system, 120 volts is recommended over a 12-volt system, especially if you intend to supply a fair amount of power.

How much resistance is there in the connecting wire? That depends on the length of the wire, the diameter of the wire, and whether the wire is made of copper or aluminum. Wires are specified by their *gauge*—the smaller the gauge (or wire number), then the bigger its diameter and, consequently, the lower its resistance. House wiring is usually No. 12 or No. 14, about the same size as the lead in an ordinary pencil. To connect your windmill to your house, you may end up having to use very heavy, very expensive wire, such as No. 0 or No. 00 (sometimes written as 1/0 or 2/0), with a diameter of around one-third of an inch. Table 3.1 gives some values of wire resistance, in ohms per 100 feet, for various gauges of copper and aluminum. Copper is standard and preferred, though aluminum is sometimes substituted because it is cheaper. Also given is the maximum allowable current for copper for the most common types of insulation, Types T, TW, RH, RHW, and THW. The most common wire in use is Type T, though Type RHW is often used for heavy currents. You must not use wire that is too small; it can overheat, damage its insulation, and perhaps even cause a fire. Moreover, wire that is too small will cause you to lose voltage in the wire so less power will be delivered to the load. To calculate the maximum voltage drop in the wire, you need to know the wire gauge, the length of the wire, and the maximum current that it will carry.

Example: Suppose our voltage source (battery, generator) is delivering 20 amps through No. 12 copper wire to a load 100 feet away. Calculate the voltage drop in the wire.

Solution: From Table 3.1, No. 12 copper wire has a resistance of 0.159 ohms per 100 feet. Since the load is 100 feet away from the source, there must be 200 feet of wire (to the load and back) with a total resistance of

$$R_w = 2 \times 0.159 = 0.32 \text{ ohms}$$

The voltage drop in the wire is therefore

$$V_w = IR_w = 20 \text{ amps} \times 0.32 \text{ ohms} = 6.4 \text{ volts}$$

If our voltage source in this example were 120 volts, then we would be losing about 5 percent of the voltage in the wires (leaving 113.6 volts for the load)—quite a bit of loss, but perhaps acceptable. If the source were only 12 volts, then more than half would be lost in the wires—clearly unacceptable!

As an aid to choosing the proper wire gauge, we have prepared Figure 3.2, based on allowing a 4 percent voltage drop in the connecting wires (electricians customarily shoot for a 2 percent drop, but we'll ease up a little here). A 4 percent voltage drop in a 120-volt system is about 5 volts; in the 12-volt system it is about half a volt. In either case, the load will receive about 8 percent less power than it would if it were located right next to the source. Figure 3.2 gives the maximum distance between source and load, for varying wire gauges, which keeps the wire voltage drop under 4 percent. The vertical axis

Table 3.1 Characteristics of Wire

Wire Gauge A.W.G.	Diameter (inches)	Resistance (ohms per 100 ft at 68°F) Copper	Aluminum	Max Current Types T,TW Copper (amps)	Max Current Types RH,RHW THW Copper (amps)
000	0.4096	0.0062	0.0101	195	200
00	0.3648	0.0078	0.0128	165	175
0	0.3249	0.0098	0.0161	125	150
2	0.2576	0.0156	0.0256	95	115
4	0.2043	0.0249	0.0408	70	85
6	0.1620	0.0395	0.0648	55	65
8	0.1285	0.0628	0.103	40	45
10	0.1019	0.0999	0.164	30	30
12	0.0808	0.1588	0.261	20	20
14	0.0641	0.2525	0.414	15	15

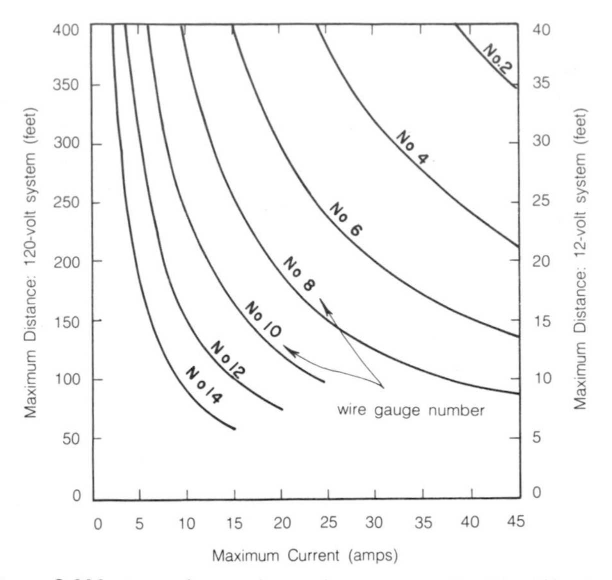

Figure 3.2 Maximum distance from voltage source to load for 4% voltage drop in connecting wires; for 12-volt and 120-volt systems using copper wire (see also Table 3.6).

on the left gives distances for a 120-volt system, and the axis on the right gives it for a 12-volt system. If you want to allow some other voltage drop you can still use this graph—for example, if you want only a 2 percent voltage drop, multiply the distances on the graph by 0.5; for a 6 percent drop, multiply distance by 1.5.

Example: Choose the right wire to allow 20 amps to be delivered to a load 100 feet away from a 120-volt source, allowing a 4 percent voltage drop.

Solution: From Figure 3.2 at 20 amps, we see that No. 12 wire is good only to about 80 feet, but that No. 10 can go to 120 feet. So choose No. 10 wire.

Calculating Your Energy Requirements

The "design" of a small-scale electrical generation system is going to be a matter of matching your needs and your money supply to equipment (which you may be able to build, but which you are likely to have to buy). When you plugged into your giant utility company, you never had to worry about overloading their generators when you turned on an appliance. The home systems we are talking about here are different. If you want to spend lots of money, you can set yourself up with a system able to meet all conceivable demands. If you want to get by, spending as little as possible, you must have a clear idea just what your energy demands are likely to be, so that you can pick the cheapest system to meet those needs adequately. It would be extravagant to install a system that can supply 600 KWH per month if you are only going to use 100 KWH. And it would be frustrating to spend a lot of money on a system that is too small, so that your ice cream keeps melting in the refrigerator.

This section, then, is on estimating how much electric energy you may need. We will then turn to choosing an appropriate wind or water system to meet those needs.

The place to start your demand estimate is with some old utility bills. They will tell you, month by month, how many kilowatt-hours you presently consume. Examine at least one full year's worth of bills since demands vary considerably with the seasons. Unless you have an air conditioner, summer demands are probably much lower than winter demands, as suggested in Figure 3.3. If your present lifestyle is comparable to that for which you are designing, then you have taken a big step towards a realistic estimate. If you don't have a year's worth of old bills lying around, phone your utility company and they will supply you with the required information from their records. But be prepared to cut a lot of waste out of your consumption. For example, on the first pass you may decide you want about 300 KWH per month to be supplied by a wind-powered system—but when you discover that wind systems cost about $3000 per 100 KWH, you may just back off a bit and start thinking small.

The second approach to figuring your monthly energy demand is to calculate it from the energy demands of each appliance. Table 3.2 lists typical power requirements for a number of electrical appliances. To figure energy demand for an appliance, you must esti-

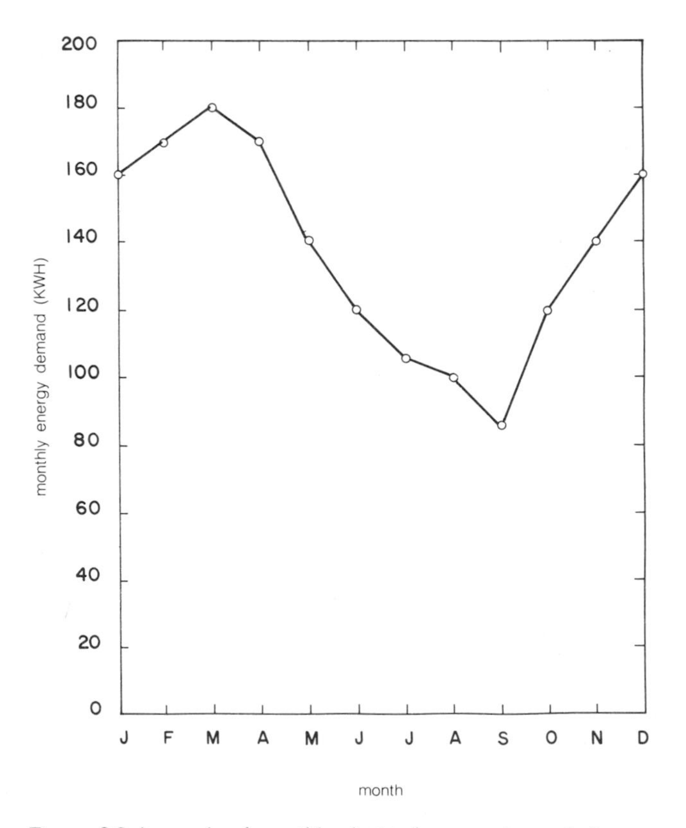

Figure 3.3 A sample of monthly electrical energy demands (summer demand is usually lower than winter).

Table 3.2 Approximate Monthly KWH Consumption of Household Appliances Under Normal Usage

Household Appliance	Rated Watts	Monthly KWH	Household Appliance	Rated Watts	Monthly KWH
Air conditioner (window)	1300	105	Oil burner or stoker	260	31
Bed Covering	170	12	Radio	80	7.5
Broiler	1375	8	Radio-phonograph	105	9
Clock	2	1.5	Range	11,720	102
Clothes dryer	4800	80	Roaster	1345	17
Coffee maker	850	8	Refrigerator	235	38
Cooker (egg)	500	1	Refrigerator-freezer	330	70.5
Deep fat fryer	1380	6	Refrigerator-freezer (frostless)	425	135
Dehumidifier	240	32	Sewing machine	75	1
Dishwasher	1190	28	Shaver	15	0.2
Electrostatic cleaner	60	22	Sun lamp	290	1
Fan (attic)	375	26	Television	255	29
Fan (circulating)	85	3	Television (color)	300	37.5
Fan (furnace)	270	30	Toaster	1100	3
Fan (roll-about)	205	9	Vacuum cleaner	540	3
Fan (window)	190	12	Vibrator	40	0.2
Floor polisher	315	1	Waffle iron	1080	2
Food blender	290	1	Washing machine (automatic)	375	5.5
Food freezer	300	76	Washing machine (nonautomatic)	280	4
Food mixer	125	1	Water heater (standard)	3000	340
Food waste disposer	420	2	Water heater (quick recovery)	4500	373
Fruit juicer	100	0.5	Water pump	335	17
Frying pan	1170	16			
Germicidal lamp	20	11			
Grill (sandwich)	1050	2.5	**In the Shop**	**Motor Size (horsepower)**	**Rated Watts**
Hair dryer	300	0.5	Bandsaw	0.50	660
Heat lamp (infrared)	250	1	Drill (portable, ⅜")	0.20	264
Heat pump	9600	—	(press)	0.50	660
Heater (radiant)	1300	13	Lathe (12-inch)	0.33	660
Heating pad	60	1	Router	0.75	720
Hot plate	1250	8	Sander (orbital)	0.20	300
Humidifier	70	12	(polisher)	1.50	1080
Incinerator	605	55	Saw (circular)	1.66	1080
Iron (hand)	1050	11	(saber)	0.25	288
Iron (mangle)	1525	13	(table)	1.60	950

mate the number of hours the appliance is used in a month and multiply that by the number of watts that the appliance draws during operation. For example, if you play a radio (80 watts) an average of 1 hour per day, then during a month's time (30 days) it would consume 30 times 80 watt-hours or about 2.4 KWH. If you do this calculation for each appliance and add up the results, you can come up with the needed figure.

There is one important exception to the above procedure. Some appliances—refrigerators or freezers, for example—may be plugged in and turned on 24 hours a day, but only consume electricity for a small fraction of that time. For such thermostatically controlled appliances, you will probably have to use the typical average monthly energy consumption given in Table 3.2 rather than perform any calculations.

As an example of how the procedure goes, consider Table 3.3, modeled for a quite modest home. Some of the big users of electricity—electric stove, water heater, or space heater—are not on this list since their energy should come from other sources (the sun or methane gas perhaps). The total of 142 KWH per month works out to about 5 KWH per day, which is about one-fourth of the American average.

So, this is the basic procedure—watts times hours equals watt-hours. But there are some refinements which we can make to this procedure. First, the wattage figures given in Table 3.2 are averages; when at all possible, you should use the actual ratings that are given on your own appliances. There is usually a nameplate somewhere with the power consumption stamped right on it. Sometimes, instead of giving the watts required, the nameplate will give the current drawn; for example, "2A" would mean 2 amps. To get power, multiply that current by 115 volts (e.g., 2 amps × 115 volts = 230 watts). Another source of power ratings for appliances not in our list are department store catalogs such as the one Sears distributes.

Second, you may not have to rely on the average *energy* information given in Table 3.2 for thermostatically controlled appliances. As you can see from the sample

Table 3.3 Example of Monthly Energy Demand Calculation

Appliance	Power (watts)	Usage (hours/month)	Energy (KWH/month)
Clock	2	720	1.5
Refrigerator-Freezer	330	—	70.5
Sewing machine	75	15	1.1
Radio-phonograph	105	60	6.3
Television	255	60	15.2
Toaster	1100	5	5.5
Washing machine	375	2	0.8
Table saw	950	5	4.8
Lights (5 at 60-watt)	300	120	36.0

Total: 142 KWH/month

calculation in Table 3.3, the refrigerator is liable to be the biggest single user of electricity in the house, which makes it particularly unfortunate to have to rely on a single "average" value given in a table. Energy consumption in refrigerators varies considerably, from brand to brand and model to model, and you may have a refrigerator that uses a lot more or a lot less than is given in the table. In 1961, the average refrigerator in the United States consumed about 70 KWH per month; but, by 1969, with increased size and such new gadgets as automatic temperature control, automatic ice-cube makers, and automatic defrosting, the average had increased more than 50 percent to about 110 KWH per month. One way to check energy requirements for new refrigerators and freezers is through consumer magazines (see, for example, the August 1974 issue of *Consumer Research* for freezers and the November 1974 issue of *Consumer Reports* for refrigerators). If you ask a dealer for this information, you'll probably get nothing but a blank stare. In general, old small refrigerators use much less energy than the big new ones.

If you're eager to get accurate information about the time your refrigerator is actually running, you might rig up a simple circuit (Figure 3.4) based on a common filament transformer that can be bought for about five dollars or scrounged out of an old tube TV. Two warnings should be noted: First, if your refrigerator is one of those big mamas drawing more than about 400 watts, this circuit may absorb too much voltage to allow the refrigerator to turn on (such a refrigerator is probably too big for a wind system anyway). And second, be sure the low-voltage side of the transformer can handle the refrigerator current (5 or 6 amps is good). In this circuit, whenever the refrigerator clicks on, the electric clock will start running. I suggest you set the clock at 12:00 and run a 24-hour test. Then read on the clock face the number of hours out of 24 that the refrigerator was actually running. For example, 8 hours on the clock would mean the refrigerator is running about 8/24 or 1/3 of the time. If it is rated at 200 watts, in one month you would then expect it to require about

$$0.2\ KW \times \tfrac{1}{3} \times 720\ hours/month = 48\ KWH/month$$

Another refinement of this procedure is to do the demand calculation as above, but do it for each month in the year or at least for each season. This will be necessary in order to match supply considerations (average monthly wind velocities vary throughout the year, as do stream flows) with demand. If you're lucky, months with higher demands (winter probably) will correspond to months of higher wind or water flow.

The next important modification to the demand calculation procedure is to separate your requirements into "essential" and "convenience" components. You

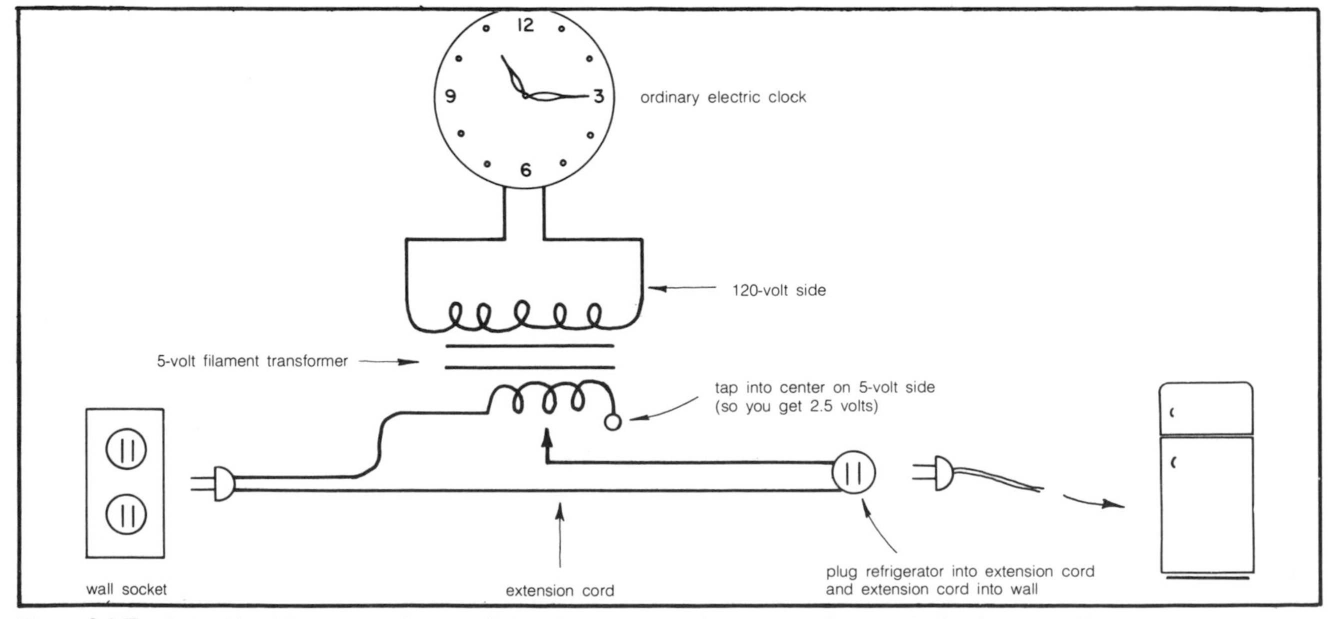

Figure 3.4 The duty cycle of thermostatically controlled appliances like a refrigerator can be measured with a circuit like this (note warning in text for large refrigerators).

may want to design your main electrical supply system to always meet the essential portion of demand and to meet the convenience portion whenever conditions are reasonably good. During those periods when the main system can't supply all of the convenience load, you can either do without or you can switch to a small auxiliary back-up unit to supply the power. Using this approach, you can get by with the smallest expenditure of money and still have a very reliable system. Later, when we work through some designs for a wind system, you'll see how to handle this calculation.

Finally—we'll mention it here and discuss it later when we look at inverters—you may have to identify separately those appliances which *can* run on dc or be modified to run on dc from those that *must* be supplied with ac. Since our energy storage mechanism is batteries and batteries supply dc only, any appliance requiring ac will require a special piece of equipment called a dc-to-ac inverter. These inverters take power themselves, so we must raise our estimate of energy required by any ac appliances anywhere from 10 to 70 percent to include losses in the inverter. Consequently, to minimize energy requirements, you may want to run as much on dc as you can. But, don't worry about this factor yet; wait until we get to inverters.

That sums up the technique for calculating energy demand. You will probably run through the calculation several times as you balance apparent needs with what is available and what you can afford.

One final step remains: to estimate the maximum power to be drawn at any one time. We need this figure to pick our wire size and also as a check on the maximum current drain from the batteries. Furthermore, the same technique will be used for picking the proper inverter. It is simple—just look through your list of appliances and estimate the total number of gadgets liable to be turned on at the same time. For example, suppose that it is evening and the TV is on (255 watts); four 60-watt light bulbs are burning (240 watts); the refrigerator is on (330 watts); the blender is blending (290 watts); and someone starts using the table saw in the shop (950 watts). The total power demand will be

$$P_{max} = 255 + 240 + 330 + 290 + 950 = 2065 \text{ watts}$$

If we have a 120-volt system, the maximum current will be

$$I_{max} = \frac{P_{max}}{V} = \frac{2065 \text{ watts}}{120 \text{ volts}} = 17.2 \text{ amps}$$

From Table 3.1 we see that No. 14 wire carries a maximum current of 15 amps, but No. 12 will carry 20; choose No. 12 wire. Checking Figure 3.2, we see that No. 12 wire will allow the source and load to be about 80 feet apart and still lose less than 4 percent of the voltage in the wires.

Notice that a 12-volt system would be entirely impractical for this much power. The wire would have to carry 172 amps, so you'd probably need to use No. 3/0 wire which is 0.4 inches in diameter—imagine wiring your house with that!

ELECTRICITY FROM THE WIND

Humankind has been utilizing the energy in the winds for thousands of years, to propel sailboats, grind grain, and pump water; and, perhaps surprisingly, we have been using it intermittently for the generation of electricity for over seventy-five years. During the 1930s and '40s, hundreds of thousands of small-capacity wind-electric systems were successfully used on farms and homesteads in the United States, before the spread of the rural electrification program in the early 1950s. And it was almost thirty-five years ago, in 1941, that the largest wind-powered system ever built went into operation at Grandpa's Knob in Vermont. Designed to produce 1250 kilowatts of electricity from a 175-foot-diameter, two-bladed prop, the unit had withstood winds as high as 115 miles an hour before it catastrophically failed in 1945 in a 25-mph wind (one of its 8-ton blades broke loose and was hurled 750 feet away).

A note here about what to call these wind-electric systems. Strictly speaking, a "windmill" is a wind-powered mill used for grinding grain into flour; calling a machine which pumps water or generates electricity a "windmill" is somewhat of a misnomer. Instead, people are using more accurate, but generally clumsier, terminology: "wind-driven generators," "wind generators," "wind turbines," "wind-driven powerplants," and "wind-driven electric power systems."

The ease and cheapness with which electricity could be generated using fossil fuels caused interest in wind-electric systems to decline; it has only been in the last few years that interest has revived—for use in both small-scale home systems and large central utilities. NASA, for example, has awarded contracts for a 125-foot, 100-KW, wind-powered system called the Mod-Zero, to be located near Sandusky, Ohio. Other studies are being made for windplants of up to three million watts.

Our interest here, however, is in small systems capable of supplying a modest amount of electricity, to one house or maybe a small cluster of houses. Before we get into details, let us state some general conclusions. First, if you live in the city or in an area where central powerplant electricity is readily available, and if cost is an important constraint, then you may as well plug into the utility. They will deliver electricity at about 2 to 4 cents per kilowatt-hour and you'll be lucky to generate your own for anything less than about 15 cents per KWH. But we all know prices are rising rapidly (electricity rates in Los Angeles have doubled in the last four years) and, in a very few years, we may very well be charged windmill prices for utility electricity.

If, on the other hand, you want to provide electricity to a rural site presently without it and your choice is between a wind-electric system, a fossil-fueled engine-generator set, or getting the power company to bring in several thousand feet of wire and poles (at roughly $5000 to $10,000 per mile), then the wind generator will not only be the most ecologically gentle system, but it will probably also be the most economical. As a rough rule of thumb, if wind speeds average at least 10 miles per hour, then wind-electric systems are economically sensible alternatives.

What about homebuilt versus off-the-shelf component systems? Homebuilt systems are reasonably cheap and provide a great deal of invaluable practical experience. They teach you to appreciate the "features" in a commercial system, but they take a great deal of time to build, are not nearly as reliable, and won't last as long as commercial units. They also provide only a very limited amount of electricity per month. If you want to provide only enough energy to run a few lights and maybe a stereo or TV a few hours a day, a homemade system assembled out of quite readily available parts (automobile alternators, telephone pole towers, hand-crafted propellers, etc.) can be sufficient. You may be able to generate 20 to 30 KWH per month for a total system cost of about $500.

To go into sufficient detail to enable you actually to build such a system takes more space than we have available here; such plans are quite readily available elsewhere (see Bibliography). We will, however, be sketching out most of the considerations that are involved in planning small homebuilt units.

If you add a refrigerator to your load, along with a few lights and some entertainment equipment, then you are out of the do-it-yourself category and into the commercial products market. We'll present later a summary of precisely what's available, along with approximate costs. For now we just note that there are two principal manufacturers of appropriately sized wind generators—Elektro of Switzerland and Dunlite of Australia. Complete systems are available which supply anywhere from a modest 75 to 100 KWH or so per month for roughly $3000 to $4000, to systems capable of supplying several hundred kilowatt-hours per month—enough for almost everything you might need, short of electric heating—that cost over $10,000. Somewhere in the middle, between a pure homebuilt and an off-the-shelf system, is the new 750-watt kit from Sencenbaugh Wind Electric in Palo Alto, California. This unit looks like it could just meet the demands of a very modest house or cabin (maybe 100 KWH in 10-mph average winds) and do so more reliably than a homebuilt and cheaper than a factory-made system.

Wind-driven Electric Power Systems

Before we begin looking in detail at each of the many components that make up a complete system, let us briefly run through the whole system as shown in Figure 3.5. Starting with the *wind*, we shall be concerned with techniques for determining average local wind conditions and estimating the amount of power and energy that can be obtained from them.

The wind strikes the *rotor* (sometimes called the turbine, fan, or propeller), causing it to rotate and thereby convert the wind's power into the power of a rotating shaft. There are many kinds of rotors to consider, ranging from the conventional horizontal-axis multibladed fans to the more unusual vertical-axis types, such as the Savonius and Darrieus rotors. The rotors usually rotate at too low a speed to be directly coupled into the generator, so most systems require some sort of *gearing*.

The *generator* (or *alternator*) converts the mechanical energy provided by the geared-up shaft into electricity. The term "generator" can refer to either a dc machine or an ac machine, though it is more common to refer to an ac machine as an "alternator." In these wind-powered systems, the ac produced in an alternator is rectified (converted to dc) by diodes in the alternator itself; viewed from the terminals, both alternators and dc generators actually "look alike" in that they both produce dc current.

The *voltage regulator* adjusts the output of the generator so that it always supplies the proper voltage to the batteries. Without this regulation, the generator voltage would vary with the speed of the rotor and the batteries could be damaged.

The *batteries*, of course, store the energy for use when the wind is insufficient for the generator to supply the load directly. Part of the design requirement is to pick the storage capacity of these batteries. The batteries supply dc only and can be directly coupled to dc loads.

The *inverter*, on the other hand, converts 12-volt or 115-volt dc to 60-cycle, 115-volt ac. Since the inverter is costly and itself consumes energy, it is usually desirable to design the system to minimize the inverter size by running as many appliances as possible on dc. The house must then be wired with separate wall plugs for dc and ac.

Finally, a *standby* system is shown to provide power during abnormally long periods of little or no wind. By incorporating this unit into the design, a certain amount of trade-off is allowed between battery storage capacity and the size of this auxiliary power source.

These, then, are the necessary components in a complete wind-electric system. In the design process that follows, we will see how to match these components to the electrical demands of your home.

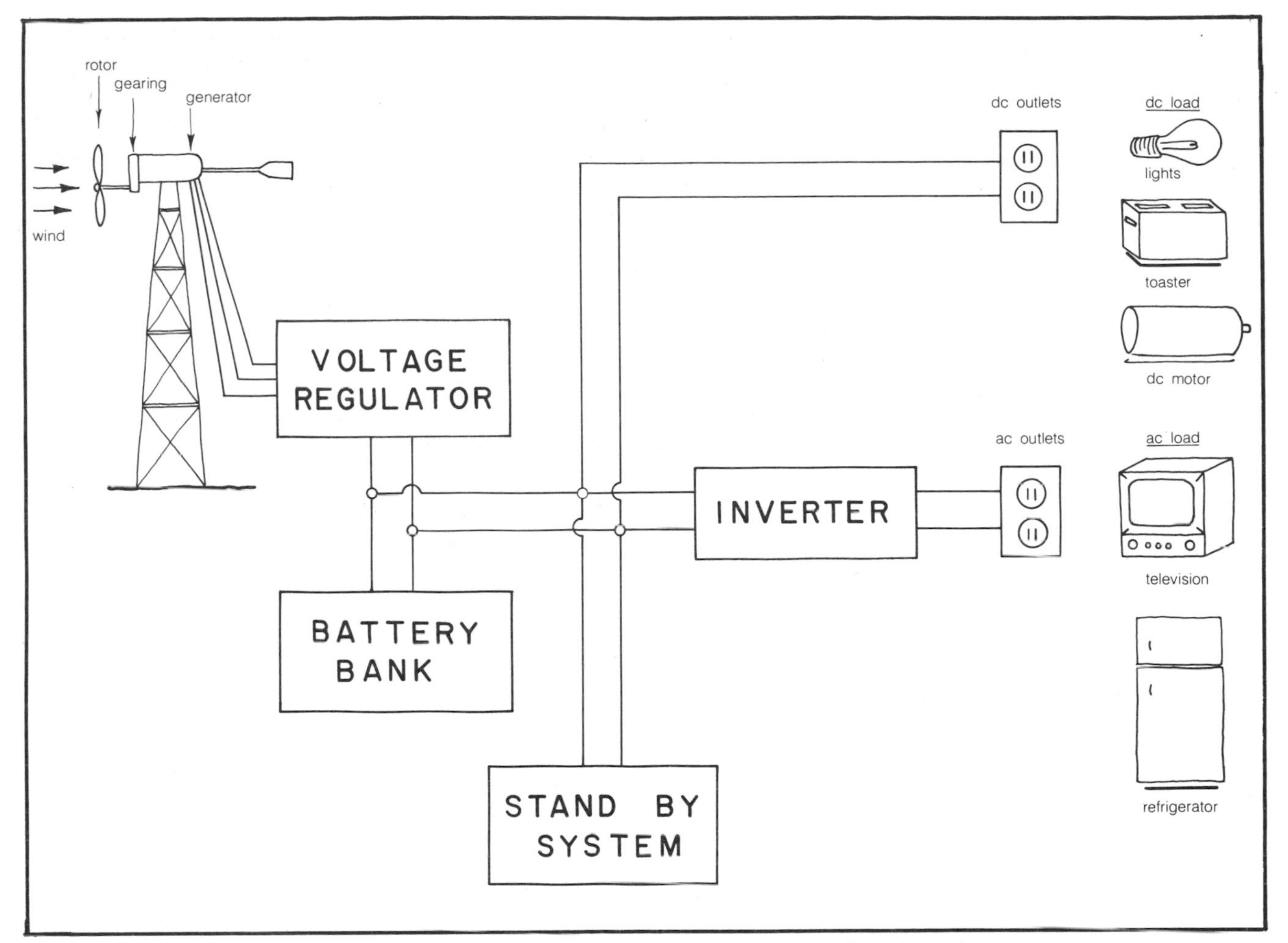

Figure 3.5 Basic schematic diagram of a wind-electric system, showing provision for ac as well as dc loads.

Power in the Wind

Since it is a relatively straightforward calculation, let us sketch out the derivation of the formula for the power contained in the wind and the amount of that power that can be extracted by a windmill. (In the equations which follow, be careful to distinguish between V for *voltage* and V for *velocity*.)

The kinetic energy (E) contained in a mass (m) of air moving at a particular velocity (V) is simply $E = mV^2/2$. If we consider the mass of air passing through an area swept out by the windmill fan (A) in an amount of time (t), then $m = \rho AVt$, where ρ is the air-mass density. Since power is energy divided by time, the final expression for the power (P) contained in the wind is

E. 3.3 $$P = \frac{\rho AV^3}{2}$$

Energy is obtained from the wind by slowing the air. However, the windmill cannot extract all of this energy from the wind or else the air behind the windmill would have to come to a complete standstill. It can be shown that a windmill derives the maximum amount of power when wind is slowed to one-third of its initial velocity, in which case the power output is 59.3 percent of the wind's power:

E. 3.4 $$P_{max} = 0.593 \times \frac{\rho AV^3}{2}$$

Using 0.08 lb/ft^3 as a reasonable value for the density of air; expressing P in watts; A in square feet (swept by the windmill); and V in miles per hour (of the wind), we can rewrite the expression for the theoretical maximum power from a windmill as

E. 3.5 $$P_{max} = 0.0031AV^3$$

Since most of the windmills of concern here will have propeller-type rotors, it is often convenient to rewrite the above equation in terms of the diameter of the propeller (D), where this diameter is expressed in feet:

E. 3.6 $$P_{max} = 0.0024D^2V^3$$

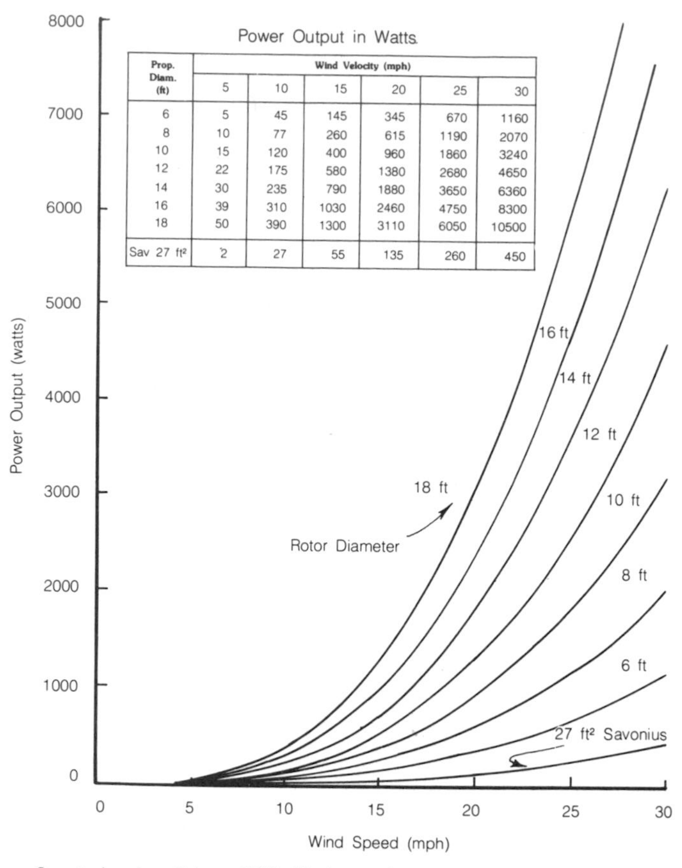

Power Output in Watts

Prop. Diam. (ft)	Wind Velocity (mph) 5	10	15	20	25	30
6	5	45	145	345	670	1160
8	10	77	260	615	1190	2070
10	15	120	400	960	1860	3240
12	22	175	580	1380	2680	4650
14	30	235	790	1880	3650	6360
16	39	310	1030	2460	4750	8300
18	50	390	1300	3110	6050	10500
Sav 27 ft²	2	27	55	135	260	450

Savonius based on efficiency of 20% of the theoretical maximum (overall efficiency 12%); prop based on efficiency of 50% theoretical max (overall 30%).

Figure 3.6 Power output for various wind speeds and rotor diameters assuming efficiency of 50% of theoretical maximum. Also included is a 27-square-foot, 20% efficient Savonius rotor.

For example, when D equals 10 feet and V equals 10 mph, the maximum power that could ever be obtained is 240 watts. Notice the power output goes up as the *square* of the prop diameter—that is, if you double the prop diameter you increase the obtainable power by a factor of four. Even more important, wind power goes up as the *cube* of the wind speed, so that doubling the wind speed gives *eight* times the available power! Thus if some cross-sectional area of wind contains 10 watts at 5 mph, then it contains 80 watts at 10 mph; 640 watts at 20 mph; and 5120 watts at 40 mph.

The power that is in the wind is one thing; the theoretical maximum power that can be extracted (59.3 percent) is another; and what a machine will actually produce is something else again. A well-designed rotor will extract about 70 percent of the theoretical maximum wind power. Furthermore, some power is lost in the gearing, which is about 90 percent efficient. And then the generator itself may be only 80 percent efficient (and that's a pretty good one). Since efficiencies are multiplicative, we can say as a reasonable estimate that wind generators are only about 30 percent efficient (0.593 × 0.7 × 0.9 × 0.8 = 0.3) in converting wind power to electrical power. Or, stated another way, windplants have efficiencies of about 50 percent of their theoretical maximum (0.7 × 0.9 × 0.8 = 0.5). Both of these ways of expressing efficiencies are commonly used, so it is important to know whether it is the percent of total wind power or of the theoretical maximum that is being referred to.

If we assume a system efficiency of 50 percent of the theoretical maximum, then the expression for power out of the generator (in watts) becomes

E. 3.7
$$P_{gen} = 0.0012D^2V^3$$

When this last equation is plotted for various prop diameters (Figure 3.6), the rapid increase in power output at higher wind speeds is clearly evident. We can also begin to see that not much power is lost if we design a system which "throws away" winds under about 8 mph.

Example: What size rotor would be required to give a 6000-watt output in 26-mph winds?

Solution: From Figure 3.6, it looks like a prop with a diameter of about 17 feet is about right. This agrees pretty well with a commercial Elektro system which is rated at 6000 watts at 26 mph and uses an 18-foot blade.

Since wind generators are so responsive to higher wind speeds, it makes sense to try to locate the plant in the best possible winds. One way is to place it on top of a tall tower, since wind speeds increase with altitude. The amount of increase is, of course, dependent on conditions at your own site; but to give you some idea of how much might be gained, *Standard Handbook for Mechanical Engineers* (Baumeister) gives the following relationship for the variation in wind velocity (V), with increasing height (H) on an unobstructed plain:

E. 3.8
$$\frac{V}{V_o} = \left(\frac{H}{H_o}\right)^n$$

V_o is the velocity at some reference height H_o, and n is an exponent that varies with wind speeds, but for our range of interest (5–35 mph) can be taken to be 0.2.

Since power varies as the cube of velocity, we can rewrite the above as a power ratio, where P is the power at height H and P_o is the power at height H_o:

E. 3.9
$$\frac{P}{P_o} = \left(\frac{H}{H_o}\right)^{3n}$$

Since it is tricky to work with fractional exponents, we have plotted this last equation in Figure 3.7, using a

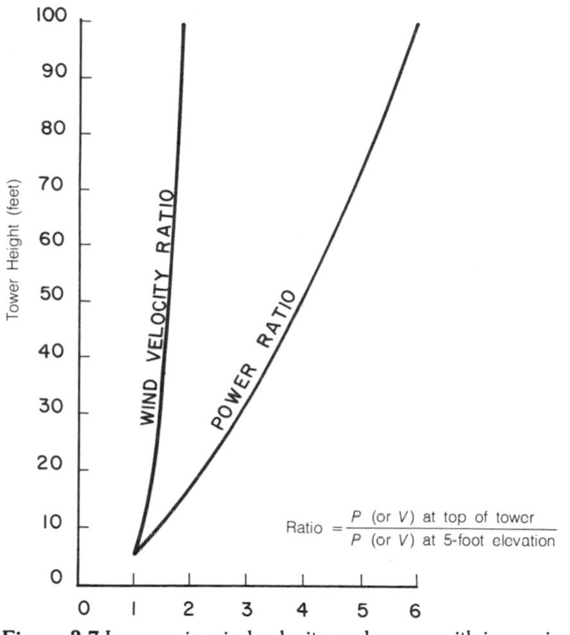

Figure 3.7 Increase in wind velocity and power with increasing height (unobstructed plain, referenced to 5-foot height).

reference height of 5 feet. The figure indicates, for example, that there is about 3 times the wind power at 30 feet than there is at 5 feet, even though the wind speed is only about 1.4 times as great. Notice that we can use this graph to help estimate the marginal gain which might be obtained in going from one height to another.

Example: Use Figure 3.7 to estimate the increase in power which might be expected by increasing tower height from 30 feet to 80 feet.

Solution: At 80 feet there is about 5.3 times the power as at 5 feet; at 30 feet there is 3 times the power. Hence by going from 30 feet to 80 feet we gain by a factor of about 5.3/3, or about 1.8.

Now for a warning and some trade-offs. The warning is that our formula is approximate and holds for an *unobstructed plain* only. Your site may be hilly and have bushes, trees, and buildings scattered around. So, before designing around this height bonus factor, you should make confirming wind-speed measurements at various heights on your own site.

The trade-offs are the increased cost of taller towers and increased losses in the connecting wires. The latter is less important and can be compensated for by using heavier (more expensive) wire. But towers are quite expensive. For example, one manufacturer lists 30-foot towers at $300 and 80-footers at about $1000.

Energy from the Wind

Now that we know how to estimate the *power* (watts) that a wind-driven generator will deliver at any given wind speed, we can move on to techniques for estimating the amount of *energy* (watt-hours) that a system will deliver each month. We have already learned how to calculate our monthly energy requirements and now we'll learn how to use local wind information to match a wind-driven generator to those requirements. This section of the chapter presents the real heart of the wind-driven generator design procedure. We are going to take a bit of a zigzag route but, by the end of this section, we'll have a simple technique.

You're going to need to know *something* about the winds at your chosen site, but precisely what information do you need and how can it be obtained? Skipping over the last half of the question for the moment, let us look first at the kind of data that would be the most useful. What you would like to have is a month-by-month indication of how many hours the wind blows at each wind speed. If you had this data, you could calculate the power output (watts) at each wind speed, multiply this figure by the number of hours the wind blows at that speed, and then sum up the watt-hours from all the winds. This would give the monthly energy that could be supplied.

This kind of data is routinely accumulated at meteorological stations all across the country and is usually summarized in either a table or graph. An example is given in Figure 3.8, based on a 5-year average for Januaries in Minneapolis. The graph may be a bit tricky to interpret if you're not used to statistics; it says, for example, that the wind speed equals or exceeds 15 mph about 20 percent of the time; 10.5 mph about 50 percent of the time, and so on.

Let us calculate the amount of energy that we might get from this wind. The exercise may be a bit tedious for some, but don't worry, we are going to have a much simpler procedure a bit later on. For those that are still

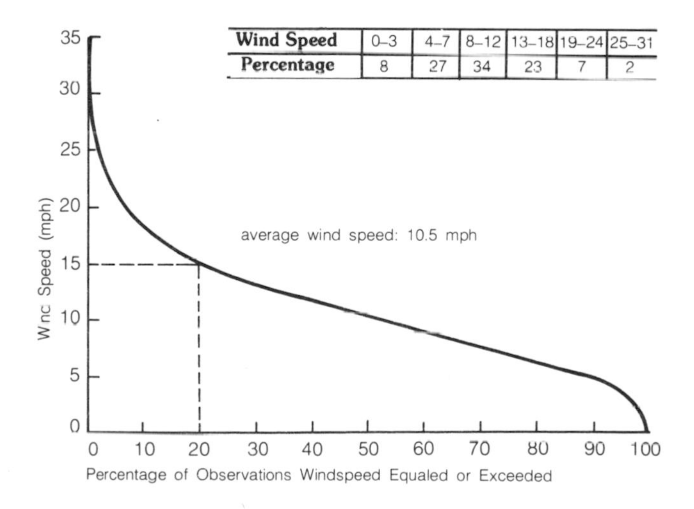

Wind Speed	0–3	4–7	8–12	13–18	19–24	25–31
Percentage	8	27	34	23	7	2

Figure 3.8 Wind-velocity duration curve and wind-frequency table for Minneapolis in January (5-year average).

interested, there are some instructive points to be gained from this example.

We need to introduce two quick facts at this point. First, real wind-driven generators produce no output power until the wind reaches a certain minimum value called the *cut-in speed*. Below this speed the generator shaft isn't turning fast enough to produce a charging current. And second, real generators are not able to produce more power than their *rated* capacity without being damaged. The speed at which the generator produces *rated* power is called the *rated wind speed*. For winds which exceed the rated speed, commercial systems are usually designed so that the rotor spills some of the wind; output power thus remains at the rated capacity, preventing damage to the generator. When people talk about a 2000-watt generator, for example, they are referring to the power output at the rated wind speed (which is usually somewhere around 25 mph).

Example: Calculate the energy that a 6000-watt wind generator, having a cut-in speed of 8 mph and a rated speed of 25 mph, would produce in January in Minneapolis. The rotor is 18 feet in diameter.

Solution: We'll start with the wind data given in the table in Figure 3.8. For each range of wind speeds, use an average speed; find the power output at the speed from Figure 3.6; multiply it by the number of hours (percent of observations) at that speed—assume 720 hours in a month; then sum up the results as has been done in Table 3.4 below. Notice that there is no output below cut-in and the output remains at 6 KW for winds higher than 25 mph.

So a little over 600 KWH could be expected. Not much energy has been lost by not picking up winds below 8 mph. The amount of this lost energy turns out to be only about 12.8 KWH, very small in comparison to 600 KWH.

Now we could do this calculation month by month,

Table 3.4 Minneapolis Wind Data (6-KW Plant)

Wind Range (mph)	% of Observations	Avg. Wind Speed (mph)	Hours per Month	Output KW (Fig. 3.6)	Energy (KWH)
0 – 3	8	1.5	57	0	0
4 – 7	27	5.5	195	0	0
8 – 12	34	10.0	245	0.4	98
13 – 18	23	15.5	165	1.4	230
19 – 24	7	21.5	50	3.9	195
25 – 31	2	28.0	14	6.0	84
32 – 38	0	35.0	0	6.0	0

Total: 607 KWH

but the drawbacks to this approach are fairly obvious: a lot of calculation is involved and everything depends on the required wind-distribution data. This data is often available from a nearby airport or weather bureau. Also, a number of years ago, the United States Department of Commerce published a series of reports called *Summary of Hourly Observations* for 112 cities across the country (that's where our Minneapolis data came from), containing month-by-month wind summaries. That information may be helpful if you live near one of the chosen cities, but even then its usefulness is limited if you live on a site where wind conditions are considerably different. You might also try to get this wind data for your site by direct measurement, but it would take years to accumulate enough data to be statistically significant and the equipment required to do the job is quite expensive.

But let us regain some perspective. There is no way that we will ever be able to *accurately* calculate the amount of energy that the wind will provide. Wind is unpredictable. We're going to have to live with a great deal of uncertainty in our estimates, so it doesn't make much sense to calculate things out to the last decimal point. There is a simple way to estimate energy obtainable when only the *average* wind speed is known. It is based on the observation that wind-velocity duration curves such as the one in Figure 3.8 tend to have very similar shapes; by knowing only the *average* wind speed, the rest of the distribution is more or less predictable.

Here is a sketchy version of the theory behind the simple procedure which we will be presenting. In the last example, we calculated that our 6-KW windplant would produce 607 KWH in a Minneapolis January. If it were running at full rated output all month (say 720 hours), then it would have delivered 4300 KWH (6KW × 720 hr). The ratio of these numbers is the *plant load factor* which, in this case, is 607/4300 or 0.14. Starting at the other end, if someone told us the plant load factor, we could calculate the energy delivered that month:

E. 3.10 *monthly KWH = load factor × 720 hr/month × rated output (KW)*

(e.g., monthly KWH = 0.14 × 720 hr/month × 6 KW = 607 KWH). The plant load factor times 720 hours per month is called the *specific output* and it has units of kilowatt-hours per month per kilowatt of installed capacity. If we analyze lots and lots of wind-velocity duration curves and, for each, calculate the specific output and note the average wind speed, we can make a plot of one versus the other. A number of researchers have done this for various spots around the world and the curves are always very similar.

The result of these calculations for each of twelve months of data at each of five sites in the United States (Albuquerque, Boston, Medford, Minneapolis, and San

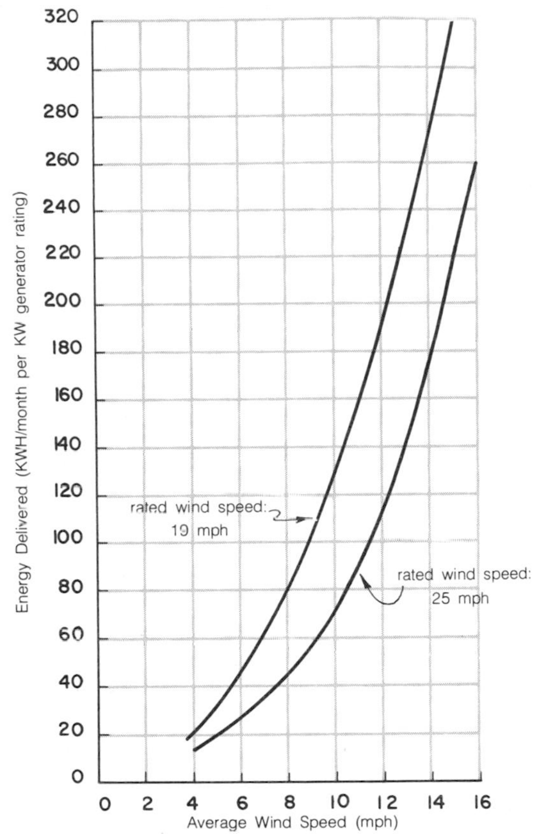

Figure 3.9 Monthly energy delivered per kilowatt of installed capacity. Cut-in wind speed 13 mph, rated wind speeds 19 and 25 mph.

Francisco) for two different values of rated wind speed (19 mph and 25 mph) is shown in Figure 3.9. This figure has been drawn in such a way as to deliberately underestimate (by maybe 20 percent) the amount of energy which you are likely to be able to obtain (by assuming a rather high cut-in speed of 13 mph). Better to guess too low than too high.

So now we have what we've been building up to—a simple technique for calculating monthly energy delivered, given only the average wind speed and the rating of the generator. From Figure 3.9 we get KWH/month per rated KW. Multiply that by your generator size (KW) and you've got monthly KWH! Moreover, this procedure is independent of the efficiency of the rotor and generator—the only assumption being that whatever efficiency the machine has at rated wind speed remains constant down to the cut-in speed. The efficiency cancels out in the calculation, as long as the rotor has been chosen to cause the generator to deliver its rated capacity at the rated wind speed. We are now ready for some examples to illustrate the procedure.

Example: Estimate the energy which a 6-KW generator, with rated wind speed of 25 mph, would supply when the average wind speed is 10.5 mph. (This is our "Minneapolis in January" example again.)

Solution: From Figure 3.9, we see that the specific output at 10.5 mph is 80 KWH/month-KW. A 6-KW generator would supply about

$$6\ KW \times 80\ KWH/month\text{-}KW = 480\ KWH/month$$

As would be expected by the conservative assumptions built into Figure 3.9, this estimate comes in somewhat below the 607 KWH/month calculated earlier.

Example: What size generator should be chosen to supply 200 KWH per month when average wind speeds are 10 mph? Assume rated wind speed is 25 mph. What size prop should be used?

Solution: From Figure 3.9, the specific output at 10 mph is about 70 KWH/month-KW.

$$\text{Generator rating} = \frac{200\ KWH/month}{70\ KWH/month\text{-}KW} = 2.9\ KW$$

or roughly a 3-KW generator. The prop diameter (obtained from Figure 3.6) at 3 KW and 25 mph is about 13 feet.

Example: What size generator rated at 25 mph would deliver the same amount of energy as a 750-watt generator rated at 19 mph, assuming average wind speeds of 10 mph?

Solution: From Figure 3.9, the 750-watt, 19-mph generator will deliver about

$$130\ KWH/month\text{-}KW \times 0.75\ KW = 97\ KWH/month$$

To get the same amount at a rated speed of 25 mph would require a generator of about (using Figure 3.9 again)

$$\frac{97\ KWH/month}{70\ KWH/month\text{-}KW} = 1.4\ KW$$

This last example points out the importance of the rated wind speed of a generator. The lower the rated wind speed, the greater will be the energy output per kilowatt of generator. So why don't manufacturers use much lower ratings? To lower the rated wind speed, a manufacturer must use a bigger (and hence more expensive) fan and must increase the gearing ratio (which also increases the cost). Also, if the average wind speed is fairly high, then a low rated speed means that much of the high-powered winds will be dumped and wasted because the generator is too small. There is, then, a problem in optimization when it comes to designing a windplant from scratch. One study indicates that, for sites with average winds around 10 mph, a low rated wind speed—around 15 to 20 mph—is best. For average wind speeds of 15 mph, it is best to use a 25-mph rating.

If you're not designing from scratch (very likely) and merely picking components, then you must simply compare windplant costs for equal monthly KWH delivered, following the technique outlined in the last example. For rated wind speeds other than 19 or 25 mph, you can use interpolation in Figure 3.9 to get a rough idea of how much variation in specific output can be expected.

Notice that you *cannot* estimate monthly energy merely by calculating the power out of the generator at the average wind speed and then multiplying this figure by 720 hours per month. The cubic relationship between power and wind speed makes speeds greater than the average have much more importance than speeds less than the average; such an estimating technique will give answers which are too low (usually about half the actual value). There is, however, a crude rule of thumb which suggests that energy per month is about double the value which would be obtained by multiplying the power output at the average wind speed by the number of hours in the month.

Measuring the Wind

While month-by-month velocity duration curves for your site (averaged over several years) would provide the best information for determining what energy the wind could supply, such data is extremely difficult to obtain. The required instrumentation is expensive and the time required to obtain statistically significant data is probably prohibitive. Moreover, the year-to-year variation in actual winds is enough to invalidate any precise calculations. So we may as well do something less precise, but easier.

We have seen that good energy estimates can be obtained from average wind speeds—data which is quite a bit easier to obtain. We recommend that you attempt to determine daily or monthly average wind speeds at your site over a specific period of time (at least several weeks) and then compare your estimates to the actual measurements made at the nearest local weather station over the same specified period. (Any number of agencies may be collecting such data—airports, U.S. Weather Service stations, military installations, National Park services, local air pollution agencies, etc.) How does your data compare to theirs? Hopefully, both sets of data will be consistent with respect to each other—that is, maybe yours is always, say, 20 percent higher than theirs. Then you may take their long-term average wind data and adjust it to your site (in this example, by raising it 20 percent). This way you get the benefit of many years' worth of measurements without having to accumulate them yourself.

By careful examination of Figure 3.9, we can see the importance of obtaining a fairly accurate estimate of average wind speeds. Very roughly, the amount of energy which can be supplied doubles for a 3-mph increase in average wind speed. The difference between a successful design and a disappointing one can therefore rest upon the difference of only a few miles per hour in our estimate.

Table 3.5 Qualitative Description of Wind Speed[a]

Wind Speed (mph)	Wind Effect
0–1	Calm; smoke rises vertically.
2–3	Direction of wind shown by smoke drift but not by wind vanes.
4–7	Wind felt on face; leaves rustle; ordinary vane moved by wind.
8–12	Leaves and twigs in constant motion; wind extends light flag.
13–18	Raises dust, loose paper; small branches are moved.
19–24	Small trees in leaf begin to sway; crested wavelets form on inland waters.
25–31	Large branches in motion; whistling heard in telegraph wires, umbrellas used with difficulty.
32–38	Whole trees in motion; inconvenience felt in walking against wind.

Notes: a. From *Standard Handbook for Mechanical Engineers* (Baumeister and Marks)

There are several approaches to measuring wind speeds, varying from crude estimates based on the observation of physical phenomena to accurate and expensive measurement equipment. Table 3.5 gives some qualitative descriptions which may be helpful to calibrate roughly your own observations.

A better approach is to buy or build a simple hand-held wind gauge. Dwyer sells one for about $7 which is available from the Whole Earth Truck Store, as well as from many boating, hobby, and scientific instrument supply houses.

The following simple gauge, which you can build yourself, was described in an issue of *Scientific American* (October 1971). A ping-pong ball attached to a length of monofilament nylon (roughly 0.08 mm to 0.2 mm in diameter) is allowed to swing in the breeze and the angle of the swing provides the measure of wind speed. The angle is measured from the vertical on a protractor to which the nylon line is attached. If the dimensions given in Figure 3.10 are carefully followed, then the instrument can use the calibration table provided in the figure.

Either of these two simple hand gauges can be used to estimate instantaneous wind speeds. If enough measurements are made throughout the day and for enough days, then a reasonable estimate of average wind speed can be determined for comparison with the measurements of a local station. Obviously this is a rather crude procedure. Its accuracy depends upon your diligence in obtaining data. Moreover, it probably won't be possible to measure the wind speed at the height at which your plant will be installed. Commercially available wind instruments, which can be installed on a pole and allow

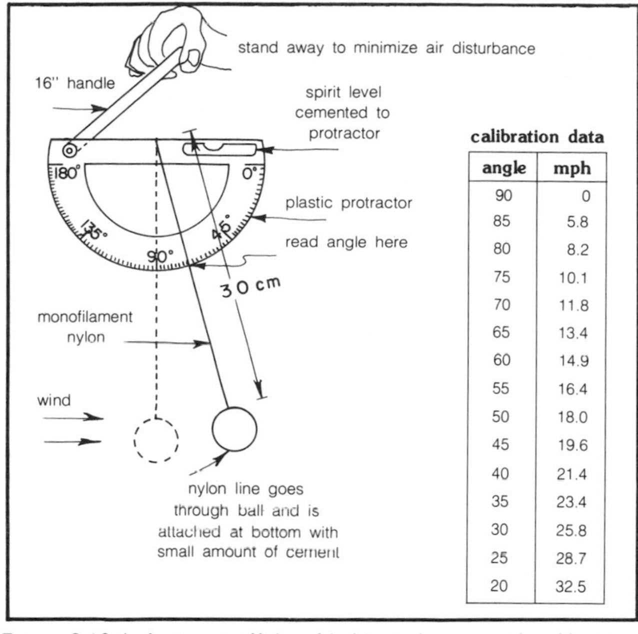

angle	mph
90	0
85	5.8
80	8.2
75	10.1
70	11.8
65	13.4
60	14.9
55	16.4
50	18.0
45	19.6
40	21.4
35	23.4
30	25.8
25	28.7
20	32.5

Figure 3.10 A do-it-yourself, hand-held wind gauge with calibration data (based on C.L. Strong, *Scientific American*, October 1971).

readouts to be taken at ground level, cost anywhere from about $25 to over $1200. The more expensive instruments include automatic recording equipment which allows you to obtain average wind speeds or even wind duration curves with accuracy. Without a chart recorder you have to make periodic measurements and so accuracy drops.

A final approach is to build a simple cup-type anemometer and attach it to a mechanical or electrical counter. If the rotational speed of the anemometer is proportional to wind speed, then the number of revolutions in a given time period—a day, for example—will be proportional to the average wind speed. Cup anemometers are reasonably linear and so this technique gives us a pretty good estimate. Moreover, we can install the device on a pole and elevate it to the desired height; we then need only record the count once a day (for daily averages) or, if the counter has sufficient capacity, once a week or month for weekly or monthly averages.

The sketch given in Figure 3.11 will help you build one. It uses cups obtained from L'eggs pantyhose containers, and bearings and shafts which can be obtained from a hobby shop. The counter should be a low-friction device which usually can be obtained from a surplus store. It should have a 6-digit capacity, to hold several days' count in pretty good winds. Or you can gear it down, which not only reduces the drag on the anemometer but also lets you accumulate several weeks' worth of data.

The only trick is to calibrate it. The best way is to install it next to a calibrated anemometer and compare your daily count to the average wind speed which the calibrated instrument indicates. Over a period of a week or so, you would be able to make your own calibration curve. Or, you can use one of the hand-held gauges to estimate an average speed for, say, 15 minutes of steady winds. Then compare your 15-minute count to that average and scale it up to one day. The more repetitions of this procedure, the more accurate the calibration becomes. Or, finally, you can attach it to a pole tied to your car, and calibrate it using a stopwatch and the car's odometer.

Besides determining average wind speeds on a month-by-month basis, it is important to know something about the highest wind speeds which you may encounter. Most windplants are designed to withstand winds as high as 80 mph and, if higher wind speeds are expected, there are extra-cost automatic controls which can be purchased. Protection for winds of up to 140 mph can be provided with these devices.

Site Selection

Choosing the best site for the windplant involves a

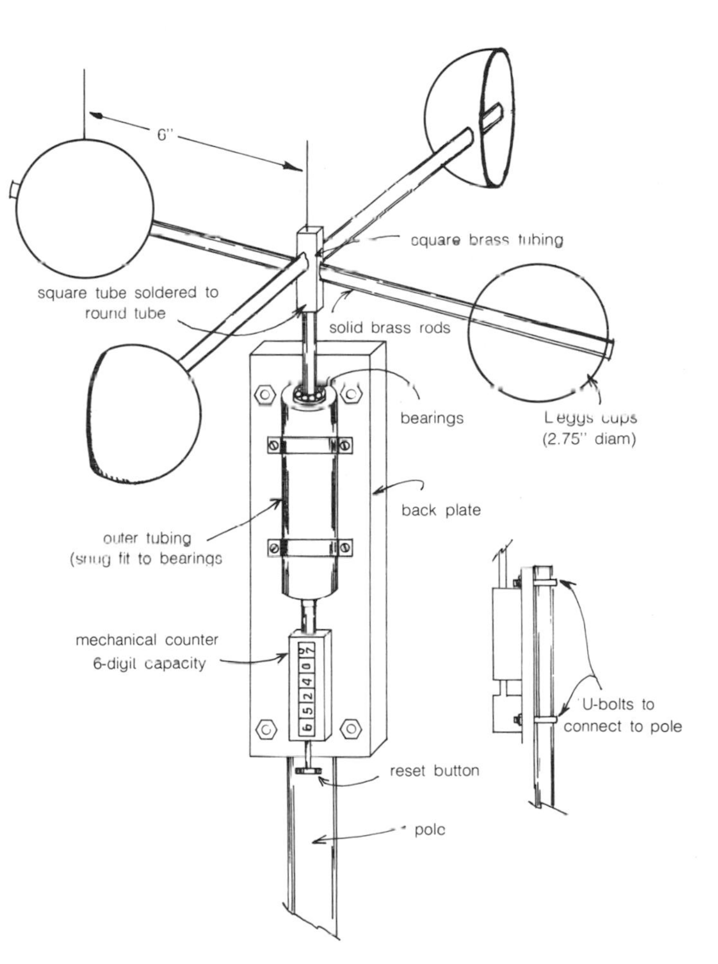

Figure 3.11 A do-it-yourself cup anemometer for measuring average wind speeds.

trade-off between conflicting demands: keeping it near enough to the house to minimize power loss in the connecting wires, yet placing it far enough away from such obstructions as buildings and trees to get it into the best winds. Obstructions which are higher than the windplant can disturb the wind for several hundred yards behind the obstacle and perhaps even 50 to 100 yards in front of the obstacle. The plant should therefore be located out of these disturbance zones. In hilly areas, the best spot is probably close to the top of a hill; but if this is impossible, it might be best to stay away from the hill altogether.

The plant should be located as high as possible and at least 15 to 20 feet higher than any nearby obstructions. It should not be placed on a roof; it could overstress the structure and probably will cause an unpleasant vibrational noise. Figure 3.7 indicated the advantages of tall towers; it is recommended that any tower be at least 40 feet high. At that height, winds may be 50 percent greater than at ground level and the power output may be several times as great.

These generalities cannot replace your own observations of the conditions that prevail on your land. Chances are either that your house is already built or that you will be picking your house site based on other criteria (solar exposure, privacy, etc.); if you keep the windplant within about 1000 feet of the house, then the number of good sites in that range is probably limited. You might hang ribbons or ping-pong balls on strings from poles erected at these considered sites and then watch their deflections with binoculars to see which site has the greatest winds. Since you are now only interested in relative conditions, you needn't worry about actual mile-per-hour comparisons.

The site should not be too far from the house, to avoid long line losses or the expense of very heavy wire. We have already seen how line losses can be computed once the current and wire gauge are known. Our earlier conclusion was that batteries and load should be kept as close together as possible to avoid losing too much battery voltage in the line. (In fact, the batteries should probably be located in or adjacent to the house.) We can perform the same calculations to estimate losses from the generator to the batteries. The maximum current to expect from your generator can be estimated by dividing the generator rated power by its voltage. It is then an easy matter to calculate line voltage drop (IR) and line power loss (I^2R) where R can be obtained from Table 3.1. To save you that trouble, we have prepared Table 3.6 which gives line voltage and power losses for various values of current for both copper and aluminum wire. Also a *rough* estimate of wire costs has been included. Notice that aluminum wire of a given gauge number is approximately equivalent (in resistance) to copper wire two gauge numbers higher and costs roughly half as much (e.g., No. 4 aluminum costs half of what No. 6 copper costs and has the same amount of resistance). Therefore, when carrying large currents over long distances, you will probably want to use aluminum wire because it is cheaper. Let's do some examples to get a feel for these numbers.

Example: Estimate the line losses for a 6-KW generator in a 120-volt system if the generator is 1000 feet from the batteries and No. 0 copper wire is used.

Solution: Maximum current will be about

$$I_{max} = \frac{6000 \text{ watts}}{120 \text{ volts}} = 50 \text{ amps}$$

From Table 3.6, we see that, at 50 amps, line losses amount to about 10 volts (meaning the generator would have to produce 130 volts to supply 120-volt batteries) and about 500 watts (out of 6000 generated). And your wallet would lose something like $2200 just to pay for the wire! If instead we choose No. 3/0 aluminum wire, losses are the same but now the price is about $950. Spending nearly $1000 on wire and still losing nearly 10 percent of our power suggests that 1000 feet is probably about the maximum distance that you could afford to place this big machine from its load.

Example: Pick a wire size which would result in no more than a 2-volt drop from generator to battery in a 750-watt, 12-volt system. The tower is 50 feet high and is 100

Table 3.6 Connecting Wire Voltage, Power Losses, and Rough Costs per 100-foot Separation between Generator and Batteries

Copper		Aluminum		Wire Voltage Drop (Volts/100 ft)							Wire Power Loss (Watts/100 ft)						
Wire No.	2-Wire $/100 ft	Wire No.	2-Wire $/100 ft	I= 10 amps	20A	30A	40A	50A	60A	70A	I=10A	20A	30A	40A	50A	60A	70A
000	360	—	—	0.12	0.24	0.38	0.5	0.62	0.74	0.86	1.2	5.0	11	20	32	44	60
00	300	0000	110	0.16	0.32	0.46	0.62	0.78	0.94	1.1	1.6	6.2	14	26	40	56	76
0	220	000	95	0.2	0.4	0.6	0.8	1.0	1.2	1.4	2.0	8.0	18	32	50	72	98
2	140	0	70	0.32	0.62	0.94	1.2	1.6	1.9	2.2	3.2	12.0	28	50	78	102	144
4	95	2	42	0.5	1.0	1.5	2.0	2.5	3.0	3.5	5.0	20.0	44	80	124	180	240
6	65	4	33	0.8	1.6	2.4	3.2	4.0	4.8	—	8.0	32.0	72	128	200	288	—
8	45	6	25	1.2	2.4	3.8	5.0	—	—	—	12.0	50.0	112	200	—	—	—
10	28	—	—	2.0	4.0	6.0	—	—	—	—	20.0	80.0	180	—	—	—	—

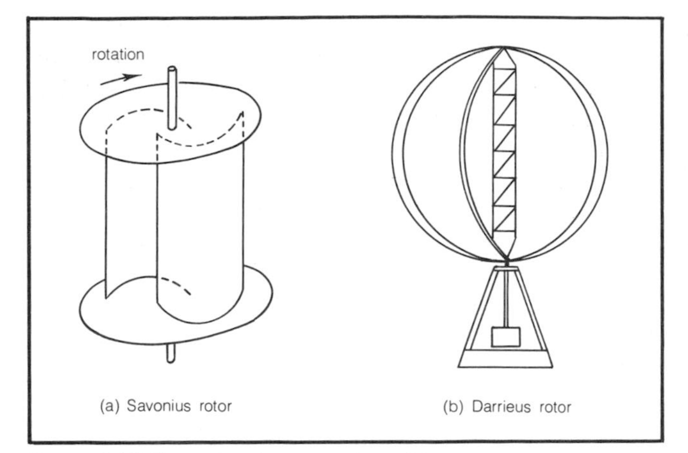

Figure 3.12 Two vertical-axis wind machines.

feet from the batteries.

Solution: The total distance from generator to batteries is 150 feet so the drop is

$$\frac{2 \text{ volts}}{150 \text{ feet}} = 1.3 \text{ volts per } 100 \text{ feet}$$

Maximum current is about

$$I_{max} = \frac{750 \text{ watts}}{14 \text{ volts}} = 54 \text{ amps}$$

In the table we see that No. 0 copper wire or No. 3/0 aluminum wire has a loss of 1.2 volts per 100 feet at 60 amps; this will be satisfactory. Our example suggests that to keep a reasonable voltage drop in 12-volt systems, they need to be much closer to the load (probably not much more than 100 feet away).

Rotors

There are many kinds of rotors which can extract power from the wind. One way to classify them is whether they rotate about a vertical or horizontal axis. The two vertical-axis machines which are receiving attention these days—the Savonius rotor and the Darrieus rotor (Figure 3.12)—are actually rather old ideas, patented about forty-five years ago. These rotors are interesting because they are always "headed into the wind"; there is no problem tracking gusty winds which constantly shift directions. Moreover, the heavy generator can be securely mounted on the ground and there is no need for slip rings to carry current from the generator to the wires connecting the load.

The Savonius, or *S*-rotor, rotates much slower than a modern two- or three-bladed propeller and is less than half as efficient in capturing the wind's energy. It must, therefore, be geared up considerably to match the speeds required by a generator and its cross-sectional area (to intersect the wind) must be quite large to make up for its inefficiency. The machines require a large surface area of material and so are heavy and hard to balance. Small do-it-yourself units can be made out of 55-gallon drums (see Hackleman's nice booklet in the Bibliography), but they deliver so little power that they hardly seem worthwhile. For example, a machine made from three 55-gallon drums stacked one on top of the other, having a cross-sectional area of about 27 square feet and an efficiency of about 20 percent of the theoretical maximum, produces about the same amount of power as a simple high-speed prop with a diameter of less than 4 feet. Hackleman's figures indicate that one such stack will probably yield less than 10 KWH per month in 10-mph average winds. Figure 3.6, in case you hadn't noticed, includes an approximate power-output curve for such a unit. *S*-rotors are not particularly appropriate for the generation of electricity, although their good-torque, slow-speed characteristics make them useful for such applications as pumping water.

The other vertical-axis machine, the Darrieus rotor, resembles a giant eggbeater. It has performance characteristics that approach a propeller-type rotor and there is reason to hope that present research efforts on these machines may soon result in a practical, economical design. However, they do have a problem with very low starting torque, so some sort of auxiliary starting system is required.

There are many different designs for horizontal-axis wind machines, but the one that probably springs to mind when you hear the word "windmill" is that romantic old, multibladed, slow-speed machine used all across the country for pumping water. Because these machines have many blades, they can develop a lot of torque (twist) in low-speed winds and are well-suited to their application. But they aren't designed to utilize high-speed winds efficiently, where the real power exists—to harness them you must have fewer blades. The fewer the blades, the faster the propeller rotates to be able to extract power from the passing wind. A high-speed prop having only two or three blades provides the best answer to the high rpm requirements of a generator; consequently, this is what most modern windplants have.

The efficiency of any rotor depends on the ratio of the speed of the tip of the rotor to the speed of the wind. This important quantity, called the *tip speed ratio*, is by definition

E. 3.11
$$\text{tip speed ratio} = \frac{2\pi r N}{V}$$

where r is the radius of the rotor, N is the rotational speed (rpm), and V is the undisturbed wind speed ahead of the rotor. Figure 3.13 shows the power coefficients of various rotors as a function of their tip speed ratio, where the

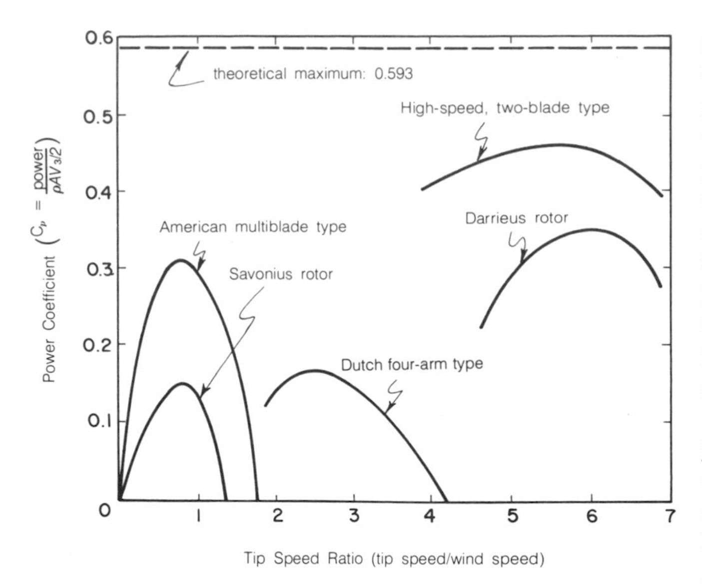

Figure 3.13 Typical power coefficients as a function of tip speed ratios for various blade designs.

power coefficient is simply that fraction of the wind's power which the rotor actually extracts. Earlier we indicated that the maximum possible power coefficient is 0.593 and this is shown in the figure. As you can see, the Savonius rotor and the multiblade windmill produce maximum power at a tip speed ratio of about 1. The high-speed prop in the figure reaches its maximum output at a tip speed ratio of over 5, and its efficiency is about three times that of a Savonius rotor. This high-speed blade has an efficiency of about 77 percent of the theoretical maximum, compared to the 70 percent we assumed for Figure 3.6.

Knowing that a well-designed prop operates with a tip speed ratio of somewhere around 4 to 6 can give us some help in estimating the gearing that is required to match the prop to a generator.

Example: Calculate the rotational speed of a 10-foot propeller in 25-mph winds if the tip speed ratio is 4. What gear ratio should be used to match it to a generator which puts out its maximum power at 2500 rpm?

Solution: From the definition of tip speed ratio (rearranged), the speed of the rotor in 25-mph winds is

$$N = \frac{V \times \text{tip speed ratio}}{2\pi r}$$

$$= \frac{25 \text{ miles/hr} \times 5280 \text{ ft/mile} \times 4}{60 \text{ min/hr} \times 2\pi \times 5 \text{ ft/rev}}$$

$$= 280 \text{ rpm}$$

The gear ratio should be 2500:280, which is about 9:1.

This example gives some rough values for rotor speed, generator speed, and gear ratios that are fairly representative of homebuilt systems using car alternators. Commercial systems are usually designed so that the generator turns much more slowly than the 2500 rpm in this example. This design feature greatly increases generator lifetime and reduces the gear ratio, but at the expense of increased generator size and cost. Dunlite, for example, uses a gear ratio of 5:1 to match a prop turning at 150 rpm to a generator which produces maximum power at 750 rpm. And Elektro, on all but its largest unit, avoids the problems of gearing altogether by running the propeller shaft directly into their very slow-speed generators.

If you are thinking of making your own blades, you just might consider using expandable paper honeycomb covered with fiberglass instead of using wood. The paper and design details are available from Windworks (see H. Meyer's listing in Bibliography). You might also check out Alan Altman's article in *Alternate Sources of Energy* (May 1974), which tells you how to design a Clark "Y" airfoil.

Whatever kind of rotor you use, provision must be made to keep the machine from reaching dangerous speeds in high winds. Although there are many ways this can be accomplished, the techniques actually used for the most part are based either on the idea of swinging the whole machine more and more off the wind as wind speeds increase or on some sort of mechanism which operates by centrifugal force. One way to have the machine swing off the wind is with an auxiliary pilot vane set parallel to the prop and perpendicular to the main vane. As the wind builds up, the force on this pilot vane eventually overcomes the guiding force on the main vane and the whole machine swings off wind.

Most bigger machines use the centrifugal force that is created at higher rpms to automatically change the pitch of the blades as the rated wind speed is neared, thereby spilling the excess wind. Dunlite (Quirk's), for example, uses spring-loaded weights attached to each prop shaft as shown in Figure 3.14. As the prop speed approaches the desired maximum, the centrifugal force on the weights overcomes the restraining effect of a magnetic latching system and the weights swing outward, thus moving the blades to a coarser pitch.

In addition to feathering or off-wind speed controls, some safety mechanism must be included so that you can halt the machine completely at any time. This can be accomplished with a folding tail vane which will place the propellers 90 degrees out of the wind.

Generators, Alternators, and Voltage Regulators

So the wind gets the rotor spinning and the rotor's shaft is coupled into the generator and the generator produces electricity. We know that a generator is characterized by the number of watts that it produces at its rated wind speed, and we know how to choose the correct generator rating for our energy needs. We also know that two different generators can put out the same amount of power at a given wind speed, but one may be loafing along at a couple of hundred rpm and the other is zooming at several thousand rpm. The slower the speed of the generator, the longer it's going to last, the heavier it is, the less gearing it requires to match the rotor, and the more it costs. In other words, a generator specifically designed for a wind machine is going to look quite a bit different than the small high-speed jobs in your car.

We also know that an alternator is a special kind of generator; but to understand why alternators are recommended over other kinds of generators, we need to know something about how they work. Whenever a conductor (e.g., copper wire) passes through a magnetic field, the electrons in the conductor experience a force which tries to push them down the wire (creating a current). Any generator requires these three components: 1) some wire windings, called the *armature*, which carry the output current; 2) a source of a magnetic field, which in very small generators may be simply a permanent magnet, but in wind generators is always some *field* windings which must be supplied with a small amount of control current (electrical currents create magnetic fields); and 3) a way

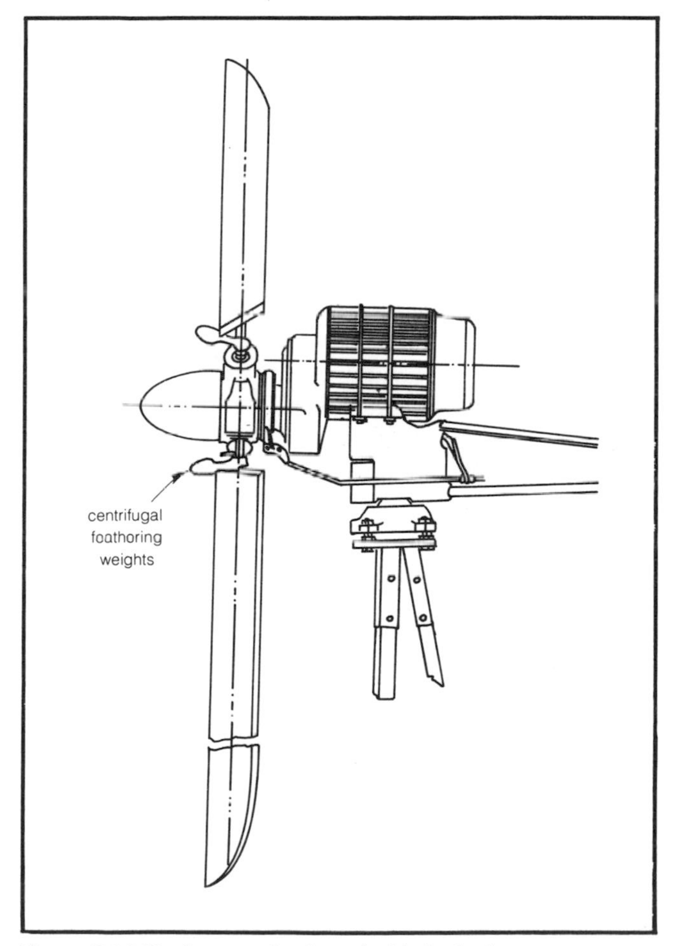

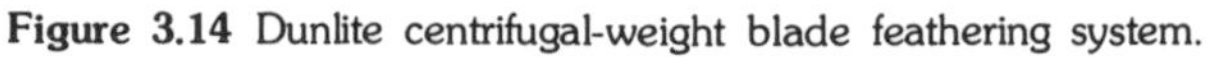
Figure 3.14 Dunlite centrifugal-weight blade feathering system.

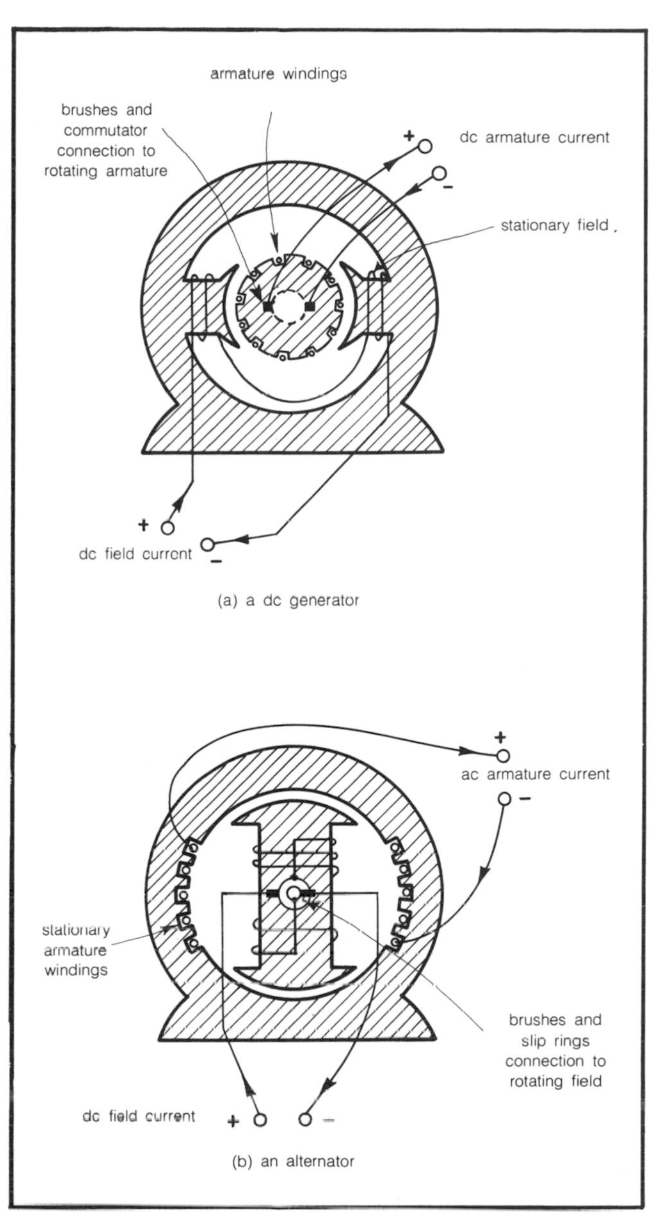

Figure 3.15 A dc generator (a) has a stationary field and a rotating armature, and an alternator (b) has a rotating field and a stationary armature.

to create relative motion between the armature and the field. That motion is created by spinning a shaft called the *rotor* within the stationary housing called the *stator*.

The big difference between an alternator and a dc generator (Figure 3.15) is in the location of the armature—whether it is spinning with the rotor (dc generator) or attached to the stationary stator (alternator). The armature, which carries maybe ten or twenty times as much current as the field, must somehow be connected to the stationary wires which carry the current to the batteries. In an alternator, the connection is direct since the armature is not rotating and this greatly simplifies the problem. In a dc generator, however, the connection must be made from the spinning rotor via commutator segments and brushes. The commutator and brushes take a lot of wear supplying the large currents from the armature; to give a reasonable lifetime between servicings, they must be well-designed—which means expensive. So for reasons of economy, lightness, and durability, alternators have the advantage. Moreover, for homebuilt systems, car alternators have further advantages: they are able to charge batteries at a lower rpm (typically 750–950 rpm, compared to over 1000 rpm for a car generator) and they also can supply more current (45–55 amps, compared to around 25–30 amps).

In an alternator, the current generated in the stationary armature is ac and its frequency varies as the rotor speed varies. Because it is variable frequency, we can't use it directly in appliances that require 60-cycle current; so we'll send it off to a battery instead. But if we sent ac directly to a battery, it would charge the battery half the time and discharge it the other half—no good. To avoid this problem, the current is *rectified* (converted to dc) by silicon diodes mounted in the housing itself so that what comes out of the alternator terminals is dc. (For you electrical engineers, Figure 3.16 is included, which shows how the 3-phase ac in the alternator is converted to dc.)

This takes care of the biggest part of the difficulty, namely picking off the large armature currents. But what about the small control current for the field which is spinning with the rotor? It can be supplied via slip rings and brushes and, since so little current is involved, this is not much of a problem; you must only check and replace the brushes periodically. Better still are the newer "brushless" alternators. (Current in stationary field windings creates a magnetic field which causes an ac current to flow in exciter windings on the rotor. This ac is rectified to dc by diodes mounted on the rotor and the dc creates the necessary magnetic field for the stationary armature.) It eliminates one of the few maintenance problems that wind generators have.

But where does the field current come from? A dc generator has enough residual magnetism in the field poles so that some current is generated even if there is no field current. This makes it possible to derive the field current right from the output of the generator itself (a shunt-connected generator). However, alternators and some dc generators get their field current from the battery; this presents a problem since, if the wind is not blowing, the batteries can simply discharge through the field windings. To prevent this discharge, expensive systems use a "reverse current relay," while in the homebuilt system some monitoring device, capable of sensing either propeller rpm or wind speed, is used to turn on the field.

The device that controls the operation of the generator and regulates transactions between generator and batteries is the *voltage regulator*. The voltage regulator regulates the output of the generator by rapidly switching the field current on and off. When the field is "on," the generator puts out as much current as it can; when it is "off," the output drops toward zero. The average output is determined by how much of the time the regulator allows the field to be on.

The regulator insures that the generator output will never be so great as to damage either the batteries or the generator itself. Batteries can be damaged and their useful lifetime shortened by charging them too rapidly or overcharging them. The rate at which batteries should be charged depends on the state of the charge already in them—they can take a lot without heating and gassing when they are discharged, but the charging rate must taper off as they approach full charge. This is handled automatically by the voltage regulator.

The other function of the regulator is to allow the generator output voltage to be adjusted to the proper level to compensate for the voltage drop in the lines connecting it to the batteries.

Energy Storage

Since we can't count on the wind to be blowing whenever we want to use electricity, and since the generator doesn't supply nice 60-cycle current at the proper voltage, the system must include some way to

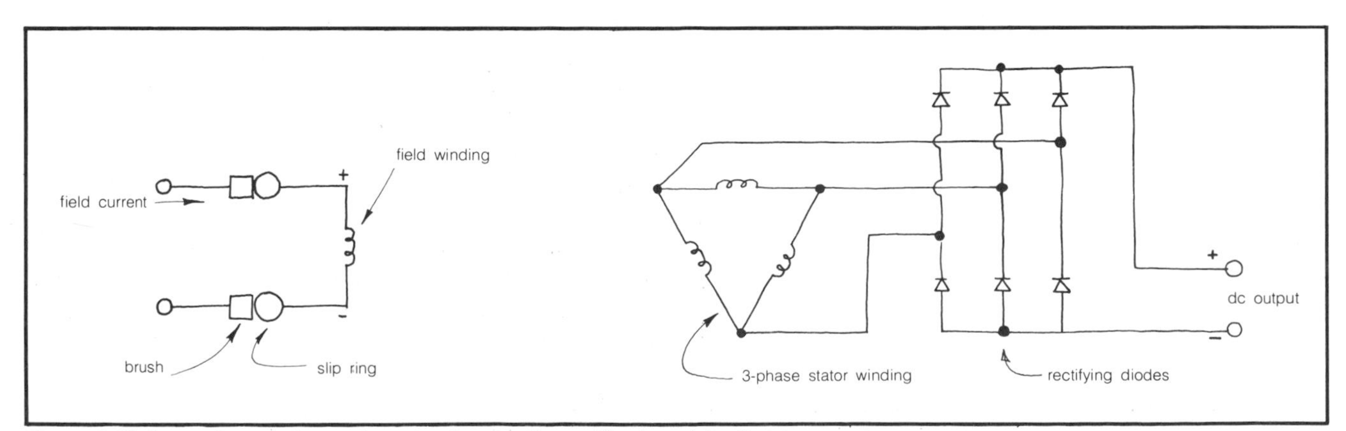

Figure 3.16 Circuit diagram for an alternator showing the rectification of the 3-phase armature current.

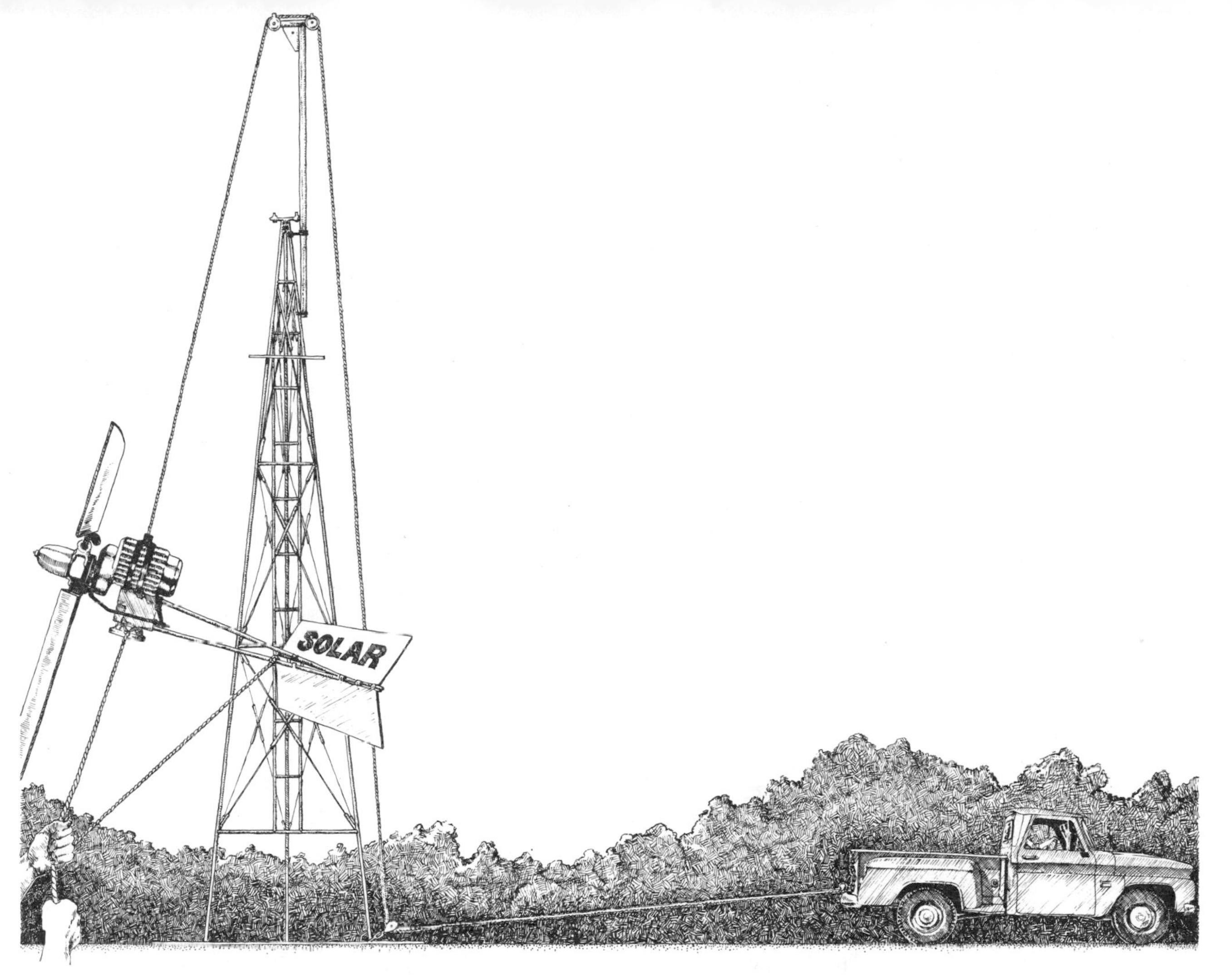

store energy. While a number of storage techniques are theoretically possible—flywheels, electrolysis of water to produce hydrogen, pumped storage, and compressed-air storage—it is the old standby, the battery, that presently offers the cheapest, most efficient, most convenient, and most readily available storage option for home usage. It is certainly to be hoped that some better storage mechanism will soon be available since batteries are a long way from being ideal; they're expensive, bulky, and require a fair amount of "tending to."

If and when the big power companies start using wind power, they won't have to worry about the storage problem: the power will be generated at the proper frequency and voltage so that it can be put directly onto their big power grid and be used immediately. Some have suggested that the small user could do the same thing—generate electricity when the wind blows, fix up its electrical characteristics, and then feed it back through your house's electric meter into the power lines. This procedure would cause the meter to run backwards, giving you "credit" for that amount of energy—credit that you would reclaim as you needed electricity in the house. While this is a neat solution to the storage problem, it is unlikely the power companies would ever go for it—and for good reason. When a repairman is working on the lines, he can disconnect the power from their end and feel safe; but if someone starts putting juice in from a windmill somewhere, it could be lethal.

So for the time being we are stuck with batteries and our choice is primarily between lead-acid batteries similar to the ones used in automobiles and nickel-cadmium (Ni-Cad) batteries like those used in aircraft. While Ni-Cads have certain advantages—they are not damaged by moderate overcharging, are smaller, lighter and more rugged, and are not affected radically by cold weather—they are so expensive that lead-acid batteries are generally recommended.

While lead-acid *car* batteries could be used, these have been designed to provide short bursts of high current; subjected to the frequent deep discharging common in a wind system, their lifetime is seriously reduced. The most practical alternative is the long-life "stationary" or "home lighting" type of battery, designed for repeated cycling from fully charged to fully discharged states. These batteries will last for more than ten years under normal windplant use and some claim lifetimes of nearly twenty years under minimal discharge rates.

Batteries are rated according to their voltage and their storage capacity. Each cell of a lead acid battery nominally produces close to 2 volts and, by arranging cells in series (connecting the terminals *plus* to *minus* to *plus* to *minus*, etc.), the cell voltages are additive so that any (even) voltage can be obtained. For example, a 12-volt car battery has six cells in series: to obtain 120 volts, we would need sixty cells in series or ten 12-volt batteries, or twenty 6-volt batteries, and so on.

The most important characteristic of a battery is its storage capacity, and determining the proper amount of storage for a home system is one of the most important parts of the design. The storage capacity of a battery is measured in amp-hours at a given discharge rate. For example, a 240 amp-hour battery with an 8-hour discharge rate is capable of delivering 30 amps for 8 hours before its output voltage drops below a specified level; it would then have to be recharged. For discharge rates longer than the specified 8 hours, the storage capacity is increased (e.g., 20 amps could be drawn for longer than 12 hours); for faster discharges, the capacity is decreased (e.g., 40 amps could not be drawn for 6 hours). Figure 3.17 shows an example of how much the capacity of a battery can change for differing discharge rates. Batteries are usually specified for discharge rates of 8, 10, or 20 hours and it is, of course, important to know which you have.

While it is the capacity in amp-hours that is usually given, we are really more interested in the amount of *energy* stored. Energy has units of watt-hours (or KWH) and, since watts are obtained by multiplying volts times amps, we can figure the energy stored in a battery by multiplying its amp-hour capacity by its rated voltage. For example a 120-volt, 240-amp-hour battery setup stores 28,800 watt-hours (120 × 240 = 28,800) or 28.8 KWH. If we were to use 7 KWH per day, the batteries would store a bit over 4 days' worth of electricity.

Example: Calculate the battery capacity required to provide 4 days of storage if we have calculated our monthly energy demand to be 150 KWH for a 120-volt system.

Solution: 150 KWH per month equals about 5 KWH per day so we want to provide

$$5 \text{ KWH/day} \times 4 \text{ days} = 20 \text{ KWH}$$

which, at 120 volts, works out to be

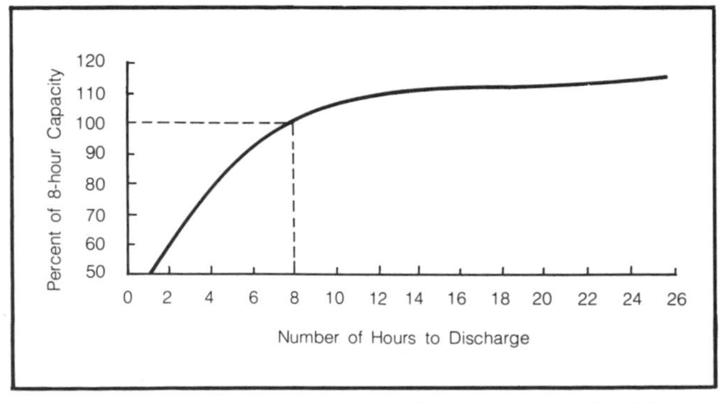

Figure 3.17 The variation in amp-hour battery capacity for differing discharge rates with the 8-hour rate equaling 100%.

$$\frac{20{,}000 \text{ watt-hours}}{120 \text{ volts}} = 167 \text{ amp-hours}$$

So we need a 120-volt battery setup rated at 167 amp-hours. If we are using 6-volt batteries, then we will need 20 of them, each rated at 167 amp-hours.

There are several guidelines concerning how many days of storage you should attempt to provide; if too small a storage capacity is provided, the batteries will be charged and discharged at so rapid a rate that their lifetime will be reduced. As a rough rule of thumb, as long as the prolonged charge or discharge current does not exceed 15 percent of the amp-hour rating, you'll get maximum battery life. So you can use the maximum current that the generator puts out to fix the minimum battery capacity. As the following example demonstrates, you simply multiply the maximum current by 7 to get minimum amp-hours.

Example: Suppose we have a 2-KW, 120-volt generator; what minimum storage capacity should be used? If our load is estimated at 150 KWH per month, how many days of storage does this correspond to?

Solution: The maximum current from the generator will be

$$\frac{2000 \text{ watts}}{120 \text{ volts}} = 16.7 \text{ amps}$$

By the rule of thumb, batteries should have a capacity of at least

$$\frac{16.7}{0.15} = 16.7 \times 7 = 117 \text{ amp-hours}$$

which, at 120 volts, is equivalent to

$$\begin{aligned} 117 \text{ amp-hours} \times 120 \text{ volts} &= 14{,}000 \text{ watt-hours} \\ &= 14 \text{ KWH} \end{aligned}$$

At the demand rate of 5 KWH per day (150/30), this would supply

$$\frac{14 \text{ KWH}}{5 \text{ KWH/day}} \cong 2.8 \text{ days}$$

This answer is fairly typical and indicates that you should probably figure on providing at least 2 or 3 days of storage. On the other hand, you should probably not figure on storing more than about 4 or 5 days' worth of energy for several reasons: 1) storage is expensive (somewhere around $60 per KWH); 2) if your winds are that bad, you should have a back-up system, a cheaper proposition than adding extra days of storage; and 3) if you have too much storage capacity, the batteries will be fully charged only infrequently and this will reduce their lifetime. They should be allowed to charge fully at least once every several weeks.

Finally, we should say something about location of the battery setup. We already know it should be as close to the load (house) as possible, to reduce line voltage losses. Moreover, batteries are sensitive to temperature—when cold, they cannot deliver as much of their rated capacity as when warm (you've experienced this on cold mornings in your car). For optimum performance they should be kept at around 75–80°F. Given line voltage losses and temperature considerations, batteries probably should be kept in the house or in a well-insulated shed adjacent to the house.

They should also be well ventilated since, during charging, batteries release hydrogen and oxygen (gassing) and hydrogen can be explosive. Finally, batteries should be kept clean and dry, off the floor, and only distilled water should be used to maintain the proper level of electrolyte.

DC-to-AC Power Inverters

Batteries supply only dc; if all of your appliances could operate with just dc, there would be no need for a dc-to-ac power inverter. And, in fact, a surprising number of typical home loads can run directly off of 110–120 volt dc: applicances which provide resistive heating (irons, toasters, electric blankets, coffee makers, fry pans, hot plates, waffle irons, and curling irons); incandescent lights; and small appliances with "universal" motors, having brushes and running on either ac or dc (vacuum cleaners, sewing machines, food mixers, shavers, and many portable hand tools). In addition, some loads can be modified to use dc: fluorescent lights (which, by the way, put out something like three times as much light per watt as incandescents) and motor-driven loads where the ac motor can be replaced with an equivalent dc motor.

There still remain some important applicances which need ac. Anything that uses a transformer (a hi-fi, TV, or those high-intensity lights) needs ac; synchronous motors (used in electric clocks and record players) and induction motors (split-phase motors, capacitor-start motors—used in heavier loads such as refrigerators and washing machines) must have ac. To handle such loads, an inverter is necessary.

There are two kinds of inverters: the older, less efficient, but decidely cheaper rotary inverters which consist of a dc motor driving an ac alternator; and the newer solid-state electronic inverters. The biggest problem with rotary inverters is their low efficiency, which may be only about 60 percent even when they are putting out their rated power; their efficiency drops even further for smaller loads. If an appliance requires, say, 100 watts of ac power which it gets from a 60 percent efficient inverter, then the actual power drain from the battery is 167 watts (100/0.60 = 167). Thus our estimate of energy demand by appliances which require ac would have to be upped by almost 70 percent! Rotary inverters are a bit hard to find except at surplus electronic supply houses—the *Energy Primer* (see Bibliography) lists some outlets.

Electronic inverters are quite a bit more efficient and require no maintenance, but they are fairly expensive—something like $100 to $150 for a fairly small 300-watt model to something like $1300 for a 3000-watt device. Prices of inverters of a given size can vary considerably, depending on how "smooth" the output is. The output of less expensive units may more closely resemble a square wave than the true sinusoidal ac that some sensitive electronic loads may require. Electronic inverters have fairly high efficiencies, usually ranging from about 80 to 95 percent. At 80 percent efficiency, the energy estimate for ac appliances should be increased by 25 percent (1/0.8 = 1.25) to account for inverter losses. Electronic inverters are available for either 12-volt or 115-volt dc inputs.

Whether you choose a rotary or electronic inverter, there is a certain amount of standby power that is lost any time an inverter is turned on—even if there are no ac appliances drawing power. Hence it is desirable to include a switch (manual or automatic) which will turn the inverter on only when ac is needed.

Now, what size inverter should you buy and should you get more than one? There are enough trade-offs here between convenience, efficiency, and costs so that no absolute solutions exist—you'll have to weigh the factors and make your own decision. The least efficient but most convenient system would be an inverter large enough to supply your entire load with ac. Simply add up the maximum number of watts that are liable to be on at one time (hopefully not more than 2000 or 3000 watts), chunk out about $1500, and you can then wire your house entirely for ac. No messing around with modified appliances or separately wired sockets for dc.

A cheaper variation is to separate your loads into those that can run directly off battery-supplied dc and those that *must* use ac. Then figure the maximum ac power that will be on at any one time and pick an inverter which can supply that power. Obviously the inverter will be smaller and cheaper than it was in the first example, and less power will be wasted because most of the load operates from dc. But you must wire the house with separate outlets; also, you must be sure never to plug an ac appliance into a dc socket, a mistake which can destroy your appliance.

A third way to design a system is with separate small inverters for each ac load or outlet. This can be the most efficient choice, since inverter efficiency is highest when it is supplying maximum power. If you have one 1000-watt inverter and most of the time it is only putting out 100 watts for your hi-fi, then its efficiency will be poor (and drain those precious KWH you worked so hard to accumulate). If instead you have two or three small inverters of a few hundred watts each, each turned on only as the demand arises, then efficiency is maximized. To know whether this latter approach is worth it, you'll need to determine—from manufacturers' spec sheets—the efficiency characteristics of inverters you are considering.

Besides such important specifications as power output, standby power drain, efficiency characteristics, voltage and frequency regulation, and wave shape, you need to determine the amount of *surge power* that the inverter can handle. Motors on such heavy-duty appliances as refrigerators and large shop tools have starting currents which may be many times greater than their normal operating currents (you've noticed your house lights dim as the refrigerator kicks on?); you must be sure your inverter can handle those transients (good ones can). Batteries, by the way, handle such surges with ease. Table 3.7 compares the surge power to running power for various sizes of induction motors and you can see the large increase during starting. Universal motors, we might note, require the same power to run as to start.

Table 3.7 Starting Power Compared to Running Power for Various Motors[a]

Motor Size (horsepower)	Watts Required to Start Motor			Running Watts
	Repulsion Induction	Capacitor	Split Phase	
1/6	600	850	2050	275
1/4	850	1050	2400	400
1/3	975	1350	2700	450
1/2	1300	1800	3600	600
3/4	1900	2600	—	850
1	2500	3300	—	1100

Notes: a. From *Electric Power From the Wind* by H. Clews.

Auxiliary Power

For a completely reliable system, you may need to include an auxiliary engine-generator to charge the batteries during prolonged periods of inadequate winds. Even if you are willing occasionally to do without electricity, this back-up unit may be necessary to return the batteries to their fully charged condition periodically; by this precaution, the batteries' useful lifetime is not decreased. Small auxiliary powerplants usually burn gasoline, but some are available which run on LP-gas, diesel, or natural gas, and it should be relatively easy to get one to run on methane from a digester.

Answering the questions of whether or not to include a back-up power generator, and if you do, how big it should be will in part depend on your analysis of month-by-month wind speeds and load requirements. For example, if your monthly winds and power demands are relatively in balance, then a small back-up unit would be all that you would need. If, however, there is a big mismatch and some months the windplant is just not going to be enough, then a larger back-up would be required.

To illustrate this point, consider the problem of meeting the month-by-month demands given in Figure 3.3 in both San Francisco and Minneapolis (both cities have the same average wind speed). In Table 3.8, monthly demand and monthly winds for each of the two cities and the corresponding KWH outputs of a 2-KW (25 mph) wind generator are given (calculated using specific output information given in Figure 3.9). The differences between supply and demand are then listed.

Table 3.8 Monthly Energy Supplies from 2-KW Wind Generator in San Francisco and Minneapolis

		San Francisco			Minneapolis		
Month	Energy Demand (KWH)	Wind Speed (mph)	Supply from 2-KW Gen. (KWH)	Supply minus Demand (KWH)	Wind Speed (mph)	Supply from 2-KW Gen. (KWH)	Supply minus Demand (KWH)
Jan	160	8.2	100	−60	10.5	160	0
Feb	170	8.8	110	−60	11.0	180	+10
Mar	180	11.2	190	+10	12.1	240	+60
April	170	12.6	260	+90	12.8	270	+100
May	140	14.0	350	+210	12.5	250	+110
June	120	15.0	440	+320	11.7	220	+100
July	110	14.2	360	+250	9.8	140	+30
Aug	100	13.2	300	+200	9.5	130	+30
Sept	90	12.5	260	+170	10.6	170	+80
Oct	120	10.0	150	+30	11.0	180	+60
Nov	140	7.6	80	−60	12.0	230	+90
Dec	160	8.4	100	−60	11.1	190	+30
Monthly Average	138	11.3	225	—	11.3	197	—

As you can see from the table, the 2-KW generator would be sufficient in every month in Minneapolis and, in most months, there is a good deal of extra power available. This "extra power" could help take the system through months when the winds are less than average. (Remember, when you see a long-term "average" wind speed for some location for some month, approximately half the time the real monthly average will be less than the long-term average—so overdesign is essential if you want to minimize the use of standby power.) Our conclusion is that, in Minneapolis, the 2-KW generator would almost always handle the entire load and an auxiliary power source, if you had one at all, could be small. A standby source of 1000 watts (about as small as they come) would boost the supply by 30 KWH per month, running only one hour per day, and that boost would probably be sufficient.

On the other hand, in San Francisco the load and supply are terribly mismatched. Winds are high in the summer, when demand is low. In the winter, demand is high and winds are low; there are four months for which the generator alone is insufficient. You could increase the size of the wind generator to meet winter demands but the 2-KW unit is already capable of supplying two to three times the summer demand. (With so much extra capacity during the summer, you might consider using some of it to heat water for your solar home. Each 100 KWH of extra capacity per month is enough to raise the temperature of 20 gallons of water by about 70°F every day—easily one person's hot-water needs.) It would be cheaper to include a fair-sized auxiliary power unit and then figure on running it an hour or so per day during the winter. You should probably not choose a standby unit with greater capacity than the wind generator: you may charge the batteries at too fast a rate. In our example, a 2-KW standby unit running one hour per day would give an extra 60 KWH per month, enough to carry us through the winter.

Gasoline-powered standby powerplants of appropriate size (1000 to 6000 watts) cost on the order of $200 to $300 per 1000 watts. Prices vary, depending on such quality factors as whether the model runs at 1800 rpm or 3600 rpm (the 1800-rpm units last longer and are more expensive) or whether it has manual or electric starting. All of these units are generally unpleasant (they are noisy, require maintenance, produce pollution, have relatively short lifespans, and consume fossil fuels) and it would seem advisable to plan your system so that the standby is needed as little as possible.

These units produce good 60-cycle, 115-volt ac power; but if you want to connect one directly to your load, remember that it must have enough capacity to handle the maximum power drawn at any one time, including the large surge currents required by some motors to start (Table 3.7). Frequently the standby will be too small to be used in this way and instead will see service recharging your battery setup. In this case, you'll need to include a full-wave bridge rectifier using silicon diodes to convert the auxiliary ac to the dc required by batteries.

What is Available?

Since the "design" of most wind-electric systems consists of matching component specifications to perceived requirements, it is important to know what's available and roughly what it costs. As interest in wind systems increases, the selection of components is correspondingly broadening; and as inflation rolls along, prices are also rapidly rising. So the following specifications—especially the prices—should be considered as only approximate.

There are only a handful of wind-generator manufacturers around the world. The two that provide the systems which are most closely matched to the needs of a small house are Dunlite of Australia (sometimes sold under the name Quirk's) and Elektro of Switzerland. Dunlite makes two sizes, 1000 watts and 2000 watts, while Elektro manufactures a variety of windplants in sizes up to 6000 watts. These systems are handled in the United States by most of the wind-energy distributors listed at the end of this section.

Dyna Technology of Sioux City, Iowa, and Bucknell Engineering of El Monte, California, both manufacture 200-watt units that are really too small for any but the most spartan of demands. Sencenbaugh Wind Electric of Palo Alto has just introduced a 750-watt kit, rated at 19 mph, that looks like it could just about handle the modest needs of a small house or cabin.

There are three more European manufacturers. Lubing Maschinenfabrik of Barnstorf, Germany, makes a 400-watt downwind unit; Enag s.a. of France produces three—400 watts, 1200 watts, and 2500 watts; and Aerowatt s.a. of Paris lists five units—28 watts, 130 watts, 350 watts, 1125 watts, and 4100 watts (these are apparently well-designed machines, but extremely expensive; the 4-KW unit costs something like $19,000!).

The characteristics of some of these wind generators are listed in Table 3.9. The prices given are West Coast, January 1975, and don't include towers, batteries, inverters, wire, or standby units. "Rough" estimates of these extras are something like: 1) Towers—maybe $300 to $600 for a 30-footer and maybe $750 to $1500 for a 70-footer (depending on size of wind generator); 2) Batteries—maybe $60 per KWH of storage; 3) Electronic Inverters—maybe $150 for a 300-watt model; maybe $900 for a 1-KW unit; maybe $1400 for a 3-KW unit; 4) Wire—prices depend on gauge, length, and material; estimates are given in Table 3.6; and 5) Standby Powerplants—about $200 to $300 per 1000 watts. Now please realize, these prices are almost ephemeral—here today, up tomorrow— and we wouldn't include them ex-

Table 3.9 Characteristics of Some Currently Available Wind-Driven Electric Generators

Manufacturer	Rated Output (watts)	Rated Wind speed (mph)	Prop Dia.	Approx.[a] Monthly Output in 10-mph avg. winds (KWH)	Usual Nominal Voltage	Approx.[b] Price ($)	Comments
Elektro G.M.b.H.	600	20	8′4″	70	12	2350	Prices include windplant, control panel with
St. Gallerstrasse 27	1200	23	9′10″	110	115	2600	metering, voltage regulator, and handbrake. High-
Winterthur, Switzerland	2500	25	12′6″	180	115	3350	speed automatic controls for shutdown of plant in
	4000	24	14′5″	320	115	4250	case of fully charged batteries or winds in excess of
	6000	26	18′	400	115	4900	80 mph are extra—about $980 for manual restart and $1170 for automatic restart. All generators are brushless and all but the 6000-watt model are direct drive.
Davey-Dunlite Co.	1000	25	11′6″	70	32	2200	Prices include windplant control panel with metering
21 Frome St.	1000	25	11′6″	70	12	3500	and voltage regulator. The less expensive 1000-watt
Adelaide, S. Australia (Marketed under the name "Quirk's")	2000	25	11′6″	140	110	3500	unit has a dc generator (brushes must be changed every 3–5 years); other two are brushless. Standard plant will withstand 80-mph winds; modifiable to 140 mph. Lubrication required every 5–7 years.
Winco Dynatech, Inc. P.O. Box 3263 Sioux City, Iowa 51102	200	20	6′	22	12	400	Includes 10′ tower and control panel.
Sencenbaugh Wind Electric P.O. Box 11174 Palo Alto, CA 94306	750	19	12′	95	12	1050/ 1400	Prices are for kit and semi-finished kit, including control panel with metering and voltage regulator.

Notes: a. Calculated from Figure 3.9. b. January 1975, West Coast prices.

cept for the fact that some feel for prices is important. As a final rough rule of thumb, for a complete wind-electric system, including all the extras, in 10-mph average winds, figure on paying about $3000 for each 100 KWH of monthly output.

The following is a list (from the *Energy Primer*; see Bibliography) of organizations or companies in the United States that act as distributors for wind-electric equipment:

Energy Alternatives
P.O. Box 223
Leverett, Maine 01054

Environmental Energies, Inc.
21243 Grand River
Detroit, Michigan 48219

Garden Way Laboratories
Charlotte, Vermont 05445

Independent Power Developers
P.O. Box 618
Noxon, Montana 59853

Penwalt Automatic Power
213 Hutcheson Street
Houston, Texas 77003

Real Gas and Electric Co.
P.O. Box A
Guerneville, CA 95446

Sencenbaugh Wind Electric
P.O. Box 11174
Palo Alto, CA 94306

Solar Energy Co.
810 8th Street, N.W.
Washington, D.C. 20006

Solar Wind Co.
P.O. Box 7
East Holden, Maine 04429

Windlite
P.O. Box 43
Anchorage, Alaska 99510

Design Summary

All the necessary design procedures have already been covered, but it may be useful to summarize the approach in one place:

1. Estimate your energy requirements using old bills or the "watts times hours" approach. Do it for at least each season and, better still, for each month. You may want to divide the demand into "essential" and "convenience" components and design the wind system to always supply at least the essential portion. For ac appliances, increase the demand according to the efficiency of the inverter. It is also recommended that all estimates be increased by about 30 percent to allow for battery inefficiencies, line losses, and less-than-average winds.
2. Acquire a month-by-month estimate of average wind speeds at the chosen site and tower height. The site must be away from obstacles, but also close enough to the load so that line losses are not excessive.
3. Using Figure 3.9 and the section "What is Available," pick a generator that will meet your needs. If you have a relatively good match between

monthly winds and monthly demands, the generator can be picked to meet at least the essential portion and perhaps most of the convenience demand. There are certain trade-offs involving auxiliary power sources and flexibility in your requirements, so hard and fast rules cannot be given here. If it seems that in some months there will be a fair amount of surplus power generated, you may want to use that surplus to augment a solar water-heating system.

4. If you are going to have an auxiliary power source, you can pick its size based on estimates of the shortfall between wind-generator output and demand. The standby should probably be smaller in watts generated than the wind generator, but large enough so that you won't have to run it more than an hour or so a day during months with light winds.
5. Pick the amp-hour storage capacity of the battery setup to be at least 7 times the maximum current the generator will supply. You may want to increase that storage figure if it doesn't give sufficient days of capacity during any given month.
6. Pick the power rating of the inverter(s) to be able to handle all the ac loads.

You will undoubtedly work through several designs as you evaluate the trade-offs between costs, convenience, and reliability; but no matter what design you settle on, there are a few things to keep in mind. Always. First, stay conservative—always overestimate the load and underestimate the supply. It is better to be surprised at how little you need to use that auxiliary power unit than to be disappointed at how often it is running. Second, when you advance to the assembly and operating stages, you must constantly be concerned with safety. The mechanical forces that build up when one of these units starts operating in high winds are very dangerous and the amounts of electrical power generated can be lethal.

Hopefully this presentation has given you enough information to get your design well underway. Final details can be checked out with the distributor from whom you purchase your components. The energy is there—it's not exactly free, but it will never run out and it's clean. Let's use it.

ELECTRICITY FROM A STREAM

If you are fortunate enough to have a small stream flowing through your property, you may be able to use it to generate electricity for your home. If your needs are

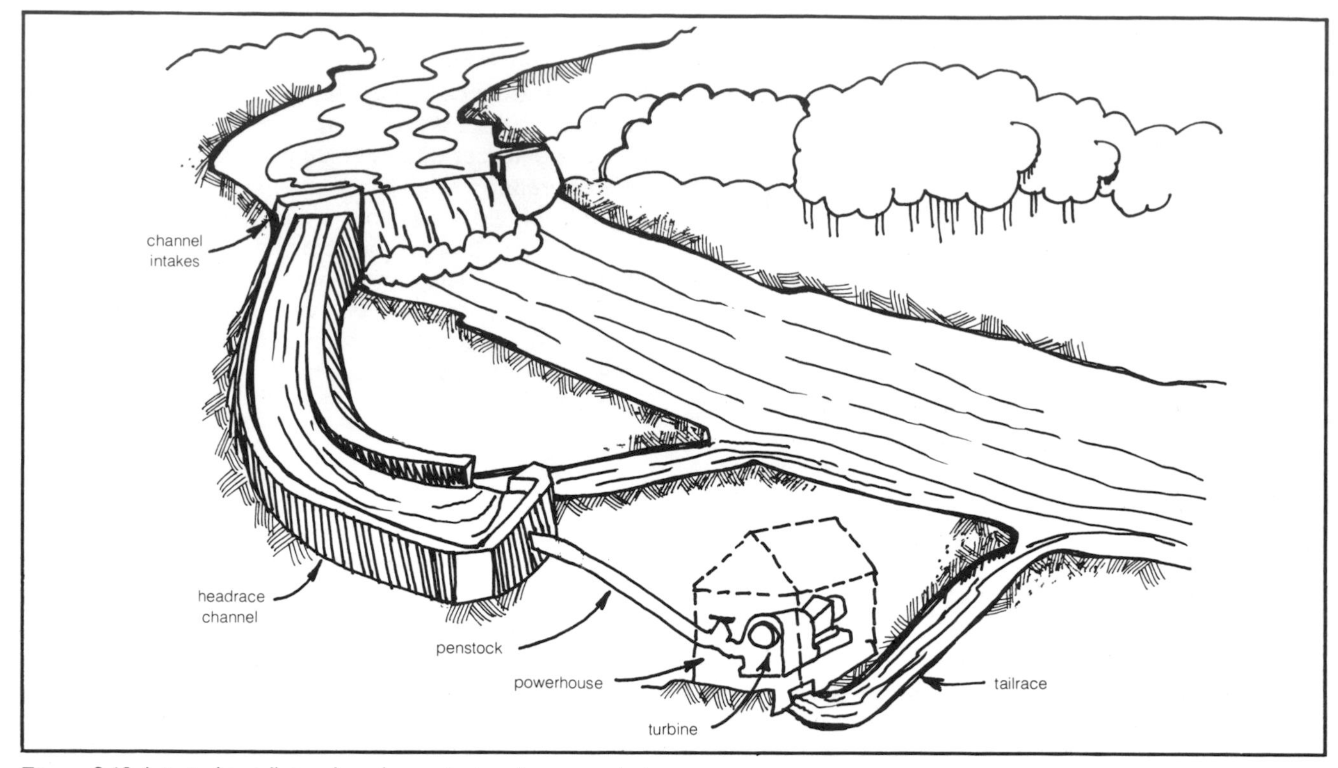

Figure 3.18 A typical installation for a low-output water-powerplant.

modest, you will probably be amazed at how small a flow is required. A meandering stream, rolling along at only a couple of miles per hour and having a cross section of only a few square feet, gives up more energy in falling several feet than you are likely to need to power *all* your electrical appliances. In this section you will learn how to estimate the energy that your stream can supply and some methods by which that energy can be captured, stored, and converted into electricity.

Power and Energy from Water

A stream contains two forms of energy: by virtue of its velocity, it has *kinetic* energy; and by virtue of its elevation it contains *potential* energy. The kinetic energy in most streams is not great enough to be useful; it is the potential energy between two sites of differing elevations that we try to exploit. Very simply, the idea is to divert some of the water from a site upstream, transport it along an elevated conduit, and then let it fall through a waterwheel or hydraulic turbine located at a lower elevation downstream. The turbine (or waterwheel) turns a generator which produces electricity. The water then returns to the stream.

Figure 3.18 shows a sample powerplant installation. The central structure is the dam which causes the stream to back up, creating a pond or reservoir to store energy and elevate the surface of the water (which increases the obtainable power). In this example, the dam includes a spillway; this feature allows the stream to overflow the dam when the pond is full, thus protecting the dam from overtopping floods.

Water can be transported from the dam to a waterwheel or a turbine in an open channel and/or in a pressure conduit called a *penstock*. The powerhouse (where the electricity is generated) may be either next to the dam as shown in Figure 3.19a, or further downstream as shown in Figure 3.19b. In the latter case, more power can be developed since the water drops further, but the increase in conduit distance is costly and some energy is lost in the conduits themselves.

The amount of power obtainable from a stream is proportional to the rate at which the water flows and the vertical distance which the water drops (called the *head*). The basic formula is

E. 3.12
$$P = \frac{QHe}{11.8} = \frac{AVHe}{11.8}$$

where P is the power obtained from the stream in kilowatts; Q is the flow of water in cubic feet per second (cfs); A is the average cross-sectional area of the stream in square feet; V is the average velocity of the stream in feet per second; H is the height the water falls (head) in feet; 11.8 is a constant which accounts for the density of water and the conversion from ft-lb/sec to KW; and e is the overall conversion efficiency.

The above relationship tells us how much power a stream has to offer. To account for the various losses which occur during the conversion to electricity, we have

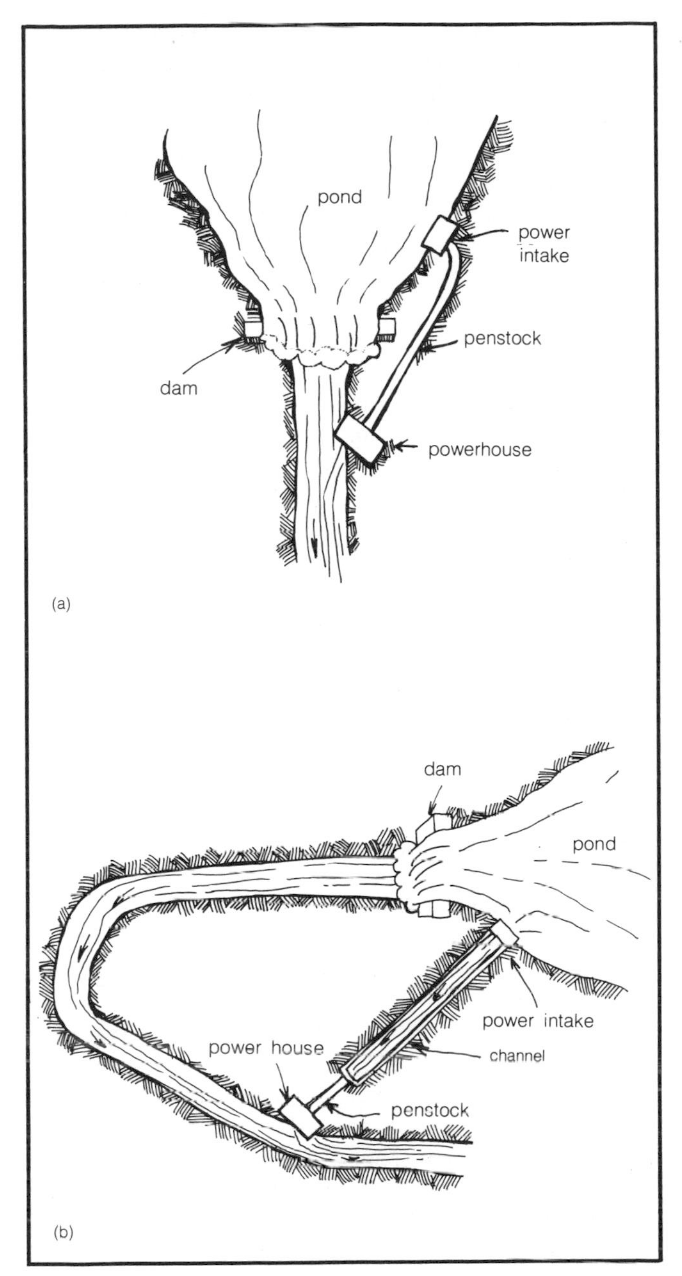

Figure 3.19 The powerhouse may be located just below the dam (a), or, for greater power, it may be located further downstream (b).

included in the above estimate an efficiency factor, *e*, which may be on the order of 0.5 to 0.7. To actually use the formula, the stream flow rate *(Q)*, or the velocity *(V)* and cross-sectional area *(A)*, must be determined. In the next section we indicate how to measure these important quantities, but for now let's look at some example calculations.

Example: At 50 percent conversion efficiency, how much power could be obtained from a flow of water having a cross-sectional area of 1 square foot, a velocity of 1 foot per second, and a fall of 10 feet?

Solution: We use our formula and find that

$$P = \frac{AVHe}{11.8} = \frac{1 \times 1 \times 10 \times 0.5}{11.8}$$

$$= 0.42\ KW = 420\ watts$$

Now 420 watts may not seem like much until you realize that this power can be obtained 24 hours a day. With this flow rate, in a month we could accumulate about

$$0.42\ KW \times 24\ hrs/day \times 30\ days/month$$
$$= 300\ KWH/month$$

This is enough electrical energy to run lights, refrigerator, freezer, power tools, pumps, hi-fi, etc.; probably all you would need unless you used an electric stove and electric space heating. (See "Calculating Your Energy Requirements" in this chapter.)

This example points out how little the *average* flow can be and yet be sufficient to meet our monthly *average* electrical-energy demand. You will recall, however, that we must be concerned not only with average electrical demand, but also with *peak* power demand. While our sample stream can deliver 420 watts continuously, what happens when we flip on a few appliances and demand jumps to several thousand watts? Either we must always have sufficient flow in the stream to provide that peak power demand or we need a way to store up energy when it is not needed to provide for these peaks. With a wind system it was recommended that such storage be provided by batteries; for a water system the most natural approach is to store potential energy in a reservoir of water behind a dam. The following examples illustrate these concepts.

Example: At 50 percent conversion efficiency, what flow rate *Q* would be required to provide 3000 watts of power given a head of 10 feet?

Solution: From our equation we have

$$Q = \frac{11.8P}{He} = \frac{11.8 \times 3}{10 \times 0.5} = 7\ cfs$$

If we know that the stream is always flowing at a rate of at least 7 cfs, then we have no problem. The stream in our first example, however, had a flow rate of only 1 cfs (using $Q = VA$; 1 ft/sec × 1 ft^2). One way to get the higher peak flow desired would be to drain off some of the water from our reservoir. The following example shows how much water would be required to satisfy a given daily demand for energy.

Example: Find the volume of water needed to generate 10 KWH per day of electricity (300 KWH/month) if the average head available is 10 feet and the overall efficiency is 50 percent.

Solution: The average power required is

$$P = \frac{10\ KWH/day}{24\ hours/day} = 0.42\ KW$$

The average flow required is

$$Q = \frac{11.8P}{He} = \frac{11.8 \times 0.42}{10 \times 0.5} = 1\ cfs$$

The amount of water to be drained from the pond in one day would be

$$\begin{aligned} Volume &= 1\ ft^3/sec \times 3600\ sec/hr \times 24\ hr \\ &= 86{,}400\ ft^3\ of\ water \end{aligned}$$

So, for example, this flow would completely drain a pond 5 feet deep with a surface area of 130 feet by 130 feet (5 × 130 × 130 = 86,000).

If you prefer a formula over the above approach, the following is an expression for pond volume required to store a given amount of energy:

E. 3.13 $$V_P = 42{,}500 \frac{E}{He}$$

where V_P equals the pond volume in cubic feet; E equals the deliverable energy stored (in KWH); and H equals the average head over the period of generation.

While the above calculation suggests that you could use the dam to store energy and then make releases only when there is an electrical demand, this may not be desirable. It would make downstream flow sporadic, which could upset the local ecology, your downstream neighbors, and could even be a violation of local water-rights laws. The pond, too, would be constantly filling and emptying. Moreover, the controls required to balance the flow with varying demands become more complicated. It is probably better to design your system to provide a constant flow large enough to handle peak demands. This requires some knowledge of year-round conditions so that you can provide for minimum expected flows.

Estimating Water Flows

It is quite important to know not only average stream flows, but also minimum and maximums to be expected. You'll need to estimate minimum flows in order to be sure you'll always have enough power; you'll need maximum flows to be able to design your structures so that they will not create a danger during peak flooding. If you have lived on the property for a long time, you may be able to recall past conditions; or you may have to gather this sort of information from other neighbors who are more familiar with the area.

Looking at the flow is one thing, but actually measuring it is something else. We will now describe two ways to go about measuring stream flows: the float method and the weir method.

Float Method

As we saw in our equation, the flow Q through the stream equals the cross-sectional area of the stream (A) measured at any site multiplied by the average velocity (V) of the water through that site ($Q = AV$). To apply the float method, first pick a section of stream about 100 feet long where the cross section is relatively constant and the stream is relatively straight.

To estimate the cross-sectional area of the stream at a given site, measure the depth of the water at a number of equally spaced points across the stream and calculate their average. The number of points needed depends on the irregularity of the cross section, but five or ten should do. The cross-sectional area is then determined by multiplying the width of the stream times the average depth. Do this at several "typical" points and average your results to obtain a more accurate answer.

To estimate the velocity, first measure the distance between two fixed points along the length of the stream where the cross-sectional area measurements were made. Then toss something that will float—a piece of wood or a small bottle—into the center of the stream and measure the length of time required for it to travel between the two fixed markers. Dividing the distance by the time gives you the average velocity of the float. But since the stream itself encounters drag along its sides and bottom, the float will travel a bit faster than the average velocity of the stream. By multiplying the float velocity by a correctional factor of about 0.8, an estimate of average stream velocity is obtained.

Example: Estimate the flow rate Q for the stream whose cross section is shown in Figure 3.20. The width is 6 feet and the five depth measurements are (in feet) 0.3, 0.5, 1.0, 1.2, and 0.5. A float traveled a distance of 100 feet along the stream in 50 seconds.

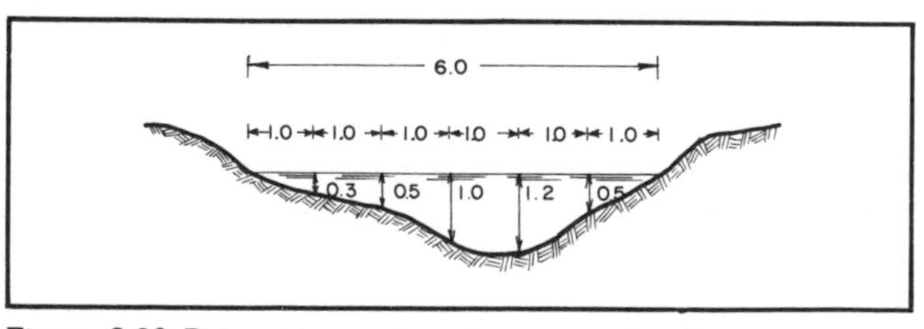

Figure 3.20 Determining a stream's cross-sectional area.

Solution: The average depth is

$$d = \frac{0.3 + 0.5 + 1.0 + 1.2 + 0.5}{5} = 0.7 \text{ feet}$$

The cross-sectional area is therefore

$$A = 6 \times 0.7 = 4.2 \text{ ft}^2$$

The stream velocity, including the correctional factor of 0.8, is

$$V = \frac{100 \text{ feet}}{50 \text{ seconds}} \times 0.8 = 1.6 \text{ feet/sec}$$

The flow rate is thus

$$Q = AV = 4.2 \times 1.6 = 6.7 \text{ cfs}$$

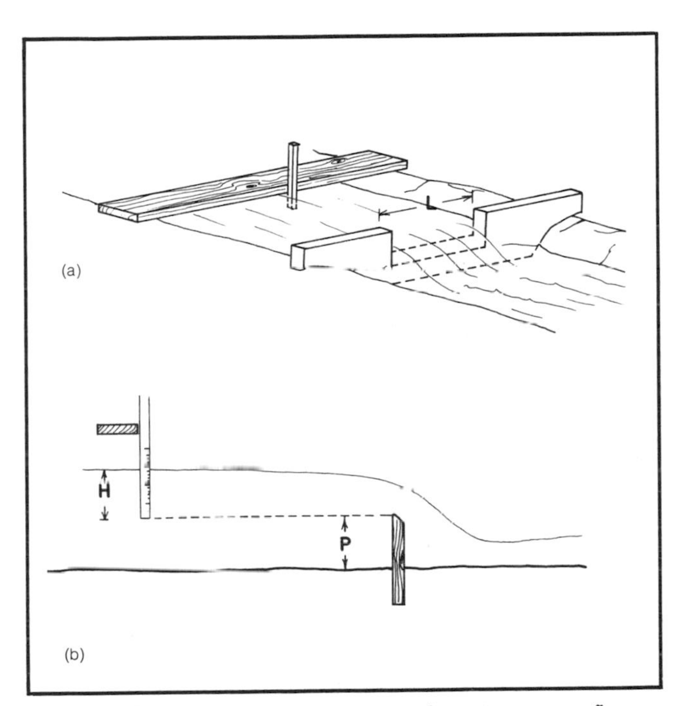

Figure 3.21 Weir and measurements to determine stream flow.

Weir Method

This method gives more accurate results than the float method, but it requires a bit more work in that a weir must be constructed. Figure 3.21 shows the basic arrangement. A water-tight, rectangular-notched dam is constructed and provision is made for measuring the height of the water surface above the notch. The weir should be located in the center of the stream, its crest should be sharp (as shown in the figure), and the streambed in front of the weir must be as flat as possible. The head H is measured several feet upstream from the weir, as shown. The dimensions of the weir and the head are all that are needed to obtain accurate flow measurements, using the following hydraulic relation:

E. 3.14 $$Q = CLH^{3/2}$$

where Q equals the flow in cfs; L equals the width of the weir opening in feet; C equals a discharge coefficient; and H equals the head above the weir in feet.

The discharge coefficient (C) can be determined from Figure 3.22, given the head (H) and the dimensions of the weir (P is the height from the streambed to the weir opening; b is the width of the stream).

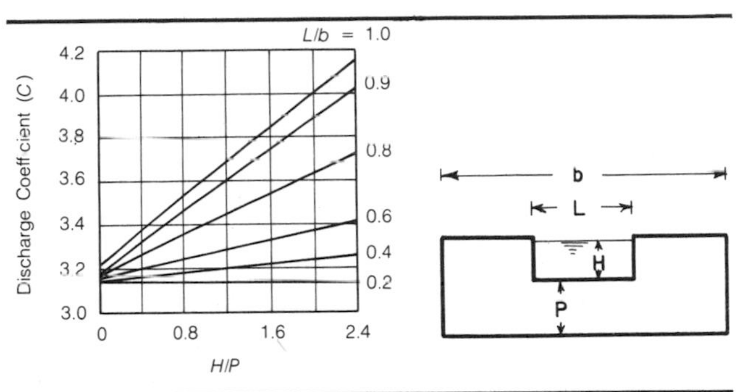

Figure 3.22 Discharge coefficient for a notched, sharp-crested weir (after King and Brater, 1963).

Example: A rectangular-notched, sharp-crested weir having a crest 0.5 feet above the streambed and a width of 2.4 feet sits in a channel 4 feet wide. The head is determined to be 0.6 feet. Calculate the flow.

Solution: To get the coefficient C from Figure 3.22, we need the following ratios:

$$\frac{L}{b} = \frac{2.4}{4.0} = 0.6 \qquad \frac{H}{P} = \frac{0.6}{0.5} = 1.2$$

From Figure 3.22, the coefficient is determined to be $C = 3.3$. So the flow is

$$Q = 3.3 \times 2.4 \times (0.6)^{3/2} = 3.7 \text{ cfs.}$$

If you install the weir permanently and put a scale on the measuring stick for the head, you can easily prepare a discharge chart. Then, any time you want to measure flow, you just read your stick and find the flow on your chart.

Dam Construction

The first step in building a dam is to select a suitable site. You should find a spot where the stream is narrow and the banks are high. Keep in mind the need to develop as much head as possible; a place where the stream is relatively steep is best. Bear in mind, too, the high cost of transmitting electrical currents at low voltages over long distances (see "Site Selection" in this chapter); you will want the site to be not much more than 1000 feet from the house.

The height of the dam will depend on the amount of water to be stored, the head needed, the topography of the site, the available materials, and your expertise. The cost, work, and materials associated with building a dam increase rapidly with dam height. If you want a dam higher than about 5 feet, you will need the advice of a qualified engineer; indeed, many states require an engineer by law.

The best season to build a dam is late summer when stream flows are lowest. You will probably need a diversion channel to allow the stream to bypass the dam during construction. Also, it is wise to set up a time schedule for your work to be sure that you can finish the job before fall rains start.

The dam must be designed properly to enable it to withstand safely the forces acting upon it: the hydrostatic pressure of water trying to overturn the dam or slide it downstream; the uplift caused by water which finds its way between the dam and its foundation; and while concrete dams can handle overflows, earthfill dams may be washed away if they are overtopped.

Water seeping through the foundation can undermine a dam and cause its failure. If you are lucky enough to have a streambed of solid rock, you don't have to worry; but most beds consist of permeable sand or gravel. One way to prevent seepage in such beds is to use sheet piles, set one right next to another to form an impermeable wall. The depth to which the piles must be driven depends on the kind of soil, the thickness of the permeable layer, and the depth of the water in the pond. If the pervious soil in the bed is not very deep, the sheet piles can be driven down to the solid rock below, as shown in Figure 3.23a. If the pervious layer is deep, then it is recommended that the foundation be sealed by covering the streambed upstream of the center of the dam with a blanket of such impervious material as clay (Figure 3.23b).

The dam itself may be any of a number of types. The most common low dams are earthfill, rockfill, crib, framed, and gravity dams.

Earthfill dams, such as those in Figure 3.23, may be constructed from clay, sand, gravel, or a combination of these elements, depending on what's available. To be on the safe side, the slope of the downstream side should be 1:2 and on the upstream side it should be about 1:2.5 or 1:3. To prevent seepage through the dam, clay is the best construction material. If you don't have enough clay for the whole dam, then a core of clay in the center (Figure 3.23) is satisfactory. The slopes on the clay core can be about 5:1. Along the sides of the core, you must have a transition zone of material of grain size intermediate between clay and the material used to construct the main body of the dam. If you don't have any clay at all, the dam can be sealed by putting a plywood or masonry wall in the core.

Since earthfill dams can be washed away if they are overtopped by water, a well-designed and well-constructed bypass spillway must be provided. Even with a good spillway, the dam should be built high enough to

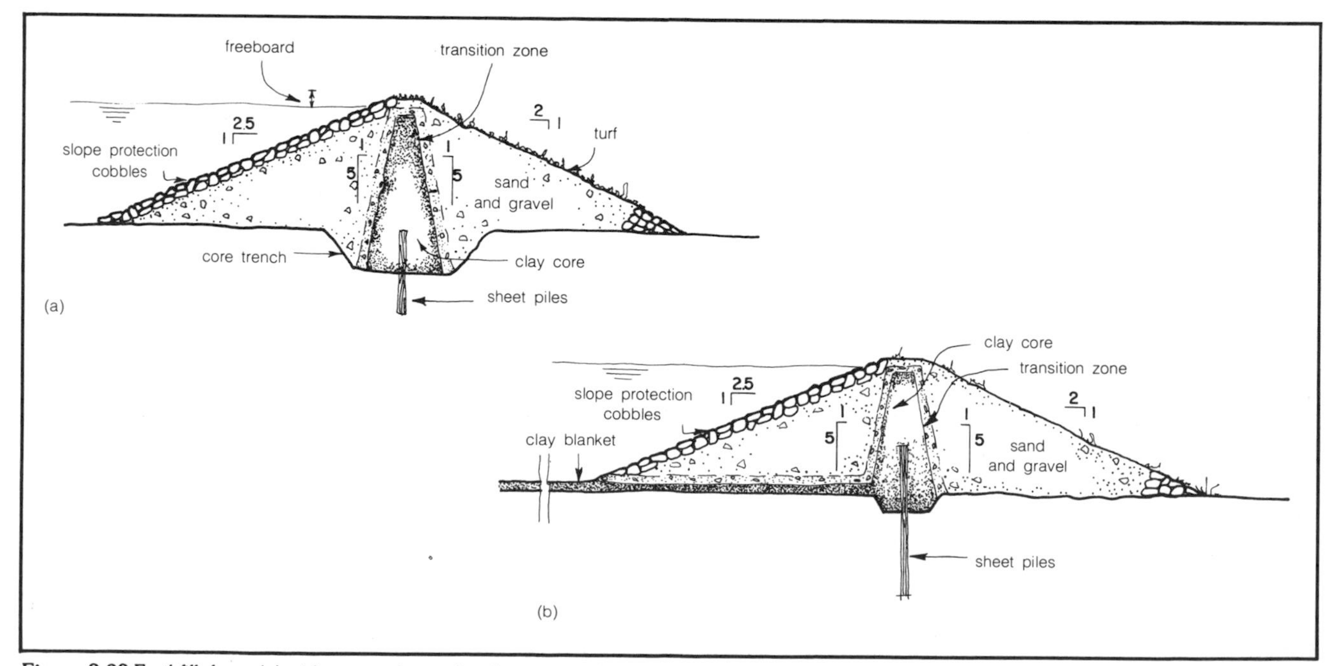

Figure 3.23 Earthfill dams (a) with center sheet piles driven to rock, and (b) sealed with an underlying layer of clay.

provide a substantial safety factor to prevent overflow during the maximum expected flood. The less information on floods that you have, the higher this safety margin should be; a washout could result in considerable destruction for which you can be (legally) liable.

Rockfill dams are similar to earthfill dams, except that the rock construction allows steeper slopes on the dam faces as shown in Figure 3.24. A slope of 1:1.3 is typical. One common way to prevent seepage through this type of dam is to cover the upstream face with concrete or asphalt.

Crib dams are probably the closest to a beaver's dam. They are made of logs or timber stacked as shown in Figure 3.25. The logs are spaced 2 or 3 feet apart, with stones or gravel to fill the openings. To prevent leakage through the dam, cover the upstream face with planks. If excess water is to be allowed to spill over the dam, then the downstream face also should be covered. Crib dams are built only to very low heights.

Framed dams are also made of wood (Figure 3.26). The sheeting which faces the water must be well supported, calked, and preserved with some water sealant. The spacing of the frame depends on the thickness of the timber and the water pressure acting on it (i.e., the height of the water in the pond).

Finally, gravity dams are made of masonry or concrete and, as shown in Figure 3.27, the spillway can be part of the dam. Discharged water from the spillway may eventually erode the streambed and start to undermine the dam foundation unless a concrete stilling basin or some rocks are used to dissipate the energy of the falling water.

Before filling the pond for the first time, some preparatory measures are recommended. It is good practice to clear the reservoir of all trees and bushes and clear a marginal strip around the water's edge. This will help prevent undesirable tastes and odors caused by decaying plants in the bottom of the pond, and enhance its usefulness for agriculture, drinking water, and safer swimming.

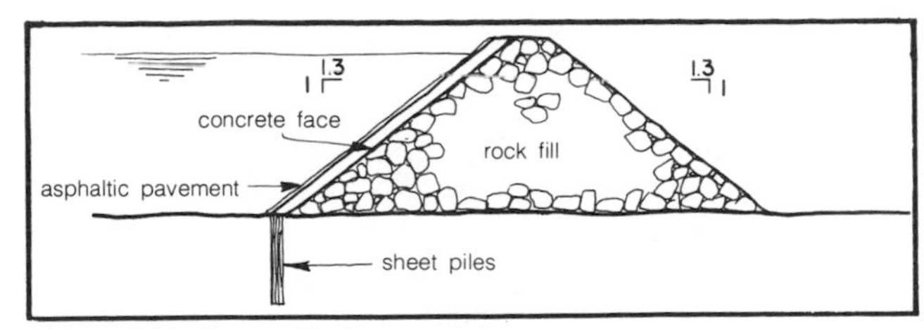

Figure 3.24 A rockfill dam.

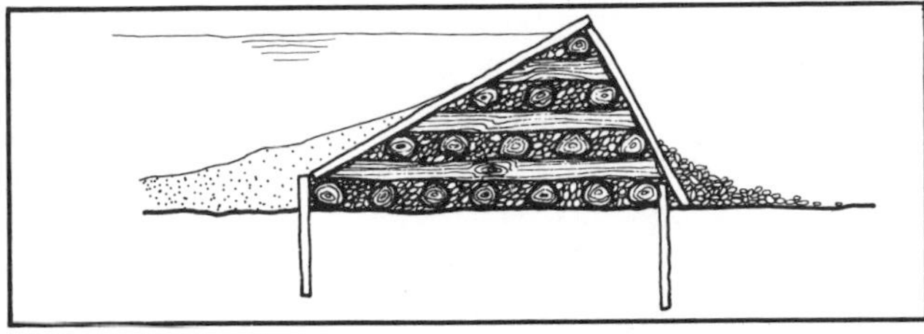

Figure 3.25 A crib dam.

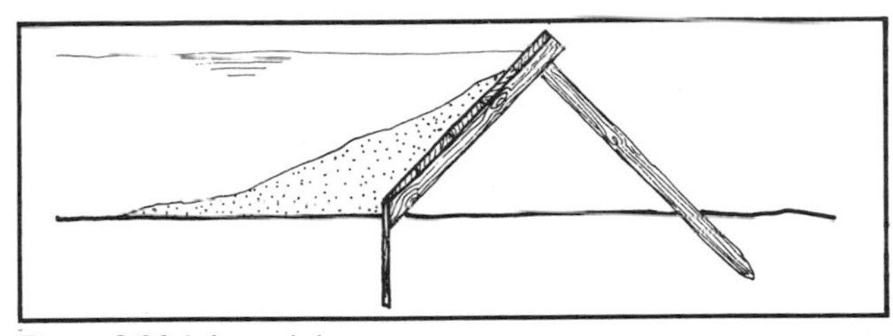

Figure 3.26 A framed dam.

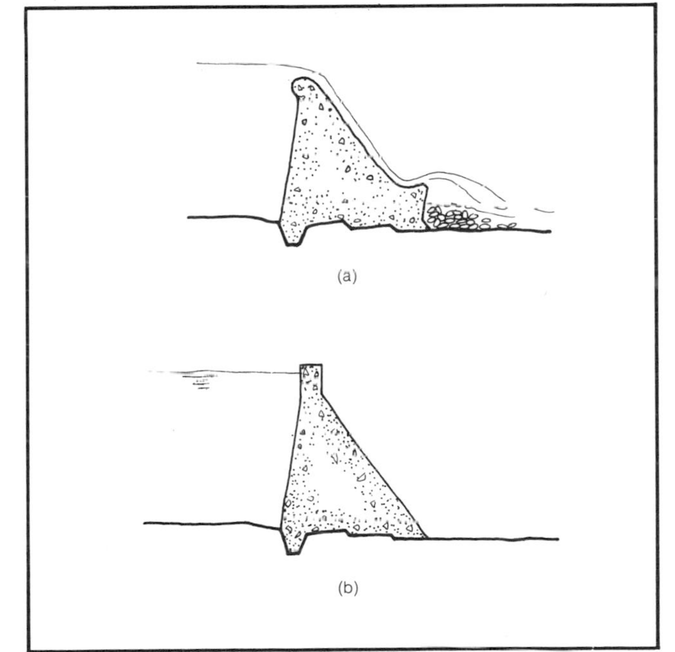

Figure 3.27 Gravity dams, with (a) and without (b) spillway.

Conduits

To utilize the water stored in the pond, it must be transported to the powerhouse in some sort of conduit. The conduit can be an open channel, as is usually the case when a waterwheel is used, or a full-pressure conduit (pipe), which is usually required for turbines. Sometimes both are used.

Water is conveyed through conduits at the expense of part of its energy, which is lost mostly because of the friction between the water and the walls of the conduit. These losses are reduced as conduit size increases, but costs also increase correspondingly. Also, reducing the slope of the channel decreases the water's velocity, thereby reducing head losses; but again, this requires larger channels to provide the same total flow rate. There are trade-offs involved in any design.

If the channel is simply a ditch in the ground, its slope will be largely determined by the terrain. But if you have a choice, a slope of about 0.1 to 0.3 percent (corresponding to a head loss of 1 to 3 feet per 1000 feet of conduit)

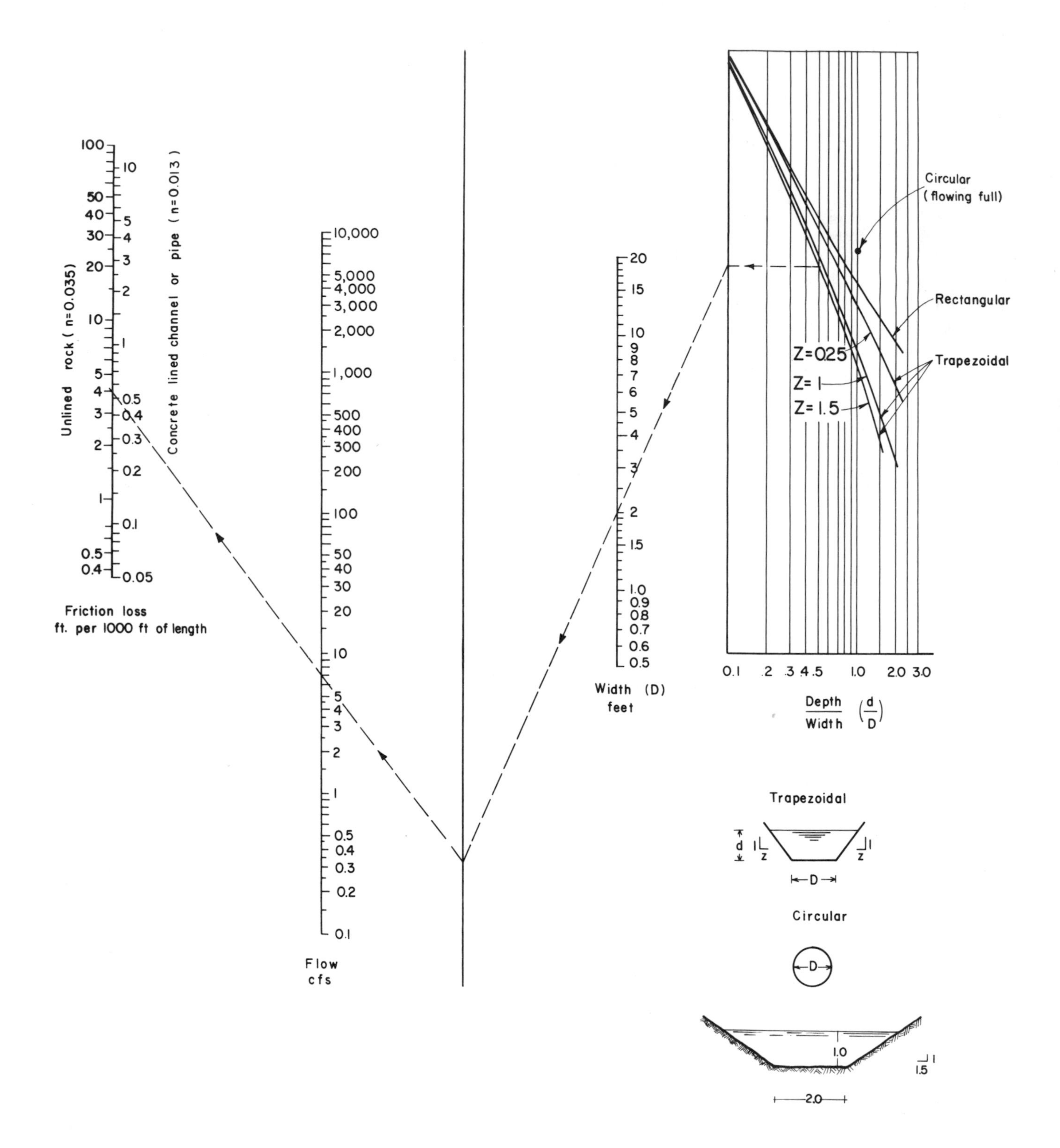

Figure 3.28 Nomograph for determining friction loss in a conduit.

is to be recommended. As a rough guideline, channels should be designed to produce flow velocities on the order of 3 to 6 cubic feet per second (less if there may be an erosion problem in the channel).

The design of a conduit consists of picking both the dimensions of the conduit cross section and the conduit's slope so that it transports a given flow of water with minimal head loss. The head loss in an open channel is equal to the slope; while in a pressure conduit, the head loss depends only on the velocity of the water and the diameter and material of the pipe. The nomograph of Figure 3.28, which applies to both the channel and pipe cases, enables us to work out graphically the required calculations.

Example: Find the slope needed to convey 7 cfs in the trapezoidal rock channel above whose base has a width (D) of 2 feet, whose sides are inclined at 1:15, with water flowing at a depth (d) of 1 foot in the channel.

Solution: The solution is worked out on the nomograph as follows: On the right-hand graph, locate the point corresponding to $(d/D) = 0.5$ for the cross section with $z = 1.5$, and project it horizontally to the left edge of the graph. From there, the line is drawn down through the point corresponding to $D = 2$ on the width line to the Center Reference Line. From the Center Reference Line, project up through the Discharge point Q= 7 cfs to the Friction Loss scale where we read off a value of about 4 feet of head for every 1000 feet of channel. So the slope is 0.4 percent.

This example indicates how to calculate the amount of head lost in conveying water from the dam site to a distant powerhouse. You will need to do this sort of calculation in order to decide whether it makes sense to locate the powerhouse near the dam or whether sufficient extra head can be gained to locate it elsewhere.

Waterwheels, Turbines, and Generators

The energy in our stream of water is converted into mechanical, rotational energy by means of a waterwheel or a turbine. Waterwheels have the advantage of being rather "low technology"; they can be built without much in the way of special skills, materials, or tools. The shaft of a waterwheel can be coupled by means of belts and pulleys directly into such low-speed mechanical loads as saws, lathes, water pumps, and mills; but their low turning speed makes it difficult to couple them into electrical generators. A typical waterwheel turns at something like 5 to 15 rpm and, since an automobile generator needs 2000 to 3000 rpm to put out much current, a gear ratio of about 300:1 would be required. Such high gear ratios are quite difficult to attain, which severely limits the usefulness of the waterwheel–automobile-generator combination. There are generators, used in commercial wind-electric plants, which put out significant power at

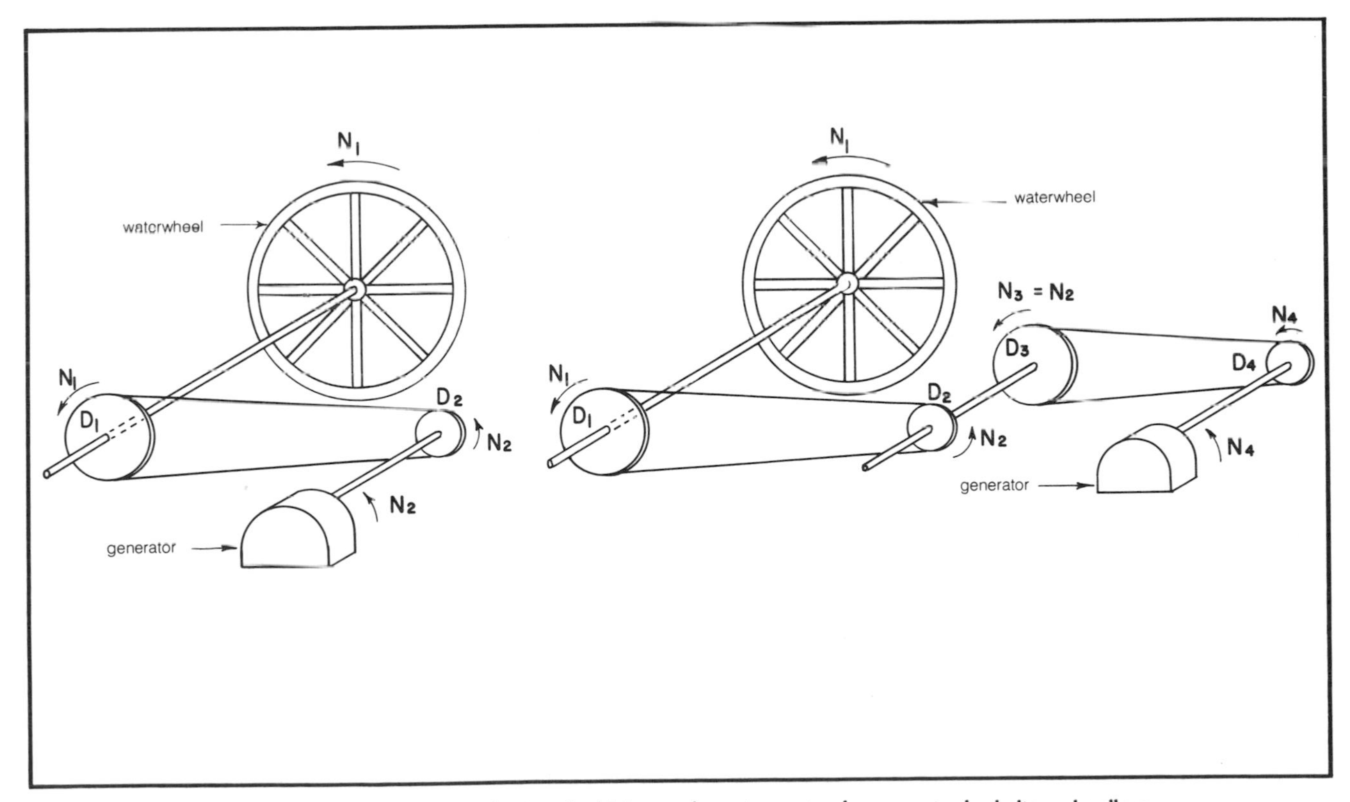

Figure 3.29 Matching the slow speed of a waterwheel to the high-speed requirements of a generator by belts and pulleys.

lower rpms; however they are considerably more expensive.

To speed up the rotation of a shaft, pulleys and belts can be used, as shown in Figure 3.29. The ratio of the speeds of the shafts is equal to the ratio of the diameters of the pulleys:

E. 3.15
$$\frac{N_2}{N_1} = \frac{D_1}{D_2}$$

To obtain a speed increase of, say, 20 to 1, the diameter of one pulley would have to be 20 times the diameter of the other. For large speed increases, the step-up should be done in stages, as shown in Figure 3.29. In this case, the total increase in speed is equal to the product of each pulley ratio:

E. 3.16
$$\frac{N_4}{N_1} = \frac{D_1}{D_2} \times \frac{D_3}{D_4}$$

In spite of the difficulty, it is possible to use waterwheels for generating electricity: Thomas Oates, for example, has described a four-stage belt-drive system which he uses to step up a waterwheel rotating at 15 rpm to a generator rotating at 800 rpm (in *Mother Earth News A Handbook of Homemade Power* — see Bibliography); Don Gilmore uses a combination of gears from an old car and some pulleys to turn a generator at 1800 rpm (see *Cloudburst* in Bibliography).

The gearing problem is greatly simplified if turbines are used since they usually rotate at a much higher speed than that of waterwheels. Turbines, however, are precision machines and must be purchased from a manufacturer. With this approach, a complete water system would end up costing several thousand dollars. One advantage of buying a system from a manufacturer is that it often includes turbine, gearing, and generator, and its output is 60-cycle ac current ready for use in your house. With a homebuilt setup, you will probably have to generate dc and include a dc-to-ac inverter if you want 60-cycle ac (see "DC-to-AC Power Inverters" in this chapter).

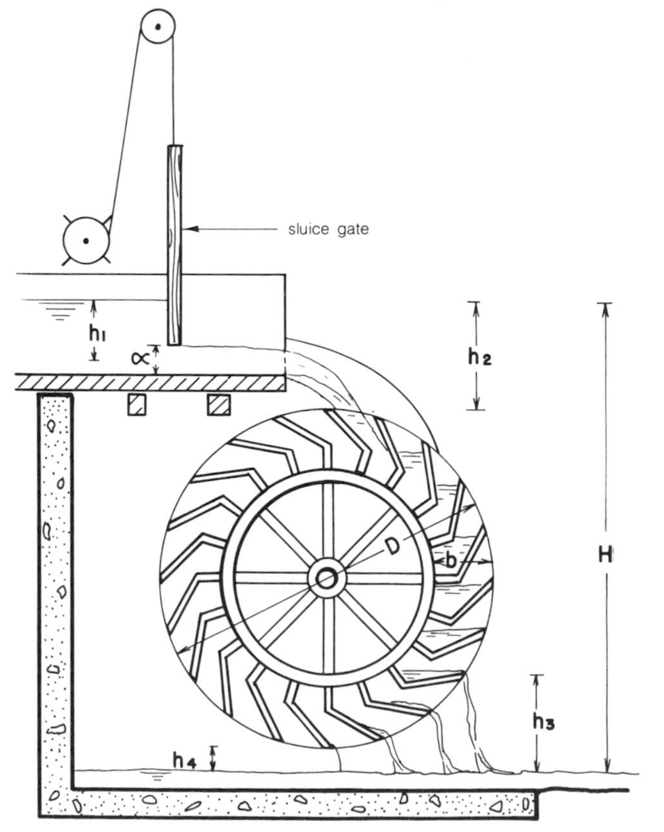

Figure 3.30 An overshot waterwheel.

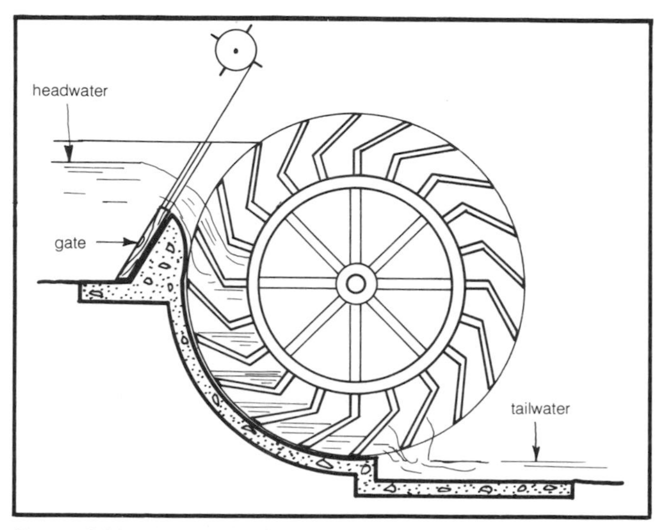

Figure 3.31 A breast wheel.

Waterwheels

Waterwheels are turned both by the weight of water in their buckets and by impulse. Their efficiency, which may be anywhere from 35 to 85 percent, is little affected by varying flow rates and they are not damaged by sand and silt in the stream. Waterwheels are categorized according to the location at which the water strikes the wheel. An *overshot* wheel (Figure 3.30) accepts water at the top of its rotation; its downward-moving side is overbalanced by the water and this overbalance keeps it in slow rotation. The *breast* wheel receives water halfway up its height (Figure 3.31) and turns in the opposite direction as an overshot type. Breast wheels are not recommended because they are relatively inefficient and are harder to build and maintain than overshot wheels. Lastly, the *undershot* wheel (Figure 3.32a) is powered by water as it passes under the wheel. One version of the undershot wheel is the *Poncelet* wheel of Figure 3.32b. By curving the vanes to reduce the shock and turbulence as the water strikes the blades, a Poncelet wheel is able to achieve much higher efficiencies than an ordinary undershot.

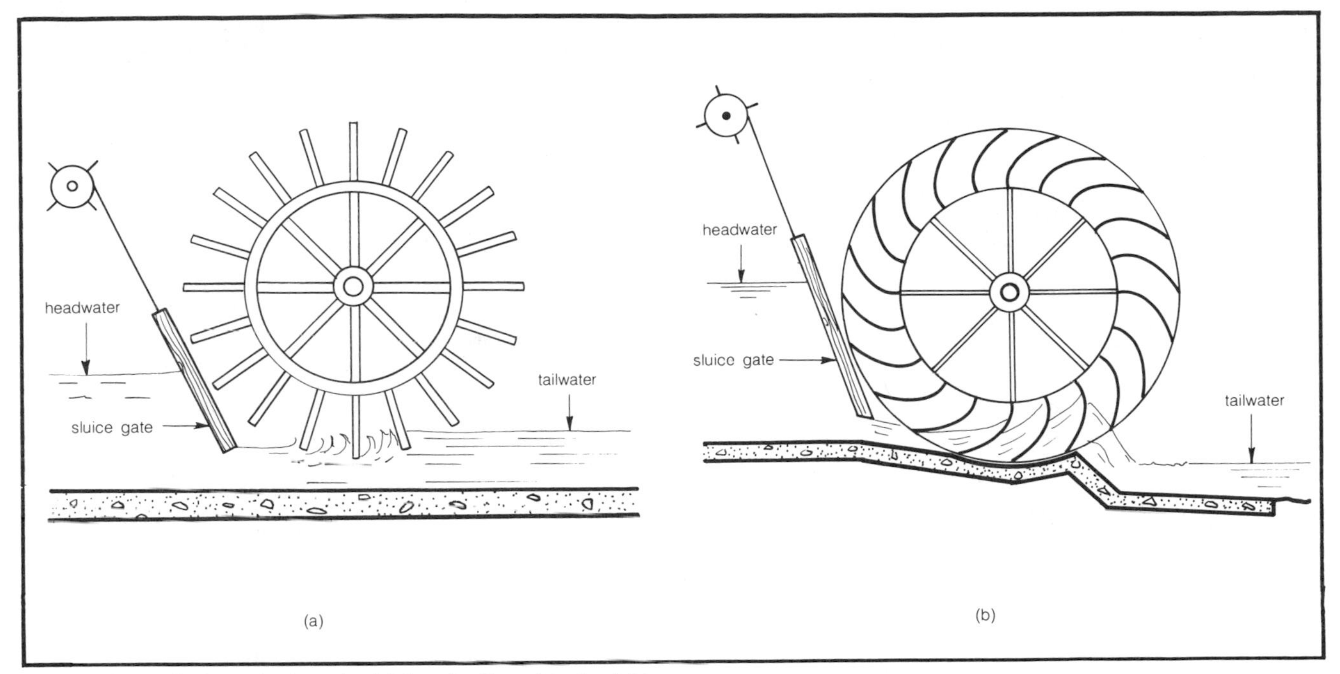

Figure 3.32 A simple undershot wheel (a) and a Poncelet wheel (b).

Overshot Wheels

The overshot wheel was one of the most widely used waterwheels in olden times. It has a number of sheet iron or wooden buckets around its periphery and is supplied with water from a flume over its top. If the buckets are smoothly curved and designed so that the water remains in them until the lowest possible point in the wheel's rotation, then relatively high efficiencies on the order of 60 to 85 percent are possible.

Overshot wheels can be used when the available head ranges from 5 to 30 feet; the diameter of the wheel is usually selected to be about three-fourths of the available fall. The flow through the flume is controlled with a sluice gate. The width of the wheel is usually 4 to 8 inches greater than the width of the sluice gate and the depth of the buckets is usually 10 to 16 inches. It is more efficient to have the buckets only about one-half to two-thirds full, to minimize water spillage.

There are several formulas which can be used to help design an overshot wheel. The variables are defined below, as well as in Figure 3.30:

D = diameter of the waterwheel (ft)

w_b = width of the buckets (ft)

w_s = width of the sluice gate (ft)

b = depth of the buckets (ft)

N = wheel rotational speed (rpm)

α = height of sluice gate opening (ft)

h_1 = head on the sluice gate, measured from water surface to center of opening (ft)

h_2 = vertical distance from water surface in flume to top of wheel (ft)

h_3 = vertical distance between tailwater surface and point where buckets start to empty (ft)

h_4 = vertical distance from bottom of wheel to tailwater surface (ft)

H = total available head, from surface of water in flume to tailwater surface (ft)

H_w = effective head for generating power (ft)

Q = flow rate (cfs)

The rotational speed of the wheel is given by

E. 3.17
$$N = \frac{70\sqrt{(h_2)}}{D}$$

The width of the buckets, assuming they are about two-thirds full, is given by

E. 3.18
$$w_b = \frac{30Q}{(D-b)bN}$$

The dimensions of the sluice gate opening, α and w_s, are related by

E. 3.19
$$w_s = \frac{Q}{6\alpha\sqrt{(h_1)}}$$

The effective head that the wheel sees is

E. 3.20
$$H_w = H - \frac{(h_3 + h_4)}{2}$$

and the power generated in kilowatts is given by

E. 3.21 $$P = \frac{QH_we}{11.8}$$

where the efficiency e is roughly 0.65 to 0.8.

Example: Design an overshot wheel for a site where the available head H is 16 feet and the flow is 8 cfs.

Solution: The wheel diameter should be about three-quarters of the fall so

$$D = 0.75 \times 16 = 12 \text{ feet}$$

Allowing a half-foot clearance between the wheel and the tailwater means h_4 equals 0.5. From the geometry

$$h_2 = H - (D + h_4) = 16 - (12 + 0.5) = 3.5 \text{ feet}$$

the rotational speed will be

$$N = \frac{70\sqrt{(h_2)}}{D} = \frac{70\sqrt{(3.5)}}{12} = 10.8 \text{ rpm}$$

Selecting the bucket depth to be 1.2 feet makes its width

$$w_b = \frac{30Q}{(D-b)bN} = \frac{30 \times 8}{(12-1.2) \times 1.2 \times 10.8} = 1.7 \text{ feet}$$

If a half-foot clearance is allowed between the top of the wheel and the bottom of the sluice gate, then the depth of water in the flume will be 3 feet. We will want the sluice gate to be narrower than the width of the buckets and we must try to find a value for α that will make this work out. Assume we open the gate 20 percent so α is 0.6 feet. Then by the geometry

$$h_1 = 3 - (0.6/2) = 2.7 \text{ feet}$$

we find that

$$w_s = \frac{Q}{6\alpha\sqrt{(h_1)}} = \frac{8}{[6 \times 0.6 \times \sqrt{(2.7)}]} = 1.35 \text{ feet}$$

That's good: it allows for some dispersion of the water as it is discharged from the gate. Finally, if we assume the buckets are shaped so that they don't start to release their water until they reach a point 2.5 feet from the tailwater, then h_3 equals 2.5 feet. The effective head is therefore

$$H_w = H - \frac{(h_3+h_4)}{2} = 16 - \frac{(2.5+0.5)}{2} = 14.5 \text{ feet}$$

Assuming the efficiency of the wheel to be 65 percent, we find the power generated to be

$$P = \frac{QH_we}{11.8} = \frac{(8 \times 14.5 \times 0.65)}{11.8} = 6.4 \text{ KW}$$

Undershot Wheels

A regular undershot waterwheel has radial paddles which are turned by the force of the water flowing beneath the wheel. The paddle design is quite inefficient and wheel efficiencies of only 20 to 40 percent are typical. They are used with heads of about 6 to 15 feet and have diameters which are typically 10 to 15 feet.

Glen Bass reported the construction of a homemade wooden undershot wheel (see *Cloudburst* in Bibliography). The wheel is 7 feet in diameter, has 24 paddles, turns at 20 rpm, and generates an estimated 3 horsepower. Using locally available materials, the builders spent only about $130. Their plan is to use the wheel to generate electricity, as well as to operate mechanically a wringer-washer, a small wood lathe, and a flour mill (connected to the shaft of the wheel).

The Poncelet wheel is an improved version of an undershot wheel. By curving the vanes to create smoother interaction with water released from the sluice gate, efficiencies of 60 to 80 percent are possible. These wheels can be used with quite low heads (about 3 to 10 feet). The wheel diameter is usually taken to be about double the head, plus 4 to 8 feet; a 14-foot diameter is usually as small as they come. The vanes are placed 10 to 14 inches apart along the periphery if they are made of metal, and 14 to 18 inches if they are wooden.

Breast Wheels

Breast wheels take their name from the arc of masonry or wood sheathing that encloses the wheel from the point where the water enters the wheel to the point it is discharged. Water is thus confined between the wheel and the breast and this close spacing can cause problems if sticks or rocks get jammed inside. The buckets of breast wheels are specially formed to allow air to escape as water flows in, which complicates construction. Because efficiencies are generally less than the overshot or Poncelet wheel and construction is also more difficult, breast wheels are not generally recommended.

Turbines

While waterwheels operate from water carried in an open channel and hence are rather slow turners, turbines receive their energy from water carried in pressure conduits; they spin much faster and so are easier to couple to electric generators. They are carefully engineered to transfer nearly all the energy from the stream of water to the generator; efficiencies of 80 to 93 percent are common.

However, a few disadvantages exist. Turbines are sophisticated pieces of machinery and therefore require a certain amount of care in the way they are installed. They wear down from the constant abrasive force of silt and small pebbles and so require a trash rack at the head of their flumes. Also, the nozzles which admit the water into the turbine must be adjusted to differing flow rates. Another factor that is important when installing a turbine is the need for a concrete structure to hold the turbine; this means building forms and pouring concrete. And finally, they are difficult to build yourself and manufactured units are expensive. But despite these qualifications, turbines are the way to go if you want electricity from water.

Turbines are classified as either *impulse* or *reaction* types. Impulse turbines use the kinetic energy of water squirting out of a nozzle at high speed to turn the turbine wheel (called the *runner*). The Pelton wheel and Michell (or Banki) turbine are both impulse turbines.

In reaction turbines, part of the available head is converted to kinetic energy and the rest remains as pressure-head. The flow takes place under pressure, which means the whole unit is enclosed in a case as opposed to the open housing of an impulse turbine. Examples of reaction turbines are the Francis, Propeller, Kaplan, Samson, and Hoppes types. The Samson and Hoppes turbines are designed for small installations, while the other three are used in larger hydroelectric projects.

Pelton Turbine

The Pelton turbine, which was first developed in America, is today the most common impulse turbine. These turbines are used only when very high heads (at least 50 feet) are available and consequently they require less flow to generate a sizable amount of power. Their efficiency can be as high as 94 percent and they turn at high speeds (up to 1000 rpm); they are well matched to the demands of an electric generator.

Table 3.10 Sample Power Ratings for Paramount Type B Impulse Waterwheels[a]

Effective Head (ft)	Range of Power Available (hp)
50	1.1– 8.6
60	1.4–11.2
70	1.7–14.0
80	2.1–17.2
90	2.5–20.6
100	3.0–24.2
110	3.5–27.8
120	4.0–31.8

Notes: a. Manufactured by Pumps and Power, Ltd.

It is unlikely that you are going to build your own Pelton turbine, since its manufacture requires quite a high level of technology. As an example of what is available, Table 3.10 gives the main characteristics of the "Paramount Type B Impulse Wheel" manufactured by Pumps and Power, Ltd. of Vancouver, Canada. This turbine is available for heads of 50 to 200 feet and in power ranges from 1 to 68 horsepower (1 hp = 746 watts).

Banki/Michell Turbine

The Banki/Michell turbine was first invented by the Australian engineer A. Michell in the beginning of this century. Later, the Hungarian professor D. Banki improved and patented it. Today it is most widely known by his name, although sometimes it is called a Michell or cross-flow turbine. Banki turbines have been built for heads from 3 to 660 feet, with rotational speeds of between 50 and 2000 rpm; they are highly recommended for a home power unit.

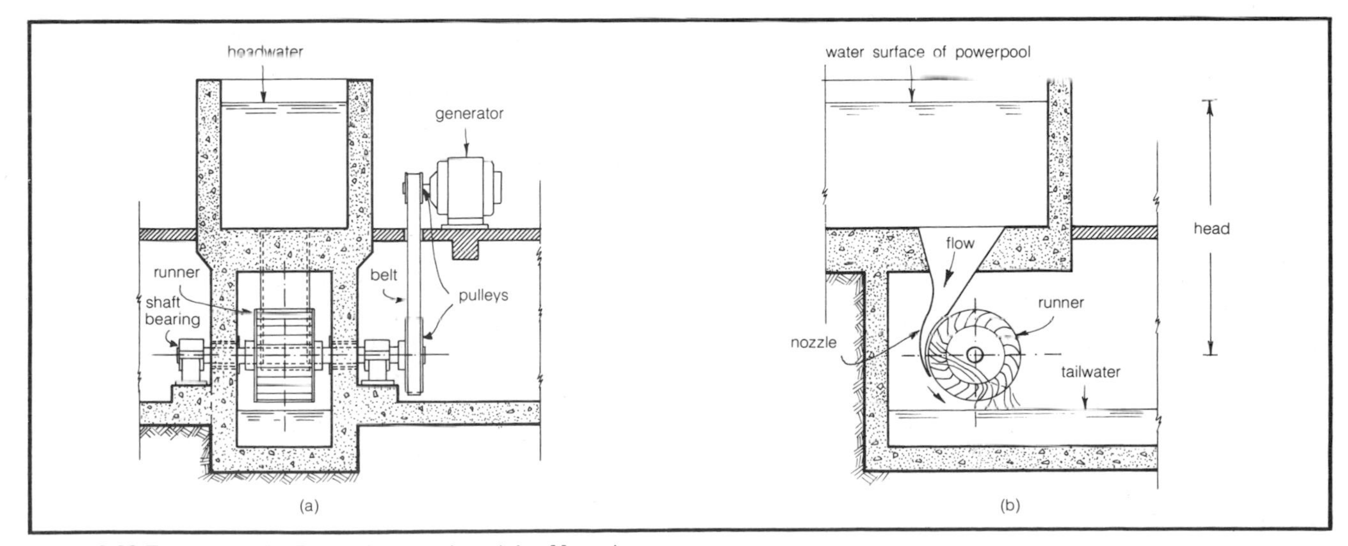

Figure 3.33 Two views of a Banki/Michell turbine (after Hamm).

The main parts of a Banki turbine are the runner and the nozzle (Figure 3.33). A jet discharged from the rectangular nozzle strikes the blades on the rim of the runner, flows over the blade, passes through the inner space of the runner, and then strikes the blades for a second time. Efficiencies of manufactured units are on the order of 80 to 88 percent and, unlike reaction types, these turbines are efficient over a wide range of flows.

Banki turbines can be built at home by a fairly skilled person. Students at Oregon State University nearly thirty years ago built such a turbine, which turned out to have an efficiency of 68 percent. And H.W. Hamm describes how to build a 12-inch Banki turbine (see Bibliography). Manufactured units are also available: Ossberger Turbinenfabrik of West Germany, for example, is very experienced in this field.

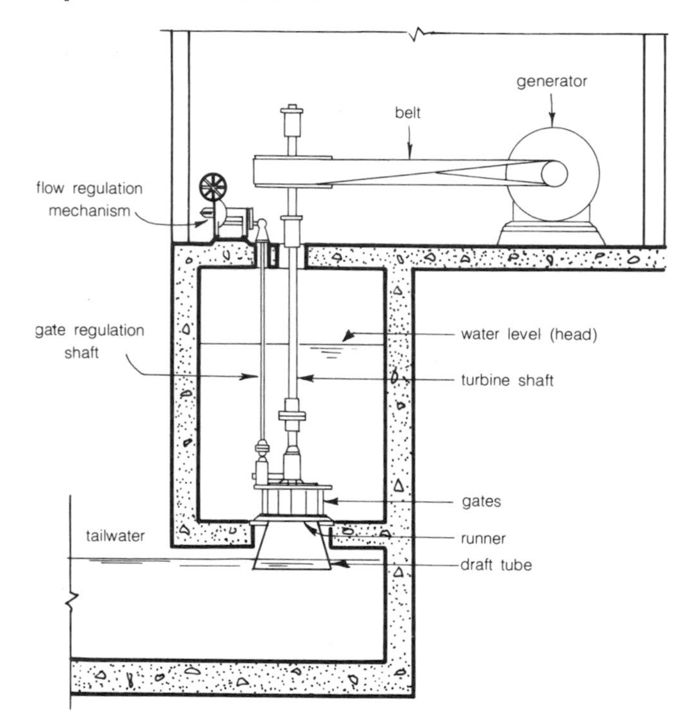

Figure 3.34 Example of a small hydroelectric facility using a Sampson turbine from James Leffel and Co.

Reaction Turbines

Reaction turbines operate under pressure with high efficiencies under reasonably low heads, but they are highly engineered units which cannot be homebuilt and which must operate within a small range of flow rates. It is important to get one that is well matched to your specific conditions; they are efficient only when operated near their design specifications.

James Leffel and Company of Springfield, Ohio, for example, markets a range of Sampson turbines designed to operate with heads of from 3 to 26 feet and with flows of from 4 to over 600 cfs. The power generated from these turbines is from 1.1 to about 1500 hp and they can be purchased with matching electrical generators. Figure 3.34 shows a typical installation.

Table 3.11 Sample Hoppes Hydroelectric Units[a]

Capacity (KW)	Range of Heads (ft)	Range of Flows (cfs)
0.5	8–12	1.7–1.1
1.0	8–25	3.2–1.1
2.0	8–25	5.5–1.8
3.0	8–25	7.8–2.6
5.0	8–25	12.7–4.3
7.5	11–25	13.3–6.3
10.0	12–25	16.3–8.0

Notes: a. Available from James Leffel and Co.

Another reaction turbine available on the market is the Hoppes hydroelectric system. It is a very compact unit with a generator mounted on the same shaft as the turbine. It comes with a governor for regulation of the rotational speed and is available with either ac or dc generators. Standard sizes deliver from 0.5 KW to 10 KW with heads of from 8 to 25 feet. Table 3.11 gives several available sizes.

Final Thoughts

As we have seen in this section, a great deal of power can be obtained from quite small stream flows, with heads that are not too difficult to obtain. Even so, not many people are likely to have the optimum set of conditions required for a hydroelectric station. For those that do, there are a number of important things to consider before damming up that nice creek that flows through the corner of your property.

You must thoroughly check out the water-rights laws which apply to your area before doing anything to alter "your" stream—and this applies to uses other than power, such as irrigation or water supply. And you should thoroughly evaluate the ecological changes that are liable to result from creating a reservoir. What is it liable to do to the water table? To the problem of mosquitoes? What's going to happen when the reservoir silts up? Are there fish that might want to swim upstream? You may have to abandon the project or radically alter your design to accommodate the needs of your local ecosystem.

On the other hand, the benefits from the project can be increased if your pond can be used for more than just the generation of electricity. Can you raise fish in the water, irrigate your land, water your animals, or float around in it all summer on your back? As they say: seek and you shall find.

Bibliography

Wind-Electric Sources

Alternative Sources of Energy. Milaca, New Mexico: Alternative Sources of Energy. The best source of current information on small-scale energy utilization. Approximately bi-monthly, for $5 per year. The May 1974 issue contains an article on windmill blade design and describes several sets of plans for wind generators.

Baumeister, T., and Marks, L.S. 1974. *Standard Handbook for Mechanical Engineers*. New York: McGraw-Hill. This book has a short (but good) chapter on wind by E.N. Fales.

Clews, H.M. 1973. *Electric Power From the Wind*. East Holden, Maine: Solar Wind Company. Probably the simplest, clearest, most informative introduction to wind-electric systems. Available from: Solar Wind Company, P.O. Box 7, East Holden, Maine 04429.

Golding, E.W. 1955. *The Generation of Electricity by Wind Power*. New York: Philosophical Library. The standard, but old, textbook.

Hackleman, M.A. 1974. *Wind and Windspinners*. Saugus, California: Earth Mind. Focus is on Savonius rotors, but packed with practical and theoretical advice useful for any homebuilt system. Available for $7.50 from: Earth Mind, 26510 Josel Drive, Saugus, California 91350.

Merrill, R., et al. 1975. *Energy Primer*. Menlo Park, California: Portola Institute. Gives a great deal of good information and tells you where to get more. Covers not only wind, but also solar power, water power, and biofuels. Available for $4.50 from: Portola Institute, 558 Santa Cruz Avenue, Menlo Park, California 94025.

Meyer, H. 1972. "Wind Generators: Here's an Advanced Design You Can Build." *Popular Science*, November 1972. Plans for the generator described in the article are available for $15 from: Windworks, P.O. Box 329, Mukwonago, Wisconsin 53149.

National Science Foundation/NASA. 1973. *Wind Energy Conversion Systems, Workshop Proceedings*. Washington, D.C. December 1973.

Sencenbaugh, J. "I Built a Wind Charger for $400!" *Mother Earth News* 20: 32–36. Detailed plans are available for $12 from: Sencenbaugh Wind Electric, P.O. Box 11174, Palo Alto, California 94306.

Syverson, C.D., and Symons, J.G. 1974. *Wind Power*. Mankato, Minnesota: Wind Power. Many good design guidelines and details. Available from: Wind Power, P.O. Box 233, Mankato, Minnesota 56001.

United Nations. 1957. *New Sources of Energy and Economic Development*. New York: United Nations Department of Economic and Social Affairs.

______. 1961. *Proceedings of the United Nations Conference on New Sources of Energy*, vol. 7 (August 21–31, 1961). New York: United Nations.

Water-Power Sources

Ball, R. 1908. *Natural Sources of Energy*. New York: D. Van Nostrand Co.

Hamm, H.W. (n.d.) *Low-Cost Development of Small Water-Power Sites*. Mt. Rainier, Maryland: Volunteers in Technical Assistance. An interesting book containing, among other things, practical instructions on how to build your own Banki/Michell turbine. Available from: VITA, 3706 Rhode Island Avenue, Mt. Rainier, Maryland 20822.

"Hydroelectric Power from a Hoppes Hydroelectric Unit." Bulletin H-49. Springfield, Ohio: James Leffel and Co.

"Improved Vertical Samson Turbines." Bulletin 38. Springfield Ohio: James Leffel and Co. The two above bulletins give information on small turbines manufactured and available from: James Leffel and Co., Springfield, Ohio 45501.

King, H.W., and Brater, E.F. 1963. *Handbook of Hydraulics*. 5th ed. New York: McGraw-Hill.

Marks, V. 1973. *Cloudburst: A Handbook of Rural Skills and Technology*. Brackendale, Canada: Cloudburst Press. This is a book born of people's experiences. It contains many valuable articles and can be ordered from: Cloudburst Press, P.O. Box 79, Brackendale, B.C., Canada.

Masoni, U. 1905. *L'Energie Hydraulique et les Recepteurs Hydrauliques*. Paris: Gauthier-Villars. This book, as well as Ball's cited above, gives information on waterwheels based on the nineteenth century's experience.

Mockmore, C.A., and Merryfield, F. 1949. *The Banki Water Turbine*. Bulletin Series no. 25. Corvallis, Oregon: Oregon State College. This book describes a Banki turbine made by students of the university and analyzes the results of experiments performed with that turbine.

Mother Earth News. 1974. *Mother Earth News: A Handbook of Homemade Power*. New York: Bantam Books. A good handbook with information on homemade waterwheels and turbines.

"Paramount Type B Impulse Waterwheels." Vancouver: Pumps and Power, Ltd. Information on this and other turbines available from: Pumps and Power, Ltd., 1380 Napier Street (P.O. Box 2048), Vancouver 6, B.C., Canada.

"Water and Power from Weissenburg Ossberger." Bavaria, Germany: Ossberger Turbinenfabrik. Information on Michell turbines and complete generating units manufactured by this company is available from: Ossberger Turbinenfabrik, 8832 Weissenburg, Bavaria, Germany.

SOLAR

Thermal comfort: what it is
and how your body maintains it

Energy from the sun: how to measure,
collect, store, and distribute it

How much heat
your house will require

Rooftop waterbeds and
south-facing barrels

Equipment: what to get
and where to get it

Solar Heating

INTRODUCTION

The technology required to heat buildings with solar energy has been with us for centuries, but the incentives to apply it have been sorely lacking. This lack of initiative stems from abundant, cheap energy; but as with so many of our natural resources, our economics failed to account for hidden costs. The fossil fuels upon which we currently rely are the result of millennia of solar energy storage by the processes of photosynthesis and biodecay. When viewed from the perspective of the time required to form these reserves, we are consuming them in the blink of a cosmic eye. The real cost of a cubic foot of natural gas reflects more than exploration and distribution costs.

These costs are beginning to reach the consumer. Gas users in California saw their bills rise by 34.7 percent in 1974 (see Figure 4.1). The underlying reason has little to do with politicians, foreign or domestic; we are simply running out of natural gas. A Federal Power Commission staff study, released in January of 1975, concluded that natural gas production from the forty-eight contiguous states has reached its peak and will decline for the indefinite future.

The same sad story is true of all of our fossil-fuel reserves—we are running out. Even coal reserves, our most abundant fossil fuel, will last but a few decades. Moreover, the mining and processing costs will be tremendous, since most of the easily obtainable reserves are already depleted. And nuclear power, derived by either fission or fusion processes, has serious implications which are only recently being addressed.

Fortunately, the sun showers us with free energy every day. We use the word "free" with some trepidation. Like wildflowers, solar energy is free only if it costs nothing to collect it. Until recently, it was cheaper in the long run to buy a conventional heating unit at an inexpensive initial cost and fuel it for fifteen or twenty years with a misleadingly cheap fossil fuel. The alternative solar heating system requires a large initial cost for equipment, although the energy will indeed be free thereafter. With recently soaring fuel costs, solar heating has become competitive with conventional systems. No doubt it will become even more attractive as fuel costs continue to climb. In fact, solar-powered *cooling* systems are now being tested for engineering and economic feasibility.

Economics alone may not be the prime reason for your interest in solar energy. In the past decade, an ecological philosophy has been emerging, and we are becoming more aware of the limitations of our natural resources. We are beginning to realize that our fossil-fuel reserves should be conserved for other uses—for the manufacture of important drugs and chemicals, for example. In this respect, heating buildings with solar energy is a particularly effective conservation technique. About 20 percent of our *total* energy usage is devoted to space heating! Others may find solar heating attractive from the standpoint of personal independence. A properly designed solar heating system can turn its back on coal strikes, natural gas prices, oil embargoes, and Alaskan pipelines.

This chapter is designed to supply you with the information and technical prowess to design and construct a workable solar heating system for a small building. It is directed at the person with very little technical background. You should do quite well if you have the spirit to read passages two or three times for full comprehension, the initiative to pursue the references for further details, and the confidence to handle simple, but sometimes tedious, algebraic computations. We assume some rudimentary knowledge of construction techniques, or else a knowledgeable friend. Finally, we suggest a willingness to experiment: learn from both your successes *and* your failures.

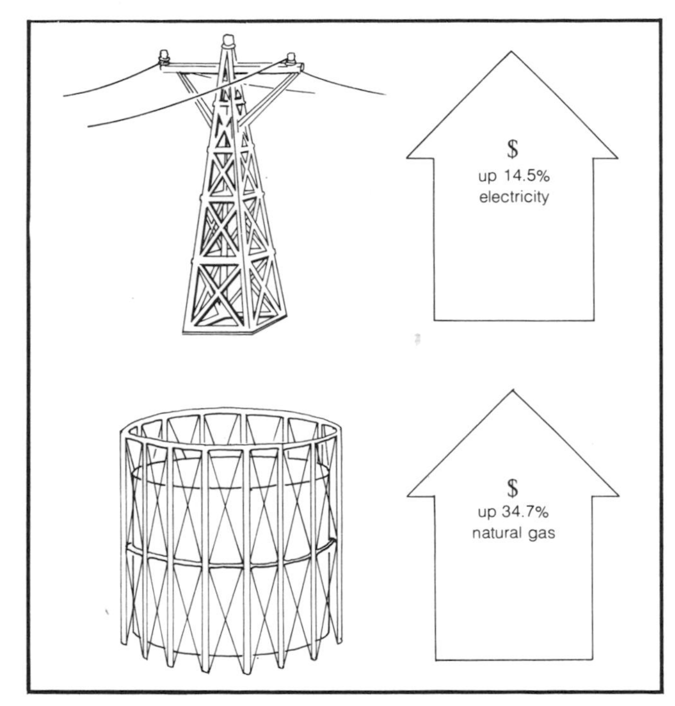

Figure 4.1 The increasing cost of energy in California (1974).

The chapter is organized into four major sections: thermal comfort—the indoor environment; climatology—the outdoor environment; heat loss—thermal communication between indoors and outdoors; and solar collection, storage, and control—the actual design process for a solar heating system.

The *thermal comfort* section is concerned with the conditions contributing to a livable indoor environment. Unfortunately, this turns out to be an elusive undertaking, for there are presently no physiological models which can accurately describe integrated human response to climatic surroundings in terms specific enough to provide design guidelines. As a result, subjective personal sensitivity (what indoor conditions make you comfortable) is still the best guide to planning for indoor comfort. We hope that this section will help establish these personal guidelines by alerting you to certain physiological ramifications to, say, indoor temperature levels. The temperature level you select has a marked effect on the energy needed to heat your home. Heating requirements, typically, drop by 25 percent if you are willing to live at 65°F rather than 75°F. This adjustment implies fewer collectors, less storage, and less initial cost. But what of the effects on comfort?

The *climatology* section includes most of the raw data necessary for the design of a solar heating system. Representative samples of Weather Service data are shown, to familiarize you with the types of information you will need to obtain for your particular locale. This section emphasizes techniques to determine the local availability of solar energy.

The *heat loss* section introduces you to the theory of heat transfer and explains techniques for minimizing heat losses from buildings. The methodology for calculating heat loss is explained in detail, with sample calculations included to help you through the process. Means to compute seasonal heating requirements, which aid in determining the economic feasibility of a particular heating design, are also discussed.

The last lengthy section, *solar collection, storage, circulation, and control,* integrates the various aspects of solar heating systems. Two categories are discussed. The first—"component systems"—examines systems comprised of individual components: different types of collectors, storage units, circulation systems, and methods of control. These components can be combined in various ways to form a complete solar heating system. The second category—"integral systems"—pertains to single structural entities which perform all three functions of collection, circulation, and control. A discussion of auxiliary heating systems, applicable to either category, is included to back up your solar heating system—just in case!

We might mention that this chapter concentrates on thermal collection systems, as opposed to photoelectric (e.g., solar cells) or photochemical techniques. Its primary emphasis is on space heating. The reasoning is quite simple: space heating requires relatively low-quality (low-temperature) energy which can be collected quite efficiently with inexpensive equipment. Photoelectric and photochemical systems simply can't compete on a cost basis.

Throughout our discussion, we concentrate on problems that deal with energy. Buildings lose *energy* in cold weather, people convert *energy* in metabolic processes, and we hope to explain just how to collect solar *energy*. In some cases, we deal with energy *rates*; for instance, we compute the energy loss *per unit time* for a building subjected to specified outside conditions. Think of an automobile trip: the distance you travel might be 100 miles; if you covered this distance in 2 hours, your average rate of travel would be 50 miles per hour (50 mi/hr).

When dealing with energy or energy rates, we will use units of *Btu* and *Btu per hour* (Btu/hr). A Btu is a measure of energy, specifically, the amount of energy necessary to raise the temperature of 1 pound of water by 1 degree Fahrenheit (1°F). The selection of the Btu as a standard energy unit was made reluctantly. The United States is the last remaining adherent to this (and other) English units. The rest of the world has adapted to the metric (Systeme International) unitary standard, and the United States will undoubtedly follow suit in the next decade. In the meantime, most of the tabular and graphic information we have are in the English system of units—hence the decision. Should the need to convert units arise, Table 4.1 should be of some help.

THERMAL COMFORT

While many contemporary homes cater to some of the more frivolous human desires, the basic function of the home remains the same—to maintain a comfortable thermal environment. But just what *is* a comfortable thermal environment? More specifically, what environmental parameters are involved, and what ranges can be tolerated to maintain a feeling of thermal comfort?

The answers to these questions are crucial, since they affect the energy consumed for space heating in a big way. Unfortunately, they can't be answered definitively. Thermal comfort is a very subjective feeling and has successfully resisted scientists' efforts to quantify it. However, certain guidelines have been established; and understanding both the basic biological aspects of thermal comfort and the environmental indices related to comfort and health can only help us to design homes that respond to the particular people who live in them.

Biology and Comfort

Human beings are mammals and, as mammals, we possess certain unique thermal systems. The body is its own source of thermal energy and can adapt to a wide range of environmental conditions. The adaption process depends upon some incredibly sensitive control mechanisms in your body. Let's look at them briefly.

To start with, you have two sets of heat-sensing organs in your skin. One set senses the outflow of heat from your body to surroundings at lower temperatures. The other set senses the inward flow of heat from objects or surroundings of higher temperatures. The outflow sensors lie very close to the skin's surface, and most of these are congregated in the fingertips, nose, and the bends of the elbow. About two-thirds of their number are proximate to the openings of sweat glands.

The inflow sensors lie deeper in the skin and are concentrated in the upper lip, nose, chin, chest, forehead, and fingers. Both sets of sensors trigger reflexes which control blood circulation through the skin. These reflexes play an essential role in heat balance, which we will talk about shortly.

The most important temperature-sensing system is located in the hypothalamus, a gland at the base of the brain, right above the pituitary. It works like a thermostat, monitoring changes in blood temperature caused by thermal events inside the body and also by temperature gradients across the skin. Like other mechanical thermostats, the one in your hypothalamus has a set point, usually close to 98.6°F. If the sensory input to your thermostat registers the body temperature at less than the set point, it will initiate physiological responses to increase the body temperature. The reverse occurs if the body temperature is too high. In this way, your body constantly strives to maintain a thermal equilibrium or heat balance, and you usually feel comfortable when this balance is achieved with the minimum amount of thermoregulation. This, then, is a simplified version of the biological basis of comfort.

Table 4.1 Useful Conversion Factors

Length
- 3.28 ft = 1 meter = 100 cm
- 1 in = 2.54 cm

Volume
- 1 ft^3 = 1728 in^3 = 0.0283 m^3
- 1 ft^3 = 7.48 gallons

Mass and Density
- 1 lbm[a] = 0.453 kilograms
- 1 lbm/ft^3 = 0.016 gm/cm^3 = 0.16 kg/m^3

Energy
- 1 Btu = 778 ft-lbf = 0.252 kcal = 252 calories
- 1 therm = 100,000 Btu
- 1 KWH = 3413 Btu

Energy Flux
- 1 langley = 1 cal/cm^2 = 3.69 Btu/ft^2

Power
- 1 KW = 3413 Btu/hr
- 1 hp = 2545 Btu/hr = 550 ft-lbf/sec

Thermal Conductivities
- 1 Btu-ft/hr-ft^2-°F = 12 Btu-in/hr-ft^2-°F
- 1 Btu-in/hr-ft^2-°F = 0.124 kcal-m/hr-m^2-°C
- = 1.44×10^{-3} watts-cm/cm^2-°C

Overall Heat Transfer Coefficients
- 1 Btu/hr-ft^2-°F = 4.88 kcal/hr-m^2-°C
- = 5.68×10^{-4} watts/m^2-°C

Heat Flow
- 1 Btu/hr-ft^2 = 3.155 watts/m^2

Pressure
- 1 lbf/in^2 = 1 psi = 27.8 in-H_2O = 2.31 ft-H_2O

Notes: a The notation "lbm" refers to a unit of *mass*, as distinguished from "lbf" which denotes a *force*. The metric system avoids the ambiguity; mass is expressed in grams or kilograms, and force in dynes or newtons.

Factors Involved in Thermal Equilibrium

Thermal equilibrium and the resulting sense of comfort are achieved by physiological and behavioral responses that control the amount of heat produced in the body and the amount of heat lost from the body. We will talk about the physiological responses which control thermoequilibrium in a later section. But, first, we should define the standard methods of heat transfer that result from these physiological responses, and discuss their relationship to thermoequilibrium in your body. The factors important to thermoregulation are: 1) metabolic rate (*M*); 2) conductive and convective exchange (*C*); 3) radiative exchange (*R*); and 4) evaporative exchange (*E*). A brief definition of each of these factors is helpful.

Metabolic Rate

Right now, the sandwich you ate for lunch is being turned into energy for the growth, regeneration, and operation of your body. This is an example of metabolic activity. Such metabolic processes are about 20 percent efficient; the other 80 percent of the energy generated is rejected as heat. The rate of metabolism is primarily controlled by the level of bodily activity. For example, if you're playing football, your metabolic rate will increase in order to supply your body with the extra energy needed to play the sport. Consequently, your metabolic energy production will increase greatly. To make things a little more complicated, your metabolic rate can differ slightly with weight, sex, age, and state of health.

Conductive and Convective Heat Exchange

Energy is lost by heat conduction through direct physical contact with objects of lower temperatures. Heat is gained by direct contact with objects of higher temperatures. When you heat up a chair by sitting in it, you are losing heat by conduction. When a heating pad warms up your aching back, you are gaining energy by conduction.

Convection has much the same physical basis as conduction, except that an additional mechanism of energy transfer is present: one of the heat-transfer agents is a fluid (air, in many cases). The air molecules exchange energy with an adjacent object, in what is initially a conductive heat-transfer process. However, in the case of convection, the air molecules are moving and thus can carry, or convect, significant amounts of energy. In this way the normal conduction process is enhanced.

There are two types of convection: natural and forced. In forced convection, the air has some significant velocity relative to the object it encounters; when you stand in a stiff breeze, run on a windless day, or sit in front of a fan, you experience forced convection. Natural convection arises due to the heating or cooling of air when it contacts an object. As the air changes temperature, it changes density and rises or falls. This "self-generated" free convection is quite common in indoor environments.

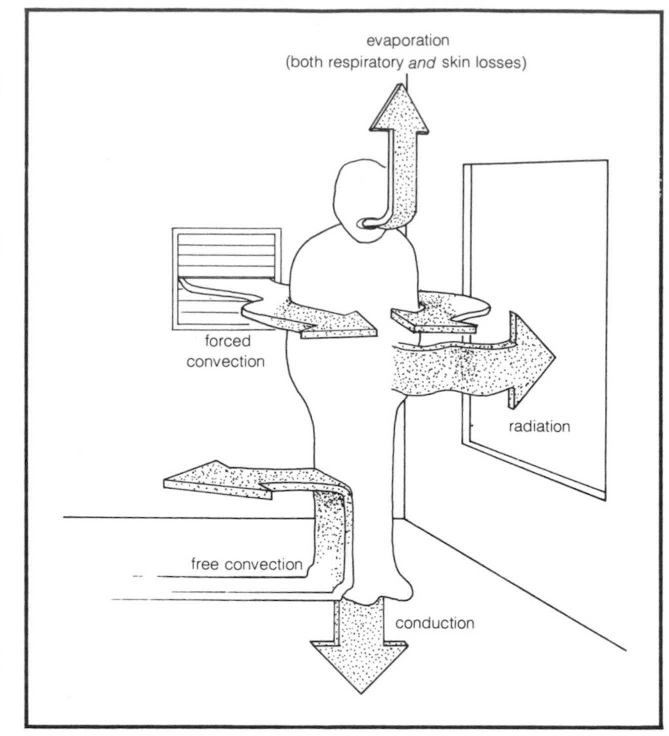

Figure 4.2 Modes of body heat loss.

Radiant Heat Exchange

Heat transfer can also arise from the exchange of electromagnetic waves. Your body will lose or gain energy depending on the temperature, texture, and geometric arrangement of the objects around you. This is probably the most complicated mechanism of heat transfer, and is quite important in establishing thermal comfort. To exchange energy by radiant means, the objects need not be in contact. Rather, they simply must be in "sight" of one another.

Light is a form of electromagnetic energy in wave lengths which are detectable by our eyes. Our interest in radiant heat transfer centers around wave lengths we find invisible—those in the infrared portion of the spectrum. You continually experience invisible radiant heat transfer. A warm electric stove element can be felt many inches away. A crackling fire produces a warm, tingly sensation in your hands and cheeks. These sensations are the result of radiant heat transfer via infrared electromagnetic waves (although the fire also produces radiant energy in the visible wave lengths).

These waves, like light, travel through the air with little or no degradation and permit your body to interact thermally with walls, windows, and other objects which make up your total environment. Direct contact with the radiating source is not necessary, which makes radiation effects particularly subtle and often mysterious. A classic

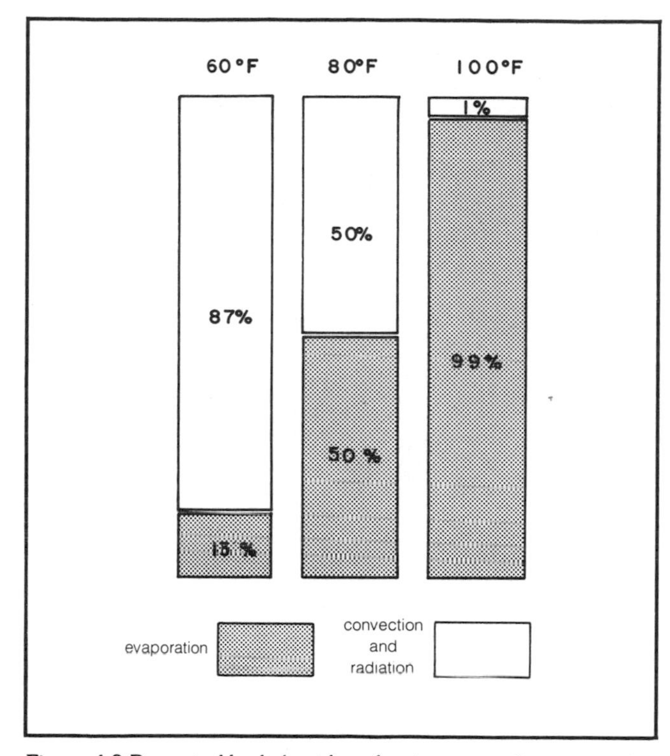

Figure 4.3 Percent of body heat loss due to evaporation, convection, and radiation (relative humidity fixed at 45%).

example is that of a room with a large window area. On a cold day, occupants can feel distinctly uncomfortable even though the air temperature in the room is 75 or 80°F. The discomfort arises from radiant heat losses to the cold window surfaces. Similarly, people can feel inexplicably hot if their environment contains a large number of hot surfaces—even if the air temperature is held at 65°F.

Evaporative Heat Loss

Respiratory passages and lungs are sites of continuous evaporative heat loss. In moderate to high temperatures, sweating of the skin is a major source of evaporative heat loss. Heat is lost in evaporation because it takes energy to turn liquid water into water vapor.

Figure 4.2 summarizes the various modes of heat transfer available to your body. These modes, along with metabolic energy production, are used to maintain thermal equilibrium. A simple relation describes the interaction between these mechanisms:

E. 4.1 $$Q = M \pm R \pm C - E$$

Here Q is the change in the thermal energy content of your body, zero in an equilibrium condition. This condition is achieved by physiological reactions which appropriately alter M, R, C, and E. (Don't confuse the "R" here—radiant heat exchange—with an "R" to come later—resistance.)

Figure 4.3 shows the interplay of these mechanisms at various room temperatures, assuming a constant relative humidity (*RH*) of 45 percent. The metabolic rate is nearly constant over the range from 60 to 100°F, but the evaporative heat loss rises rapidly to dominate at high temperatures. At lower temperatures, convection and radiation play the dominant roles. Pure conduction usually has little effect on bodily heat loss.

Describing the Indoor Environment

Air temperature is a well-known environmental index. When combined with three other indices, relative humidity, mean radiant temperature (*MRT*), and air velocity, a fairly complete description of the indoor environment results.

As shown in Figure 4.3, *air temperature* affects two methods of heat exchange which are essential to the energy balance of your body—convection and evaporation. In fact, it is the principal parameter that can affect your state of comfort when your body is close to its optimal comfort zone. Most heating systems are designed with this fact in mind and concentrate single-mindedly on air-temperature control. *Relative humidity* is an oft-neglected, but important, index of environmental comfort. Simply stated, it is a measure of the quantity of water vapor in the air. It is typically reported in percent, with the percentage referenced to a "saturated" state in which the air holds all the vapor it can without some condensation occurring. The relative humidity is closely tied to air temperature, since warmer air can retain more moisture. Thus, if air with a fixed quantity of water vapor is heated, the relative humidity drops.

Relative humidity can be measured with a wet bulb/dry bulb thermometer combination. The dry bulb thermometer has an everyday mercury or alcohol design which measures the air temperature. The wet bulb thermometer is identical in construction, except that a water-soaked wick is placed around the bulb. The wick promotes evaporation of the water surrounding the bulb, thereby depressing the thermometer reading. The lower the relative humidity, the higher the evaporation rate, and the greater the temperature depression. The difference in readings between the wet and dry bulb thermometers thus can be directly correlated with relative humidity and both specifications are used interchangeably.

High relative humidity results in a muggy atmosphere and stifles the evaporative cooling mechanism. Low-humidity conditions tend to make people very uncomfortable; dry skin, dry mucous membranes, and contact lens discomfort are common symptoms. Excessive drying can also weaken wooden structural members and furniture. While high-humidity conditions are often difficult to rectify, low-humidity situations are easily handled with indoor plants or inexpensive humidifiers.

The *mean radiant temperature (MRT)* is a measure of radiative effects arising in a room. We mentioned in the preceding section that walls and windows which communicate with the outdoors can have inside surface temperatures well below the room air temperature. The cold surfaces can cause significant discomfort, and later sections describe insulating techniques to alleviate this problem.

Air velocity refers to the speed of the air moving through the room. High velocities tend to increase convective heat losses; we say it feels "drafty." Very low velocities or stagnant conditions are undesirable in that odors and moisture can rapidly accumulate. Indeed, these problems arise long before any significant oxygen depletion occurs. The odor problem is tied to both health considerations and aesthetics. Excessive moisture can be downright destructive to furniture and the interior portions of walls, ceilings, and floors. Unless suitable vapor barriers are incorporated into a structure, walls literally can rot from moisture.

Generally, the minimum air velocity needed to avoid stagnation is 10 feet per minute, with 20 to 45 feet per minute preferable. This is often translated to an air-flow or air-exchange criterion of 25 cubic feet per minute per person, or one complete exchange of room air every hour. We will explain these concepts more fully in ensuing sections.

Optimal Settings for Indoor Environments

Now that we have looked at the factors which affect thermal comfort, we ask the question, "Are there really optimal settings for these factors that will produce the most desirable indoor climate?" Researchers have conducted many studies on optimum comfort conditions, and have come up with general comfort indices like "operative temperature," "resultant temperature," "effective warmth," "effective temperature," "revised effective temperature," and so on.

The most widely used comfort criteria are set by the American Society of Heating, Refrigerating, and Air-Conditioning Engineers (ASHRAE, for short). ASHRAE standards serve as a guide for architects and engineers in the space-conditioning trade. The latest comfort standards are given in Table 4.2 and are formally referred to as Standard 55-66. These criteria are referenced to

Table 4.2 Thermal Comfort Conditions (ASHRAE Standard 55-66)

Air Temperatures	73–77°F
Relative Humidity	Below 60%
Mean Radiant Temperature	Equal to air temperature[a]
Air Velocity	10–45 ft/min

Notes: a. For every degree the MRT is below 70°F, increase the air temperature by 1°F.

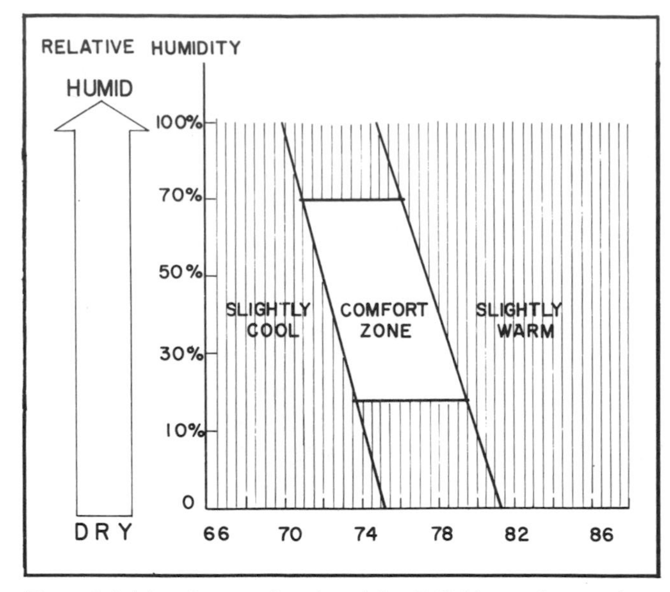

Figure 4.4 A baseline comfort chart (after R.G. Nevins, Institute for Environmental Research, Kansas State University).

persons performing light office work and wearing appropriate clothing.

By way of comparison, Figure 4.4 presents some experimental results compiled by R.G. Nevins of the Institute for Environmental Research. A large sample of sedentary people was exposed to varying air temperature and humidity and questioned as to their personal comfort. If we consider "slightly warm" and "slightly cool" as acceptable states, we have extended that comfort range significantly.

It is interesting to note that recent comfort criteria established in the United States differ from those reported in the 1920s. These, in turn, differ substantially from those obtained in England. Why is this? A host of factors could account for the differences, including physical ability to acclimatize to lower temperatures, physical activity, psychological state, and type of clothing worn.

The first factor—the body's ability to acclimatize—is dependent upon general physical health. If you move from one set of environmental extremes to another, your body will undergo pronounced physiological changes. Blood volume and viscosity alter, heart rate varies, and blood circulation undergoes appropriate changes. Frequent exposure to various environments strengthens the body's ability to acclimatize quickly and effectively.

Physical activity will also affect the temperature at which a person feels comfortable. Metabolic rate (probably the most important parameter in coping with temperature extremes) can range widely: 220 Btu/hr while sleeping; 325 Btr/hr while reading or resting; 550 Btu/hr when sweeping the floor; and even higher rates for intense physical exercise. The temperature of our environment must be adjusted according to the state of physical activity: a person skiing in 40°F temperatures might be perfectly comfortable, while a person trying to read or write at similar temperatures would find it intolerable. Consequently, the average American today—bound to either office, auto, or home (all heated or cooled to maintain a very limited temperature range) and getting precious little outdoor physical exercise—might well require higher temperatures to be comfortable than an American of the 1920s, who lived in a house subject to much greater temperature variation and engaged in more physical work, without benefit of a car or various other labor-saving machines.

Of course, psychological conditioning is another real consideration. If you're not used to (or can't produce) a high indoor temperature, you don't expect it. In such places as England, where central heating is not as common as in the United States, persons normally become acclimatized to lower indoor temperatures. The same is undoubtedly true with air conditioning; the more it is available, the more people demand its use in summer months.

Psychological conditioning also greatly influences what people perceive as their alternatives to discomfort. The fourth factor, the type of clothing worn indoors, is an obvious example. It is always possible to put on warmer clothes or additional layers of clothing when we feel chilly. But for most people with access to a thermostat, such an alternative is not always considered.

Perhaps it is time to re-evaluate our indoor dress habits and to start weaving garments that insulate our bodies instead of merely decorating them. Maybe it is time for people to toughen up a little, to get our bodies in shape, to quit pampering ourselves psychologically, and to put on warmer clothes when relaxing around the house instead of cranking up the thermostat. But it's not necessarily that simple. Important health and efficiency considerations *do* remain as we decide what indoor conditions to design for in a structure using solar heating: these are the considerations that determine our needs for heat storage and auxiliary heating should our solar system break down or an extended cloud cover cause us to use up all our stored solar energy.

First, consider your likely indoor activities. It's all well and good to say we should always be physically active; but if you're a guitarist, or an avid reader, or a writer, chances are you're going to need warmth available on a fairly constant basis to keep the fingers movable. Second, determine the low extreme of outdoor temperatures in the area you plan to build. You may be able to withstand temporary 30°F temperatures quite well when you're healthy, active, and warmly dressed; but it could have severe consequences for any sudden sickness or other emergency, particularly if accompanying weather conditions make medical assistance difficult to acquire.

Remember, too, that individuals have susceptibilities to cold temperatures and humidity which no comfort index can pinpoint. When your fingers are turning blue, an expert's statement that your environment is at a comfortable temperature is going to be no comfort at all. Your best guide is your own body. Practice being aware of various indoor climatic conditions and how they make you feel—when you're healthy or sick, active or just hanging out. Ultimately, this self-study will be your best guide.

CLIMATOLOGY

Before *any* heating system can be designed, the local climate must be thoroughly investigated. Climatological data is primarily needed to estimate the most severe heating load, although related data can be used to predict the monthly or annual energy expenditures for space heating. When we design a solar-heated home, our investigation must be more detailed. Specifically, the availability of solar energy must be ascertained.

Climate and Energy Requirements for Space Heating

Specific climatological factors affecting heating requirements include the temperature difference between the inside and outside of the dwelling, the wind conditions, and shading. The temperature difference is most important, since the rate of heat loss is directly related to this quantity. The overall effect of temperature difference is gauged in two ways: using minimum design temperatures and heating degree-day tabulations.

The most rapid *rate* of heat loss experienced by your building will occur, as you might imagine, when the outside temperature is lowest. This maximum rate may only occur for several hours a year; nonetheless, your heating system must be designed to cope with this situation. The problem is analogous to buying an automobile with a 200-horsepower engine. Driving normally, you seldom use more than 30 or 40 horsepower—the rest of the engine capacity is held in reserve.

Minimum outside design temperatures have been recommended by ASHRAE, and are tabulated for major cities and towns across the country. Table 4.3 is an abbreviated version for Alabama. Note that three temperatures are listed for each locale. The median of annual extremes is computed from thirty or forty years of temperature data, using the coldest temperature recorded each year. The values in columns labeled *99* or *97.5* percent are temperatures which are *exceeded* for that percentage of the time during the three months of December, January, and February. For example, in Auburn, Alabama, the 99 percent design temperature is 21°F, which means that it was colder than 21°F for 1 percent of the time covered in the three-month period. Design temperatures for your particular city or region can be found in Appendix 4A, or in a copy of the ASHRAE *Handbook of Fundamentals* (found in any library). Local chapters of ASHRAE publish even more complete design tables for areas in their jurisdiction. Another way to estimate the outside minimum design temperature is to contact a local office of the National Weather Service. They can supply you with average minimum temperatures and the record low temperature measured in your area.

As you choose lower and lower design temperatures for your locale, your design will become more conservative. This is precisely the implication of the ASHRAE percentage levels—the 99 percent level is more conservative than the 97.5 percent level. Using the former figure, your heating system will be able to handle more severe situations than if you choose the latter. Our conventional practice usually has been to choose very low, ultraconservative design temperatures. This philosophy was encouraged by the small increase in the "initial cost" of conventional, fossil-fired heating systems as their capacity was increased. But in solar heating systems, the initial cost is significant, as are incremental costs for added capacity. Thus the design temperature should be selected carefully.

While the minimum design temperature aids in determining the maximum *rate* of heat loss, degree-day information can tell us about energy requirements over the long term. A degree-day accrues for every degree the average outside temperature is below 65°F for a 24-hour period. Days with temperatures above 65°F are not counted. For example, if the outside air temperature is

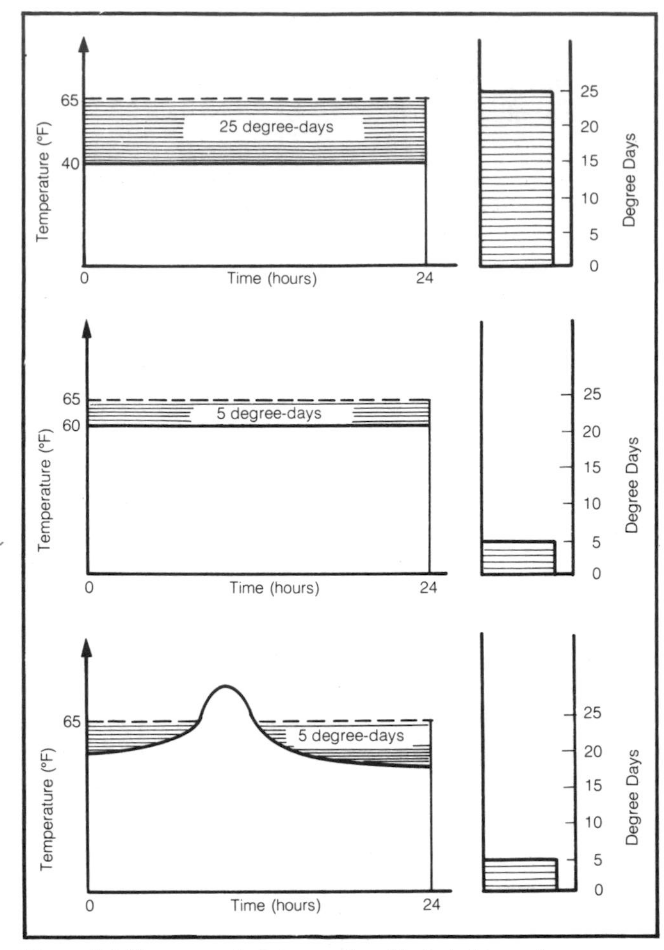

Figure 4.5 The concept of degree-days illustrated in various ways.

constant at 40°F for a 24-hour period, we would accumulate

$$(65 - 40)°F \times 1 \text{ day} = 25 \text{ degree-days}$$

If this condition persisted for one month, we would accumulate

$$(65 - 40)°F \times 30 \text{ days} = 750 \text{ degree-days}$$

during that month. Naturally, outside air temperatures are seldom constant over a 24-hour period—much less for 30 days—and professionally compiled tabulations take this fact into account. Figure 4.5 illustrates the concept graphically. Suffice it to say that the degree-day is a measure of the severity of weather in a given location, incorporating both temperature levels and durations.

Annual degree-day figures, which are useful in computing energy requirements for heating, can be estimated from the map in Figure 4.6. If you live in Oklahoma City, for example, you are situated about halfway between the lines for 3000 and 4000 degree-days annually. As a rough estimate, you might use 3500 degree-days per year.

Table 4.3 Winter Weather Data and Design Conditions for Major Cities in Alabama[a]

Col. 1 State and Station[b]	Col. 2 Latitude[c] ° ′		Col. 3 Elev,[d] Ft	Winter Col. 4 Median of Annual Extremes	99%	97½%	Col. 5 Coincident Wind Velocity[e]
ALABAMA							
Alexander City..	33	0	660	12	16	20	L
Anniston AP....	33	4	599	12	17	19	L
Auburn.........	32	4	730	17	21	25	L
Birmingham AP.	33	3	610	14	19	22	L
Decatur........	34	4	580	10	15	19	L
Dothan AP.....	31	2	321	19	23	27	L
Florence AP....	34	5	528	8	13	17	L
Gadsden........	34	0	570	11	16	20	L
Huntsville AP...	34	4	619	8	13	17	L
Mobile AP......	30	4	211	21	26	29	M
Mobile CO......	30	4	119	24	28	32	M
Montgomery AP.	32	2	195	18	22	26	L
Selma-Craig AFB	32	2	207	18	23	27	L
Talladega.......	33	3	565	11	15	19	L
Tuscaloosa AP..	33	1	170r	14	19	23	L

Notes: a. See Appendix 4A for complete listings.

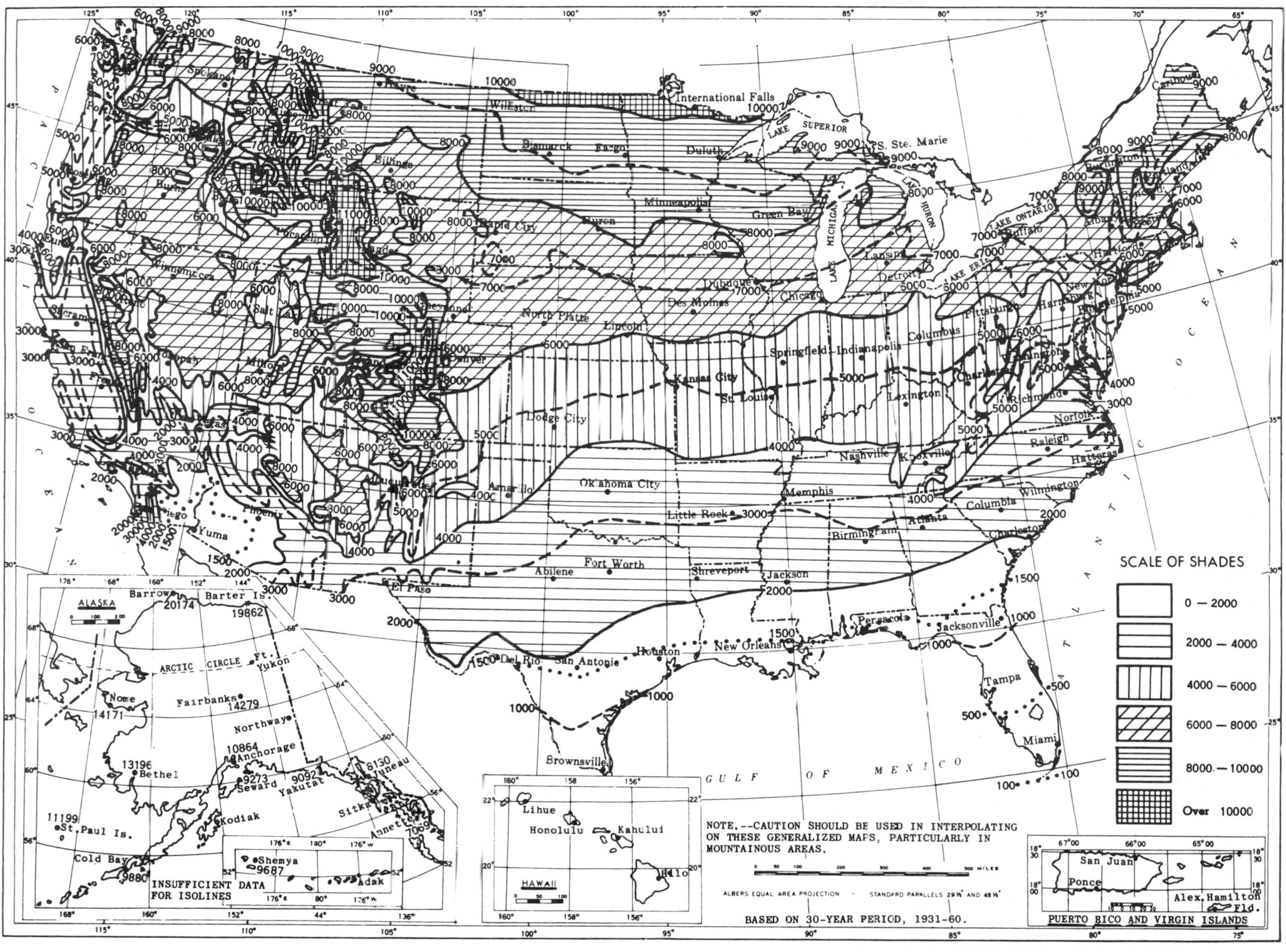

Figure 4.6 Normal total heating degree-days, annual (base 65°F).

More useful degree-day data are available in tabular form, on a monthly basis. The United States Department of Commerce publishes this information, state by state, in a pamphlet entitled "Monthly Normals of Temperature, Precipitation, and Heating and Cooling Degree-Days." Pamphlets for your state may be obtained by writing the National Climatic Center, Federal Building, Asheville, North Carolina 28801. Each request should be accompanied by 25 cents. Alternatively, if you are interested in just one specific locale, you might telephone the nearest office of the National Weather Service and ask for this information. The ASHRAE *Guide and Data Book* also lists degree-day information for major cities across the country. Table 4.4 is excerpted from this source; the complete listing is presented in Appendix 4B.

Monthly degree-day data are useful in determining the range of heating conditions. For instance, the monthly degree-day figures for the San Francisco area (see Table 4.4) suggest year-round moderate heating requirements. In the city itself (San Francisco—C), the maximum number of degree-days is accrued in January. However, the remaining months have significant entries—even during the summer. On the other hand, Mount Shasta, California, experiences extremely severe weather in December and January but very mild weather in the summer months.

There are two possible siting problems that may cause undue heat loss if a solar house (or any house, for that matter) is not properly located. The first problem is wind, which can significantly increase the heat loss from a house. On most sites, wind is not a major difficulty; however, in some mountainous or coastal areas, continuous high winds may exist. Appendix 4A lists the relative wind velocities for many American cities. The wind-velocity entries are a rough indication of average wind speeds during winter months: *VL* signifies very light; *L*, light; *M*, medium; and *H*, high. However, this information is of limited value, since wind conditions are so closely tied to local topography.

If you have a fair amount of land, check your site carefully; a portion of it may get less wind because of topography or thick vegetation. If so, that's probably a good location for your house (assuming it doesn't complicate aspects of solar-collector design or home ventilation; see Chapter 2). If no effective windbreaks exist on the site, you can create them by planting trees, particularly conifers (pines, firs, etc.) or other tall, dense trees and shrubs to form hedgerows. You can consult nurseries or knowledgeable gardeners in your area for specific information.

The second potential problem is shade from such site features as rock outcroppings or trees. Shading of a solar-heated house in winter is detrimental for several reasons. First, shading the collectors decreases the available solar radiation; hence less energy is collected. Also, sunlight falling on any other noncollector part of a house helps to warm the house, and this free energy should be utilized as much as possible. Finally, shading means lower local air temperatures and, therefore, greater heat loss from the house.

However, during the summer, shade can be used for cooling, for the same reasons stated above. A nice way to get the benefits of both winter sun and summer shade is to grow such deciduous trees as oaks, elms, and maples around the house. As seen in Figure 4.7, such trees lose their leaves in the winter, letting sunlight pass through.

Table 4.4 Average Monthly and Yearly Degree-Day Totals for Major Cities in California[a]

Station	Avg. Winter Temp	July	Aug.	Sept.	Oct.	Nov.	Dec.	Jan.	Feb.	Mar.	Apr.	May	June	Yearly Total
Bakersfield.............A	55.4	0	0	0	37	282	502	546	364	267	105	19	0	2122
Bishop..................A	46.0	0	0	48	260	576	797	874	680	555	306	143	36	4275
Blue Canyon............A	42.2	28	37	108	347	594	781	896	795	806	597	412	195	5596
Burbank................A	58.6	0	0	6	43	177	301	366	277	239	138	81	18	1646
Eureka..................C	49.9	270	257	258	329	414	499	546	470	505	438	372	285	4643
Fresno..................A	53.3	0	0	0	84	354	577	605	426	335	162	62	6	2611
Long Beach.............A	57.8	0	0	9	47	171	316	397	311	264	171	93	24	1803
Los Angeles.............A	57.4	28	28	42	78	180	291	372	302	288	219	158	81	2061
Los Angeles.............C	60.3	0	0	6	31	132	229	310	230	202	123	68	18	1349
Mt. Shasta..............C	41.2	25	34	123	406	696	902	983	784	738	525	347	159	5722
Oakland................A	53.5	53	50	45	127	309	481	527	400	353	255	180	90	2870
Red Bluff...............A	53.8	0	0	0	53	318	555	605	428	341	168	47	0	2515
Sacramento.............A	53.9	0	0	0	56	321	546	583	414	332	178	72	0	2502
Sacramento.............C	54.4	0	0	0	62	312	533	561	392	310	173	76	0	2419
Sandberg...............C	46.8	0	0	30	202	480	691	778	661	620	426	264	57	4209
San Diego...............A	59.5	9	0	21	43	135	236	298	235	214	135	90	42	1458
San Francisco...........A	53.4	81	78	60	143	306	462	508	395	363	279	214	126	3015
San Francisco...........C	55.1	192	174	102	118	231	388	443	336	319	279	239	180	3001
Santa Maria.............A	54.3	99	93	96	146	270	391	459	370	363	282	233	165	2967

Notes: a. See Appendix 4B for complete listings.

Figure 4.7 Natural shading techniques for buildings.

During the summer, their leafy branches block direct sunshine and shade the house. Most of these trees are slow growers, so unless you can plant the trees several years before you build your house, only the existing trees will be of much help for seasonal shading purposes.

The Nature of Solar Energy

The thermal exchange between the sun and earth is a superb example of radiation heat transfer. The sun radiates inconceivable amounts of energy derived from continuous nuclear fusion reactions. Due to the tremendous distances involved, only a small portion of this energy is intercepted by the earth. Even then, only about 80 percent of this energy reaches the earth's surface. The short, or ultraviolet, wave lengths are largely absorbed in the upper atmosphere, while other wave lengths are attenuated to a lesser extent. Most of the radiation that reaches us, eight minutes after leaving the sun, is either visible or else of infrared (long) wave length. The attenuating effects of the atmosphere can be seen in Figure 4.8.

Even with attenuation, the amount of energy reaching the earth's surface is staggering. The solar energy falling on the continental United States in a year is about a thousand times that of our annual energy consumption! The problem, of course, is harnessing this widely dispersed and intermittent energy resource. The first step is to identify locations which receive large amounts of radiation.

The availability of solar energy depends upon latitude, season, and weather patterns. Dependency upon the first two variables becomes obvious when one understands the trajectory of the earth around the sun. Figure 4.9 depicts a highly simplified model of this relationship.

The path of the earth around the sun is not a circle but an ellipse (though not a very pronounced ellipse). Since the sun is at one focus of the ellipse, the distance between the sun and the earth varies throughout the orbit. The earth is closest to the sun in December (89.8 million miles) and farthest in June (95.9 million miles). This relatively small variation makes an appreciable difference in the radiation intensity, since the intercepted radiation decreases with the square of the distance. Those of us in the northern hemisphere may thus ask, "Why isn't it warmer in January?" This is related to the earth's tilt, which is also depicted in Figure 4.9. The tilt, combined with orbital factors, gives rise to our seasons.

Figure 4.10 looks at the earth's tilt in a slightly different (and ancient!) perspective. Here we consider the earth fixed and the sun as a moving entity. With this viewpoint, we see the sun making its northernmost

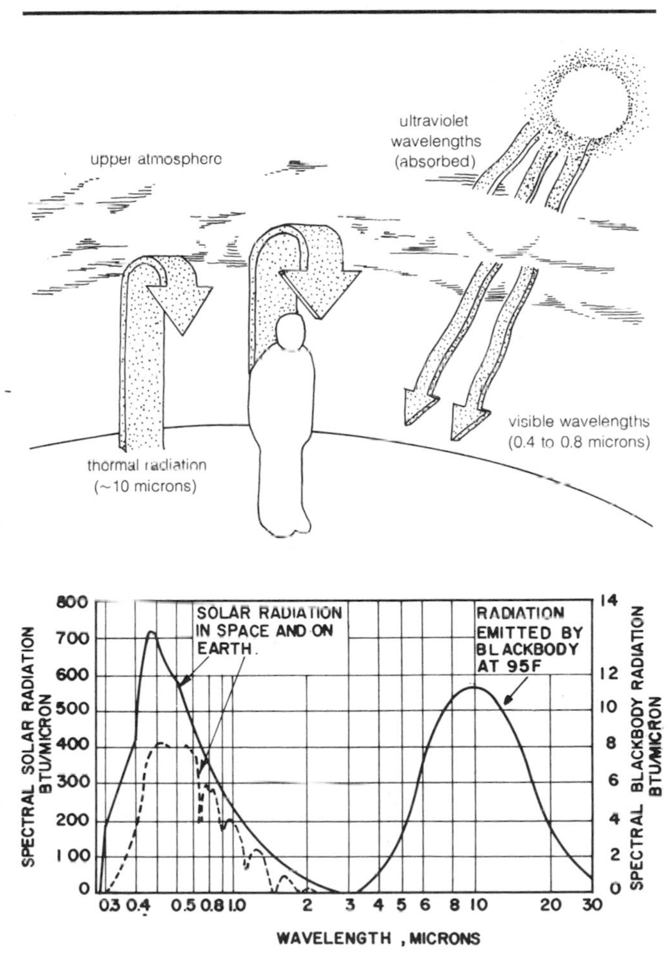

Figure 4.8 Spectral characteristics of solar and infrared radiant energy (1 micron = 0.0001 centimeter).

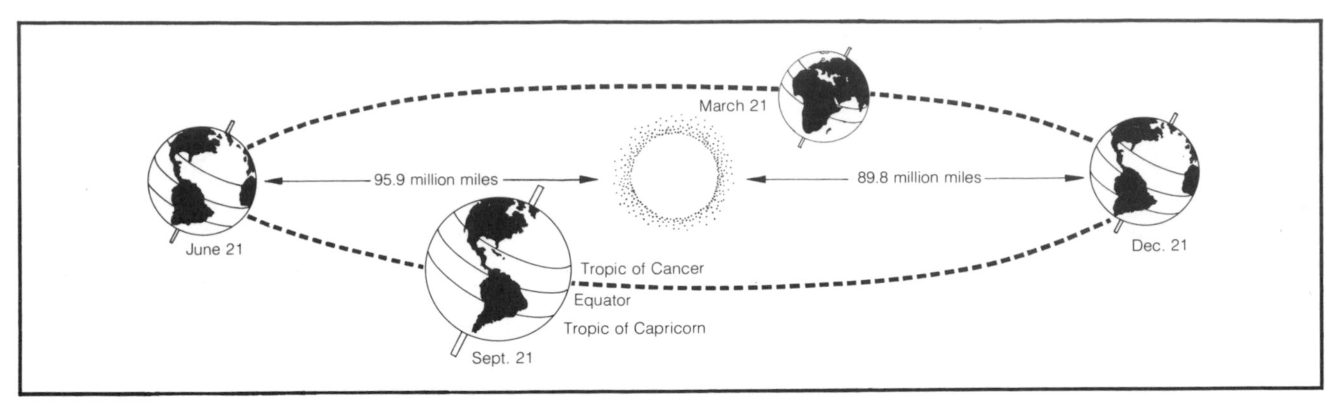

Figure 4.9 The motion of the earth around the sun.

excursion on June 22. At this time the sun is directly overhead at the Tropic of Cancer. The autumnal and vernal equinoxes occur when the sun crosses the equator on September 23 and March 21, respectively. The sun is never *directly* above any portion of the continental United States, although it reaches a maximum solar altitude at the summer solstice.

The solar altitude, which is the angle the sun rises above the horizon, is important in two respects. First, a higher solar altitude means that the solar radiation travels a shorter distance to traverse the atmosphere. A low solar altitude forces the radiation to travel through a great deal more air mass; the attenuating effects are proportionally greater. (The stunning visual effects at sunrise or sunset are thus possible—when else can you look directly at the sun and enjoy it so?) Second, higher solar altitudes imply more daylight hours.

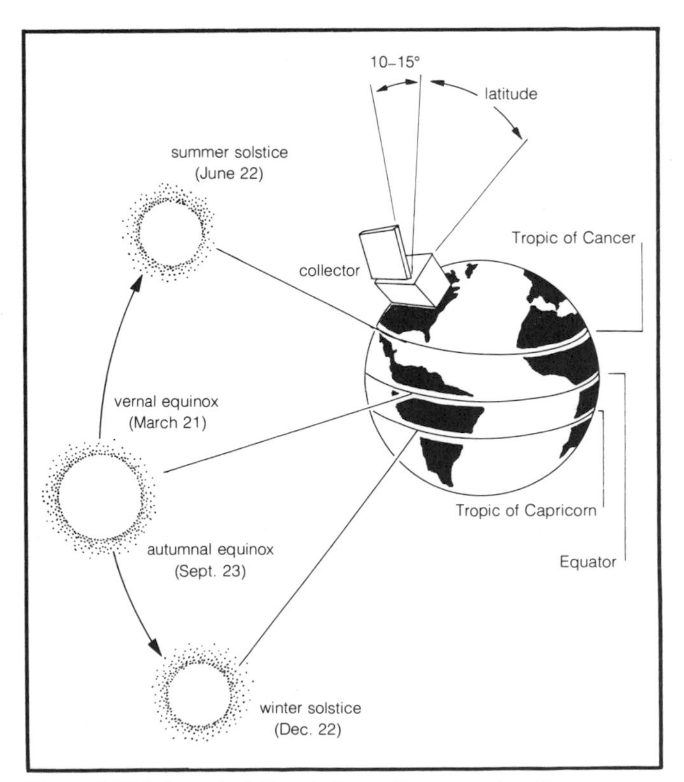

Figure 4.10 Seasonal relations between the sun and earth.

The relationship between seasons, solar altitude, and incident solar radiation is shown in Figure 4.11. The solar altitude is shown as a function of the daylight hour (by the clock) for a number of days during the year. An accompanying graph shows the amount of solar energy which would fall on a horizontal surface. These values assume there are no local atmospheric anomalies; no clouds, smog, or mist. Note that the impinging solar energy, or *insolation*, is given in Btu per hour per square foot (Btu/hr-ft^2). This is a set of units which can be used to express the rate of energy falling on each square foot of surface area—in this case, horizontal surface area. The curves show us what we've suspected all along—more solar energy is available during the summer solstice than during any other time of the year. The insolation values are higher and the exposure time is longer.

ASHRAE publishes values of solar radiation computed for a variety of latitudes and months (see Appendix 4C). If cloud cover and other meteorological wonders were never present, this information would be a sufficient (and welcome) indicator of solar availability for any locale. Unfortunately, things are a bit more complicated.

Availability of Solar Energy

The presence of fog, cloud cover, storms, or heavy smog can appreciably reduce the incoming solar radiation. Thus, the ASHRAE insolation values, based on cloudless sky conditions, are only a limited help. There are two basic methods to estimate the actual availability of solar energy. First, you can obtain actual data taken over some significant time period. The National Weather Service has a few stations which perform this type of measurement, and sometimes local universities or pollution control districts may monitor insolation. The Weather Service stations which record this data are listed in Table 4.5. Table 4.6 illustrates a typical format for the data, and Figure 4.12 presents an interesting overview of national

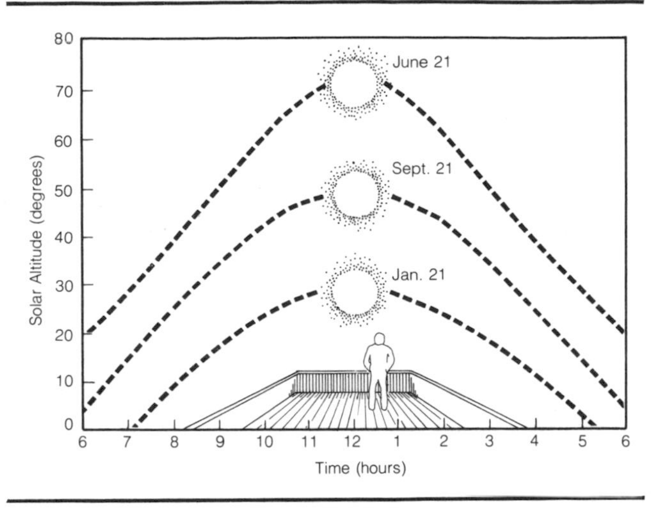

Figure 4.11a Solar altitude at 40°N latitude (June, September, and January).

patterns. Monthly average insolation figures, where the average is taken over several years, are quite suitable for design calculations. Unfortunately, there are all too few stations equipped to supply this information.

A second technique to estimate average insolation is presented in a later section—"Sizing a Flat Plate Collector." The method is based on the fact that most of the variation in insolation arises from differences in cloud cover between various locales. If we could count on perfectly clear days all year long, the only major factors causing variance in insolation would be the season of the year and the latitude of the site. By knowing the percentage of cloud-free days each month, we can adjust the cloudless insolation data accordingly.

The percent of possible sunshine is tabulated by month for most locations in the United States. Table 4.7 presents some representative values. San Jose, California, for example, has sunshine during 54 percent of the daylight hours during January. The local branch of the National Weather Service can supply this figure for your

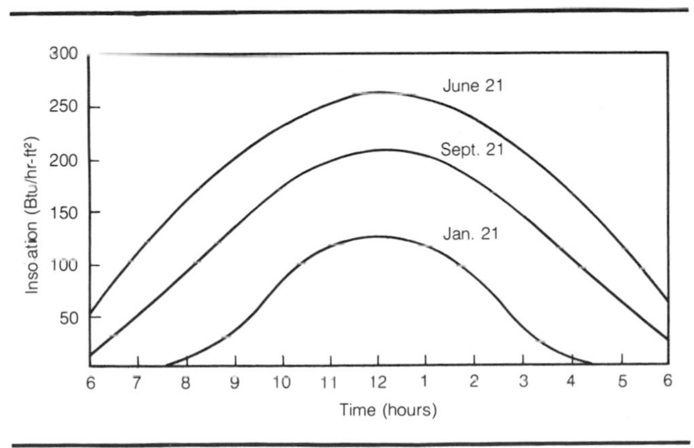

Figure 4.11b Cloudless insolation values for a horizontal surface located at 40°N latitude.

town. The actual calculation of insolation estimates will be deferred until we deal with solar collectors.

On-site Measurements

It is conceivable that certain locales may be so unique that no reliable published climatological data are available.

Table 4.5 U.S. Weather Service Stations Currently Recording Insolation

Alaska
Annette
Barrow
Bethel
Fairbanks
Matanuska
Arizona
Page
Phoenix
Tucson
Arkansas
Little Rock
California
Davis
El Centro
Fresno
Inyohern
Los Angeles
Riverside
Santa Maria
Colorado
Grand Junction
District of Columbia
Washington
Florida
Apalachicola
Miami
Tallahassee
Idaho
Boise
Illinois
Argonne
Indiana
Indianapolis
Iowa
Ames
Kansas
Dodge City
Manhattan
Louisiana
Lake Charles
Maine
Caribou
Michigan
East Lansing
Sault Ste. Marie
Mississippi
Columbia
Montana
Glasgow
Great Falls
Nebraska
North Omaha
Nevada
Ely
Las Vegas
Reno
New Mexico
Albuquerque
New York
Astoria
Geneva
Ithaca
New York Central Park
North Carolina
Cape Hatteras
Greensboro
North Dakota
Bismarck
Ohio
Cleveland
Oklahoma
Oklahoma City
Oregon
Medford
Portland
Pennsylvania
State College
Rhode Island
Newport
South Dakota
Rapid City
Tennessee
Nashville
Oak Ridge
Texas
Brownsville
El Paso
Fort Worth
Midland
San Antonio
Vermont
Burlington
Washington
Seattle
Tacoma
Wisconsin
Madison
Wyoming
Lander
Laramie
"Proposed" Stations
Boulder, Colorado
Columbia, Missouri
Salt Lake City, Utah
Sterling, Virginia

Table 4.6 Daily Averages of Solar Energy Received on a Horizontal Surface[a,b]

City	Jan	Feb	Mar	Apr	May	Jun	Jul	Aug	Sep	Oct	Nov	Dec
Miami, Fla.	1100	1284	1535	1756	1852	1771	1749	1716	1528	1351	1232	1085
Brownsville, Tex.	1240	1465	1782	1919	2144	2590	2472	2336	1934	1786	1221	1192
Gainesville, Fla.	1011	1255	1587	1937	2066	1904	1823	1624	1446	1181	974	841
Apalachicola, Fla.	1192	1542	1616	1941	2074	2192	2103	2007	1646	1513	1317	959
New Orleans, La.	756	915	1207	1487	1590	1697	1491	1439	1402	1258	952	804
Lake Charles, La.	1041	1114	1531	1771	2118	2151	1926	2052	1594	1439	1262	819
El Paso, Tex.	1328	1546	2125	2524	2716	2731	2531	2435	2103	1786	1428	1207
Fort Worth, Tex.	1015	1218	1734	1875	2140	2332	2303	2258	1779	1587	1277	867
La Jolla, Calif.	930	1181	1550	1882	2007	2066	2015	1830	1572	1273	1107	904
Charleston, S. C.	923	1232	1664	2059	2288	2166	1989	1945	1509	1203	1137	786
Griffin, Ga.	1063	1107	1181	2103	2288	2273	2170	2066	1546	1358	1044	738
Riverside, Calif.	974	1151	1506	1823	2007	2207	2184	2011	1712	1321	1018	782
Santa Maria, Calif.	1070	1380	1882	2251	2506	2399	2428	2369	1945	1594	1114	867
Albuquerque, N. M.	1133	1354	1834	2236	2494	2749	2502	2299	2018	1712	1284	1085
Hatteras, N. C.	941	1063	1550	2103	2229	2266	2229	2125	1587	1269	1015	756
Oak Ridge, Tenn.	642	852	1055	1483	2103	2000	1838	1708	1517	1129	753	598
Las Vegas, Nev.	963	1292	1956	2111	2362	2771	2539	2332	2044	1483	1166	845
Nashville, Tenn.	524	753	1089	1557	1838	1934	1867	1668	1439	1125	779	465
Stillwater, Okla.	923	1004	1520	1801	1838	2196	1889	1937	1565	1255	900	775
Fresno, Calif.	664	1037	1539	2096	2406	2642	2576	2288	1889	1380	915	605
Davis, Calif.	738	1026	1417	1982	2387	2642	2605	2303	1834	1336	745	576
Washington, D. C.	568	738	1225	1513	1716	1867	1808	1631	1373	1085	745	539
Columbia, Mo.	598	930	1229	1631	1801	2077	2221	1856	1624	1299	731	664
Seabrook, N. J.	686	908	1321	1668	1897	2007	1838	1771	1336	1052	771	535
Grand Lake, Colo.	790	1144	1624	2030	2177	2362	2236	1989	1720	1328	863	613
Salt Lake City, Utah	572	908	1424	1882	2015	2192	2303	2247	1631	1026	661	443
New York, N. Y.	450	705	956	1339	1572	1646	1620	1351	1166	897	546	395
Sayville, N. Y.	635	904	1236	1631	1970	2066	1812	1694	1292	1070	686	498
State College, Pa.	506	642	1015	1428	1572	1845	1889	1683	1321	915	605	424
Lincoln, Neb.	686	930	1247	1576	1852	2052	2122	1775	1509	1114	771	613
Upton, N. Y.	568	790	1181	1550	2030	2081	1779	1683	1328	1111	642	487
Cleveland, Ohio	373	675	1030	1550	2140	2214	2192	1934	1705	1041	487	406
Newport, R. I.	583	845	1196	1535	1786	1963	1860	1683	1358	1092	668	524
Put-in-Bay, Ohio	509	756	1137	1476	1734	2092	2000	1830	1387	1026	557	399
East Wareham, Mass.	557	790	1085	1498	1720	1897	1768	1668	1299	996	627	542
Blue Hill, Mass.	601	923	1196	1465	1756	1911	1838	1708	1343	1077	601	498
Boston, Mass.	454	745	1085	1328	1661	1823	1690	1502	1184	937	502	406
Medford, Ore.	391	768	1232	2107	2790	2590	2804	2494	1753	1063	550	362
Ithaca, N. Y.	435	760	926	1173	1624	1867	1845	1697	1299	940	476	391
Twin Falls, Idaho	613	827	1269	1705	2184	2303	2280	1985	1646	1255	738	450
E. Lansing, Mich.	384	649	945	1284	1395	1638	1653	1432	1048	819	380	343
Madison, Wis.	539	797	1166	1498	1727	1904	1993	1690	1321	959	557	443
Toronto, Ont.	351	605	1004	1317	1668	1926	1756	1627	1144	797	399	347
St. Cloud, Minn.	627	878	1461	1734	2070	2066	2118	1667	1343	1048	646	561
Caribou, Me.	531	745	1192	1697	1771	1963	2015	1690	1343	937	450	391
Spokane, Wash.	443	731	1240	2111	1782	2269	2487	2144	1587	923	491	280
Seattle, Wash.	229	328	1033	1823	1867	2280	2170	1753	1255	627	325	229
Glasgow, Mont.	576	900	1446	1852	2362	2494	2435	2011	1395	900	642	450
Winnipeg, Man.	524	745	1225	1461	1945	2048	2052	1808	1166	841	417	365

Notes: a. Values from direct measurements or inferred from "% cloudiness" data; from Strock's *Handbook of Air Conditioning, Heating and Ventilating*. b. Units are Btu/ft^3-day.

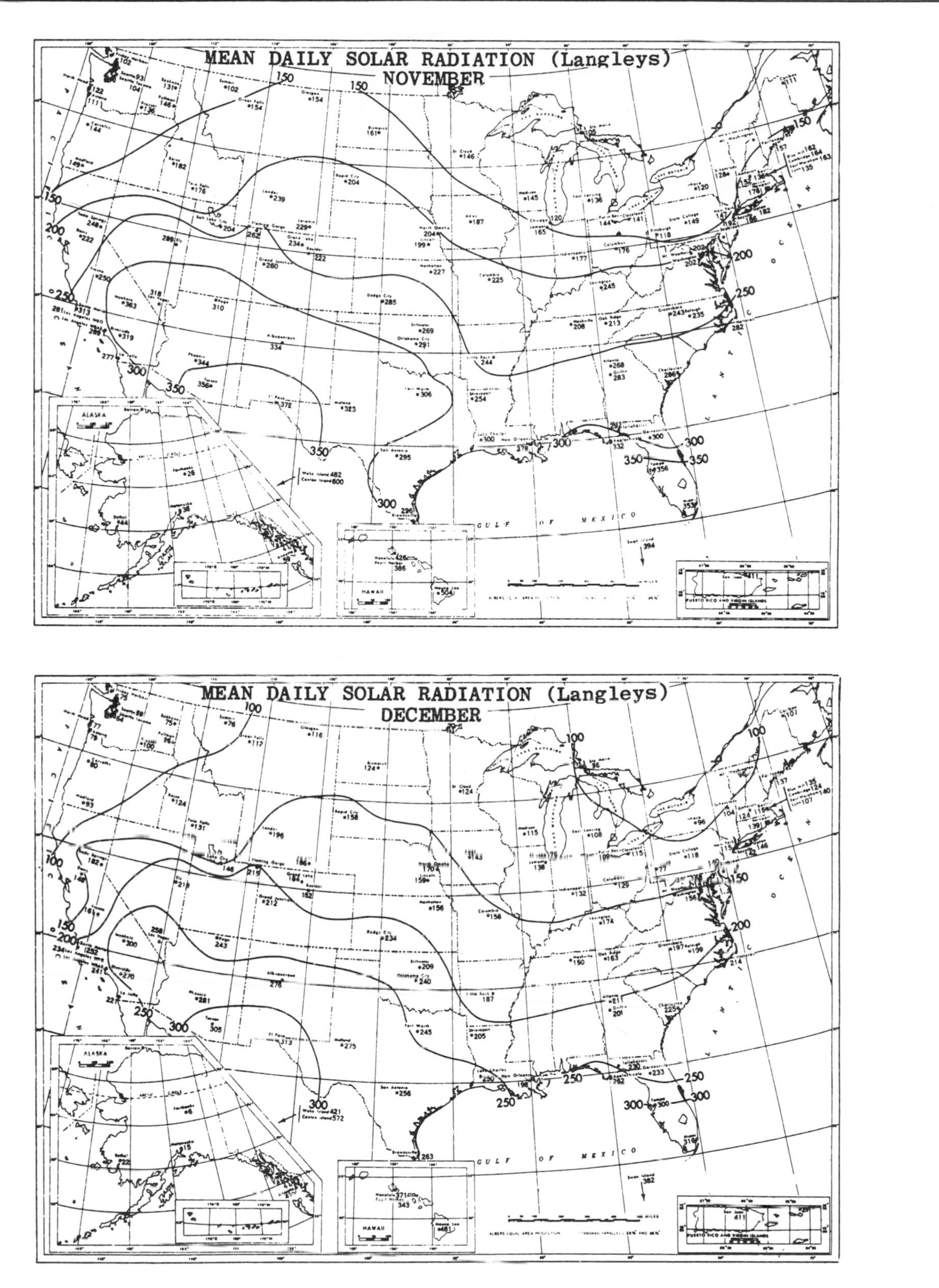

Figure 4.12 Mean daily solar radiation in November and December (one langley equals 3.69 Btu/ft²).

Notes: a. 1 Langley equals 3.69 Btu/ft².

Table 4.7 Percent of Possible Sunshine in January (Selected Major Cities)[a]

State and City	%
Alabama	
Anniston	50
Birmingham	45
Mobile	51
Montgomery	51
Arizona	
Flagstaff	—
Phoenix	76
Tucson	78
Winslow	—
Yuma	84
Arkansas	
Fort Smith	42
Little Rock	42
California	
Bakersfield	—
El Centro	—
Eureka	39
Fresno	49
Long Beach	—
Los Angeles	71
Needles	—
Oakland	—
Pasadena	—
Sacramento	39
San Bernardino	—
San Diego	67
San Francisco	53
San Jose	54
Colorado	
Denver	67
Durango	—
Grand Junction	53
Pueblo	71
Connecticut	
Bridgeport	—
Hartford	46
New Haven	52
Delaware	
Wilmington	—
District of Columbia	
Washington	46
Florida	
Apalachicola	58
Jacksonville	57
Key West	68
Miami	66
Pensacola	55
Tampa	60
Georgia	
Atlanta	48
Augusta	56
Brunswick	—
Columbus	—
Macon	54
Savannah	54
Idaho	
Boise	39
Lewiston	—
Pocatello	35
Illinois	
Cairo	42
Chicago	42
Moline	42
Peoria	47
Springfield	38
Indiana	
Evansville	38
Fort Wayne	35
Indianapolis	36
Terre Haute	44
Iowa	
Davenport	—
Des Moines	54
Dubuque	48
Keokuk	—
Sioux City	56
Kansas	
Concordia	60
Dodge City	61
Topeka	51
Wichita	61
Kentucky	
Louisville	41
Louisiana	
New Orleans	49
Shreveport	47
Maine	
Augusta	—
Bangor	—
Eastport	45
Portland	55
Maryland	
Baltimore	48
Cumberland	—
Massachusetts	
Boston	48
Fitchburg	—
Nantucket	43
Springfield	—
Worcester	—
Michigan	
Alpena	28
Detroit	31
Escanaba	41
Flint	—
Grand Rapids	26
Lansing	—
Marquette	28
Saginaw	—
Sault Ste. Marie	32
Minnesota	
Duluth	46
Minneapolis	49
St. Paul	—
Mississippi	
Meridian	—
Vicksburg	45
Missouri	
Columbia	52
Kansas City	51
St. Louis	46
Montana	
Billings	47
Butte	—
Havre	49
Helena	51
Kalispell	—
Missoula	29
Nebraska	
Lincoln	57
North Platte	62
Omaha	54
Valentine	62
Nevada	
Reno	64
Winnemucca	53
New Hampshire	
Concord	48
Manchester	—
Portsmouth	—
New Jersey	
Atlantic City	51
Jersey City	—
Newark	—
Trenton	48
New Mexico	
Albuquerque	69
Roswell	—
Santa Fe	—
New York	
Albany	42
Binghamton	32
Buffalo	32
Canton	—
New York	51
Oswego	19
Rochester	32
Syracuse	31
North Carolina	
Asheville	48
Charlotte	52
Greensboro	49
Raleigh	49
Wilmington	57
North Dakota	
Bismarck	54
Devils Lake	52
Fargo	48
Williston	51
Ohio	
Akron	—
Cincinnati	40
Cleveland	28
Columbus	36
Dayton	38
Sandusky	34
Toledo	32
Youngstown	—
Oklahoma	
Oklahoma City	57
Tulsa	49
Oregon	
Baker	41
Eugene	—
Medford	—
Portland	19
Roseburg	25
Pennsylvania	
Altoona	—
Erie	22
Harrisburg	44
Oil City	—
Philadelphia	45
Pittsburgh	32
Reading	44
Scranton	37
Rhode Island	
Block Island	45
Pawtucket	—
Providence	49
South Carolina	
Charleston	58
Columbia	53
Greenville	53
South Dakota	
Huron	52
Rapid City	55
Sioux Falls	—
Tennessee	
Chattanooga	43
Knoxville	42
Memphis	45
Nashville	42
Texas	
Abilene	59
Amarillo	65
Austin	48
Brownsville	48
Corpus Christi	49
Dallas	46
Del Rio	53
El Paso	75
Fort Worth	54
Galveston	50
Houston	45
Palestine	46
Port Arthur	47
San Antonio	46
Utah	
Modena	—
Salt Lake City	46
Vermont	
Burlington	34
Rutland	—
Virginia	
Cape Henry	40
Lynchburg	48
Norfolk	50
Richmond	49
Roanoke	—
Washington	
North Head	28
Seattle	28
Spokane	26
Tacoma	21
Walla Walla	24
Wenatchee	—
Yakima	34
West Virginia	
Bluefield	—
Charleston	—
Huntington	—
Parkersburg	29
Wheeling	—
Wisconsin	
Green Bay	46
La Crosse	—
Madison	46
Milwaukee	38
Wyoming	
Cheyenne	61
Lander	63

Notes: a. From U.S. Weather Service.

In such a case, you might consult local airports, ranchers, lighthouses, the U.S. Soil Conservation Service, the Department of Agriculture, pollution-control agencies, little old ladies with bunions, or any other local people who might be concerned with weather in their business or profession. Or, if you have the time and patience, you may want to make your own climatological measurements on-site over a period of time. The necessary instruments (thermometers, anemometers, insolation meters, etc.) are commercially available. Many of these instruments are incredibly expensive, particularly those designed for professional meteorologists or shipboard applications. (The Taylor Instrument Company does have a relatively inexpensive series of instruments available. For instance, a seven-day recording thermometer is priced around $90 [1974]. For catalog information, write the company, % the Sybron Corporation, Arden, North Carolina 28704.)

Figure 4.13 shows a simple device which can be used to measure insolation. It is comprised of a simple volt-ohm meter (VOM), which can be purchased for under $20 in any electronics supply house, and a silicon solar cell. The silicon cell need only be a square inch or so in area, and should cost $1 or $2. The VOM is set on one of the milliampere (ma) current scales, and the output current of the cell can be read directly. Different intensities may require switching to different current ranges.

The output is nearly linear with respect to incident energy intensity, which is all we need to know if only relative measurements are required. For instance, if we wished to compare the insolation on a horizontal surface with one at some tilt angle, we would simply read the meter in both positions and compare the outputs.

Absolute calibration is a bit of a problem unless you have an accurate standard nearby. If a calibrated system is nearby, you need only visit the station once to standardize your instrument. Or, you might refer to the Solar Position and Intensity charts in Appendix 4C. Choose the table representing your latitude and the proper month. Insolation figures are given hourly, assuming cloudless skies. The figures can be used in the following way to obtain a rough calibration: On a perfectly clear day, place your meter in a horizontal position and read the current output when the sun reaches its highest position in the sky. This position corresponds to a "solar time" of noon (although your watch may not *read* noon). Record the meter current and then consult the Solar Intensity charts in Appendix 4C. The horizontal insolation value for solar noon is very close to the value your meter is reading.

As an example, suppose you performed your measurements at 40°N latitude on May 21. Consulting Appendix 4C, we find the noon insolation to be 301 Btu/hr-ft^2 for a horizontal surface. If the meter is reading 150 ma, then each ma (301/150) corresponds to approximately 2 Btu/hr-ft^2. If at a later time your meter reads 50

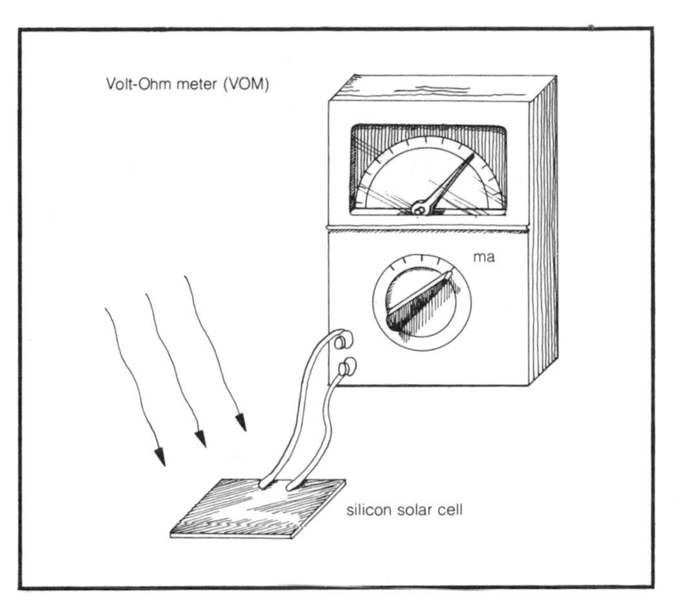

Figure 4.13 A simple scheme to measure incident solar energy.

ma, then the insolation would be (50 × 2) 100 Btu/hr-ft^2.

We might insert a general word of warning. You should understand that on-site collection of *meaningful* weather data is a substantial undertaking; measurements should extend through an entire year or over several years. Unless expensive automatic recording equipment is used, two or more visits each day to your site would be necessary for you to read the instruments and record the data. This may not be feasible until you are actually living there.

An alternative (and certainly an easier) approach is to collect the best climatological data available from local sources and then use a conservative design program to compensate for any errors. For instance, if the insolation data comes from a station many miles from your site, you might reduce the insolation figure by 5 to 10 percent, to accommodate any suspected differences in smog or cloud cover.

HEAT LOSS IN DWELLINGS

A dwelling is designed to create an artificial environment—a living space climatically detached from the outdoors. The heat loss experienced by a building reflects the degree of separation. Our goal is to minimize heat losses in order to reduce the energy requirements for heating. The more successful we are, the more attractive solar heating becomes. We begin with a more detailed discussion of the heat-transfer modes introduced earlier, when we considered the thermal aspects of physical comfort.

Basically, heat is thermal energy which is transferred between two objects because of a temperature difference between them. The direction of energy transfer is from the higher to the lower temperature (Figure 4.14). No net

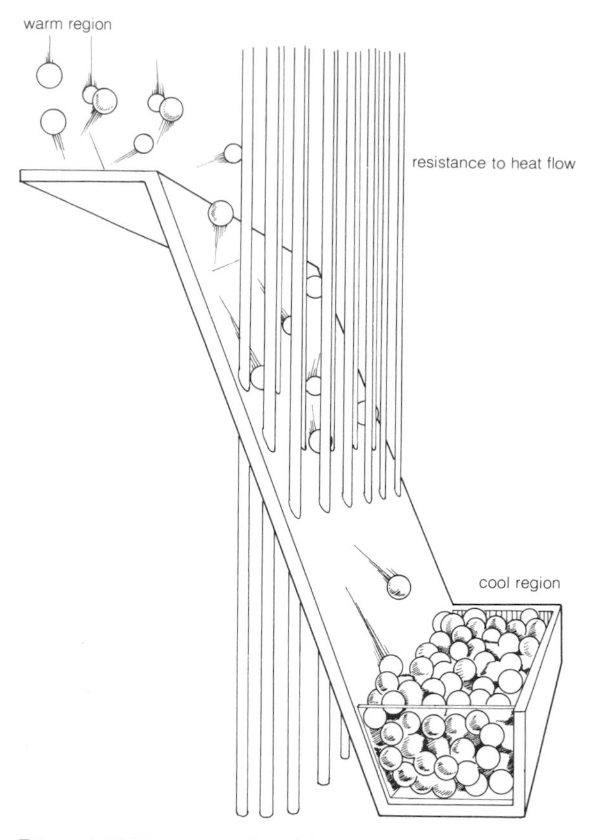

Figure 4.14 Heat is transferred from warmer regions to cooler ones.

heat transfer occurs between two objects at the same temperature. Obviously, all our basic heat-loss problems arise when outside temperature is lower than the inside temperature of our home. Furthermore, the amount of heat transferred in a given time interval (the rate of heat transfer) is generally proportional to the difference in temperature between the two bodies in question: the greater the temperature difference, the greater the rate of heat transfer. So in heat transfer, temperature differentials are all-important. The rate of heat transfer is generally expressed in Btu per hour (Btu/hr). When dealing with building materials, it is sometimes useful to consider the heat transfer rate per square foot of surface area. The units are then Btu/hr-ft^2. The usefulness of the latter notation will become evident shortly.

Mechanisms of Heat Loss

Given the concepts of heat and temperature difference, we can begin to quantify the aspects of heat transfer and other mechanisms of energy loss. As we mentioned in the section on thermal comfort, there are three classic mechanisms of heat transfer: conduction, convection, and radiation. An additional energy loss will occur when the living space exchanges air with the environment. This air exchange is called either ventilation or infiltration, depending on whether it is intentional or not.

Conduction

Conduction is the transfer of energy through a medium by direct molecular interaction. Excited (hot) molecules transfer some of their vibrational energy to their cooler neighbors. This energy transfer on the molecular level results in a large-scale flow of energy from higher to lower temperature regions. Cooking with a frying pan is a good introduction to conduction heat transfer. The energy from the pan is conducted along the handle and eventually reaches your hand. This effect is accentuated if the handle is short and/or made of metal.

A poor conductor is one which retards heat transfer; it is known as an insulator. The most important measure of an insulator is its thermal conductivity (k). Thermal insulators are materials which have low values of k; porcelain, glass, wood, and dry soil are good examples. By comparison, metals (e.g., copper and aluminum) have high thermal conductivity and so are poor thermal insulators. This is one reason why, in cooking, metal handles get very hot while wooden handles don't. To insulate your home, we seek a material which is economical, practical, and has a low k-value.

Table 4.8 lists the thermal conductivities of common substances. Scan it for familiar materials, recalling that low conductivities imply good insulators. Note that the units are rather awkward: k is given in Btu-in/hr-ft^2-°F. This is no cause for concern; we will explain how to deal with these units in due time.

Two other quantities enter into conduction heat transfer: the area and thickness of the insulating materials. The rate of heat flow (Btu/hr) is directly proportional to the area through which the heat energy can move. For instance, if 100 Btu/hr is transferred through 1 square foot of a given insulation, then we know that 200 Btu/hr will be transferred through 2 square feet of the same insulation. This concept becomes clear if we think of water flowing through a hole—the bigger the hole, the greater the amount of water flowing through it (other things being equal). More to the point, the larger the surface area of your house, the greater the heat-loss problem. The rate of heat flow is inversely proportional to the thickness of the insulating material, which is just a way of saying that thicker insulation reduces heat loss. If we insulate with 6 inches of fiberglass instead of 3 inches, then we will reduce the wall's heat-loss rate by about 50 percent.

In summary, four factors affect the rate at which heat energy conducts through a substance: the temperature differential, the thermal conductivity of the substance, the heat-transfer area, and the thickness of the substance. This information is related by a simple conduction equation:

E. 4.2 $$q_c = A\frac{k}{t}(T_i - T_o) = A\frac{k}{t}\Delta T$$

where q_c equals the rate of heat transfer by conduction (Btu/hr); A equals the area (ft²); k equals the thermal conductivity (Btu-in/hr-ft²-°F); t equals the thickness (in); and T_i, T_o equal the inside and outside temperatures (°F). Note that the equation reaffirms our discussion: q_c increases if A, k, or $(T_i - T_o)$ increases. The inverse relationship between q_c and t is also verified: as t increases, q_c will decrease. Let's perform a simple calculation to get a feel for the numbers and units involved.

Example: Suppose we had a wall of dry clay 4 inches thick, with a surface area of 50 square feet. If one face is held at 75°F and the other at 35°F, let's find the rate of heat transfer.

Solution: Using Table 4.8, we find a thermal conductivity of 3.5 to 4.0 Btu-in/hr-ft²-°F. We'll choose the higher value and estimate the maximum heat-transfer rate. Using Equation 4.2, we find

$$q_c = (50\ ft^2)\left(\frac{4.0\ Btu\text{-}in/hr\text{-}ft^2\text{-}°F}{4\ in}\right)(75 - 35)°F$$

$$= 2000\ Btu/hr$$

As we mentioned earlier, it is often handy to have the conductive heat-transfer rate per square foot of surface area. We will denote this quantity by q_c''. In our example

Table 4.8 Thermal Conductivity (k) of Miscellaneous Substances at Room Temperature

Material	Density at 68°F (lbm per cu ft)	Conductivity k (Btu-in/hr-ft²-°F)
Air, still	—	0.169–0.215
Aluminum	168.0	1404–1439
Asbestos board with cement	123	2.7
Asbestos, wool	25.0	0.62
Brass, red	536.0	715.0
Brick		
Common	112.0	5.0
Face	125.0	9.2
Fire	115.0	6.96
Bronze	509.0	522.0
Cellulose, dry	94.0	1.66
Cinders	40-45	1.1
Clay		
Dry	63.0	3.5-4.0
Wet	110.0	4.5-9.5
Concrete		
Cinder	97.0	4.9
Stone	140.0	12.0
Corkboard	8.3	0.28
Cornstalk, insulating board	15.0	0.33
Cotton	5.06	0.39
Foamglas	10.5	0.40
Glass wool	1.5	0.27
Glass		
Common thermometer	164.0	5.5
Flint	247.0	5.1
Pyrex	140.0	7.56
Gold	1205.0	2028.0
Granite	159.0	15.4
Gypsum, solid	78.0	3.0
Hair felt	13.0	0.26
Ice	57.5[a]	15.6
Iron, cast	442.0	326.0
Lead	710.0	240.0
Leather, sole	54.0	1.1
Lime		
Mortar	106.0	2.42
Slaked	81–87	—
Limestone	132.0	10.8
Marble	162.0	20.6
Mineral wool		
Board	15.0	0.33
Fill-type	9.4	0.27
Nickel	537.0	406.5
Paper	58.0	0.9
Paraffin	55.6	1.68
Plaster		
Cement	73.8	8.0
Gypsum	46.2	3.3
Redwood bark	5.0	0.26
Rock wool	10.0	0.27
Rubber, hard	74.3	11.0
Sand, dry	94.6	2.28
Sandstone	143.0	12.6
Silver	656.0	2905.0
Soil		
Crushed quartz		
(4% moisture)	100.0	11.5
Fairbanks sand		
(4% moisture)	100.0	8.5
(10% moisture)	110.0	15.0
Dakota sandy loam		
(4% moisture)	110.0	6.5
(10% moisture)	110.0	13.0
Healy clay		
(10% moisture)	90.0	5.5
(20% moisture)	100.0	10.0
Steel		
1% C	487.0	310.0
Stainless	515.0	200.0
Tar, bituminous	75.0	—
Water, fresh	62.4	4.1
Wood		
Fir	34.0	0.8
Maple	40.0	1.2
Red oak	48.0	1.1
White pine	31.2	0.78
Wood fiber board	16.9	0.34
Wool	4.99	0.264

Notes: a. At 32°F.

we would simply divide our result by 50 square feet.

E. 4.3 $$q''_c = \frac{q_c}{A} = \frac{k}{t}(T_i - T_o)$$

We can modify Equation 4.2 for a more general expression:

$$q''_c = \frac{2000\ Btu/hr}{50\ ft^2} = 40\ Btu/hr\text{-}ft^2$$

If we had several distinct sections of clay wall exposed to the same temperature difference, we would simply add their areas and multiply by q''_c to obtain the overall heat-loss rate.

Convection

Convection is the transfer of heat by fluids and gases in contact with solid surfaces. We have already mentioned that there are two types of convection: free and forced. In buildings, free convection is of primary importance for interior walls. Forced convection occurs when fluid motion is induced by "external" factors—a fan in a room or a strong wind rushing by an outside wall. Forced convection differs little in principle from free convection; it simply involves higher fluid or gas speeds. A fluid at high speed will cause more heat transfer than the same fluid at a lower speed. This is why you feel colder on a windy day at 20°F than you do on a day with no wind at the same temperature ("chill factor").

The rate of heat transfer due to convection is directly proportional to the temperature difference between the surface and the adjacent fluid ($T_s - T_f$); the heat transfer area (A); and a convective (or "film") coefficient (f), which describes the specific fluid properties and flow characteristics. In equation form, this relationship is

E. 4.4 $$q_{cv} = Af(T_s - T_f)$$

where q_{cv} equals the heat transfer due to convection (Btu/hr); T_f, T_s equal the fluid and surface temperatures (°F); f equals the convective, or film, coefficient (Btu/ft²-hr-°F); and A equals the area (ft²). The direction of heat transfer is determined by the relative size of T_f and T_s. The factor f increases with fluid velocity. For instance, f equals approximately 1 Btu/hr-ft²-°F for free convection of air next to a wall, but can increase to 10 Btu/hr-ft²-°F in very windy conditions.

An area-based heat-transfer rate can be defined for convective situations. The rationale is the same as that for conduction heat transfer:

E. 4.5 $$q''_{cv} = \frac{q_{cv}}{A} = f\ (T_s - T_f)$$

Example: Compute the convective heat loss from a wall whose surface is held at 50°F while exposed to moving air with a temperature of 10°F, given the convective coefficient (f) of 5 Btu/hr-ft²-°F.

Solution: The heat transfer per square foot would be

$$q''_{cv} = \frac{5\ Btu}{hr\text{-}ft^2\text{-}°F}(50 - 10)°F$$

$$= 200\ Btu/hr\text{-}ft^2$$

If we had 50 square feet of wall surface, the total heat-loss rate would be

$$q_{cv} = q''_{cv}\ A = \frac{200\ Btu}{hr\text{-}ft^2}(50\ ft^2)$$

$$= 10{,}000\ Btu/hr$$

Radiation

Radiation is the transfer of heat by electromagnetic waves. Solar energy was already given as an example of radiant energy in transit; hence, we see that radiant energy does not need a medium through which to move. If we have two bodies at different temperatures in a closed, evacuated system (i.e., a vacuum, which eliminates the other transfer modes of conduction and convection), their temperatures will eventually equalize by the exchange of radiant energy. We also mentioned that radiant energy exists at varying electromagnetic wave lengths— some visible, most not. (Visible-light energy is conceptually similar to infrared energy, but infrared wave lengths are not part of the visible spectrum.)

Referring back to Figure 4.8, we can see the spectral character of radiant energy. Incident solar energy is concentrated around the visible wave lengths (0.4 to 0.8 microns). These relatively short wave lengths are characteristic of a high-temperature source—indeed, the sun's effective temperature is about 10,000°F. All objects actually emit radiant energy. Relatively low-temperature objects like your body emit infrared energy of long wave length. This is shown qualitatively in Figure 4.8 by the curve labeled "Radiation Emitted by a Black Body at 95°F." Note the wave lengths are on the order of 10 microns.

If you are getting the feeling that radiant heat transfer can get complicated, you're right. It is very difficult to estimate quantitatively radiation heat-exchange rates. We merely wish to reiterate that your body exchanges radiant energy with surrounding surfaces and objects; and so we must keep all interior wall surfaces at "reasonable" temperatures in order to prevent discomfort, through proper insulating techniques.

On a related note, we might call your attention to a common method for increasing a wall's insulating characteristic, based on radiant heat transfer. You may have seen fiberglass batting or insulating material which has a layer of shiny aluminum foil on one or both sides. The foil reflects the thermal radiation that would normally cross an air gap in a wall. Again, the analysis is hard to treat quantitatively, but this method is effective and you should be aware of its underlying principle. We should note that materials reflect differing amounts of radiation depending on the wave length of radiation. For instance, one might think that white paint would reflect the thermal radiation just as well as aluminum foil. Our reasoning is based on our visual experience: white paint reflects radiation in the visible, or short, wave lengths (light). However, it *absorbs* most of the energy in the thermal, or long, wave lengths! White paint, it turns out, would be a poor substitute for aluminum-foil backing.

Air Exchange

There is another way a house can lose energy. Whenever air leaves a heated space, it takes thermal energy with it. The warm air is replaced by cold air which must be heated in order to maintain a comfortable temperature. This air exchange can take place in two ways—by infiltration and by ventilation. Infiltration is unintentional air exchange which occurs because of various leaks in the house; the shabbier the construction, the greater the infiltration. Large amounts of infiltration take place around the edges and through the joints of windows and doors. Wall-floor joints and corner joints are also potential leakage areas.

Ventilation is the intentional exchange of air to avoid stuffiness and to rid the house of odors. Ventilation requires that our structure "exhale" warm air, and so a loss of energy necessarily results. During the heating season, we seldom worry about ventilation in small buildings because normal infiltration rates have a sufficient ventilating effect. Natural ventilation is deliberately encouraged through open doors or windows during the summer.

Later in this section, we will introduce a way to quantify the heat loss due to infiltration. For now, simply note that common sense and good construction techniques are your best guides for minimizing infiltration losses.

Thermal Resistance and Coefficient of Transmission

In the previous section, we examined the individual heat-loss mechanisms and the equations that govern two of them: conduction and convection. More frequently these mechanisms act in concert, and we need to develop methods to quantify their combined effects. The concepts of *thermal resistance* and *coefficient of transmission* enable us to do this.

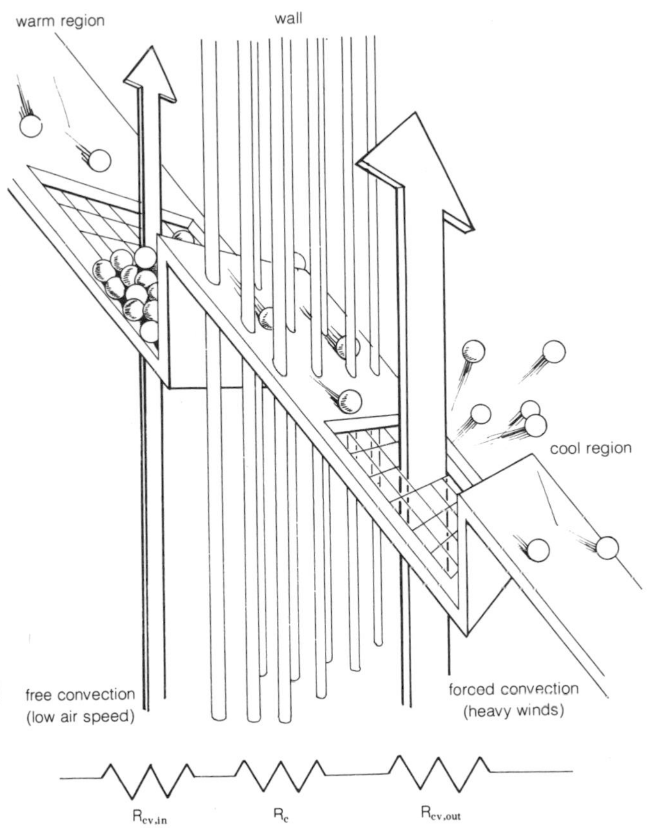

Figure 4.15 The thermal resistance of a simple wall.

Figure 4.15 shows a simple wall subjected to inside and outside convective processes. The wall has an overall thermal resistance due to the combined retarding effects of the wall and convective boundaries. The higher the overall resistance, the lower the heat transfer for an overall temperature difference ($T_i - T_o$). Each mode of heat transfer has an associated expression for thermal resistance. For instance, the wall in Figure 4.15 has two "convective resistances" and one "conductive resistance." We can see the functional form of these resistances in the equations (4.2 and 4.4) introduced earlier. Using expressions for heat flow per square foot, for conduction we find

E. 4.3 $$q''_c = \frac{k}{t}(\Delta T_c)$$

and for convection,

E. 4.5 $$q''_{cv} = f\ (\Delta T_{cv})$$

The ΔT's represent temperature differences over the regions experiencing the conductive or convective heat flow. The temperature differences are analogous to voltages (V) in an electrical circuit. The heat flow (q_c or q_{cv})

is then analogous to the flow of electrical current (I). From Ohm's Law, we know voltage (V), current (I), and resistance (R) are related by $V = IR$ or

E. 4.6 $$I = \left(\frac{1}{R}\right)V$$

By direct analogy to Equation 4.2 and Equation 4.4, we find

E. 4.7 $$q'' = \left(\frac{1}{R_{th}}\right)\Delta T$$

where q'' equals the heat-transfer rate and R_{th} equals thermal resistance. Then, for convection,

E. 4.8 $$R_{th} = R_{cv} = \frac{1}{f}$$

while for conduction,

E. 4.9 $$R_{th} = R_{c} = \frac{t}{k}$$

These resistance expressions make sense if you examine them. First, small values of f, the convective coefficient, yield *large* resistances. Recall that f is small when the air velocity is small, and large when air velocity is large. *Translation*: Hollow walls with drafty air spaces lose much more heat than sealed walls where air flow is restricted. Second, the conduction resistance varies directly with thickness (t) and inversely with thermal conductivity (k). A large thickness combined with a low conductivity yields a very high thermal resistance. *Full translation*: Use insulating materials with low conductivities, and the thicker, the better.

The concept of thermal resistance is particularly useful in computing heat loss over a composite wall, one made up of layers of different materials. In many cases, the same amount of heat must flow through each resistive member, so the situation is similar to an electrical circuit with elements connected in series. Series circuits have the property that the overall resistance is the sum of the individual resistances. For the simple wall in Figure 4.15, the overall resistance can be calculated as follows:

E. 4.10 $$R_{total} = R_{cv,\ out} + R_c + R_{cv,\ in} = \frac{1}{f_{out}} + \frac{t_{wall}}{k_{wall}} + \frac{1}{f_{in}}$$

so that

E. 4.11 $$\frac{q}{A} = q'' = \frac{(T_i - T_o)}{R_{total}} = \frac{(\Delta T)_{overall}}{R_{total}}$$

Note that the heat flows and resistances are based on one square foot of wall surface. The advantage of this method is that it can be used for very complicated wall structures. In these complicated cases, more thermal resistances are involved, but the overall effect is just the sum of the individual components.

We will see shortly that much of this work has already been done for us. ASHRAE publishes complete tables for many types of wall construction. These tables sometimes report overall thermal resistances, but, more frequently, they report the *overall coefficient of transmittance* (U) which is simply the reciprocal of R_{total}:

E. 4.12 $$U = \frac{1}{R_{total}}$$

With this definition, Equation 4.11 becomes

E. 4.13 $$q'' = \frac{q}{A} = U(\Delta T)_{overall}$$

on a square foot basis, and the overall rate of heat transfer is given by

E. 4.14 $$q = UA(\Delta T)_{overall}$$

Note that the convective and conductive subscripts have been dropped, since we are now dealing with combined effects.

If you're the average nontechnical person, you are probably somewhat confused at this point. That's fine—this is the most difficult section in the chapter. Take a break, read the section again, and then proceed to the examples.

Incidentally, you might be asking yourself why we bothered to introduce thermal resistances and conductances if U-values are already available. Good question. The main reason is that you might very well be experimenting with new wall constructions that are not covered by the ASHRAE tables. For instance, many builders are experimenting with expanded polyurethane walls. No U-values are published for these walls, but they are easily computed with Equations 4.10 and 4.12.

Example: How can we figure heat loss across a window?

Solution: Figure 4.16a is a pane of quarter-inch glass. The inside and outside convective coefficients can be taken as 1.4 and 6.0 Btu/hr-ft²-°F, respectively. These f-values are typical, assuming free convection on the inside surface and a 15-mph wind on the outer surface. Referring to Table 4.8, the thermal conductivity of glass is about 5 Btu-in/hr-ft²-°F. The overall resistance is the sum of the three individual resistances (Equation 4.10):

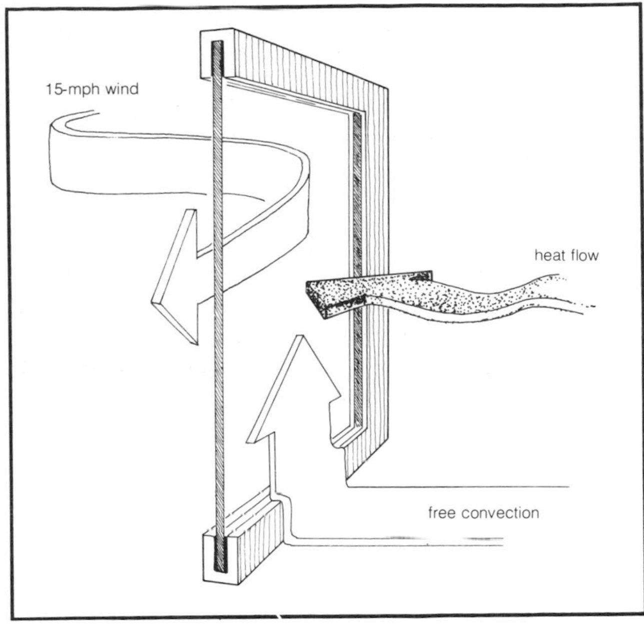

Figure 4.16a A simple window.

$$R_{total} = \frac{1}{6} + \frac{0.25}{5} + \frac{1}{1.4} = 0.93 \frac{\text{hr-ft}^2\text{-°F}}{\text{Btu}}$$

The overall coefficient of transmission from Equation 4.12 is simply

$$U = \frac{1}{R_{total}} = 1.07 \frac{\text{Btu}}{\text{hr-ft}^2\text{-°F}}$$

Note that our result is independent of specific outside or inside temperatures. Also, our calculations are very general in that they are based on one square foot of window pane.

Example: Now let's treat a more specific problem. What would be the total heat transfer for 25 square feet of window surface subjected to a 70°F indoor air temperature and a 30°F outdoor air temperature?

Solution: Using our results in Equation 4.14, we find

$$\begin{aligned} q &= UA(\Delta T) \\ &= \left(\frac{1.07 \text{ Btu}}{\text{hr-ft}^2\text{-°F}}\right)(25\text{ft}^2)\ (70 - 30)\text{°F} \\ &= 1070 \text{ Btu/hr} \end{aligned}$$

Heat Transfer Through a Composite Wall

Next consider the composite wall shown as a cutaway in Figure 4.16b. We will assume that the inside air temperature is 70°F and the outside air temperature is 30°F. Our job is to estimate the rate of heat transfer across the wall.

Table 4.9 Film Coefficients (f) and Resistances (R) for Vertical Surfaces[a]

	Film Coefficient (f)	R[b]
Free Convection	1.46	0.68
7.5 mph	4.00	0.25
15 mph	6.00	0.17

Notes: a. From ASHRAE *Handbook of Fundamentals*.
b. $R = 1/f$; units are hr-ft²-°F/Btu.

Since heat must flow through each element in turn (including the convective resistances), the system can again be represented by a series circuit. Let's start from the inside and move outward. The first resistive element is the convective (or film) resistance. We would expect free convection on this inside wall, so f_i would be on the order of 1 Btu/hr-ft²-°F. More precise values are given in Table 4.9. Using Equation 4.8, the resistance of the element is given by

$$\begin{aligned} R_{cv} &= \frac{1}{f_i} = \frac{1}{1.46} \\ &= 0.68 \frac{\text{hr-ft}^2\text{-°F}}{\text{Btu}} \end{aligned}$$

The half-inch plywood sheet (designated by the number "1") represents a conductive resistance, and the thermal conductivity of plywood can be found in a

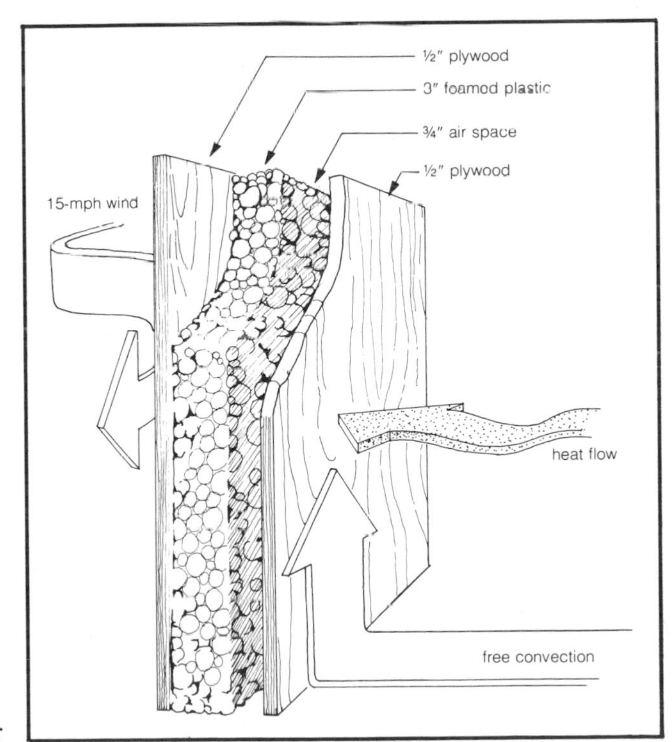

Figure 4.16b A simple composite wall.

Table 4.10 Air-Space Resistances (R) for 50°F Mean Temperature[a]

Position of Air Space	Direction of Heat Flow	Air Space Bounded by Ordinary Materials		Air Space Bounded by Aluminum Foil	
		0.75-inch R	4-inch R	0.75-inch R	4-inch R
Horizontal	Upward	0.78	0.85	1.67	2.06
Horizontal	Downward	1.02	1.23	3.55	8.94
Vertical	Horizontal	0.96	0.94	2.80	2.62

Notes: a. From ASHRAE *Handbook of Fundamentals*; units of hr-ft²-°F/Btu.

number of references. Table 4.8 lists values for wood, but information more pertinent to the construction trade is listed in Appendix 4D. This appendix makes our job particularly easy in that it already includes the thermal resistance as well as the thermal conductance (C). The thermal conductance is related to the thermal resistance in the following way:

E. 4.15 $$C = \frac{1}{R}$$

For half-inch plywood we find (under the section of Appendix 4D labeled "Building Board")

$$R_1 = \frac{1}{C} = 0.62 \frac{hr\text{-}ft^2\text{-}°F}{Btu}$$

Next, we encounter an enclosed air space. In reality, the air space involves two additional convective (or film) resistances and a single term to account for conduction across the air gap. The combined effect of these resistances is tabulated in Table 4.10. For a three-quarter-inch vertical air space with no reflective foil, we find the resistance to be

$$R_2 = 0.96 \frac{hr\text{-}ft^2\ °F}{Btu}$$

Incidentally, did you notice that a foil-lined air space has nearly three times the resistance of an ordinary air space?

The 3-inch foam plastic represents another conductive resistance and the values in Appendix 4D are applicable ("Block and Board Insulation"). We have a number of plastics to choose from, all listed under the subheading "rigid polystyrene." The entries differ in density or manufacturing technique; some are expanded using freon gas, R-11 or R-12, while others simply use air. For our example, let's choose "polyurethane, R-11 exp." Conductivities are given for a wide range of temperatures, but for home insulation, 50°F is reasonable and k equals 0.16 Btu-in/hr-ft²-°F. Using Equation 4.9, we compute the resistance for the 3-inch thickness:

$$R_3 = \frac{t}{k} = \frac{3\ in}{0.16\ Btu\text{-}in/hr\text{-}ft^2\text{-}°F}$$

$$= 18.75 \frac{hr\text{-}ft^2\text{-}°F}{Btu}$$

Note the relatively large size of this thermal resistance; foamed plastic is an excellent insulating material!

The remaining sheet of plywood can be handled in the usual way ($R_4 = R_1 = 0.62$). The outside convective coefficient is obtained assuming a 15-mph wind (Table 4.9) and $R_{cv,\ out}$ comes to 0.17. Total resistance is simply the sum of the individual elements.

$$R_{total} = 0.68 + 0.62 + 0.96 + 18.75 + 0.62 + 0.17$$

$$= 21.8 \frac{hr\text{-}ft^2\text{-}°F}{Btu}$$

The overall coefficient of transmission (Equation 4.12) is simply

$$U = \frac{1}{R_{total}} = 0.046 \frac{Btu}{hr\text{-}ft^2\text{-}°F}$$

The heat transfer per square foot of wall area is given by Equation 4.13:

$$q'' = U(T_i - T_o) = 0.046 \frac{Btu}{hr\text{-}ft^2\text{-}°F} (70\text{—}30°F)$$

$$= 1.83\ Btu/hr\text{-}ft^2$$

To obtain the total heat transfer in, say, a 200-square-foot wall, we simply multiply our result by 200:

$$q = q'' \times 200\ ft^2 = 367\ Btu/hr$$

Simplified Heat-Transfer Calculations

The previous examples illustrate a step-by-step method to compute the total thermal resistance of a wall. The details will be particularly useful to those who are experimenting with new building techniques and materials. In many situations, however, wall construction follows a set pattern, and thermal resistance calculations for these "standard" walls have been computed. The results are reported in terms of the coefficient of transmission (U).

U-values are given for several window configurations in Table 4.11. Note that a single pane with an outdoor (exterior) exposure has a coefficient of 1.13, nearly in agreement with the calculated value in the last section. Also note that double-pane windows (that is, modular

Table 4.11 *U*-Values of Windows, Skylights, and Light-transmitting Partitions[a]

PART A—VERTICAL PANELS (EXTERIOR WINDOWS AND PARTITIONS)—FLAT GLASS, GLASS BLOCK, AND PLASTIC SHEET

Description	Exterior		Interior
	Winter	Summer	
Flat Glass			
single glass	1.13	1.06	0.73
insulating glass—double			
3/16 in. air space	0.69	0.64	0.51
1/4 in. air space	0.65	0.61	0.49
1/2 in. air space	0.58	0.56	0.46
insulating glass—triple			
1/4 in. air spaces	0.47	0.45	0.38
1/2 in. air spaces	0.36	0.35	0.30
storm windows			
1 in.–4 in. air space	0.56	0.54	0.44
Glass Block			
6 × 6 × 4 in. thick	0.60	0.57	0.46
8 × 8 × 4 in. thick	0.56	0.54	0.44
—with cavity divider	0.48	0.46	0.38
12 × 12 × 4 in. thick	0.52	0.50	0.41
—with cavity divider	0.44	0.42	0.36
12 × 12 × 2 in. thick	0.60	0.57	0.46
Single Plastic Sheet.	1.09	1.00	0.70

PART B—HORIZONTAL PANELS (SKYLIGHTS)—FLAT GLASS, GLASS BLOCK, AND PLASTIC BUBBLES

Description	Exterior		Interior
	Winter	Summer	
Flat Glass			
single glass	1.22	0.83	0.96
insulating glass—double —			
3/16 in. air space	0.75	0.49	0.62
1/4 in. air space	0.70	0.46	0.59
1/2 in. air space	0.66	0.44	0.56
Glass Block			
11 × 11 × 3 in. thick with cavity divider	0.53	0.35	0.44
12 × 12 × 4 in. thick with cavity divider	0.51	0.34	0.42
Plastic Bubbles			
single walled	1.15	0.80	—
double walled	0.70	0.46	—

PART C—ADJUSTMENT FACTORS FOR VARIOUS WINDOW TYPES (MULTIPLY U VALUES IN PARTS A AND B BY THESE FACTORS)

Window Description	Single Glass	Double or Triple Glass	Storm Windows
All Glass	1.00	1.00	1.00
Wood Sash—80% Glass	0.90	0.95	0.90
Wood Sash—60% Glass	0.80	0.85	0.80
Metal Sash—80% Glass	1.00	1.20	1.20

Notes: a. From ASHRAE *Handbook of Fundamentals*; units are Btu/hr-ft²-°F.

Table 4.12 *U*-Values of Solid Wood Doors[a]

Thickness[b]	Winter No Storm Door	Winter Storm Door[c] Wood	Winter Storm Door[c] Metal	Summer No Storm Door
1 in.	0.64	0.30	0.39	0.61
1¼ in.	0.55	0.28	0.34	0.53
1½ in.	0.49	0.27	0.33	0.47
2 in.	0.43	0.24	0.29	0.42

Notes: a. Units are Btu/hr-ft²-°F, from ASHRAE *Handbook of Fundamentals*.
b. Nominal thickness.
c. Values for wood storm doors are for approximately 50 percent glass; for metal storm doors values apply for any percent of glass.

windows with two sheets of glass separated by an air space) have a significantly lower *U*-value. Essentially, the thermal resistance of the double-pane window is twice that of a single pane.

Analogous values for solid wooden doors are given in Table 4.12. You can see that the addition of a storm door lowers the overall conductance significantly. Doors of nonstandard construction can be handled with the basic thermal resistance techniques discussed earlier. For instance, a hollow door is treated as a sort of composite wall.

Overall conductances for walls, ceilings, and elevated floors are given in Appendix 4E. Care should be taken to use the proper tables. Each is clearly marked to account for various circumstances. For instance, if we had a masonry wall with 4-inch brick face, 12-inch cinder blocks, and five-eighths-inch plaster on the interior wall, we would make our way to Table VII in Appendix 4E and find our wall as the third listed. Column "B" would correspond to a plaster wall, and the *U*-value would be 0.30 Btu/hr-ft²-°F. Similar searches can account for all exterior facing surfaces.

A correction factor is often applied to the tabulated *U*-values to compensate for additional insulation or added air spaces. These factors are given in Table 4.13. You enter the table with the *U*-value of the uninsulated wall, and read to the right until you reach the appropriate column. Suppose, for example, we added 3 inches of insulation to the masonry wall described in the preceding paragraph. Turning to "Part A" of Table 4.13, we find that 3 inches of insulation added to our original wall reduces the *U*-value from 0.30 to 0.069 Btu/hr-ft²-°F. Thus the thermal resistance is increased by about four times!

In a move to promote energy conservation in buildings, ASHRAE recently proposed guidelines for transmission coefficients. These are summarized in Table 4.14. Separate guidelines have been suggested for roofs and ceilings, floors, and "gross exterior walls." The latter term represents all vertical exterior surfaces, including walls, windows, and doors. It is computed using the following formula:

Table 4.13 Correction of *U*-Values with Addition of Insulation or Air Spaces to Uninsulated Building Sections[a]

U Value Without Added Insulation	*Fibrous Insulation Thickness—Inches*				*One Air Space of Effective Emissivity* E			*Two Air Spaces of Effective Emissivity* E			*Three Air Spaces of Effective Emissivity* E		
	½	1	2	3	0.82[e]	0.20	0.05	0.82	0.20	0.05	0.82	0.20	0.05
Col. 1	2	3	4	5	6	7	8	9	10	11	12	13	14
Part A—Walls													
0.70	0.304	0.194	0.113	0.080	0.752	0.463	0.380	0.437	0.240	0.189	0.298	0.150	0.112
0.60	0.284	0.186	0.110	0.078	0.630	0.412	0.341	0.392	0.225	0.177	0.276	0.144	0.108
0.45	0.246	0.168	0.104	0.075	0.460.	0.331	0.280	0.318	0.195	0.158	0.237	0.130	0.098
0.40	0.230	0.161	0.101	0.074	0.409	0.299	0.258	0.291	0.185	0.149	0.222	0.125	0.095
0.35	0.212	0.152	0.097	0.072	0.354	0.267	0.234	0.262	0.172	0.140	0.205	0.119	0.092
0.30	0.192	0.142	0.093	0.069	0.30	0.234	0.207	0.232	0.158	0.130	0.186	0.112	0.087
0.28	0.184	0.138	0.091	0.068	0.28	0.221	0.196	0.220	0.151	0.125	0.178	0.108	0.084
0.26	0.175	0.133	0.089	0.066	0.26	0.208	0.185	0.207	0.144	0.120	0.169	0.104	0.082
0.24	0.166	0.127	0.087	0.065	0.24	0.194	0.173	0.194	0.137	0.115	0.160	0.100	0.079
0.22	0.156	0.121	0.084	0.064	0.22	0.180	0.161	0.180	0.129	0.110	0.150	0.096	0.076
0.20	0.145	0.115	0.081	0.062	0.20	0.165	0.149	0.166	0.120	0.104	0.140	0.091	0.073
0.18	0.134	0.108	0.078	0.060	0.18	0.150	0.137	0.152	0.112	0.098	0.129	0.086	0.069
0.16	0.123	0.100	0.074	0.057	0.16	0.136	0.124	0.137	0.103	0.090	0.118	0.080	0.065
0.14	0.111	0.092	0.069	0.054	0.14	0.120	0.111	0.122	0.094	0.083	0.106	0.075	0.061
0.12	0.098	0.083	0.064	0.051	0.12	0.105	0.098	0.107	0.084	0.075	0.094	0.068	0.056
0.10	0.085	0.073	0.058	0.047	0.10	0.089	0.084	0.091	0.074	0.066	0.082	0.061	0.051
0.08	0.070	0.062	0.050	0.042	0.08	0.073	0.068	0.074	0.062	0.056	0.068	0.053	0.045
Part B—Floors (Heat Flow Down)													
0.70	0.305	0.195	0.113	0.080	0.70	0.240	0.114	0.377	0.122	0.057	0.262	0.086	0.042
0.60	0.284	0.186	0.110	0.078	0.60	0.236	0.111	0.346	0.118	0.056	0.246	0.084	0.041
0.50	0.260	0.175	0.106	0.076	0.50	0.210	0.106	0.310	0.114	0.055	0.228	0.082	0.041
0.45	0.246	0.168	0.104	0.075	0.45	0.200	0.103	0.290	0.111	0.055	0.217	0.081	0.040
0.40	0.230	0.161	0.101	0.074	0.40	0.189	0.100	0.268	0.107	0.054	0.205	0.079	0.040
0.35	0.212	0.152	0.097	0.072	0.35	0.176	0.096	0.244	0.103	0.052	0.192	0.077	0.040
0.30	0.192	0.142	0.093	0.069	0.30	0.162	0.091	0.219	0.098	0.051	0.175	0.074	0.039
0.28	0.184	0.138	0.091	0.068	0.28	0.156	0.089	0.208	0.096	0.050	0.168	0.073	0.039
0.26	0.175	0.133	0.089	0.066	0.26	0.150	0.087	0.197	0.094	0.049	0.161	0.072	0.038
0.24	0.166	0.127	0.087	0.065	0.24	0.143	0.084	0.185	0.091	0.048	0.153	0.070	0.038
0.22	0.156	0.121	0.084	0.064	0.22	0.136	0.081	0.173	0.088	0.047	0.145	0.068	0.037
0.20	0.145	0.115	0.081	0.062	0.20	0.128	0.078	0.160	0.084	0.046	0.136	0.066	0.036
0.18	0.134	0.108	0.078	0.060	0.18	0.119	0.074	0.148	0.080	0.045	0.126	0.064	0.036
0.16	0.123	0.100	0.074	0.057	0.16	0.109	0.070	0.133	0.076	0.044	0.116	0.061	0.035
0.14	0.111	0.092	0.069	0.054	0.14	0.099	0.065	0.118	0.071	0.042	0.105	0.058	0.034
0.12	0.098	0.083	0.064	0.051	0.12	0.088	0.060	0.103	0.066	0.040	0.094	0.054	0.033
0.10	0.085	0.073	0.058	0.047	0.10	0.076	0.054	0.089	0.058	0.037	0.081	0.049	0.031
0.08	0.070	0.062	0.050	0.042	0.08	0.063	0.047	0.072	0.050	0.033	0.068	0.044	0.028
Part C—Ceilings (Heat Flow Up)													
0.70	0.305	0.195	0.113	0.080	0.690	0.472	0.403	0.427	0.262	0.216	0.307	0.180	0.146
0.60	0.284	0.186	0.110	0.078	0.588	0.417	0.362	0.385	0.244	0.204	0.284	0.171	0.139
0.50	0.260	0.175	0.106	0.076	0.488	0.361	0.318	0.339	0.224	0.189	0.258	0.160	0.131
0.45	0.246	0.168	0.104	0.075	0.438	0.331	0.295	0.316	0.212	0.180	0.243	0.154	0.126
0.40	0.230	0.161	0.101	0.074	0.389	0.300	0.270	0.288	0.199	0.170	0.227	0.146	0.121
0.35	0.212	0.152	0.097	0.072	0.340	0.269	0.244	0.260	0.185	0.158	0.209	0.138	0.115
0.30	0.192	0.142	0.093	0.069	0.292	0.237	0.215	0.230	0.168	0.145	0.189	0.129	0.108
0.28	0.184	0.138	0.091	0.068	0.272	0.224	0.203	0.217	0.161	0.140	0.181	0.125	0.104
0.26	0.175	0.133	0.089	0.066	0.253	0.211	0.191	0.204	0.154	0.134	0.172	0.120	0.101
0.24	0.166	0.127	0.087	0.065	0.234	0.199	0.179	0.191	0.146	0.128	0.163	0.115	0.097
0.22	0.156	0.121	0.084	0.064	0.214	0.186	0.166	0.178	0.137	0.120	0.153	0.109	0.093
0.20	0.145	0.115	0.081	0.062	0.195	0.173	0.154	0.164	0.128	0.114	0.143	0.104	0.088
0.18	0.134	0.108	0.078	0.060	0.176	0.159	0.141	0.150	0.119	0.106	0.132	0.097	0.084
0.16	0.123	0.100	0.074	0.057	0.156	0.146	0.128	0.136	0.109	0.098	0.121	0.090	0.079
0.14	0.111	0.092	0.069	0.054	0.137	0.132	0.115	0.120	0.099	0.090	0.109	0.083	0.073
0.12	0.098	0.083	0.064	0.051	0.118	0.118	0.101	0.105	0.088	0.080	0.096	0.075	0.068
0.10	0.085	0.073	0.058	0.047	0.099	0.105	0.088	0.090	0.076	0.071	0.082	0.067	0.062
0.08	0.070	0.062	0.050	0.042	0.079	0.091	0.074	0.073	0.064	0.061	0.067	0.058	0.056

Notes: a. From ASHRAE *Handbook of Fundamentals*; units are Btu/hr-ft²-°F.

Table 4.14 Maximum Heat Flow at Design Conditions (U_o)[a]

Degree-Days		U-Value
Gross Exterior Walls[b]		
0-999		0.60
1000-2999		0.36
3000-4999		0.30
5000-9999		0.26
10,000 or greater		0.17
Floors		
0-2500	NR[c]	NR[c]
2501-4500	0.13[d]	0.26[e]
4501-8000	0.10[d]	0.20[e]
8000 or greater	0.08[d]	0.20[e]
Opaque Roofs and Ceilings		
0-1000	0.12[f]	0.14[g]
1001-2000	0.08[f]	0.14[g]
2001-5000	0.05[f]	0.09[g]

Notes: a. ASHRAE recommendations from *Proposal 90P*; *U*-values in Btu/hr-ft²-°F.
b. One- and two-family living units.
c. No recommendation.
d. Floors facing unheated spaces.
e. Slabs on grades.
f. Ceiling.
g. Flat deck with rigid roof insulation and exposed structural system.

E. 4.16
$$U'' = \frac{U_{w1}A_{w1} + U_wA_w + U_dA_d}{A_{w1} + A_w + A_d}$$

Here, U_{w1}, U_w, and U_d refer to the *U*-values of the walls, windows, and doors, respectively; the areas are similarly subscripted. U'' is simply a weighted average.

The guidelines become more stringent (as they should be) in regions with high degree-day accumulations. When constructing or re-insulating a home, we suggest these values be used as upper limits. Typically, floor and ceiling guidelines are easy to satisfy. However, the average value of a gross exterior wall is raised substantially by any window area included in the design. You will probably want to consider minimizing window area and/or installing double-pane insulating glass (see Table 4.11).

When you consider insulation for walls, ceilings, or floors, don't be stingy. Insulation is a great investment. It quickly pays for itself in fuel savings and helps make your home more comfortable.

Table 4.15 U^*-Values for Ground-Water Temperatures[a]

T_{gw}	Basement Floor U^*	Subterranean Walls U^*
40°F	3.0	6.0
50°F	2.0	4.0
60°F	1.0	2.0

Notes: a. From ASHRAE *Handbook of Fundamentals*; units are Btu/hr-ft².

Heat Loss from Ground Floors and Underground Walls

Heat loss from basement floors and subterranean walls can be treated with a modified form of the heat-loss equation:

E. 4.17
$$q = U^*A$$

Values of U^* are given for various groundwater temperatures (T_{gw}) in Table 4.15. The inclusion of the water temperature eliminates the need for a temperature-difference term in Equation 4.17. As a result, the units of U^* are different from the other *U*-values in the preceding sections. In most localities, the groundwater temperature remains around 50°F, independent of season. More accurate data can be obtained from your local water utility.

A popular method to compute heat loss from concrete floors at grade (ground) level utilizes the perimeter of the floor as an indicator of the potential heat loss. The reasoning here is that horizontal conduction losses from the inner area of the floor to the outer perimeter can be significantly higher than the losses "straight down" if the edges of the concrete slab are "exposed." (The method *does*, however, include a factor for losses "straight down.") The governing equation is given by

E. 4.18
$$q_{floor} = F_2P_e(T_i - T_o)$$

where P_e is the exposed perimeter of the floor (in feet). A special heat-loss factor is denoted by F_2 and has units of Btu/hr-ft-°F. For this equation to apply, two conditions must be satisfied. First, the floor slab must be properly insulated and waterproofed. Figure 4.17 shows some recommended techniques and materials. Zonolite polystyrene foam is a popular slab insulating material (Grace/Zonolite, 62 Whittemore Avenue, Cambridge, Massachusetts 02140). It should cover the vertical edges of the floor and extend several feet under the edge. The vapor barrier or waterproofing can be either sheets of thin plastic film or a coating which is applied on-site. The UniRoyal Corporation (Mishawaka, Indiana 46544) markets such a coating "system" which combines special sealants and thin plastic membranes. Many other coatings are available, and we suggest you contact a building supply house for details.

As a second condition, the ratio of the slab area to the *exposed* slab perimeter must be *less* than 12 to 1. The edge of the slab is considered to be *exposed* if it is in direct contact with the outside air. Values of F_2 are given in the figure.

Example: Compute the heat loss experienced by a 16-by-20-foot concrete slab edged with 1.5 feet of insulation. The conductance of the insulation is 0.35 Btu/hr-ft²-°F and the temperature differential is 50°F. Only two adjoining sides of the slab are exposed to the outside air.

Solution: Equation 4.18 and the F_2 data in Figure 4.17 are valid if the ratio of the slab area divided by exposed slab perimeter (A/P_e) is less than 12. Two adjoining sides are exposed so that

$$P_e = 16\ ft + 20\ ft = 36\ ft$$

and

$$A = 16\ ft \times 20\ ft = 320\ ft^2$$

so that the ratio is

$$\frac{A}{P_e} = \frac{320}{36} = 8.9$$

which is less than 12. Thus, from Figure 4.17 we can take the value

$$F_2 = 0.61 \frac{Btu}{hr\text{-}ft\text{-}°F}$$

and consequently Equation 4.18 gives us the heat loss from the floor:

$$q = F_2 P_e (T_i - T_o)$$

$$= \left(\frac{0.61\ Btu}{hr\text{-}ft\text{-}°F}\right)(36\ ft)\ (50°F)$$

$$= 1098\ Btu/hr$$

Heat Loss Due to Infiltration

There are two ways to calculate infiltration losses: the air-exchange and the crack-estimation methods. The air-exchange procedure is less accurate and should be used only when the crack-estimation technique is impractical. The *air-exchange equation* is based on the number of total air changes per hour for a dwelling of a given volume. This equation is written as

E. 4.19 $$H = 0.018(T_i - T_o)nV$$

where H equals the heat loss in Btu/hr due to infiltration; n equals the number of air changes per hour; V equals the volume of the dwelling (ft³); T_i equals the inside temperature (°F); T_o equals the outside temperature (°F); and 0.018 is a factor combining the density and heat capacity of air. For a normal residence, n equals 1 air change per hour. It is not so hard to figure the volume of our house; then we can easily determine H.

The *crack-estimation* method requires the determination of the total linear feet of window and door "cracks" present in the dwelling (the perimeters of each door and window). Using a table to find the approximate air flow through these cracks, we can determine our heat

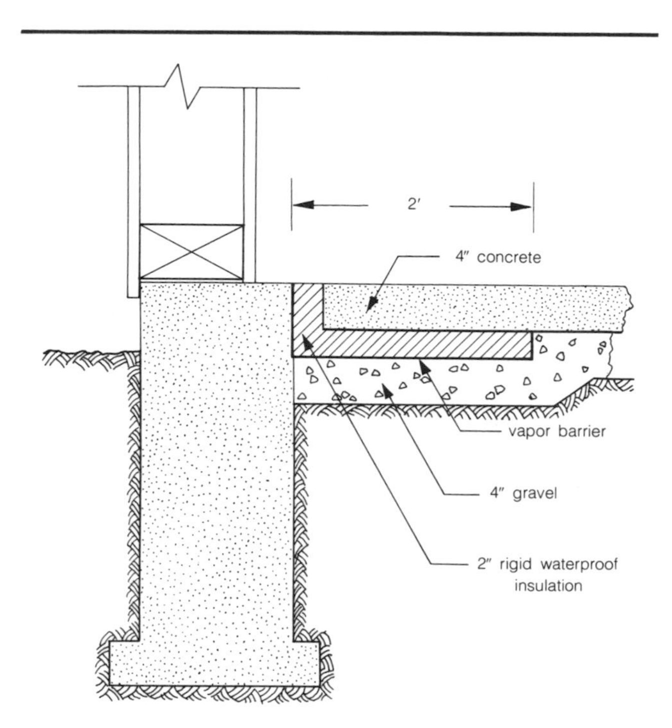

Conductance of Insulation[a]	Total Horizontal and/or Vertical Width of Insulation		
	Values of F_2[b]		
	1 foot	1.5 feet	2 feet
	Unheated Slabs[c]		
0.15	0.29	0.26	0.25
0.20	0.38	0.35	0.33
0.25	0.49	0.44	0.42
0.30	0.58	0.52	0.50
0.35	0.67	0.61	0.59
0.40	0.77	0.70	0.67
	Heated Slabs[c]		
0.15	0.32	0.28	0.27
0.20	0.43	0.39	0.37
0.25	0.57	0.58	0.48
0.30	0.70	0.62	0.59
0.35	0.83	0.75	0.71
0.40	1.00	0.88	0.83

Notes: a. In Btu/hr-ft²-°F.
b. In Btu/hr-ft-°F.
c. In the case of unheated slabs, temperature difference is design indoor minus design outdoor temperature. In the case of heated slabs, temperature difference is the temperature of the heating medium on the slab minus the outside design temperature.

Figure 4.17 Typical floor construction and values of F_2 for use in Equation 4.18.

loss. Because air which enters on one side of the house must push air out from the other side, and since the wind usually blows from one direction at any given time, we normally use only *half* of the total feet of crack length in our calculation. The equation we use is

E. 4.20 $$H = 0.018(T_i - T_o)IL$$

where H equals the heat loss in Btu/hr by infiltration; I equals the tabulated cubic feet per hour of leakage per foot of crack (ft³/hr-ft); and L equals the linear feet of crack (ft). Values of I are given in Table 4.16.

Example: Suppose a building has two 3-by-5-foot weatherstripped, double-hung, wood-sash windows of average construction, and three 2-by-4-foot non-weatherstripped windows of the same construction. The dwelling also has two well-fitted 3-by-7-foot doors on opposite sides of the building. The windows are reasonably (if not evenly) distributed around the house so we can use half of the total crack length. The wind is at 10 mph, the temperature is 70°F inside and 25°F outside. Assuming there are no windbreaks around the building, let's estimate the heat loss by infiltration.

Solution: First, we find the total crack length, or perimeter of the three types of openings and the appropriate values of I to use in Equation 4.20. Adding up the total window and door perimeters, we find the results listed below:

	Feet of Crack	*I (ft³/hr-ft)*
Weatherstripped windows	32	13
Nonweatherstripped windows	36	21
Doors	40	69

The values of I are taken directly from Table 4.16. Note the difference that weatherstripping makes on the windows! We can now compute the heat loss due to infiltration (deleting units to simplify the presentation of the math):

Table 4.16 Infiltration (I) through Cracks of Windows and Doors[a]

Type of Window	*Remarks*	*Wind Velocity, Miles per Hour* 5	10	15	20	25	30
Double-Hung Wood Sash Windows (Unlocked)	Around frame in masonry wall—not calked	3	8	14	20	27	35
	Around frame in masonry wall—calked	1	2	3	4	5	6
	Around frame in wood frame construction	2	6	11	17	23	30
	Total for average window, non-weatherstripped, 1/16-in. crack and 3/64-in. clearance. Includes wood frame leakage	7	21	39	59	80	104
	Ditto, weatherstripped	4	13	24	36	49	63
	Total for poorly fitted window, non-weatherstripped, 3/32-in. crack and 3/32-in. clearance. Includes wood frame leakage	27	69	111	154	199	249
	Ditto, weatherstripped	6	19	34	51	71	92
Double-Hung Metal Windows	Non-weatherstripped, locked	20	45	70	96	125	154
	Non-weatherstripped, unlocked	20	47	74	104	137	170
	Weatherstripped, unlocked	6	19	32	46	60	76
Rolled Section Steel Sash Windows	Industrial pivoted, 1/16-in. crack	52	108	176	244	304	372
	Architectural projected, 1/32-in. crack	15	36	62	86	112	139
	Architectural projected, 3/64-in. crack	20	52	88	116	152	182
	Residential casement, 1/64-in. crack	6	18	33	47	60	74
	Residential casement, 1/32-in. crack	14	32	52	76	100	128
	Heavy casement section, projected, 1/64-in. crack	3	10	18	26	36	48
	Heavy casement section, projected, 1/32-in. crack	8	24	38	54	72	92
Hollow Metal, Vertically Pivoted Window		30	88	145	186	221	242
Doors	Well fitted	27	69	110	154	199	—
	Poorly fitted	54	138	220	308	398	—

Notes: a. Units are ft³/hr-ft; from ASHRAE *Handbook of Fundamentals*.

$$H = H_{windows} + H_{doors}$$

$$= 0.018\ (70 - 25)(13)\frac{32}{2}$$

$$+ 0.018\ (70 - 25)(21)\frac{36}{2}$$

$$+ 0.018\ (70 - 25)(69)\frac{40}{2}$$

$$= 168 + 306 + 1118 = 1592\ Btu/hr$$

The actual infiltration rate, in cubic feet per hour, is easily computed. Using one-half of the crack lengths listed above, we find

E. 4.21 *Infiltration Rate* $= \left(\frac{L}{2}I\right)_{windows} + \left(\frac{L}{2}I\right)_{doors}$

If you are alert, you will recognize that Equations 4.20 and 4.21 can be combined. If you compute the total infiltration rate with Equation 4.21, the heat loss is simply

E. 4.22 $H = 0.018(T_i - T_o)$ *(Infiltration Rate)*

If we had the total heat-transfer loss for the building in our example, we could add that figure to the infiltration loss and find the total heat load for the building (if we had no special heat losses to consider, like a fireplace). Of course, we know that there is some infiltration through the walls, but if we are careful during construction, we needn't calculate these losses. Later, we'll add a safety factor to our heat load to take wall infiltration and other things into account.

Figure 4.18 A dome in Chico, California.

Infiltration can be reduced significantly by sealing and weatherstripping potential leakage areas. Windows and doors are prime targets for this treatment. If at all possible, the builder should use weatherstripped windows. Also, wooden sashes and frames have an advantage over their metal counterparts because of wood's low thermal conductivity. When the window is installed, care should be taken to insulate the void between the frame and the wall and to calk or otherwise seal the gap between outside frame and wall. Similar precautions apply to doors. A thick door insulates better and should be well fitted in its jamb. Remember that some doors are hollow. Unavoidable heat loss results every time we open a door, so in extremely cold environments you might think about double doors (one is always closed, similar to the air locks in spacecraft).

Two Sample Heat-Load Calculations

Thus far we have been concerned with heat loss for specific parts of a dwelling—walls, windows, doors, etc. Now we can reconstruct the house and determine the *overall heat losses*, or *heat load*. When calculating heat losses, we are concerned with *maximum* heating loads, those associated with the most adverse probable climatological conditions of temperature and wind. These conditions occur during the winter season, and usually we calculate heat losses using the *winter design temperature* and the *maximum coincident wind speed*. To determine the appropriate data for temperature and wind conditions, reconsult the climatology section of this chapter.

A Dome

As our first example, let's consider a hypothetical 30-foot diameter, hemispherical dome to be built in Chico, California. The dome is constructed of 4-inch-thick polyurethane and has a wooden floor with a ventilated crawl space. The plans call for 100 square feet of single-pane plastic bubbles, but the builder is willing to consider double-pane bubbles if we can show her the benefits. A single 7-by-4-foot door is used. Figure 4.18 illustrates the proposed design.

As a first step, we define the design conditions. Our client desires a 70°F indoor temperature and Appendix 4A lists temperatures of 23, 29, and 33°F for the median of extremes, 99 percent level, and 97.5 percent level, respectively. A conservative choice might be 25°F, which is also a nice round number! The wind is seen to be very light (*VL*), so we may take 7.5 mph as a reasonable design criterion.

Now we obtain *U*-values for the various structural components. Referring to Table 4.11, the winter *U*-value of a single-walled plastic bubble is 1.15. We will assume that the door is solid wood, 1.5 inches thick, and has no storm door. Referring to Table 4.12, we find a U-value of 0.49. The construction details of the wooden frame floor haven't been specified, but we can obtain representative values from Appendix 4E. Table I is most pertinent, as it deals with frame-construction ceilings and floor. (Often one person's ceiling is another person's floor.) Table I also happens to be the most confusing to read.

The left-hand column in Table I describes the ceiling material (e.g., acoustical tile). The type of floor construction above the ceiling is described in the headings across the top of the table. By now you have undoubtedly guessed that the table is designed for multistory dwellings, and our dome floor is a very special case. Unless we intend to provide acoustical tile for chipmunks living in the crawl space, we should opt for no ceiling material (line 1). The heat flow is downward through the floor in the winter, which limits us to floor designs "H" through "L." Design "L1" looks particularly sound. Successive layers of insulating board, hard board, and linoleum yield a U-value of 0.24.

The floor we have chosen is much like the example pictured at the top of Table I. A 7.5-inch air space is created between the studs, and it wouldn't hurt to add some more insulation. Let's use 3 inches of additional fibrous insulation and revise our U-value using Table 4.13. In "Part B" of the table, we read down the first column to the raw U-value of 0.24. Reading to the right, we find 3 inches of fibrous insulation (column 5) reduces the U-value to 0.065. This corresponds to a three-fold reduction in heat loss.

We are left with the polyurethane walls. Ordinarily, we would check through Appendix 4E to find a wall of similar construction. In this particular example, our search would be futile. Polyurethane walls are relatively new and, in fact, have not yet been approved for construction in many localities. In situations such as these, we must fall back to the elaborate technique outlined in the "Thermal Resistance" section of this chapter. Our situation is much like the simple wall shown in Figure 4.15, but this time the wall is constructed of 4-inch polyurethane. (The hemispherical shape of the wall can be neglected—with such a large radius we can treat it as a flat wall.) For this situation, we use Equation 4.10:

$$R_{total} = R_{cv,\ out} + R_c + R_{cv,\ in}$$

$$= \frac{1}{f_{out}} + \frac{t_{wall}}{k_{wall}} + \frac{1}{f_{in}}$$

The convective resistances are easy enough to handle (using Table 4.9 for our f-values), but we must locate a value of k for polyurethane. If you'll recall, such information appears in Appendix 4D under "Blocks, Boards, and Pipe Insulation." If the foam is expanded with freon (R-11), we look under "polyurethane, R-11 exp" and find k equals 0.16 Btu-in/hr-ft²-°F for 50 to 75°F. (You might wish to use the k-value 0.17, tabulated for 25°F, if you are designing for a very cold climate. However, this is Chico and, as you can see, it *really* doesn't make much difference in the final result since all the values in the table are about the same.) We now compute the overall resistance of the wall:

$$R_{total} = \frac{1}{4.0} + \frac{4}{0.16} + \frac{1}{1.46} = 25.93\ \frac{hr\text{-}ft^2\text{-}°F}{Btu}$$

The U-value is simply the reciprocal:

$$U = \frac{1}{R_{total}} = 0.039\ \frac{Btu}{hr\text{-}ft^2\text{-}°F}$$

Now we need the appropriate areas for each structural member. The windows encompass 100 square feet, and the door is 28 square feet ($7\ ft \times 4\ ft = 28\ ft^2$). The floor has an area equivalent to a circle 30 feet in diameter:

$$A_{floor} = \pi R^2 = \pi(15\ ft)^2 = 706\ ft^2$$

To find the area of the hemispherical wall, we might have to dust off an old geometry book. The surface area of a sphere is $4\pi R^2$, so a hemisphere has an area of $2\pi R^2$. In our case, we must subtract the window and door areas to obtain the true wall area.

$$A_{wall} = 2\pi R^2 - 100\ ft^2 - 28\ ft^2$$

$$= 2\pi(15)^2 - 128$$

$$= 1286\ ft^2$$

Finally, we must estimate the heat loss due to infiltration. We have one door with a perimeter of 22 feet and 100 square feet of plastic bubbles of unknown size. For purposes of analysis, let's assume there are four windows of 25 square feet each. If square, they would each be 5 feet on a side, yielding a total perimeter of 80 feet. The I-values would normally be found in Table 4.16. Our design wind velocity is 7.5 mph, which would require interpolation between the 5- and 10-mph values. But why not use the 10-mph values and be extra safe? There's nothing listed for infiltration around plastic bubble windows, but you will notice values ranging from 2 to 69 ft³/hr-ft for "double-hung, wood sash" windows. We would suggest that 10 ft³/hr-ft is reasonable for a well-installed bubble, and that's the value we'll use. (This is an example of the type of judgments *you* will be making.) We will require that the door be well fitted, which

Table 4.17 Summary of Dome Heat Loss[a]				
	Area (ft²)	*U*-Value	Heat-Loss Rate (Btu/hr)	% Total
Windows	100	1.15	5175	45
Walls	1286	0.039	2257	19
Floor	706	0.065	2065	17
Door	28	0.49	617	6
Infiltration	—	—	1571	13
Total:			11,685 Btu/hr	

Notes: a. Based on (70-25)°F design temperature difference and wind velocity of 7.5 mph.

corresponds to an *I*-value of 69 ft³/hr-ft at 10 mph. Let's use 70 as a round number.

We may now use Equation 4.21 to compute the total infiltration rate. The windows are reasonably distributed around the dome, so we'll use half the window crack length. However, there is only one door so we will use the entire crack length to be on the safe side:

$$\text{Infiltration Rate} = \left(\frac{80}{2}\right)10 + (22)70$$

$$= 1940 \ ft^3/hr$$

$$= 32 \ ft^3/min \ (cfm)$$

You might recall that ventilation requirements are typically 25 cfm per person. The above result emphasizes a common fact: often infiltration provides sufficient ventilation by itself.

We can now summarize the heat loss in Table 4.17. Equation 4.22 allows us to compute the heat loss due to infiltration, and Equation 4.14 will give us the heat-loss rate across the various structural members. Probably the most striking aspect of our result is the percentage of heat loss through the plastic bubbles. A puny 100 square feet of window area accounts for 45 percent of the total heat loss! This is a *classic* situation: in order to reduce heat loss one must either reduce the total window area or revert to insulating windows. For instance, Table 4.11 shows that double-walled plastic bubbles have *U*-values of 0.70. If we employed them, our total heat-loss rate would decrease to 9628 Btu/hr at design conditions—a 17 percent saving!

A Conventional Dwelling

Consider the simple house illustrated in Figure 4.19. It rests on an unheated concrete floor with a completely exposed perimeter, and is of standard stud-wall construction. The wall construction corresponds to wall number "D4" in Table IX (Appendix 4E). Once again, we'll add 3 inches of insulation, which changes the *U*-value from 0.24 to 0.065 (see Table 4.13). Interior walls need not be considered if they merely serve as room dividers. There is no interior ceiling, and the pitched roof is of insulated wooden-rafter construction corresponding to roof number "C4" in Table XI (Appendix 4E). The roof is inclined at 45 degrees. There are solar collectors on the southern half of the roof and normally 2 or 3 inches of insulation is used in collector construction. This insulation should be counted when computing the heat loss from the roof area; the appropriate adjustments could be found in Table 4.13 (*U*-value corrections). For this example, let's *assume* that the corrected *U*-value has been found to be 0.070, and that the northern half of the roof has been insulated to achieve the same effect. Positions and sizes of windows and doors are indicated in the figure.

We wish to find the heating load of this dwelling. For design conditions, we'll assume an average wind speed of 15 mph, an inside temperature of 65°F, and an outside

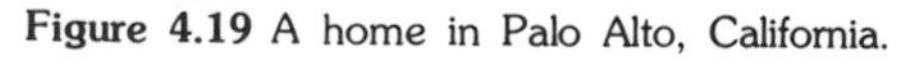
Figure 4.19 A home in Palo Alto, California.

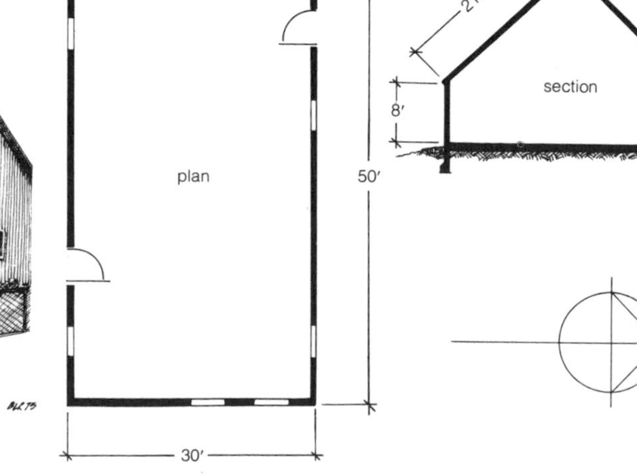

design temperature of 30°F.

We have five separate calculations to make, one each for walls, roof, doors, windows, and floor. The actual wall area can be calculated as follows: For the north and south faces we have

$$A_{NS} = wall\ area - door\ area - window\ area$$

$$= 2(8 \times 50) - 2(3 \times 7) - 4(2 \times 5)$$

$$= 718\ ft^2$$

The east and west faces must include the triangular gables (area for a triangle is one-half the base times the height—$A = bh/2$):

$$A_{EW} = wall\ area + gable\ area$$

$$- door\ area - window\ area$$

$$= 2(8 \times 30) + 2\left(\frac{30 \times 15}{2}\right)$$

$$- 0 - 4(2 \times 5)$$

$$= 890\ ft^2$$

The total actual wall area is then 1608 square feet (890 + 718).

The roof, consisting of two rectangular surfaces, has an area of 2120 square feet (2 × 21.2 × 50). The two doors (which are to be 2 inches thick) have a total area of 42 square feet (2 × 3 × 7). Eight 2-by-5-foot windows encompass 80 square feet (and they will be of double-pane construction).

The concrete floor must be handled with Equation 4.18:

$$q_{floor} = F_2 P_e (T_i - T_o)$$

Table 4.18 Heat-Loss Summary for Conventional House in Figure 4.19[a]

	Area (ft²)	*U*-Value	Heat-Loss Rate (Btu/hr)	% Total
Windows	80	0.650	1820	11
Walls	1608	0.065	3658	23
Roof	2120	0.070	5194	33
Doors	42	0.430	632	4
Floor	1500	—	1400	9
Infiltration	—	—	3220	20
Total:			15,924 Btu/hr	

Notes: a. Based on (65-30)°F design temperature difference.

Let us say we are using the highest resistance (lowest conductance) insulating material in a two-foot width; we find that F_2 equals 0.25 (see table of Figure 4.17). The exposed perimeter of the concrete slab is 160 feet and the area (50 × 30) is 1500 square feet. The ratio of slab area to exposed perimeter (1500/160) is 9.4; remember, this ratio must be less than 12 for us to use Equation 4.18. The heat loss from the floor is then

$$q_{floor} = \left(\frac{0.25\ Btu}{hr\text{-}ft\text{-}°F}\right)(160\ ft)\ (65 - 30)°F$$

$$= 1400\ Btu/hr$$

Infiltration losses are handled with the crack-estimation method. By simple arithmetic, the total crack lengths for the eight windows and two doors are 112 and 40 feet, respectively. The windows and doors are equally distributed around the house, so these values will be halved. The appropriate value of I for a well-fitted door at 15-mph wind is 110 ft³/hr-ft (see Table 4.16). The double-paned windows are undoubtedly well fitted, so we might model them as "rolled section steel sash windows: residential casement," as listed in Table 4.16. The corresponding I-value for 15-mph wind is 52 ft³/hr-ft. From Equation 4.21, the total infiltration rate is then

$$Infiltration\ Rate = \left(\frac{112}{2}\right)52 + \left(\frac{40}{2}\right)110$$

$$= 5112\ ft^3/hr$$

$$= 85\ cfm$$

Once again, ventilation requirements will probably be satisfied by infiltration. The heat loss associated with the infiltration is given by Equation 4.22:

$$H = 0.018(65 - 30)°F \times 5112\ ft^3/hr$$

$$= 3220\ Btu/hr$$

Now we look up the U-values for our 2-inch doors in Table 4.12 ($U = 0.43$) and for our double-pane windows (0.25-inch space; $U = 0.65$) in Table 4.11; and suddenly, using Equation 4.14 to equate our areas and U-values, we are in a position to summarize the heat loss for the house. These results are found in Table 4.18. Our example is quite simplified, but if you understand where all the entries are found, you should be able to handle almost any residential calculation.

It is often good practice to add a "safety factor" of around 10 percent to the computed heat load. This accounts for incalculable losses and unforeseen circumstances (e.g., outside temperatures lower than the design temperature). Thus, we might revise our last figure of 15,924 to 17,500 Btu/hr. Our heating system (perhaps with a supplementary auxiliary system) should be designed to handle this peak heating load, assuming that 30°F is our winter design temperature.

Seasonal Heating Requirements

Seasonal energy requirements can be computed with the heating-load result (the figure immediately above) combined with degree-day data from the climatology section. You'll recall that a degree-day accrues for every degree the outside temperature is below 65°F during a 24-hour period. To find our heating requirements for any period of time, we need only know the number of degree-days in that period and our heating load in Btu per hour. Then we use the following equations:

E. 4.23 $$S = 24Q(\text{Degree-Days})$$

where S equals the season heating requirement (Btu); Q equals the heat loss per degree of temperature difference for a dwelling (Btu/hr-°F); 24 equals the hours in a day; and Q is defined as follows:

E. 4.24 $$Q = \frac{\text{Total Heat Load}}{T_i - T_o}$$

Let's find the yearly heating requirement for the house in our last example, supposing it were sited in Palo Alto, California. Palo Alto has roughly 3000 degree-days per year (check Appendix 4B for your area), and T_o, the design temperature in the last example, is 30°F. First, we find Q:

$$Q = \frac{17{,}500 \text{ Btu/hr}}{(65 - 30)°F} = 500 \text{ Btu/hr-°F}$$

Then, using Equation 4.23, we find that

$$S = 24(500)(3000) = 36{,}000{,}000 \text{ Btu}$$

You should remember that, in this calculation, the quantity $(T_i - T_o)$ in Equation 4.24 must be the *same* temperature differential that was used in computing the heat load.

This last manipulation may appear to be a step backwards. First we multiply by the design temperature differences, and now we are dividing by the same quantity! Q, the heat loss per degree of temperature difference, is independent of the design conditions selected. In contrast, the design heat loss (q) is very much dependent upon the design temperatures. The latter value is used primarily to size the heating components so that they can handle the coldest day that can reasonably be expected. We will see much more of Q and S in our solar heating calculations.

You also might have noticed a potential discrepancy between the calculation of the seasonal heat loss, S, and the way in which degree-days are defined. You will recall that degree-days accrue only when temperatures dip below 65°F. The implication is that heating will be required only when the outside temperatures are below 65°F. This is a good approximation for indoor temperatures in the range of 65 to 70°F. In this range, "natural" energy sources in the dwelling will carry the heat load: metabolic heat production, energy from lights, cooking, and general appliance use. If, for any reason, you need higher indoor temperature (e.g., for tropical plants), you will have to adjust the published degree-days values to compensate for the higher indoor temperature. As a rule of thumb, you should increase the degree-day figure for any winter month by 30 degree-days for each degree over a 70°F indoor design temperature. For instance, if you design for a 78°F indoor temperature, increase the monthly degree-day figure (8×30) by 240 degree-days. For warmer months, in which the outdoor temperature remains above 65°F for a significant amount of time, the correction factor is less obvious and we recommend that you use the degree-day figure with no corrections. (In reality, you'll seldom design for temperatures in excess of 70°F and so our discussion is largely academic; relax.)

We can utilize the seasonal heating requirement to determine the economic feasibility of a particular solar house design. For example, using the above figure of approximately 36 million Btu per year in terms of the present cost of electricity (about $6 per million Btu), we predict a savings of $216 per year with solar heating. Over a fifteen-year period, fuel savings would be about $3200, assuming a constant fuel price. This is enough money to build a suitable collector system. Some conventional forms of energy (such as natural gas) are much cheaper than electricity; however, they are rapidly becoming more expensive. Most people consider solar heating practical at the present time, provided one has the money for the initial capital outlay. We can also use the heating figures to determine fuel costs for any auxiliary heating systems we might employ. To compute the cost of this supplementary energy (excluding capitalization of the equipment), we need know only the *percentage* of the heat load handled by the auxiliary system and the cost of the fuel.

You can obtain local fuel prices from your utility company or compute them yourself using your monthly energy bill. Electrical consumption is reported in kilowatt-hours (KWH), where one KWH is 3413 Btu. Gas consumption is tallied in "therms," where a therm is

100,000 Btu. Simple algebra will yield your dollar-per-Btu costs.

As an example, suppose you were billed $15.00 for the use of 90 therms of gas. Ninety therms corresponds to 90 times 100,000, or 9 million Btu: the cost would be $1.67 per million Btu. Or, suppose your electric bill was $25.00 and your consumption was 1200 KWH (4.1 million Btu). Dividing the dollar figure by the Btu value (25/4.1), we find a rate of $6.10 per million Btu.

Some Concluding Thoughts

By now you should have a good feel for heat-loss calculations. Your expertise can be used to size *any* heating system, not only solar systems. With regard to minimizing heat loss, let common sense be your guide. We would, however, like to make a few closing comments.

As you know, a poorly insulated house will require large amounts of energy to maintain a given environment. Many builders choose this option to save a few dollars on insulation; the owner is then saddled with horrendous fuel bills for the life of the home. Aside from fuel costs, these poorly insulated homes can be very uncomfortable. One subtle effect is related to the mean radiant temperature of the room. As you recall, your body can transfer significant amounts of energy by radiation. A poorly insulated house has cold interior walls and your body continually loses energy to these surfaces. To compensate, the room air temperature must be raised significantly. We've often seen temperatures close to 80°F required before the occupants felt comfortable. The higher air temperature results in more heat loss and even higher fuel bills! Even then, conditions are far from satisfactory; the air is drier and one often feels stifled. Remember: insulate to save energy *and* be comfortable.

Even in a home with good wall insulation, windows pose a problem for much the same reasons outlined above. They not only permit the escape of great amounts of energy, but they create low mean radiant temperatures. Architects usually compensate by keeping furniture away from the windows or placing a heating unit just below the window surface. The heater blankets the window with warm air, increasing the temperature of the inside surface. Of course, this compensation costs a lot in the way of energy.

Double-pane windows alleviate this situation significantly, since they have twice the thermal resistance of single-pane windows. Unfortunately, the cost is also about double that of single-pane glass—which is expensive if you've priced glass lately. Nonetheless, it's worth serious consideration. Any building-goods outlet can direct you to a supplier. For example, one company in San Francisco, Weathermaster, Inc., (255 Channel Street, San Francisco, California 94107) specializes in replacement units for existing single-pane windows. Special molding which minimizes infiltration is included in the design. PPG Industries also markets double-pane glass under the trademark "Twindow." More complete listings of double-pane window and skylight manufacturers can be found in Sweet's *Architectural Catalog*. (This multivolume set can be found in any major library; it contains every conceivable component used in building construction. We *highly* recommend this reference.)

We might say a few words about *quality* construction. Insulation will be purposeless if your house is carelessly constructed. Resulting infiltration losses can ruin the best-laid plans; window and door frames are classic culprits. If you are not absolutely certain of construction techniques, you might pick up a copy of *The Home Building Book* (D. Brown) or *Your Engineered House* (R. Roberts). Both cover all the technical details of house construction.

If you should run into a special heat-loss problem not covered here, we recommend the ASHRAE *Handbook of Fundamentals* or any of the heating and ventilating texts listed in the Bibliography. You now should have sufficient background to understand and apply anything they present.

Summary of Methodology

We will attempt to summarize the essential heat-loss information in a more condensed, usable form.

Calculating Heat Loss

1. *Determine design criteria*: Determine the type of construction to be used in the house. Determine the design temperatures, inside and outside (T_i and T_o). Determine the design wind speed (usually 15 mph).
2. *Calculate heat-transfer losses*: Find areas and U-values, and use Equation 4.14 to compute heat loss from walls, roof, doors, and windows. Find floor perimeter, the factor F_2, and then use Equation 4.18. For unoccupied ventilated attics, ignore the uppermost roof structure and compute the heat loss through the ceiling only. Assume the attic air temperature is equal to the outside design temperature.
3. *Calculate infiltration losses*: Use the crack estimation method and Equations 4.20, 4.21, and 4.22. For each kind of window and each variety of door, find the total feet of crack and the appropriate I-value (Table 4.16). Usually the total feet of crack is halved for this calculation. Add up all the separate infiltration losses to find H_{total}.
4. *Find the heat load*: Add q_{total} and H_{total} to find the raw (unadjusted) heat load. Add a safety factor of about 10 percent to the raw heat load to find the

total adjusted heat load (or demand). If desired, use degree-day information and Equations 4.23 and 4.24 to determine the energy requirements for space heating.

Constructional Considerations

1. Don't skimp on insulation—fill all appropriate constructional voids.
2. Seal around windows and doors—try to use weatherstripped units. Seal walls where necessary. Draperies help reduce window losses.
3. Minimize window area and the number of doors in the house.
4. Double-paned windows cut heat losses in half. Storm doors can also cut heat losses. Both are good investments in colder climates.
5. By fitting south-facing windows with overhangs, we can trap solar energy during the winter, when the sun is low in the sky, and also keep out the high, hot summer sun.
6. Underground or subsurface construction has a heat-loss advantage, because ground temperatures below the freeze line are higher than ambient air temperatures. Disadvantages are higher cost, more construction labor, and the necessity of a stronger constructional design.
7. It is nearly impossible to *over*-insulate, because excess insulation costs (relatively) so little money.
8. Avoid large metal fasteners which extend completely through a wall. Because of their low k-value, they are heat-transfer "freeways."
9. Externally exposed flooring edges can be sources of substantial heat loss, so insulate the flooring perimeter.
10. Don't neglect ceiling (or roof) insulation; warm air rises, and the temperature differential will be greater near the ceiling, which means more potential heat loss.
11. Many insulations should not be packed, because packing destroys dead air space. Follow directions when you install insulation.

Site and Orientation Considerations

1. Minimize wind and shading during the heating season. Wind tends to reduce insulation effectiveness, and shade unnecessarily lowers the outside temperature, increasing temperature differential and hence heat load. Utilize natural windbreaks and deciduous trees.
2. To reduce infiltration losses, don't place large glass areas or poorly fitted doors on the windward side of the building.
3. Try to maximize the insolation available to your home by judicious orientation and location.

SOLAR HEATING SYSTEMS

At this point, you should have an idea of the indoor environment you desire and also the ability to compute the energy you need to maintain it. Now we can discuss the application of solar energy to space heating.

A solar heating system must perform four important functions: energy collection; storage; circulation and distribution; and control (see Figure 4.20). *Collection* is the process of absorbing the incident solar radiation. There are a number of ways that this absorption can be accomplished, but we will concentrate on devices which convert the radiant energy directly to thermal energy forms (heat). Such *heliothermal* collectors fall into two main categories—flat and focused. *Flat collectors* make no attempt to concentrate the incoming solar rays; they can take the form of *collector plates* or massive structural members integral to the dwelling. *Focused collectors* are able to achieve higher operating temperatures due to optical concentration of the incoming rays. Both types often rely upon the *greenhouse effect*—that is, the capture of solar energy by some sort of glazing material—to enhance their collection efficiencies.

The intermittent nature of solar energy demands that some form of thermal energy *storage* be considered. This typically involves materials which are able to store substantial amounts of energy in a reasonably small space. Water, rocks, concrete, and certain chemical salts all have certain advantages (and disadvantages) for this purpose.

The *circulation and distribution system* transfers energy among the collectors, storage system, and house interior, as conditions warrant. During the day, for example, energy must be removed from the collectors and

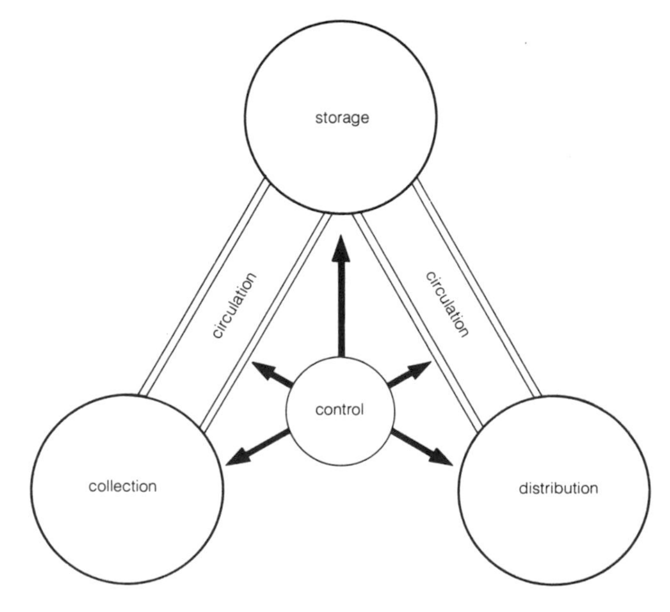

Figure 4.20 The basic elements of a solar heating system.

transferred both to the storage device and to the interior environment. However, in the evening or during cloudy days, the collectors must be isolated from the overall system. Energy transfer then occurs between the storage system and the house interior.

As you might imagine, even simple circulation schemes require some measure of *control*. A typical control system would employ valves, pumps, thermostats, and/or simple manual adjustments, striving for a harmonious interaction among all components of the heating system.

Solar heating systems can be divided into two broad categories. *Component systems* combine separate entities (e.g., collectors, storage units, circulation schemes, and control mechanisms) to form a complete, functioning system. *Integral systems* depend upon a single structural unit for the functions of collection, circulation, and storage. A thick concrete or water-filled southern wall, with appropriate solar design, would fall into the latter category. This chapter is somewhat biased toward the component approach, although detailed descriptions of integral systems are presented. The emphasis reflects the relative difficulty of predicting the thermal performance of integral designs.

Flat Plate Collectors

Flat plate collectors are stationary, intermediate-temperature (usually around 140°F), fluid-flow systems which can collect from 40 to 70 percent of the incident solar energy. They are probably the most common solar collection devices for space and water heating. Collectors are normally designed for rooftop installation and require about one-half to one-and-a-half times the available roof area to heat a home adequately, depending upon house design and climate. The flat plate operates best when exposed to direct sunshine, but, unlike focusing collectors, it also operates well with diffused (indirect) radiation and under hazy conditions.

Virtually all flat plates have the same general design features, depicted in Figure 4.21. The construction is "sandwich style." The bottom layer is a stiff backing material, often plywood, well insulated with about 3 inches of fiberglass. In new buildings, the collectors can be integrated with the normal wall or roof construction to save materials. For instance, if the collectors are integral to the roof, the insulation retards heat loss from both the collectors *and* the roof area. Similarly, when installing collectors on existing roofs, don't forget to adjust the U-value to account for the added insulation (see Table 4.13).

The next layer is the collector itself—a sheet of metal or plastic which absorbs the incident solar energy on its surface. A portion of this energy travels by conduction to the circulating fluid, which flows in passageways integral to the collector. Aluminum, copper, and thin galvanized steel are acceptable collector materials. The key selection criterion is the material's ability to transfer the energy absorbed on the surface to the circulating fluid. This means that the collector materials must have a high thermal conductivity and that the average distance the energy must travel to reach the circulating fluid must be kept to a minimum. To express these two requirements quantitatively, we resurrect the concept of thermal resistance (Equation 4.9) from our discussion of heat loss:

$$R = \frac{t}{k}$$

You will recall that R is thermal resistance, t is the thickness of the material (or, in this case, the length of the conduction path), and k is its thermal conductivity. Low values of thermal resistance, which are desirable in collector design, occur when t is small and/or thermal conductivity (k) is large. Referring back to Table 4.8, we see that copper and aluminum are two materials with high thermal conductivities, and thus make good collectors. Materials with lower k-values (e.g., steel, plastics) can also be used successfully, provided that conduction paths (t) are especially short.

We'll call the circulating fluid (very often water, with some additives occasionally mixed in) the third layer up, although the exact location of the fluid varies according to individual collector design. For example, referring to Figure 4.22, you can see that in collector "a" (integral construction), the water flows through pipes extending equally below (into the insulation) and above (into the air space, described later) the collector plate surface. In

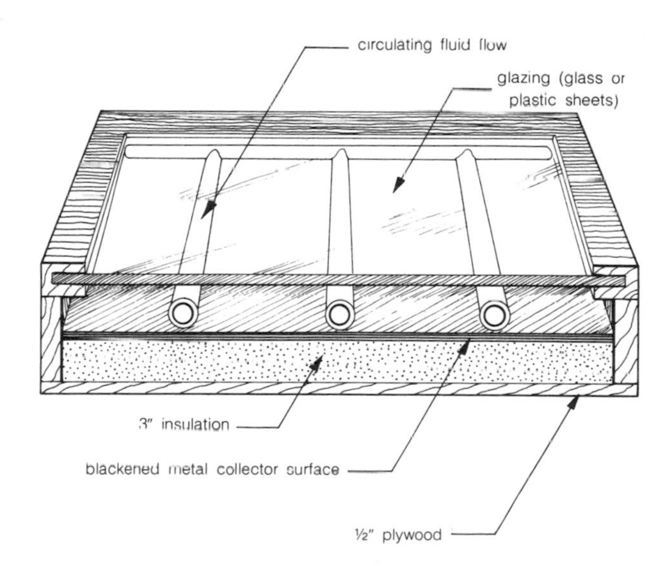

Figure 4.21 A typical flat plate collector.

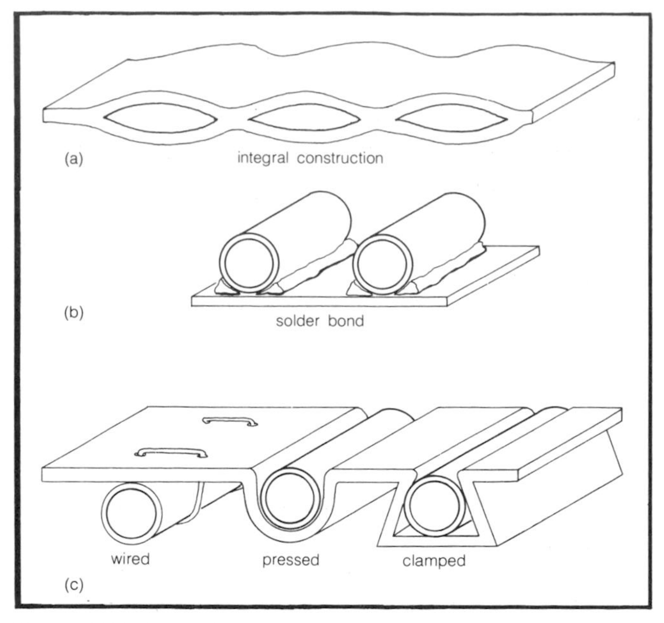

Figure 4.22 Various bonding schemes.

collector "b," the pipes lie on and extend above the collector plate; in collector "c," they are recessed below the plate surface.

The next "layer" is the upper surface of the collector plate. This surface is usually coated with a material which enhances the absorption process. Black or dark-colored paints with suitable temperature capacities are often used. Infrequently, exotic "selective" coatings are employed which discriminately absorb solar radiation and minimize losses by reradiation.

The uppermost layer of the sandwich consists of one or more sheets of glazing—glass or rigid, specially fabricated plastic. Between the glazing and the collector is an air space which can vary in thickness. The glazing and air space(s) jointly give rise to the greenhouse effect. You've probably noticed this phenomenon if you've ever jumped into an automobile that has been sitting in the hot sun with its windows closed. Simply stated, the windows permit the inward passage of solar radiation, but don't allow long-wave, outbound radiation to escape.

The energy we receive from the sun is in the form of "high-quality" radiation of short wave length (some of which is visible to the eye). Materials chosen for collector glazing are nearly transparent to this short-wave radiation; almost all of it passes through. For example, glass—probably the best glazing material—transmits about 90 percent of the incident radiation. Of the remaining 10 percent, half is reflected and half absorbed by the glass itself. The radiation transmitted through the glazing is absorbed by the metal collector and partially transferred to the circulating fluid. The metal collector plate necessarily becomes warm and energy is reradiated from the plate outward. This outbound radiation is of a longer wave length (in the infrared portions of the spectrum) to which the glazing is "opaque." Thus the radiant energy is trapped between the glazing and the collector plate. To summarize, we use the greenhouse effect to increase collector efficiency by trapping radiation which would otherwise be reradiated and lost.

It should be noted that plastics, while tempting choices for glazing materials, are often unsuitable because of transmission bands in the infrared region. Mylar and Tedlar can be classified as marginal in this regard; a good portion of the reradiated thermal energy will pass through these materials. However, some fiberglass-based materials are now entering the market with optical properties quite similar to those of glass. The Kal-Lite division of Kalwall Corporation (1111 Candia Road, Manchester, New Hampshire 03103) markets such a material under the trade name of "Sunlite." This material is significantly lighter and cheaper than glass (about 30 cents per square foot) and has nearly identical transmission characteristics.

The air space between the glazing and the collector is a crucial element, but the collector actually is rather insensitive to the thickness of this space. The minimum gap should be on the order of an inch. Larger gaps are acceptable but have a limited effect (refer back to Table 4.10, where typical air-space resistances are listed). Note that the thermal resistances for a 0.75-inch and a 4-inch gap differ only slightly. But the thermal resistance can be increased significantly if *successive* layers of air and glazing are used. This technique, however, has some practical limitations, cost among them.

The air space within the collector assembly should be reasonably well sealed to minimize convective heat losses. Sealing also reduces the amount of dust that can infiltrate the collector assembly and adversely affect performance. (On the other hand, small accumulations of dust on the exterior surface of the glazing have been shown to have a minimal effect on collector performance.) An absolute seal is *not* advisable, however, since heating and cooling of the air requires some expansion space. Finally, many designs include desiccants to eliminate moisture problems.

Some Construction Details

Now that we have a general idea how flat plate collectors work, we can look at the design aspects in more detail. Hopefully, we can alert you to some common pitfalls which lurk for unwary collector designers and also help you evaluate commercially available units.

Flat collector plates can be classified according to the flow configuration of the circulating fluid. A typical *series* configuration uses a length of tubing, with a single inflow and outflow point, which is curved a number of times across the collector face, like human intestines. A typical *parallel* configuration diverts the inflow into many separate flow paths. The flow recombines at the outflow end

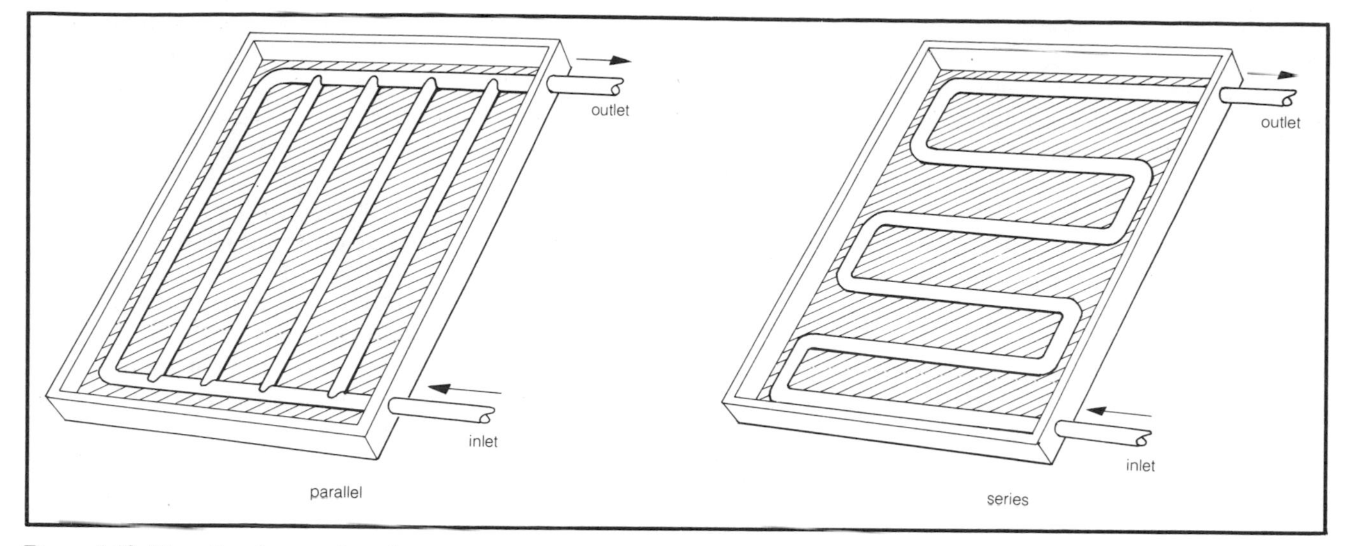

Figure 4.23 Alternative flow configurations.

of the collector. Figure 4.23 illustrates the two flow configurations.

Probably the most common type of series solar collector employs metal tubes bonded in some way to the collector plate. Tubing materials (in order of decreasing cost and thermal conductivity) include copper, aluminum, and steel. The tubes are spaced about 2 to 8 inches apart; closer packing in metal systems is much more expensive without a significant increase in performance. Since the tube surface occupies a relatively small portion of the total collector area, a considerable amount of heat must be transferred to the tubes via conduction along the collector plate. It is therefore critical that the tubes be in good thermal contact with the collector. The thermal resistance of a secure bond can be hundreds of times less than a badly soldered or poorly clamped joint. Experimentation has shown that if good thermal contact is made, steel pipes perform nearly as well as copper at a much lower cost.

Figure 4.22 illustrates various schemes for bonding the tubing to the collector. In collector "a" the tubes are constructed integrally with the backing. This is the most costly (but most effective) means of bonding. Collector "b" shows the most common bonding method, in which the tube is soldered or similarly connected to the collector. Collector "c" shows various other connection methods. Strapping is easy, but it usually results in poor thermal contact. Pressing and clamping both require considerable working of the collector surface. One series scheme that eliminates the bonding problem simply dribbles water through channels attached to the collector plate, shown in Figure 4.24.

An important advantage of a series system is ease of inflow and outflow *manifolding*. A manifold is simply a connecting pipe (or pipes) which distributes the flow from a single input or output line into the various flow paths across the collector. Since a series system has only one collector flow path, the manifolding is particularly simple. The disadvantages of a series system include high tubing costs and bonding problems.

In *parallel* systems, the input is diverted into many paths. Each path makes one pass over the collector plate and must be recollected at the output end. Parallel systems can be built using the bonded-tube construction characteristic of series systems. An advantage of parallel-tube construction over series-tube construction is that pressure loss (from friction) within the pipe is reduced because of the shorter path between inflow and outflow. In theory, a smaller pressure loss is desirable, since this factor determines the pump power required to circulate the collector fluid. However, solar systems have such low circulation rates that frictional effects seldom become dominant design criteria. Figure 4.25 illustrates some parallel-flow design cross sections using corrugated steel or aluminum roofing material. In example "a," two

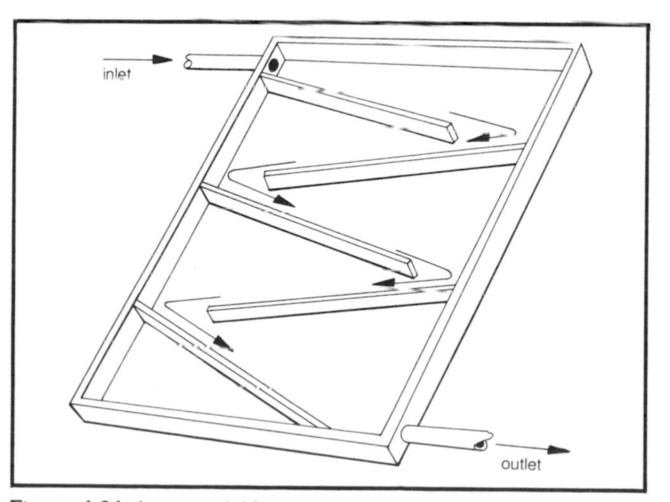

Figure 4.24 A series dribble system.

opposing sheets of corrugated steel are tack-welded together. In example "b," the corrugated-steel sections are offset slightly. The amount of offset and fluid flow rate are matched so the fluid will contact the upper collector surface and thereby increase heat transfer. In example "c," sheet metal and corrugated steel are tack-welded.

The major problem encountered with parallel-flow systems arises when manifolding the input/output flow streams. Each additional flow passage requires one more leakproof manifold connection. So it is probably advantageous to use a series configuration if water is circulated. Air systems lend themselves to parallel flow because leaking manifolds do not pose a critical problem.

To reiterate, the collector plate must exhibit a low thermal resistance to heat flow and should be carefully constructed to minimize leakage. In practice, the second criterion is most difficult to meet. Each and every connection must be carefully soldered or clamped with hose material. The situation is aggravated by the daily stressing

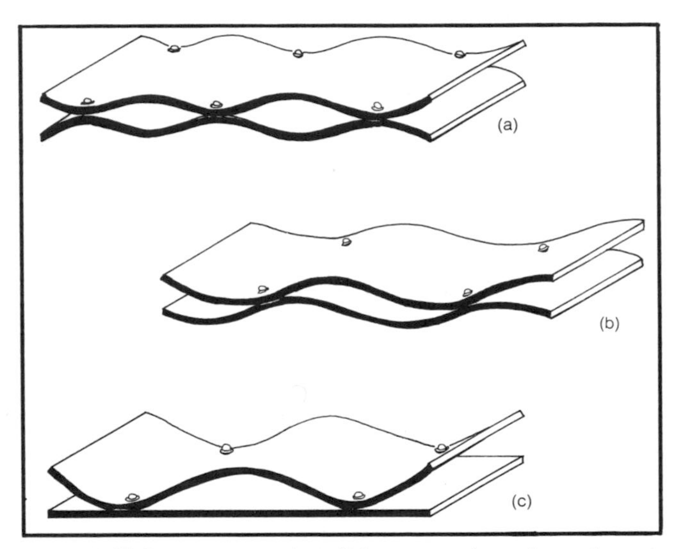

Figure 4.25 Cross sections of parallel systems utilizing flat and corrugated metal sheets.

of the joints due to thermal expansion—the stretch of metal components as they heat up. All joints and pipe runs should be designed with this expansion problem in mind, since an unyielding system will eventually develop cracks.

One of the best sources we've seen for metalworking details is a pamphlet entitled "How to Design and Build a Solar Swimming-Pool Heater," by Francis de Winter. Copies can be obtained from the Copper Development Association, (405 Lexington Avenue, New York, New York 10017). As the title and source imply, the emphasis is on the design of swimming-pool heaters using copper. However, most of the material is applicable to any type of metal collector. (De Winter also includes an excellent discussion on the physics of the collector, detailed economic analyses, and suggested collector designs—right down to the amount of solder required per joint!) Another excellent publication, entitled the *Copper Tube Handbook*, is also available from the Copper Development Association. Step-by-step soldering and pipe installation techniques are presented in a series of photographs.

The surface of the collector plate should be coated with a material which strongly absorbs energy in the solar wave lengths. Flat black or dark-colored paints perform admirably, provided they can withstand the 200°F temperatures which are often experienced by the plate. A paint store can provide you with the temperature tolerances of various paints, or you might investigate the spray paints used on automobile engine blocks. These paints can withstand very high surface temperatures and are commonly available in automotive stores. Also, 3-M produces an acceptable, albeit expensive, ultraflat black paint known as "Nextel."

You may see commerical collectors with selective surface coatings. As we mentioned, these coatings absorb solar energy very strongly, but reradiate thermal radiation in far smaller quantities. Unfortunately, we have yet to see an easy-to-apply, do-it-yourself selective coating. The chemical processes involved are complex and currently quite expensive. However, the results are dramatic; you should be alert for new selective coatings as they enter the market.

The insulation beneath the collector plate should be equivalent to 2 or 3 inches of glass wool. Polyurethane is a tempting choice due to its excellent insulating properties. However, polyurethane tends to deform severely at temperatures above 160°F; we do not recommend that it be put in direct contact with the collector plate. Instead, you might consider a multilayer insulating affair, with fiberglass in contact with the plate, followed by a layer of polyurethane.

Probably the most vulnerable part of the collector is the glazing seal and support system. Once again, thermal expansion is a major problem. The continuous movement of the glazing relative to its sealing system often produces leaks in the weatherproofing. In extreme cases, the thermal expansion can actually crack glass. Plastics have an advantage here since they have a degree of give.

Mounting procedures for vertical glass panes are well established. These techniques, which are summarized in Figure 4.26, can be used with reasonable success on inclined systems provided the unsupported glass area is not outrageous. Large spans of unsupported glass tend to flex under their own weight and the edges dig into the sealing hardware. This problem is closely tied to the *bite*, the amount of edge material supported by the seal system. We would recommend that you increase the bite as the collector inclination approaches the horizontal or the span area increases. Do this by increasing the *rabbet* depth (*A*); don't skimp on the clearance dimension (*B*).

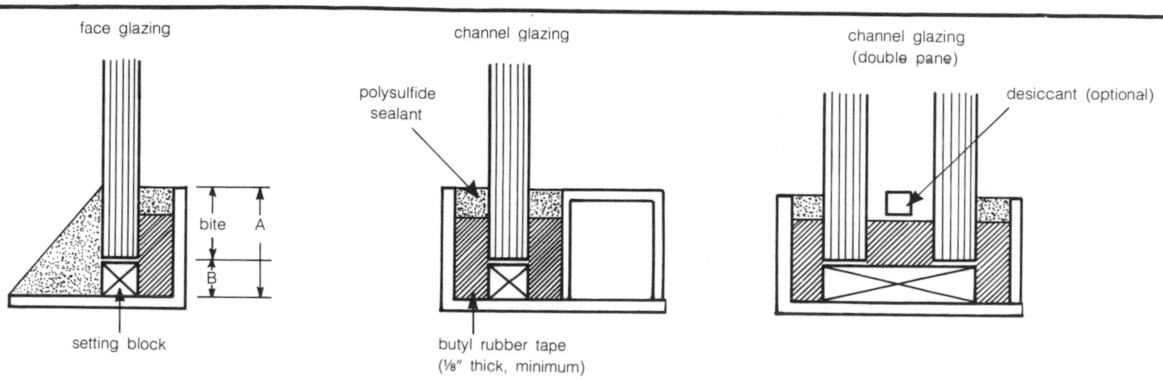

DOUBLE PANE Type Glass and Thickness		Glass Size		Tolerances	A	B	
Glass Thick (in)	Total Thick (in)	Area (ft²)	Width or Height (in)	Unit Size (in)	Rabbet Depth (min) (in)	Clearance at Head Sill & Jambs (allow) (in)	Setting Block Height (range) (in)
1/8–1/8	9/16 or 13/16	15	80	± 1/16	9/16	3/32	1/16–1/8
1/8–1/8	9/16 or 13/16	15	80	± 1/16	5/8	1/8	1/16–3/16
3/16–3/16	11/16 or 15/16	16	48	+ 1/3–1/16	9/16	3/32	1/16–1/8
3/16–3/16	11/16 or 15/16	30	120	+ 3/16–1/16	5/8	1/8	1/16–3/16
3/16–3/16	11/16 or 15/16	30	120	+ 3/16–1/16	3/4	1/4	3/32–5/16
1/4–1/4	13/16 or 1 1/16	50	144	+ 3/16–1/16	3/4	1/8	1/16–3/16
1/4–1/4	13/16 or 1 1/16	70	144	+ 3/16–1/16	3/4	1/4	3/32–5/16
1/4–1/4	13/16 or 1 1/16	70	144	+ 3/16–1/16	7/8	1/4	1/8–3/8

Figure 4.26 Glazing details for single-pane and double-pane glass.

SINGLE PANE	Glass Size		Tolerances	A	B	
Glass Type and Thickness (in)	Area (ft²)	Width or Height (in)	Glass Cutting Size (in)	Rabbet Depth (min) (in)	Clearance at Head Sill & Jambs (allow) (in)	Setting Block Height (range) (in)
1/8	25	128	± 1/16	1/2	11/64	3/32–1/4
1/8	67	128	± 1/16	5/8	15/64	3/32–3/8
1/4	100	120	± 1/16	1/2	11/64	3/16–1/4
1/4	140	156	± 1/16	5/8	1/4	1/8–3/8
1/4	207	229	± 3/32	3/4	11/32	3/16–1/2
5/16	207	229	± 3/32	3/4	11/32	3/16–1/2

Two types of glazing configurations are shown in Figure 4.26. A face-glaze is probably the simplest in terms of collector construction. Channel glazing requires a channel-shaped molding, which means that you are in for a lot of woodworking or that you will be buying prefabricated aluminum channel stock for your collectors. Double-pane windows can be face-glazed, but channel glazing is much more common. Note that a desiccant can be placed between the two windows to prevent moisture from condensing on the inner window surfaces.

The sealant system is comprised of three main components. A setting block is used to position and support the pane while the sealant is put in place. These blocks are spaced evenly around the window channel. Butyl rubber tape is used to maintain a clearance between the window and sharp edges which might be present in the support system. A final seal is achieved with a silicon or (more commonly) polysulfide compound which can be applied with a calking gun. One sealant looks particularly attractive for collector construction since it is designed to flex in systems beset with thermal-expansion problems. The polymer is marketed by DAP under the trade name "Flexiseal." DAP also manufactures butyl rubber tape. Sweet's *Architectural Catalog* lists a number of other manufacturers offering similar materials.

There are some cardinal rules which apply to glazing installation. Cleanliness is Godliness in this business, and some manufacturers even suggest the application of a special primer before the sealant is put in place. Also, it's inadvisable to work with these sealants in cold weather. All in all, follow the manufacturer's directions.

The overall *size* of the collector is worth some thought. Large collectors are heavy and unwieldy, and usually contain large pieces of glazing. Most designers favor modular construction—units of 20 or 30 square feet—which can be connected to one another in either series or parallel configuration. These modules typically weigh less than 100 pounds and can be transported and installed easily. If repairs are needed, only one module need be taken out of service; the rest of the array remains in operation.

To conclude, if you intend to construct your own collectors, the key word is *care*. In addition, you will need access to metal-working and carpentry facilities. Materials can be purchased through the usual outlets, but check with surplus houses, junk yards, and demolition sites for cheaper prices.

Commercially available flat plate collectors are generally sold at prices ranging from $6 to $10 per square foot of surface area. Naturally, the material costs are significantly less. To get a feel for the raw costs, let's consider a typical collector system. The design utilizes copper plate with copper tubing affixed to the surface. A single layer of fiberglass glazing is employed. Prices in Table 4.19 reflect the retail situation in San Francisco during the spring of 1975.

Costs will be different with each individual design, depending upon availability of materials, local prices, and your access to used or junk materials. As an example of the variations you might encounter, copper prices were close to $2 per pound in 1974 but in early 1975, prices plummeted to 63 cents per pound! This price is competitive with many aluminum alloys, and copper is somewhat better suited for collector applications.

Commercially Available Flat Plate Collectors

There are a number of commercial collectors already on the market, and new designs are introduced almost weekly. Some are sold as complete units, including the collector plate, backing, and glazing. Other companies simply sell the collector plate itself.

The flat plate designs we have seen so far differ only slightly, and we suspect the same can be said for thermal performance. There does, however, seem to be a clear difference in manufacturing quality and materials—items which affect the life and overall performance of the collector. Appendix 4F contains a fairly complete description of collectors on the market at this time.

Many collector manufacturers rely on the Olin Corporation, of East Alton, Illinois, to supply collector plates. Olin produces a "Rollbond" panel, fabricated of aluminum or copper, which has flow passages integral to the collector. This is accomplished by silk-screening a flow design on a thin metal sheet and then bonding a second sheet to the first. The bond "takes" on all but the silk-screened area. High-pressure air is then applied to a crevice between the two sheets, to expand the unbonded metal into an integral flow passage. (This technology was originally developed for the heat exchangers used in refrigerator freezers.)

Olin (and similar sources like the Tranter Manufacturing Company in Lansing, Michigan) is geared for high-production runs of many thousand square feet. Panels are made to the customer's specifications, with setup charges running close to $700. Thereafter, production costs are based on a "per pound of material" basis. Very recently, Olin announced a line of predesigned copper panels for retail purchase. Detailed information is included in Appendix 4F.

When purchasing panels or constructing those of your own design, you should be aware of some potential interfacing problems that involve the rest of your solar heating system. Aluminum, for instance, is prone to corrosive reactions when "sharing" a system with other metals. Thus, while aluminum collector plates are often cheaper than copper plates, more care must go into the system design to prevent debilitating corrosion. This will be discussed in more detail when we treat circulation systems.

Collector Efficiency

The collector efficiency is the ratio of the energy actually absorbed by the circulating fluid (i.e., the energy we can use to heat our home) to the amount of solar energy incident on the collector assembly:

E. 4.25 $$\text{Collector Efficiency, } e = \frac{\text{Energy Absorbed}}{\text{Energy Incident}}$$

Table 4.19 Costs for Raw Materials of a Typical Flat Plate Collector[a]

Material	Cost per sq ft	500 sq ft
Copper collector, flat stock 500 sq ft of 10 oz. material gives 315 lb. @ $1.00 per lb. (0.0135" thick)	$.63	$315
Copper tubing, Type L 1.5 linear ft per sq ft of collector gives 750 ft @ 52¢ per ft	.78	391
Insulation	.10	50
Solder @ $2.50 per lb., 15¢ per ft of tubing for 750 ft	.23	113
Manifold fittings 50 @ 24¢ ea.	.02	12
Valves 2 @ $15.00 ea.	.06	30
Other fittings	.03	15
Glazing (plastic, single layer)	.30	150
Paint and odds and ends	.11	55
Collector coating	.024	10
Total:	about $2.29	$1141

Notes: a. Taken to be 500 ft^2.

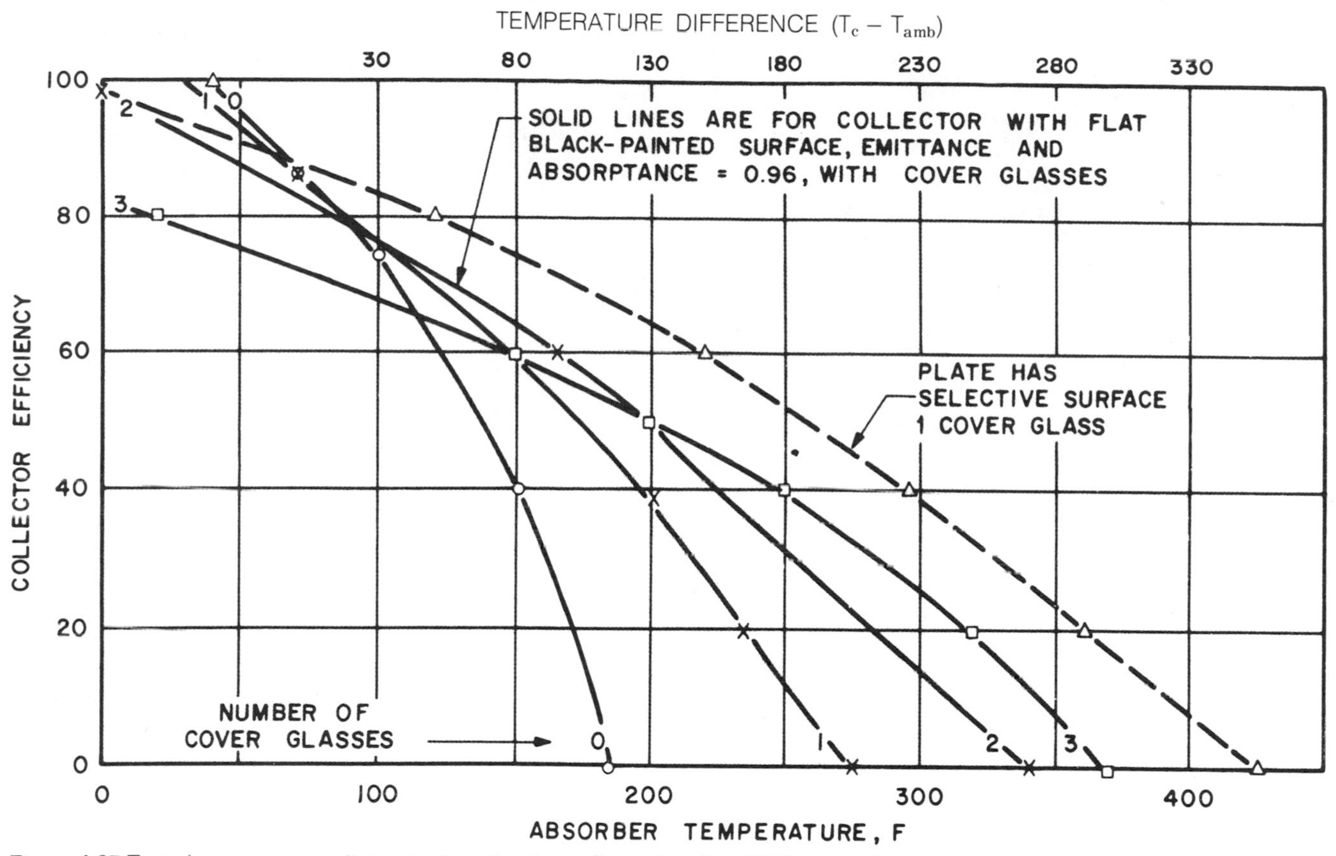

Figure 4.27 Typical instantaneous efficiencies for a flat plate collector based on 70°F air and sky temperature. Use the top scale to estimate efficiencies at other air temperatures. Nominal insolation of 300 Btu/ft^2-hr. (Reprinted with permission of McGraw-Hill Publishing)

For instance, if 1000 units of solar energy fall on a collector assembly in one hour and 600 units are collected by the circulating fluid in the same time period, we would say that

$$\text{Collector Efficiency, } e = \frac{600}{1000} = 0.60 = 60\%$$

Figure 4.27 demonstrates typical relationships between collector efficiences, average absorber (or loosely, collector) temperature, and the number of sheets of glazing used in construction. Looking at the graph, you will see that with moderate collector temperatures between 140 and 160°F, an unglazed collector (the curve labeled "0") is about 40 percent efficient. One sheet of glazing increases the efficiency to 60 percent—probably a worthwhile improvement. An additional sheet of glazing (two sheets, total) increases the efficiency by only 5 percent and probably isn't worth the added expense. A third sheet of glazing actually *reduces* the efficiency, compared to the two-sheet configuration! (The reflective and absorptive losses associated with the glazing begin to take their toll.) You should note that our conclusion applies only to the range of temperatures from 140 to 160°F. At higher temperatures, the triple-glazed collector exhibits higher efficiencies than single- or double-glazed units; if you desire higher operating temperatures, additional glazing might be economically justified.

Since Figure 4.27 will prove quite useful to us in later sections, some further comments are warranted. First, the curves are based on 70°F ambient conditions. At lower ambient temperatures, efficiencies suffer since thermal losses are greater. We can correct for this, approximately, by modifying the temperature scale to indicate the temperature *difference* between the the absorber plate and the environment. Thus an absorber temperature of 100°F corresponds to an absorber-environment temperature difference (100 − 70) of 30°F, and so on. This modified scale appears at the top of the figure.

The new scale can be used in the following way. Suppose the average winter temperature (see Appendix 4B) in your locale is 30°F and you desire an estimate of the collector efficiency operated at an average of 160°F. The temperature difference in this case (160 − 30) would be 130°F and the efficiency, say for a double-glazed unit, would be about 50 percent. Thus our scale change has extended the utility of the efficiency curves significantly.

Our modified curve is quite valuable in another respect—it's the only efficiency curve we are likely to have! Commercial collector manufacturers seldom specify efficiencies in their literature, largely because standardized efficiency measurement methods have been lacking. Very recently, the National Bureau of Standards issued proposed testing standards, but manufacturers have had little time to implement them. Until they do, Figure 4.27 will have to suffice for our calculations. (One problem is that the efficiency curves differ somewhat when measured at different insolation values. Figure 4.27 is based on an insolation of 300 Btu/ft^2-hr; however, we will take it to be a representative average over a range of insolation values.)

You may be wondering how the average absorber temperature is defined and what parameters control it. Truly representative temperatures are difficult to assign, since fluids entering the collector are cooler than those exiting. Necessarily, the portions of the plate near the inlet fitting are cooler during normal operation, and suffer fewer thermal losses. Just the opposite is true for portions near the fluid exits. The representative absorber temperature is *usually* taken as a simple average between the inlet and outlet fluid temperatures. For example, if water entered a collector at 100°F and exited at 150°F, we would say that the average absorber temperature was about 125°F.

The absorber temperature is affected in several ways. Naturally, the amount of incident solar energy is a key element. Another important factor is the flow or circulation rate of the collector fluid. The effects of flow rate will be discussed later in more detail, but for now we can simply observe that a slower fluid flow rate causes a larger temperature rise across the absorber plate. The greater temperature rise is attributable to its longer exposure to incoming solar energy. Imagine covering a fixed distance on a rainy day; the slower you travel, the wetter you get. The same principle applies to a fluid traversing an absorber plate; the slower it goes, the hotter it gets (and the higher the average plate temperature becomes). Thus, by controlling the speed of your circulation system, you can achieve virtually any average plate temperature you desire.

A final and obvious parameter is the inlet fluid temperature. Often this is equal to the temperature of your storage system. If your storage system is charged up after several balmy, sunny days, you might have inlet fluid temperatures of 150°F or more. Clearly, the average collector temperature will be higher still, and your efficiency will be on the low side. On the other hand, if weather has been bad, the temperature of your storage system might be near 100°F. The inlet collector temperature would be about the same, and the average plate temperature would also be relatively low—at least compared to the previous situation. Efficiencies would be comparatively high, however, and just in the nick of time! Thus, the interplay with the storage facility is a subtle but civilized one; collection efficiencies tend to be high when you need them high and taper off as the demand diminishes.

Table 4.20 Collector Interception Efficiency for Various Misalignments

Misalignment Angle	Percent of Maximum Solar Energy Intercepted[a]
0°	100.0
5°	99.6
10°	98.5
15°	96.5
20°	94.0
25°	90.6
30°	86.6
35°	81.9
40°	76.6
45°	70.7
50°	64.3

Notes: a. Cosine of the angle of misalignment multiplied by 100.

Collector Orientation

Naturally, your collector should be oriented to intercept as much of the sun's radiation as possible. For best results, we should always have the panel perpendicular to the incoming sunlight, but this arrangement is difficult to achieve on a continuous basis. As you recall, the sun sweeps across the sky each day and shifts to the north and south seasonally (see Figure 4.10). Therefore, a stationary collector will operate at less than optimum efficiency for much of the time.

How much energy is lost because of collector "misalignment"? For small angles, the loss is proportional to the cosine of the angle. At very steep angles, the reflective properties of the glazing become more important, and the cosine rule no longer applies. A zero misalignment angle corresponds to a perfectly aligned collector, one perpendicular to the incoming rays. Table 4.20 gives you some idea of the cosine behavior.

Evidently, for orienting flat surfaces to receive sunlight, "pretty close" is close enough. Even if your collector is misaligned by as much as 25 degrees, you are still intercepting over 90 percent of the possible sunlight. This

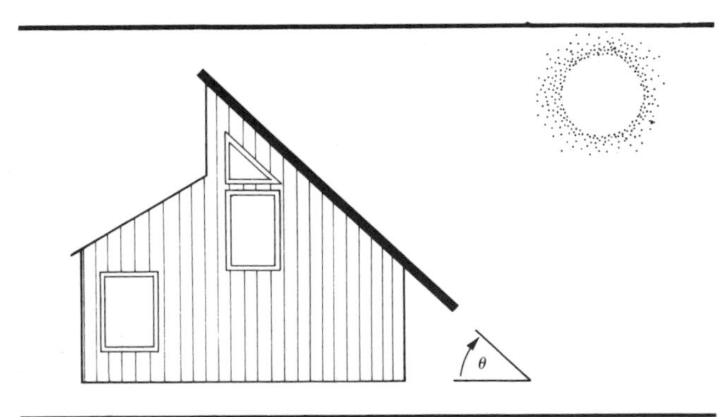

Figure 4.28 Collector inclination angle (θ).

is acceptable for flat plate collectors, which can use heat from diffuse solar radiation as well as from direct radiation, and can operate on bright cloudy days. Thus, flat plate collectors (and integral variations described later) can be constructed with vertical, horizontal, or tilted orientations and, if sized properly, can collect enough solar energy to heat a house. In contrast, focused collectors can only operate with direct radiation; they require additional equipment to track the sun continuously.

Collector orientation preferably should be selected for operation during the winter months, when heating demand is high. During these months, the sun moves lower in the southern sky (in the northern hemisphere); one should therefore orient the collectors more vertically than horizontally to receive sunlight from nearer the horizon. If your roof doesn't lend itself to a direct southerly exposure, a southwestern exposure should be favored.

Many studies have been undertaken to determine the optimum tilt angle for a south-facing stationary collector. The consensus seems to be that the optimum winter tilt angle should be equal to the local latitude plus 15 degrees. The inclination angle is measured from the horizontal, as shown in Figure 4.28. Note that if one desires to mount the collectors directly on the roof, a relatively steep roof line is required. Fortunately, as we see from Table 4.20, one need not strictly adhere to the consensus guideline.

We might close this section by discussing a method to augment the available solar energy by means of reflective panels. The technique is clearly illustrated in Figure 4.29. Fixed or movable panels can be fabricated with aluminized mylar sheet, which has excellent reflective properties. In some designs, the panels double as insulators which swing up to cover the collectors at night—thereby preventing heat loss and/or freezing. Unfortunately, it is difficult to analyze these reflective schemes in a quantitative fashion (in fact, the following section on collector sizing will not consider such configurations). However, the method should be recognized as an effective augmentation scheme. At least one of the example homes described later, in our discussion of integral systems, uses reflective panels.

Sizing a Flat Plate Collector

You should now have a good idea of how a collector works and of basic construction techniques. Next we need to determine the required collector area for a particular application. To do this, we must have estimates of the heat loss (heat load) from the house, the collector efficiency, and the available solar energy. The first two factors have already been discussed: the heat loss can be computed using techniques presented in the "Heat Loss" section of this chapter, and the efficiency can be estimated from Figure 4.27 or manufacturers' data.

The available solar energy is more difficult to estimate. The insolation data presented in the section on climatology is of limited use since it fails to take into account all of the important variables—collector orientation, inclination, and local weather phenomena. For instance, the solar position and insolation data in Appendix 4C account for collector angle and inclination, but neglect local weather effects. The U.S. Weather Service insolation measurements (see Table 4.6) reflect local climatic conditions but are reported for horizontal surfaces only. Complete data, encompassing all of these variables, may be available for many localities in the next few years. In the meantime, we must make do with some approximate methods.

We will describe these methods in a moment, but a basic statement on the limitations of our method is in order. Throughout the rest of this chapter, we will be introducing a variety of "approximate techniques" for sizing components in a solar heating system. Some of these techniques can incur errors of 20 percent or more. So why do we use them?

First, treating the problem in an exact way is very difficult, even for a trained engineer. For example, the efficiency of a solar collector is related to its average temperature (see Figure 4.27), which in turn is related to both the hour of the day and the season. In addition, the

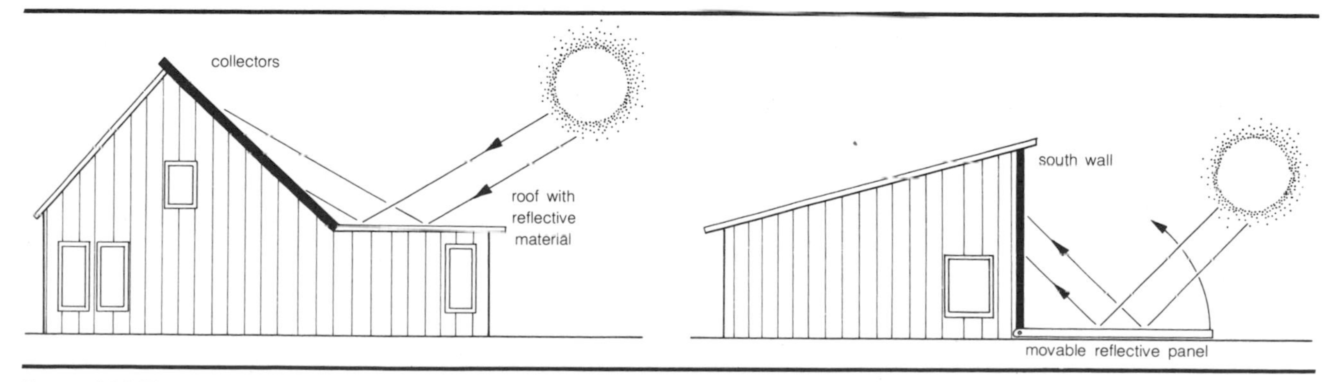

Figure 4.29 The use of reflective materials to augment solar energy available to collectors.

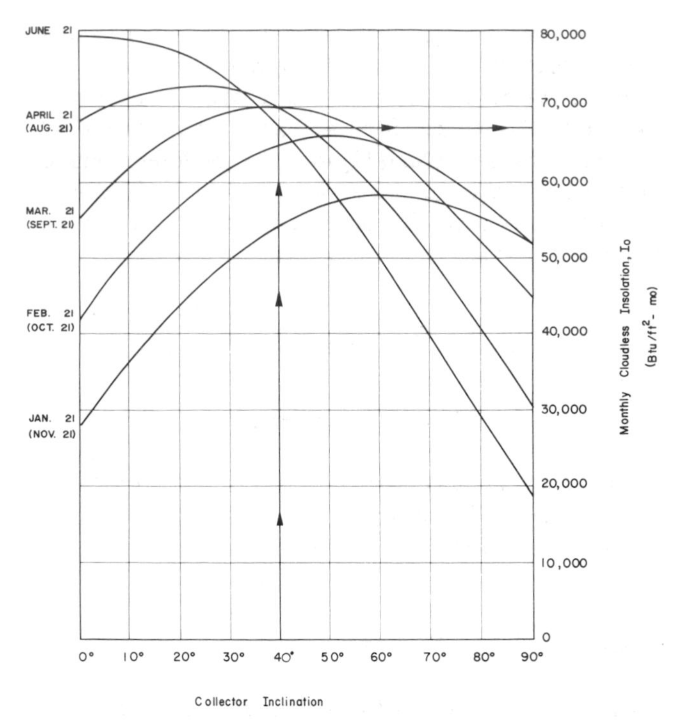

Figure 4.30 Monthly solar availability for cloudless conditions at 40°N latitude (data taken from Appendix 4C).

quantity of solar radiation striking the collector is also changing with time. What a mess to analyze exactly! Fortunately, the monthly *average* insolation, *average* collection efficiencies, and other approximations help us bypass these problems and still arrive at reasonable estimates of energy collection.

Second, it is futile even to attempt any sort of exact analysis. The weather, which plays such an important role in all of the calculations, is too unpredictable to design for every possible contingency. For instance, the Weather Service might report that a certain locale had an average January insolation of 30,000 Btu per square foot per month over the last thirty years. If you carefully design your solar heating system based on this figure, you can be sure that the very next January will be "extra cloudy." This is one of the many corollaries to Murphy's Law, well known to engineers, which states that if anything can go wrong, it will. Unfortunately, Mother Nature and Mr. Murphy are particularly good friends; but despite this friendship, the point here is that our methods, albeit approximate, will get you reasonably close to a good design. One warning: in light of this discussion, you should see the need for an auxiliary heating system in your design.

The first method we describe makes use of the ASHRAE insolation data presented in Appendix 4C. The "daily surface totals" have been converted to monthly figures and plotted for a variety of collector inclination angles in Figure 4.30. We have used data for a 40°N latitude to construct this curve but similar plots can be sketched for any latitude.

The curves enable us to find the *maximum* amount of energy that will strike your collector for a given inclination angle and month. The entire month is assumed *cloudless* and the only variation in solar intensity is due to the seasonal and daily motion of the sun. With this assumption, you can readily see that the curves will be valid for all geographic localities at 40°N latitude. (In actuality, little error is introduced by using this curve for any part of the continental United States except Alaska.)

Observe that a symmetry exists for dates equally spaced about June 21. For instance, March 21 occurs three months before June 21, and September 21 is three months afterwards. Thus the insolation values for March and September are essentially identical and one curve represents them both. Other symmetric pairs are shown on the graph.

To incorporate the effects of cloud cover, we use an approximation scheme based on published values of "percent possible sunshine." (You may recall that this quantity was introduced in our discussion of climate. Table 4.7 contains some representative values.) The cloudless insolation value (I_o) is adjusted using the following empirical relation:

E. 4.26 $$I = I_o\left[0.35 + 0.61\left(\frac{\%\ Sunshine}{100}\right)\right]$$

The numbers 0.35 and 0.61 are known as *climatic constants* and vary with specific location; the values in Equation 4.26 are average figures. This formula attempts to account for energy which reaches us even on partially cloudy days; thus weighting is not totally dependent upon the percent-sunshine figure. Remember, this "I" is not the infiltration factor of earlier formulas, but rather the adjusted monthly insolation.

Example: We wish to estimate the available solar energy in the month of June for a collector inclined at 40 degrees. We will also assume that the percentage of sunshine in that month is 60 percent for our particular locale.

Solution: Referring to Figure 4.30, we first draw a vertical line from 40 degrees until it strikes the curve for June. We then hang a right and continue until we reach the vertical axis. This would be the total energy striking our collector that month *if it was sunny every day*. You can see the I_o would be about 68,000 Btu/ft² per month, or 2260 Btu/ft² per day. This amount is obviously too high since it is sunny only 60 percent of the time. To account for the cloudiness, we apply Equation 4.26:

$$I = I_o[0.35 + 0.61 \left(\frac{60}{100}\right)]$$

$$= 68,000\ (0.35 + 0.37)$$

$$= 48,600\ Btu/month\text{-}ft^2$$

Using this method, collector sizing becomes straightforward. Essentially, we seek to balance the energy lost from the house over a given time interval with the energy collected from the sun (when shining) during the same interval. Often this time interval is taken as a day, but the calculations can also be done on a monthly basis.

Example: Let's compute the amount of collector area for a typical house located in Fresno, California. Suppose the heat loss from the house (including a 10 percent safety factor) is 25,000 Btu/hr when based on an outside design temperature of 30°F and an inside design temperature of 65°F. The collectors are double-glazed and operate at an average temperature of 150°F. They face to the south at an angle of inclination of 50 degrees.

Solution: First we obtain an estimate of the average collector efficiency. Consulting Table 4.4 (excerpted from Appendix 4B), we find the average winter temperature in Fresno to be 53°F. This means that the average temperature differential between the collector and its environment (150 − 53) will be about 100°F. Using the top temperature scale in Figure 4.27, we can estimate the efficiency to be about 60 percent.

Next we compute the heat loss for the most severe month, typically January. Referring to the degree-day information in Table 4.4, we find that 605 degree-days accrue during January in Fresno. Now Equation 4.23 may be applied, from our discussion of heat loss:

$$S_{month} = 24Q(Degree\text{-}Days)_{month}$$

where S is the total energy requirement and Q is the heat loss per degree of temperature difference. We first derive Q from Equation 4.24:

$$Q = \frac{Heat\ Load}{T_i - T_o} = \frac{25,000\ Btu/hr}{(65 - 30)°F}$$

$$= 714\ Btu/hr\text{-}°F$$

Note that Q is based on the *design* temperature difference (65 − 30)°F, so the energy loss in January (S) will be (24 × 714 × 605) about 10,371,000 Btu. This calculation accounts for the fact that the outside temperature varies daily and isn't always 30°F.

The solar energy available that month, per square foot of collector, can be found with Equation 4.26. First, however, we need to know the percentage of sunshine in the month of January. Table 4.7 indicates a figure of 49 percent for Fresno. Using Figure 4.30, the total monthly cloudless insolation (I_o) for a collector tilt of 50 degrees is found to be 58,000 Btu per square foot. Applying the cloudiness correction of Equation 4.26:

$$I = 58,000\ \left[0.35 + 0.61 \left(\frac{49}{100}\right)\right]$$

$$= 37,700\ Btu/month\text{-}ft^2$$

We are now finally prepared to compute the required collector area. The basic relation is simple. On the average, heat lost per month must equal heat gained per month.

E. 4.27 $$S = I \times e \times A$$

where S is the heat loss for the month, I is the adjusted monthly insolation, e is the collector efficiency, and A the area of the collector. Then

$$10,371,000\ Btu = 37,700 \frac{Btu}{ft^2} \times 0.60 \times A$$

Therefore, we find that the required collector area is

$$A = \frac{10,371,000}{(37,700)\ (0.60)} = 458\ ft^2$$

This is a reasonable number, considering the moderate climate chosen and our assumption of 100 percent solar heating. (In many cases, it will not be feasible to provide for complete solar heating in the coldest months, because of economic or space constraints.)

If the month is abnormally cold and cloudy weather occurs simultaneously, or if either lasts for several consecutive days, you are likely to require some auxiliary heating. Alternatively, you may wish to increase the collector area by 5 or 10 percent, thereby adding an additional safety factor. By and large, however, the use of monthly data will give sufficiently accurate results to avoid embarrassingly frigid situations.

There is a second method to determine I if estimates of insolation on a horizontal surface are available (see Table 4.6). Here the problem is translating data referenced to a horizontal surface to data of an inclined plane with a particular orientation. The effects of cloud cover are already accounted for in the horizontal values.

Consider the Fresno site once again. Table 4.6 indicates that the average *daily* insolation in January is

664 Btu/ft². This corresponds to 19,920 Btu/ft² per month (664 × 30), referenced to a *horizontal surface.*

In order to estimate the energy intercepted by the south-facing, 50-degree-tilt collector in the previous example, we examine Figure 4.30 (the information can also be taken directly from Appendix 4C). This graph excludes the effects of weather phenomena—it is based only on the geometric relations between the earth and the sun. We can use this information in the following way. Taking the curve labeled January, we can read cloudless insolation values for 0-degree and 50-degree inclination angles. The 0-degree inclination angle corresponds to a horizontal surface. We find, for January at 40°N latitude:

Inclination Angle	*Cloudless Insolation per Month*
0°	29,000 Btu/ft²
50°	58,000 Btu/ft²

Thus, averaged over a day or month, a panel inclined to 50 degrees will receive about twice the amount of energy (58,000/29,000) a horizontal surface receives. This figure will vary with inclination angle and month. But in this case, we can multiply our measured horizontal insolation, which includes the effect of cloud cover, by 2:

$$I_{January} = (2)\,(19{,}920) = 39{,}840 \text{ Btu/ft}^2$$

Our result is higher than the 37,700 Btu/ft² per month (January) obtained using the percent-sunshine method. In fact, Figure 4.30 and the percent-sunshine adjustment yield consistently lower insolation estimates and often the discrepancy is more substantial. We suspect the error lies in Figure 4.30. The values are taken from Appendix 4C and were calculated assuming *zero ground reflectivity*; that is, energy was assumed to come directly from the sun or diffuse sky conditions. In reality, the environment surrounding any given surface will reflect solar energy to that surface This effect can be neglected for any average surface, since it, too, will reflect much of the incident solar energy. But for collectors, which are coated to absorb as much sunlight as possible, we can count on a net gain from the reflective environment. Thus, the zero-reflectivity assumption probably explains the lower results generated by Figure 4.30 and the percent-sunshine adjustment.

You're undoubtedly asking, "Which method should I use?" Actually, either is satisfactory. The graphical technique seems to overplay the effect of cloudiness, while the method using horizontal insolation measurements seems to underplay it. We might suggest you use both methods (if, indeed, horizontal insolation data are available) and then choose a value somewhere in between. The closer you adhere to the lower estimate, the more conservative your design will be. Such a design will also entail larger collector areas and initial costs.

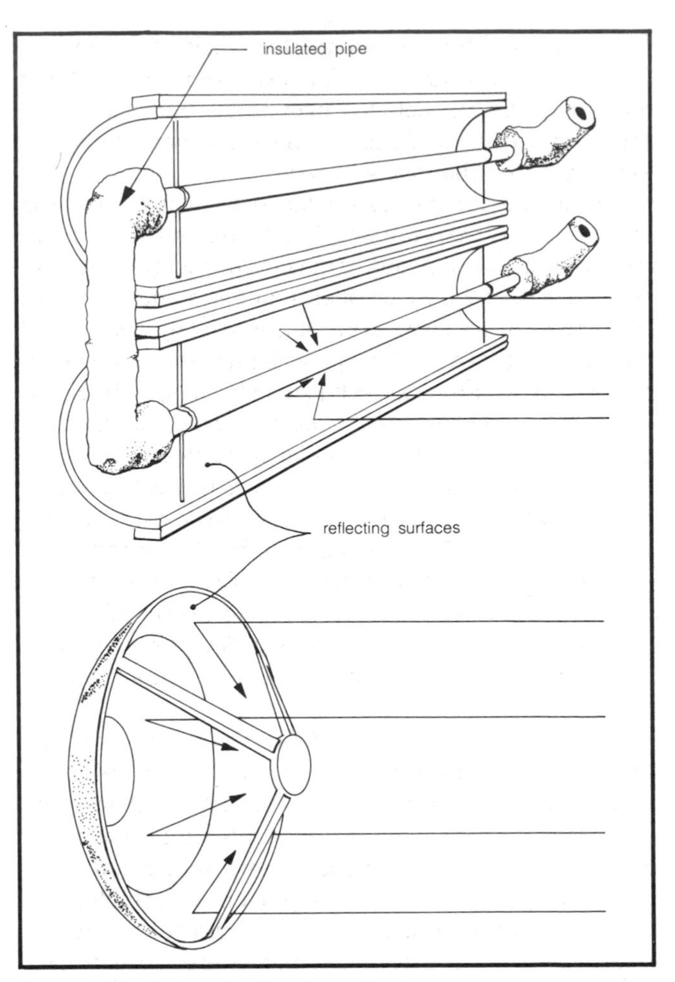

Figure 4.31 Designs for focusing collectors.

Hopefully, more useful insolation data will be available to the public in the near future, information based on actual radiation measurements around the country, reported on a monthly basis for a variety of inclinations and orientations. Until then, you will have to plod along with these two methods or else instrument your proposed site and record your own insolation data.

Focusing Collectors

Focusing collectors produce high temperatures (300–1000°F) by directing all incident sunlight to a point (the focal point) or line (the focal axis) through which an enclosed, circulating medium is passed (see Figure 4.31). Although the ideal focuser has a parabolic cross section, collectors with hemispherical or cylindrical shapes are easier to make and can be almost as efficient.

There are three basic components to any focusing collector: the focusing unit; the pipe containing the circulation medium which actually collects the heat; and insulation for the interconnecting pipes. The focusing unit consists of a frame covered with a highly reflective material such as thin aluminized mylar. Searchlight mirrors purchased from a surplus store make excellent focusers, or you can easily fabricate your own frame or mold. It is also possible to integrate the frame and reflective surface by flexing shiny sheet metal such as aluminum into the desired shape.

When the focus or axis of the focusing unit is located, the circulation pipes are attached to the unit so that they pass through these points of maximum energy concentration. (Locating the focus often requires some degree of mathematical sophistication.) It is advantageous to coat the collecting areas of the pipe with paint that can absorb high temperatures. Connecting lengths of pipe should be well insulated because the circulating fluid leaves the collector either as very hot water or as steam; the potential heat loss from these sources is much greater. Several collector units are necessary for most heating uses, so a circulation scheme must be developed.

An outer covering of plastic or glass can protect the collectors from weathering, prevent dust accumulation on the reflective surface, and reduce collector heat losses. The collector units may be mounted just about anywhere.

The greatest advantage of focusing collectors is their production of high temperatures and, thus, high-quality heat. In fact, solar cooking is quite feasible with focusing units. However, the high temperatures imply larger heat losses, a notable disadvantage. There are two other disadvantages to focusers. The production of high temperatures requires virtually unclouded direct sunlight; if direct sunlight is not available, the focuser collects little usable energy. Consequently, the use of focusers is now limited to areas with very few days of cloud cover during the heating season. And second, most focusers require sun-tracking (movement which follows the sun's path for continual focusing). Continuous tracking requires automatic equipment, which can be very expensive.

Several commercially available focusing collectors (or kits) are listed in Appendix 4F. The Solergy Collector (Solergy, Incorporated, 150 Green Street, San Francisco 94133) is perhaps the most interesting. Its unique design may obviate the need for tracking, which makes it quite attractive for solar heating applications. We feel the tracking requirements of other focusing collectors make them impractical for space heating.

Energy Storage System

Since periods without sunshine are inevitable, it is necessary to include a storage system for a solar-heated building. In the short term, stored energy is needed during the night, when the collectors are inoperative. In the long term, you will occasionally experience several days or weeks of sunless conditions due to storms or heavy cloud cover. A properly designed energy storage system, perhaps combined with some auxiliary heating, will carry you through these periods.

Clearly, the larger the storage unit, the less likely you are of running out of energy during a given sunless period. On the other hand, a small unit is desirable from the standpoint of space utilization and cost. The actual sizing decision depends on the particular situation. You can base the decision on your knowledge of the local climate; for example, you might conclude that the probability of having more than three consecutive sunless days is acceptably small and risk being cold on a fourth sunless day. The size of your storage unit can also reflect the availability of auxiliary heating systems. If you are planning to install gas or electric back-up heaters in any event, you might size your storage system for only one day of sunless operation. On the other hand, if you're determined to do without auxiliary systems, you might try to provide for a five- or six-day storage capacity.

Means of Energy Storage

There are several techniques of energy storage at your disposal. The simplest (and recommended) technique involves energy storage in a *sensible* form (commonly called "sensible heat"). In this scheme, energy is stored internally in a suitable substance by increasing its temperature. To increase the storage capacity, you either use more of the substance or increase its temperature further. The former alternative is limited by costs (both for the material itself and the container to hold it) and the amount of space you wish to commit to energy storage. The latter alternative also has its drawbacks. As the temperature increases, the prevention of unwanted heat loss from the storage facility becomes more difficult. High temperatures also limit your choice of construction materials for the container.

The ability of common substances to store energy *sensibly* is indicated in Table 4.21. The important quantities are density (ρ) which has units of mass per unit volume, and the specific heat of the substance (C_p) which has the units of energy per unit mass ("pound-mass" or "lbm") per degree of temperature difference. The product of ρ times C_p gives the amount of energy stored per cubic foot of material and per degree of temperature rise. For water,

E. 4.28
$$\text{Stored Energy} = \rho C_p = \left(\frac{1.0\ Btu}{lbm\text{-}°F}\right)\left(\frac{62.4\ lbm}{ft^3}\right) = 62.4\ Btu/ft^3\text{-}°F$$

As an example, suppose we had a storage tank containing 100 cubic feet (748 gallons) of water and raised its temperature 10°F. Taking V as the volume and

Table 4.21 Thermal Properties of Potential Storage Media[a]

Media	Temp. Range, °F	Melting Temp., °F	C_p Btu/lbm-°F	ρ lbm/ft³	k Btu/hr-ft-°F	α ft²/hr	L Btu/lbm	Material Cost, $/lbm
Water-ice	32	32	0.49	58	1.3	0.046	144	Nil
Water	90–130	32	1.0	62	0.38	0.0061	—	Nil
Steel (scrap iron)	90–130	—	0.12	489	36.0	0.44	—	0.003
Basalt (lava rock)	90–130	—	0.20	184	1.14	0.031	—	0.002
Limestone	90–130	—	0.22	156	0.73	0.021	—	0.002
Paraffin wax	90–130	100	0.7	55	0.13	0.004	65	0.08
Salt hydrates								
$NaSO_4 \cdot 10H_2O$	90–130	90	0.4	90	—	—	108	0.012
$Na_2S_2O_3 \cdot 5H_2O$	90–130	120	0.4	104	—	—	90	0.05
$Na_2HPO_4 \cdot 12H_2O$	90–130	97	0.4	94	—	—	120	0.034
Water	110–300	—	1.0	62	0.38	0.0061	—	Nil
Fire brick	110–800	—	0.22	198	0.70	—	—	0.01
Ceramic oxides MgO	110–800	—	0.35	224	20.0	0.06	—	0.14
Fused salts $NaNO_3$	110–800	510	0.38	140	—	—	80	0.024
Lithium	110–800	370	1.0±	33	—	0.000435	286	10.00
Carbon	110–800	6750	0.2	140	2.4	0.103	—	0.20
Lithium hydride	1260	1260	1.0±	36	4.2	—	1200	10.00
Sodium chloride	1480	1480	0.21	135	3.65	0.13	223	0.02
Silicon	2605	2605	—	146	—	—	607	—

Notes: a. By mixing fused salts, any desired melting point between 250 and 1000°F can be attained.

ΔT as the temperature rise, the total increase in stored energy would be:

E. 4.29
$$\text{Increased in Stored Energy} = V_p C_p \Delta T$$

$$\times \left(\frac{1.0\ Btu}{lbm\text{-}°F}\right) \times (10°F)$$

$$= 62{,}400\ Btu$$

As another example, let's look at the same volume of basalt rock experiencing the same temperature increase. Referring to Table 4.21, the specific heat is one-fifth that of water (C_p = 0.2 Btu/lbm-°F), but the density is much higher (ρ = 184 lbm/ft³). The important quantity, $\rho \times C_p$, is 36.8 Btu/ft³-°F, which is about half that of water. We would then expect that the same volume of rock would store about half as much energy as water for the same temperature increase. Using Equation 4.29 with the appropriate numbers, we find

$$\text{Increase in Stored Energy} = (100)(184)(0.2)(10)$$

$$= 36{,}800\ Btu$$

which, indeed, proves our supposition. (Our calculation is based on *solid* basalt rock. In an actual application, crushed rock would be used to permit air or water circulation through the storage unit. Typically, the rocks are about 2 inches in diameter. With an air circulation system the product of $\rho \times C_p$ for crushed rock is less than for solid rock (around 20 Btu/ft³-lbm). This is due to the air spaces between the rocks, spaces which contribute little to the system's storage capacity.)

Alternate storage techniques capitalize on the energy released or absorbed when a substance *changes phase* (e.g., "energy of vaporization," often called "latent heat"). When we speak of a phase, we are referring to the liquid, vapor, or solid state of a substance. For example, the solid phase of water is ice, and the vapor phase is steam. The freezing of water or the melting of ice are examples of phase changes between the liquid and solid states of water. Boiling and condensation involve the phase changes between liquid and vapor states.

The amount of energy involved in a phase change is substantial. If we had for example, a pound of liquid water at 62°F, about 150 Btu of energy would be required to raise it to the boiling point (212°F) and an additional 1000 Btu to vaporize (or boil) it. The 1150 Btu would then be stored in the vapor or steam. By contrast, to store the same amount of energy sensibly, we would have to increase the temperature of *ten* pounds of water by 115°F! (This difference in storage capacity accounts for the severity of steam burns as opposed to hot-water burns: as the steam touches your skin and condenses, a tremendous amount of heat energy is transferred.) Another unique aspect of phase change involves the temperature behavior of the substance. During the phase change, the addition or release of energy occurs with *no* change in the material's temperature. For instance, a heated pan of water at normal room pressure will remain at 212°F during the entire boiling process. This is quite different from sensible heat addition or release, where an accompanying temperature change is experienced.

While storage schemes involving phase changes enable us to store large amounts of energy with small quantities of material, there are some drawbacks. Liquid-to-vapor phase changes involve large changes in density. For example, a pound of steam at 212°F and room pressure requires about two thousand times the space required by a pound of liquid at the same pressure and temperature. These tremendous volume changes render most vapor-liquid systems impractical for energy storage systems. Solid-liquid phase changes are more attractive in that they display less drastic volumetric fluctuation. For example, the densities of liquid water and ice differ by only a few percent. The main problem with solid-liquid phase changes is that they generally occur at relatively low temperatures. Water melts (or freezes) at 32°F, which is obviously too low a storage temperature to heat a living space. Paraffin is a bit better, since it melts at 100°F (see Table 4.21); but even 100°F is marginal for use in a heating system for a home. At the other end of the spectrum, common table salt melts at 1460°F and would make a fine high-temperature storage medium. Unfortunately, flat plate collectors seldom supply temperatures in excess of 200°F. We feel that such phase-change systems are impractical for the average builder at this time.

Substances which undergo reversible chemical reactions also have aroused considerable interest as storage media. Certain salts undergo hydration reactions which release or absorb energy depending on the "direction" of the reaction, often referred to as a "freezing" or a "melting" process because it has physical similarities to a simple phase change. Several examples of hydrated salts are given in Table 4.21. For example, a pound of hydrated sodium sulfate ($NaSO_4 \bullet 10\ H_2O$) absorbs 108 Btu (column "L") when it gives up its water of hydration. The same amount of energy is released when the action reverses. Since the density of sodium sulfate is 90 lb/ft^3, this is equivalent to 9720 Btu/ft^3. The process occurs at a relatively low temperature; its "melting point" is shown as 90°F. But other salts with higher reaction temperatures are available.

While salt hydrates offer substantial economies in size and weight, their successful utilization requires a well-developed chemistry background. Problems arise with nonuniform reaction rates throughout the storage unit, primarily due to density variations between the hydrated and unhydrated salt. The density variations cause a separation or "stratification" of the mixture. Also, the reaction rates themselves are limited, which means that you are limited as to how quickly you can store or retrieve energy. Finally, periodic chemical treatments must be administered to recharge the storage unit. All in all, it's a hassle.

Our basic recommendation is to stay with sensible storage systems which employ water or rock. If you have special talents (as a chemist, perhaps) then feel free to experiment with fused salts or the like. These methods *do* work—they just require considerable expertise.

Sizing the Storage Unit

Sizing the storage unit is relatively straightforward. Assuming you have calculated the heat loss from your house, you can then compute the heat loss per degree of temperature difference (see Equation 4.24). Suppose we take our sample house in Fresno, California, which has a heat-loss rate of 25,000 Btu/hr when the inside temperature is 65°F and the outside is 30°F. The heat loss per degree of temperature difference was computed to be

$$Q = \frac{25{,}000\ Btu/hr}{(65 - 30)°F} = \frac{25{,}000\ Btu/hr}{35°F}$$

$$= 714\ Btu/hr\text{-}°F$$

Next, choose the "maximum" continuous time period you would expect to be without sun. As we mentioned earlier, this choice depends upon many factors, which you alone properly can consider. For illustrative purposes, we will design for a three-day period.

Now find the degree-days per month accumulated during the most severe month of the year. Continuing with our example, Fresno, California, accrues 605 degree-days in the month of January (see Table 4.4). As an *estimate* of the degree-days accumulated in any three days of that month, we would multiply by a tenth (3/30). Thus, on the average, we would expect 60.5 degree-days to accumulate in a three-day period. We can then use a related form of Equation 4.23 to estimate the total heat loss in the three-day period:

$$S_{3\text{-}day} = 24Q\ (\text{degree-days for 3 days})$$

$$= (24)\ (714)\ (60.5)$$

$$= 1{,}036{,}700\ Btu$$

This is the amount of energy we must store.

To find the amount of storage material needed, we use a modified form of Equation 4.29, where volume is equal to the total storage requirement divided by the product of density, specific heat, and temperature difference:

$$V = \frac{S_{3\text{-}day}}{\rho C_p \Delta T} \qquad \text{E. 4.30}$$

Example: Assume that we are designing a water storage system: how big should our tank be?

Solution: If we assume that the maximum water storage temperature will be 150°F, and the minimum useful temperature is 80°F, the effective temperature difference (150 − 80) is 70°F. For water, ρ equals 62.4 lbm/ft³ and C_p equals 1.0 Btu/lbm-°F. Using Equation 4.30, we find the required volume of water is

$$V = \frac{1{,}036{,}700 \text{ Btu}}{\left(\frac{62.4 \text{ lbm}}{\text{ft}^3}\right)\left(\frac{1 \text{ Btu}}{\text{lbm-°F}}\right)(70°\text{F})} = 237 \text{ ft}^3$$

Since there are 7.48 gallons in each cubic foot, we require a storage tank for 1775 gallons (237 × 7.48). This amount of water can be stored in a cylindrical storage container 6.7 feet high and 6.7 feet in diameter ($V = \pi R^2 h$). Calculations for rocks or other storage materials follow a similar procedure.

Our method suffers from some inaccuracy, primarily due to the way we compute the number of degree-days accumulated in three days. If the month accrues 605 degree-days, we cannot be certain that they accrue evenly over the month. We could experience particularly severe weather in the first half of the month followed by balmy weather for the latter half. Clearly, our sample storage system could not cope with such a situation. This problem isn't as severe as it may appear. First, we *are* using monthly data, which certainly has more resolution than, say, seasonal or yearly data. Second, the calculations for heat loss and collector area should each include safety factors of 10 percent. Third, and perhaps most important, the three-day design criterion is purely arbitrary. No matter what solar system you design, there is always a chance that you will run out of energy sometime during the operating life of the system. Systems with small storage capacity simply have a larger probability of running out than do larger systems.

The Realities of a Storage Device

Probably the most important design feature in the storage unit is the insulation. The amount used largely depends upon the location of the storage unit, with larger amounts needed for outdoor or underground installations. Locating the storage facility *in* the dwelling is a considerable advantage since any heat loss serves to warm the house. However, indoor units should contain substantial insulation anyway, so you have some measure of control over when the heat enters the living space. Heat loss from the storage unit also decreases if the surface area is kept to a minimum. For a given volume, a spherical shape gives the smallest surface area. The second-best configuration is a right circular cylinder with its diameter equal to its height, the tank shape we used in the preceding example.

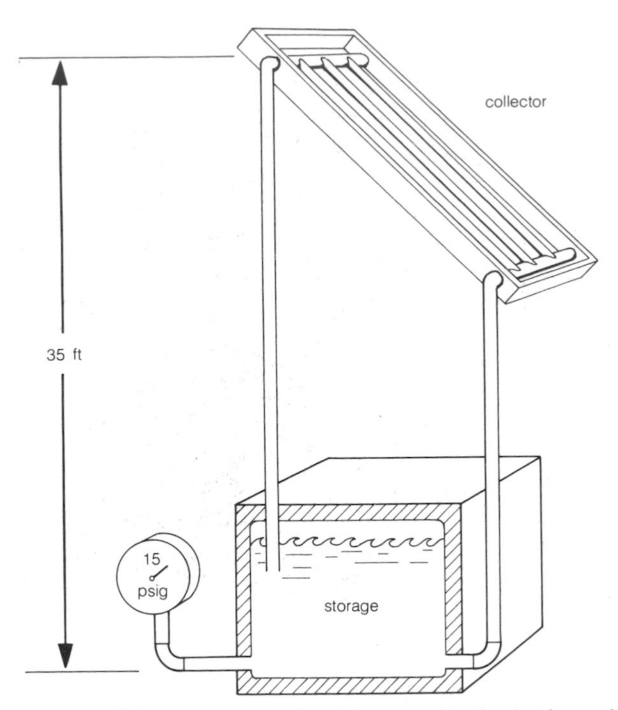

Figure 4.32 Hydrostatic pressure-head developed in the first loop of a collector system utilizing water (pump omitted for clarity).

If you're considering a water system, you might be wondering where you can *get* a large tank. Why, at a tank store, of course! If you look in the Yellow Pages of your phone book, you will probably find several dealers who deal in new and used tanks. New steel tanks sell for about a dollar per gallon of capacity. New fiberglass tanks can also be purchased for slightly more; it is also possible to fabricate your own at considerable savings. The only requirement is that the unit withstand the planned operating temperatures and pressures. There is no difficulty with steel in the temperature department, but some plastics are questionable. Fiberglass is acceptable for temperatures under 200°F. The tank will be under a slight (but significant) pressure—particularly if the collectors are mounted atop your roof and the storage tank is below. A hydrostatic pressure head of about 15 psi (pounds force per square inch) will develop for each 35 feet of elevation. This is illustrated in Figure 4.32. Twenty or thirty psi doesn't sound like much, but try multiplying that number by the area of your tank to estimate the magnitude of the forces involved! So make sure your tank can withstand the operating pressures and temperatures.

You can also buy used tanks from salvage yards. A typical used 1000-gallon steel tank might cost around $300. Often these tanks have been used for chemical storage, so you'll want to be sure they are carefully cleaned (look in the Yellow Pages under "Tank Cleaners").

Propane tanks, which are plentiful in rural areas, can

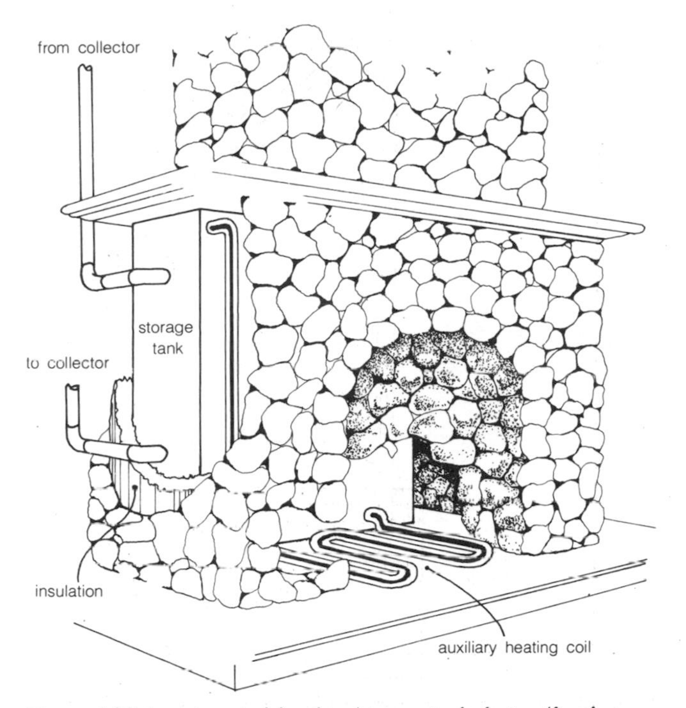

Figure 4.33 An integrated fireplace/storage-tank design (fireplace can be used as auxiliary heat source).

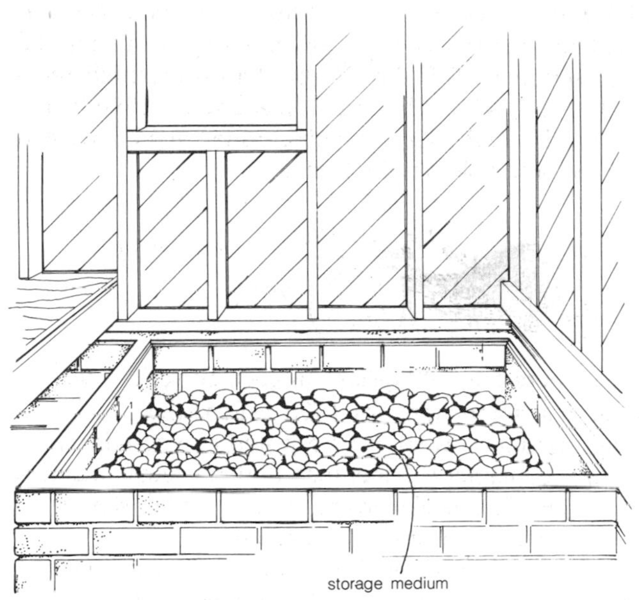

Figure 4.34 A basement/foundation construction with thermal storage area.

also be considered. Or, a very acceptable storage tank can be made from a piece of concrete pipe. You often can get chipped pipes at bargain prices from local highway or public works departments. You then pour a concrete pad and cement the pipe vertically. An insulated wooden top completes the unit.

It's probably a good idea to coat the inside of your storage tank with a corrosion-resistant material. Swimming-pool supply houses have special epoxy-based paints that fill the bill nicely. For instance, PPG Industries (Paint Division) markets a metal coating system consisting of a zinc primer known as "Aquapon" and an epoxy known as "Coolcat Coal Tar." The primer is quite expensive, around $50 per gallon, but the epoxy is much less. The surface must first be prepared by sandblasting; one coat of primer and two coats of epoxy finishes the job. Commercial tank coaters use this treatment regularly, and you may want to leave the work to them. As a final alternative, you might install a plastic bladder along the interior walls of the storage device.

Note that you will have to get into your tank to apply any of these coatings. This requires some sort of access hole. Probably the most simple arrangement is to have a removable top.

Two practical considerations remain: how to get the tank to your site and where to put it when you get it there. A steel or fiberglass tank is easily transported in a rental truck. Bring along a couple of friends and some metal pipes to help roll the tank on and off the truck. A concrete pipe probably requires a small crane or "cherry-picker" to be loaded and unloaded. You might contact a local hauler who specializes in moving such material.

Basically, there are three locations you can consider for the tank: underground, above ground and outside, or inside the dwelling. Any outside installation, whether above or below ground, must be heavily insulated with 4 to 8 inches of high-performance, weatherproofed insulation. Pipes leading to and from the unit must be as short as possible and heavily insulated. Placing tanks inside dwellings is difficult unless you are building from scratch (an existing garage might be an exception).

If you *are* building from scratch, you have some interesting options. For instance, you might make the tank the centerpiece of your main room—perhaps with an integral fireplace! Figure 4.33 illustrates this concept, which is particularly attractive since the fireplace can be used as an auxiliary heating system. Or, you might construct a *square* storage tank in one corner of the building, perhaps in the basement. As shown in Figure 4.34, this would be a simple addition if you are using a cinder-block foundation. Naturally, the interior surface would have to be properly waterproofed if you are planning a water storage system. A storage system based on rock (a rock bin) would probably not require a special coating. (By the way, a rock bin has no particular construction oddities, but it does require insulation.)

Regardless of the tank design, you should be aware of the stratification present in fluid storage systems—particularly water. The warmer water rises to the top of the tank while the colder material sinks to the bottom. This important process affects the placement of inlet and outlet pipes, a topic we will discuss later.

Circulation Systems

In the next few sections, we are concerned with the transfer of energy between the components of a heating system. Referring to Figure 4.35, you can see that there are basically two energy transfer (or circulation) loops—one between the collectors and the storage unit, and another connecting the storage system and the living space. Often (but not always) these loops are physically separated, since different collection media may be used in the respective loops.

The *collector-to-storage* or *first loop* simply transfers energy from the collector to the storage device. Of course, there *are* some subtleties involved. We must take steps to prevent freezing and minimize corrosion, and define operating strategies to optimize energy collection. The major function of the *storage-to-interior* (*second loop*) is the effective distribution of stored thermal energy throughout the interior environment. In many ways, this loop is more complex than the collector/storage loop. The design of a balanced and responsive distribution system requires considerable expertise. Fortunately, there is a good deal of prior art in the field of heating and ventilating, so that we can use standardized design techniques. But before discussing either loop in detail, we need to address some more general considerations.

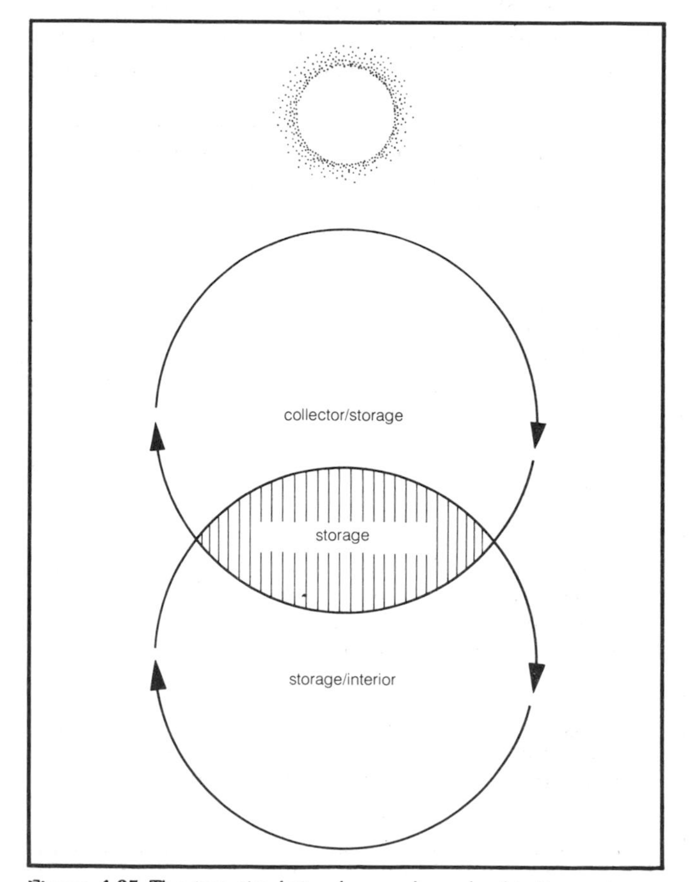

Figure 4.35 The two circulation loops of a solar heating system.

Circulation Fluids

The two most common and versatile circulation media are air and water. Water is perhaps the most popular of the two. It has been used for years in hydronic heating systems, and the technology is well developed. In solar heating systems, there is the added advantage that the circulating medium and storage material can be one and the same substance. Further, where space is limited, water pipes have an inherent advantage over more voluminous air ducts. Disadvantages of hydronic systems include the expense of leakproof storage facilities and piping, the danger of freezing, and general corrosion problems. However, these difficulties can be surmounted with thoughtful planning.

Freezing of the circulation water, particularly in the collector system, deserves serious attention since the associated expansion can rupture your expensive collectors. Damage can occur whenever freezing temperatures are approached, and the problem is aggravated when the circulation system is turned off. There are several ways to avoid freezing. First, simply be sure that noncirculating water is never exposed to freezing weather. For instance, you might drain the collector sections on particularly cold nights. Or, you can maintain a minimum circulation rate throughout the collector system, thus preventing any substantial ice formation. The latter solution is not highly recommended, since you will experience a net loss of energy from the storage system as you circulate.

A more popular method involves lowering the freezing point of the circulation fluid. This is commonly done by adding antifreeze to the system. Several commercial antifreezes are available; however, since many solar heating systems utilize components of differing metallic composition (e.g., aluminum, copper, steel), only mixtures which are compatible with multimetal systems should be used. Most antifreezes have an ethylene glycol base with various buffers and inhibitors added to achieve multimetal compatibility and pH (acid-base) stability. Such antifreezes undergo decomposition in the course of normal service, and organic acids are produced in the process. Solar heating systems using antifreeze probably will require periodic draining and replacement of the circulation mixture in order to prevent the accumulation of harmful acids. In automotive cooling systems, coolant changes are usually recommended yearly; however, because solar heating systems differ appreciably from automobile systems, appropriate replacement intervals remain to be established. You should monitor the pH of your antifreeze/water mixture (with litmus paper, for example) and replace the mixture when pH levels exceed the manufacturer's recommendations.

The freezing point of the antifreeze/water mixture depends upon the relative concentration of glycol. Table 4.22 lists the temperature specifications for one popular commercial brand. You might use the "median of annual extremes" listed in Appendix 4A to determine the right concentration for your locality.

Large quantities of glycol can run into considerable expense, a strong incentive to separate the collector circulation medium from the fluid used in the storage unit. (There should be little need for an antifreeze solution in the storage unit, and the cost is prohibitive anyway.) Figure 4.36 illustrates a way to isolate the collector circuit by using a series of tubes (*heat exchanger*) to transfer energy into the storage unit. Note that an *expansion tank* is included in this circuit, since we require room for liquid expansion as it is heated during the day. Naturally, all of this hardware adds to the complexity of the system, but it may be a necessary addition in colder climates.

There is one other method to reduce the probability of freezing. Lightweight insulating covers, perhaps of polyurethane, can be placed over the collectors at night. The practicality of this method is largely determined by the physical location of the collectors.

Corrosion problems can be handled in numerous ways. First, you should use de-ionized or distilled water in your system. Chemical corrosion inhibitors can be added; sodium di-chromate is popular, with typical concentrations on the order of a pound per 50 gallons. Many antifreeze compounds already include corrosion inhibitors (e.g., Mobil Chemical Virco-Pet 30).

Systems which have two or more types of metal are often subject to debilitating galvanic corrosion. Aluminum components are particularly susceptible when used in loops with copper or steel components. To prevent these reactions between the aluminum and other metal ions, all aluminum components should be galvanically (electrically) isolated from other metal components. Short lengths of plastic or rubber pipe, interspersed strategically, can effect this electrical isolation. Also, the use of a *getter column* is often advantageous. A *getter* is a cylinder which contains a number of aluminum sheets, over which the circulation water passes before entering the aluminum portion of the heating system. The column traps metal ions in the aluminum sheets, which are intended to corrode; at intervals you replace the column. An in-line water de-ionizer provides much the same protection.

While water enjoys a great deal of popularity, many homes have successfully utilized hot-air systems. One advantage of air as a circulating medium is that flow passages need not be leakproof. Further, if air is used in the collector/storage loop, it is more likely to be used in the interior loop, which allows for some savings in component costs. Freezing and corrosion pose no problems, although storage units for air systems are usually quite massive—usually crushed rocks. However, the ready availability of these special materials can make air an attractive medium. A final advantage is that ducting for heating is equally applicable for cooling.

Some of the disadvantages of air systems have already been noted. For one, the ducting is quite bulky. Flow noise is often a problem, and the power require-

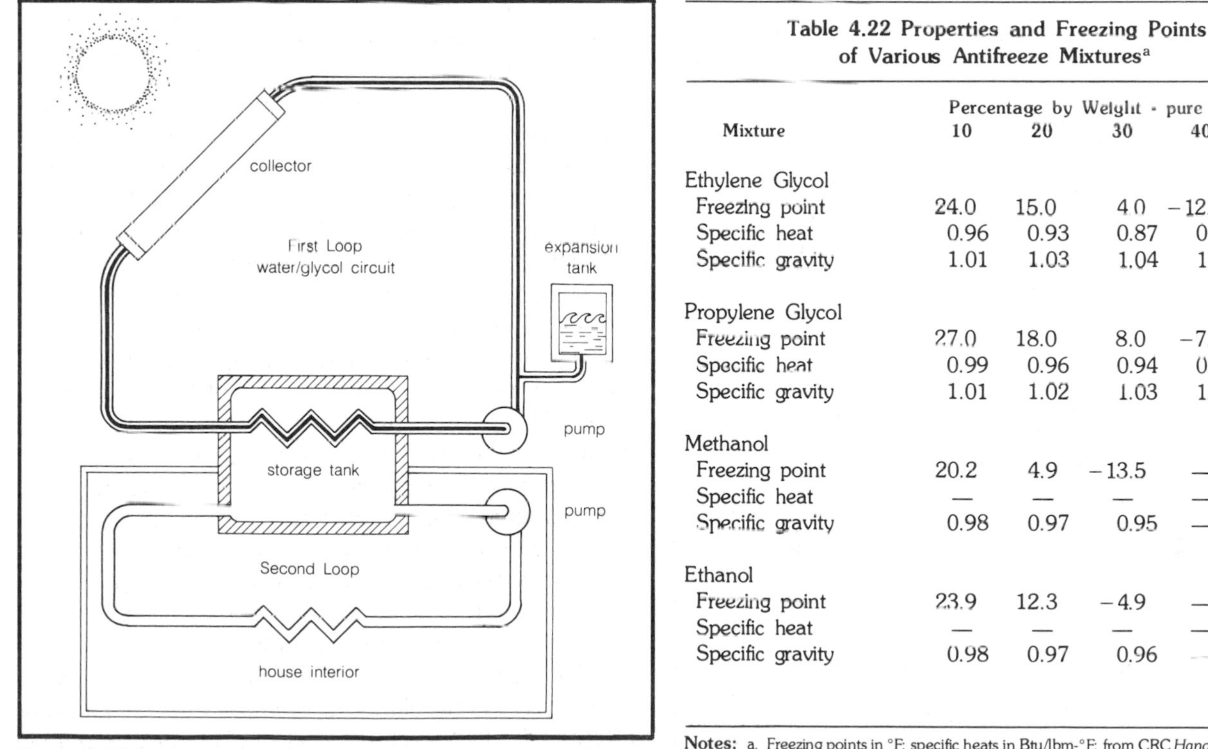

Figure 4.36 A separate water/glycol circuit for use in freezing climates.

Table 4.22 Properties and Freezing Points of Various Antifreeze Mixtures[a]

Mixture	Percentage by Weight - pure antifreeze 10	20	30	40	50
Ethylene Glycol					
Freezing point	24.0	15.0	4.0	−12.0	−32.0
Specific heat	0.96	0.93	0.87	0.81	0.76
Specific gravity	1.01	1.03	1.04	1.06	1.07
Propylene Glycol					
Freezing point	27.0	18.0	8.0	−7.0	−25.0
Specific heat	0.99	0.96	0.94	0.89	0.85
Specific gravity	1.01	1.02	1.03	1.04	1.04
Methanol					
Freezing point	20.2	4.9	−13.5	—	—
Specific heat	—	—	—	—	—
Specific gravity	0.98	0.97	0.95	—	—
Ethanol					
Freezing point	23.9	12.3	−4.9	—	—
Specific heat	—	—	—	—	—
Specific gravity	0.98	0.97	0.96	—	—

Notes: a. Freezing points in °F; specific heats in Btu/lbm-°F; from CRC *Handbook of Applied Engineering Science*.

ment for the circulating blower is substantially higher than for an equivalent water system. This is particularly true in solar heating systems where, due to lower source temperatures, flow rates must be about twice that of conventional hot-air furnace systems.

Natural and Forced Circulation

There are two basic methods of fluid circulation: free and forced. Circulation loops using the former technique are called *free-convection*, *gravity*, or *thermosyphon* systems. Fluid movement is induced by differences in density between different parts of the fluid. Because hot fluid is less dense than colder fluid and thus tends to rise, a naturally circulating collector/storage loop is possible with proper positioning of components. Figure 4.37 illustrates a typical thermosyphon circuit used in a collector/storage loop. To achieve thermosyphoning action, it is necessary that the storage tank be several feet above the collector unit. This requirement makes rooftop collectors unfeasible since the heavy storage tank would have to be supported above the roofline.

Note the relative position of the inlet and outlet pipes in Figure 4.37, taking advantage of the stratification of temperature in the storage tank. Relatively cold water is taken from the bottom of the tank, heated in the collector, and deposited in the upper portion of the tank. Circulation is sustained by the differing densities of the cold *downcomer line* and the warm collector/riser combination.

The thermosyphon principle can also be used in the second or interior loop. Decades ago, many hot-water heating systems were based on this design; in fact, the ASHRAE *Guide and Data Book* still contains substantial design information. In such systems, the storage tank must be located *below* the heated space, for the same reasons it must be located above the collectors. (You can see that a thermosyphon effect in *both* the collector/storage and storage/interior loops would require some very strange architectural geometry.)

In recent years, thermosyphon heating systems have all but disappeared. Considerable care is required to design one properly: friction effects in various lines must be balanced, reverse flow prevented, and trapped air and drainage problems faced. Forced circulation systems, using pumps and blowers, can be planned in a fraction of the time. Of course, pumps and blowers require electrical or mechanical energy which may be at a premium (if available at all). When dealing with water systems we suggest that you rely on small, low-power circulation pumps. A possible exception would be the collector/storage loop. In air systems, the added pumping requirements are significant enough to consider seriously a total or partial thermosyphon system. Soon we will present more detailed information on the sizing and design of these circulation systems, but first we must lay some additional groundwork.

Piping and Ducting

We have already suggested that the circulation systems are comprised of either pipe or ducting. Pipes for fluid circulation come in a variety of diameters, running from one-quarter to several inches in outside diameter. Copper is probably the most popular material, although galvanized steel is used in many applications. The wall thickness of the pipe varies; it is denoted by "Schedule" numbers or, in the case of copper, by Type. Common copper pipes for plumbing are Types "K," "L," and "M." A variety of fittings, elbows, and tees are available in addition to straight pipes.

Certain kinds of plastic piping are also available. They are relatively cheap and lightweight and are easily assembled with special cement and simple tools. The major drawback of plastic has to do with specific temperature limitations. ABS (acrylonitrile butadiene-styrene), RS (rubber styrene), and PVC (poly-vinyl chloride) pipes are *not* recommended for hot-water service. However, CPVC (chlorinated poly-vinyl chloride) pipe can withstand temperatures up to 180°F under operating pressures of 100 psi. In solar heating systems, pressures seldom exceed 15 or 20 psi, so this temperature limit might be extended somewhat beyond 180°F. All in all, plastic piping is quite tempting. But because the pipe has had limited service in hot-water systems, you should proceed with some caution.

Ducts are round, square, or rectangular and are constructed of galvanized sheet metal. As with pipes, right-angle bends, tees, and other fittings are available. But due to the large size of duct systems, a large amount of custom fabrication must be done for particular jobs.

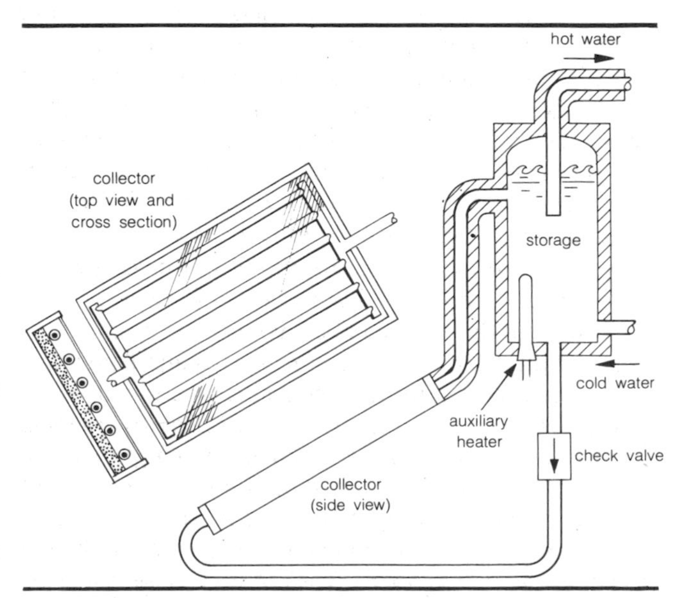

Figure 4.37 A thermosyphon system with collector and storage.

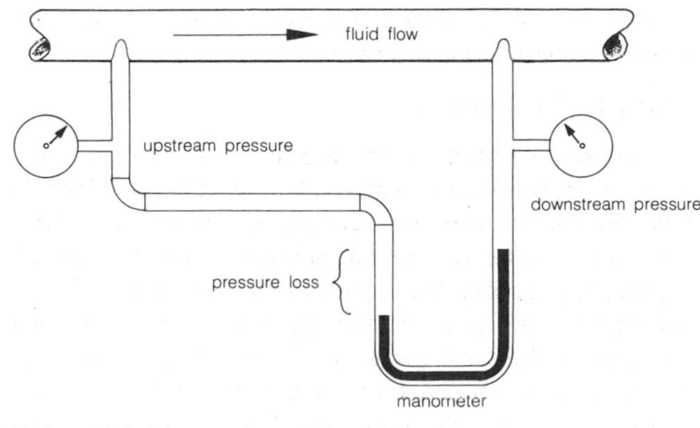

Figure 4.38 Pressure drop along a pipe.

Duct sizes vary from six inches up to several feet on a side.

Ducting and pipe supplies can be obtained at almost any hardware outlet. You might consult a Sears catalog to get an idea of what's available and current prices. Sears also sells some inexpensive booklets outlining installation techniques.

Flow rate is an important quantity related both to pipes and ducts. For liquids, the flow rate is frequently expressed in gallons per minute (gpm). For air, flow rates are given in cubic feet per minute (cfm). The flow rate is of particular interest to us since we can use it to estimate the retarding friction experienced by the flow. And knowing the flow rate and retarding friction enables us to select a proper blower or pump for our circulation system.

The retarding friction in a passage is expressed in terms of pressure drop. That is, the friction causes a loss of pressure in the direction of the flow. You might expect to see units of pounds per square inch (psi), but typically pressure drop (or head loss) is expressed in inches or feet of water. The terminology originates from the methods used to measure the pressure drop. Losses along several feet of pipe usually amount to only a few hundredths of a psi. While these subtle effects do add up, such small pressures are best measured with water manometers like the one shown in Figure 4.38. One psi corresponds to about 2.3 feet-H_2O or about 27.7 inches-H_2O. You can imagine, then, that one inch of water corresponds to a very small pressure drop! (For instance, drinking with a straw requires only a "few inches" of vacuum.)

Pressure drop or head loss along straight pipe ducting is tabulated for a variety of flow sizes and flow rates. Losses due to flow through bends, tees, and valves can also be calculated. Frequently, losses in fittings are tabulated in terms of "equivalent length of straight pipe or duct." Tables 4.23 and 4.24 present information for tubing (similar data can be obtained for duct systems by consulting one of the heating and ventilating handbooks listed in the Bibliography). Table 4.23 reports the pressure loss in feet-H_2O per 100 feet of pipe length. In

Table 4.23 Pressure Loss Due to Friction[a]

Flow (gpm)	Copper or Steel Tube ½″	¾″	1″
1	1.8	0.5	—
2	6.4	1.2	0.5
3	13.1	2.3	0.7
4	21.7	4.2	1.2
5	31.9	6.0	1.6
10	108.0	19.9	5.8
15	—	40.7	11.6

Notes: a. Units of ft-H_2O per 100 ft of tube.

FITTINGS

Fitting	Connection	Material
REGULAR 90° ELL	SCREWED	STEEL
		CAST IRON
	FLANGED	STEEL
		CAST IRON
LONG RADIUS 90° ELL	SCREWED	STEEL
		CAST IRON
	FLANGED	STEEL
		CAST IRON
REGULAR 45° ELL	SCREWED	STEEL
		CAST IRON
	FLANGED	STEEL
		CAST IRON
TEE-LINE FLOW	SCREWED	STEEL
		CAST IRON
	FLANGED	STEEL
		CAST IRON
TEE-BRANCH FLOW	SCREWED	STEEL
		CAST IRON
	FLANGED	STEEL
		CAST IRON
GLOBE VALVE	SCREWED	STEEL
		CAST IRON
	FLANGED	STEEL
		CAST IRON
GATE VALVE	SCREWED	STEEL
		CAST IRON
	FLANGED	STEEL
		CAST IRON
ANGLE VALVE	SCREWED	STEEL
		CAST IRON
	FLANGED	STEEL
		CAST IRON

Table 4.24 Equivalent Pipe Lengths of Common Hydronic Fittings[a]

	Steel Pipe	Copper Tubing
90° Elbow	25	25
45° Elbow	18	18
90° Elbow (long radius)	13	13
90° Elbow (welded)	13	13
Reducer coupling	10	10
Gate valve (open)	13	18
Globe valve (open)	300	425
Angle radiator valve	50	150
Radiator or convector	75	100
Boiler	75	100
Tee (typical)	100	100

Notes: a. Given in diameters of pipe.

contrast, Table 4.24 lists the equivalent length of straight pipe for a variety of fittings. For instance, a 90-degree elbow (right-angle bend) is equivalent to 25 "diameters" of straight pipe loss.

Example: Suppose we have 150 feet of three-quarter-inch copper pipe. The line contains sixteen 90-degree elbows and four open gate valves. Let's find the pressure drop if the pipe has a flow rate of 5 gpm of water.

Solution: Using Table 4.24, we find:

Length of ¾-inch tube	150 feet
16 elbows @ 25 diameters each (1 diameter = ¾ inch)	25 feet
4 gate valves @ 18 diameters each (1 diameter = ¾ inch)	5 feet
Total equivalent length:	180 feet

Referring to Table 4.23, we see that a three-quarter-inch tube carrying 5 gpm loses 6.0 feet-H_2O per 100 feet of tube. The equivalent length is 180 feet, so the total pressure loss is

$$\text{Pressure Loss} = (180\ ft)\left(\frac{6.0\ ft\text{-}H_2O}{100\ ft}\right)$$

$$= 10.8\ ft\text{-}H_2O$$

Don't confuse the units of "ft-H_2O," a measurement of pressure, with the "feet" units of pipe length.

In many cases, the above pressure loss and flow rate would be sufficient to size the pump needed to sustain the assumed flow rate. We must make an exception for pipes which experience a *net* change in elevation. An additional foot of head loss accrues for each foot of elevation. Suppose, for instance, that the pipe in the previous example underwent a 5-foot rise in elevation from the inlet to the exit. Then the total pressure or head loss (10.8 + 5.0) would be 15.8 ft-H_2O.

The methodology for computing pressure drop in ducts is identical, except that different tables must be used. Elevation changes can be neglected, since air is so much lighter than water. Head loss is usually in inches-H_2O rather than ft-H_2O. The details and appropriate tables can be found in any of the reference texts dealing with heating and ventilating (see Bibliography).

Pumps and Blowers

Probably the most difficult task in sizing components in a circulation system is the computation of head loss. But once that information is available, along with the flow rate, we can then select a pump (for liquids) or blower (for air). Typical performance curves for several circulating pumps are shown in Figure 4.39. Each curve represents a different size pump. Any given size has a distinct relation to both the flow rate it handles and the pressure behind that flow. If the pump has little pressure drop (head) to deal with, the flow rate will be high (right-hand portion of the curve). However, as the required head increases, the flow capacity decreases (moving left on the curve). We can see that model number *1″-PR* would probably handle the 5-gpm flow of our previous example (a head of 15.8 ft-H_2O).

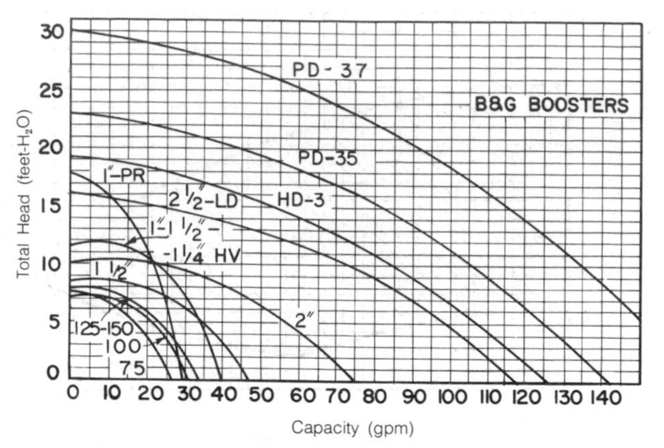

Figure 4.39 A family of pump curves for line-mounted pumps (with permission from Bell & Gossett, Inc.).

The curves in Figure 4.39 are based on pumps running continuously at a constant speed. But variable-speed pumps are also available which permit operation over a much wider range. Figure 4.40 depicts the range of performance for a typical variable-speed unit. Constant-speed pumps can be converted to this type of operation by driving the motor with an electronic speed controller. Today these controllers are small and relatively cheap (often less than $10). They are used to control the speed of drills and other machine tools that use alternating-current motors and can be purchased at any hardware or electronics store. We have been quite pleased with an inexpensive, solid-state device available from Omnetics (P.O. Box 113, Syracuse, New York 13211).

Aside from flexibility, speed controllers offer significant energy savings. Suppose you experience a moderately cool day and wish to circulate warm water or fluid through the living space, but at a reduced flow rate. (The reduced flow rate will lower the normal heat ration added to the living space and keep temperatures comfortable, not too hot.) With a fixed-speed pump, you must run at full speed and retard the flow by partially closing a valve. The closing of the valve increases the total head loss in the system and thus reduces the flow rate (see Figure 4.39). It also wastes a considerable amount of electrical energy for the pump. A variable-speed pump, on the other hand, can be adjusted to give the desired flow rate with negligible power waste.

Air blowers (or *movers*) are selected in much the

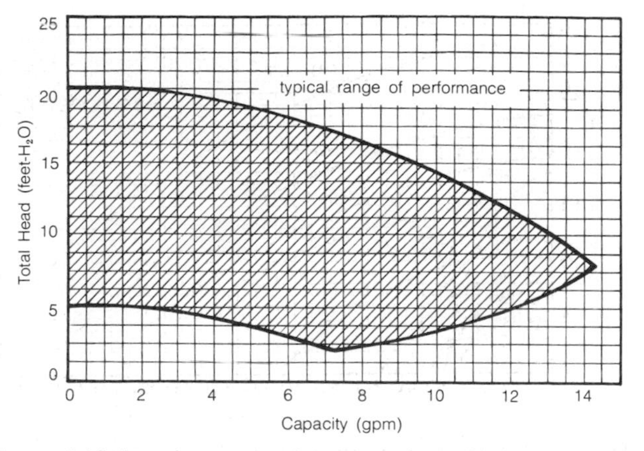

Figure 4.40 A capacity curve for a typical variable-speed pump.

same way as pumps. Curves similar to Figure 4.39 are available. These curves have much the same shape, although the coordinates generally read cfm instead of gpm, and inches-H_2O instead of feet-H_2O.

We will have more to say about pump and blower selection when we discuss the collector/storage and storage/interior loops in more detail. But now we would like to offer some suggestions that may be helpful in acquiring your pump or blower.

The *pumps* you require are probably going to be small. In fact, it's often difficult to find pumps small enough! Water flow rates in solar heating systems seldom exceed 5 gpm, and the resulting pressure drops are quite small. You can purchase reasonably sized pumps from a heating and ventilating supply house which carries hydronics systems or from Sears. Pumps from Bell and Gossett are also quite popular. Calvert Engineering (7051 Havenhurst Avenue, Van Nuys, California 91406), too, carries a reasonable line of inexpensive pumps ($15 and up). Other brands are available, but be sure that the considered pump can withstand your operating temperatures. Also, if you plan to use antifreeze or other additives, make sure the pump material is compatible.

Most pumps come with an integral motor, with typical ratings on the order of 0.1 horsepower. You can compute the minimum power required in water systems by multiplying the gpm-rating by the head in ft-H_2O and dividing by 3960. The result will be in horsepower; you should double your answer to account for the inefficiencies in small pumps. Using this formula, you will find that most solar circulation schemes require ridiculously small pump motors. Direct-current or alternating-current motors can be used, provided their speed and torque characteristics match the pump. Of course, direct mechanical drive of the pump by windmill or waterwheel cannot be ruled out.

Blowers can be obtained from most of the sources already mentioned. These are quite common in home heating systems, and you should be able to obtain a used unit without too much difficulty. Drive motors will be around 0.5 to 1.0 horsepower for small systems. Once again, we see that blowers must be disproportionately more powerful than their water-circulating counterparts.

Insulation

If the heating system is operating properly, the circulating fluid will always be warmer than the immediate surroundings. Heat transfer is unavoidable, but we can take steps both to minimize unwanted losses and to control the rate of energy addition to the inside environment.

The remedies are quite straightforward. The storage tank and all transfer piping should be heavily insulated. Standard pipe insulation is available in a variety of sizes and materials. Lately, people have been turning to new, foamed-plastic insulation which is lightweight and easily installed. The Armstrong Cork Company (home office in Lancaster, Pennsylvania) markets Armaflex in flexible pipe, ribbon, and sheet form. The sheet form can be applied to the large curved surfaces you may encounter on a storage tank. Similar products are marketed by the APCO Company (1350 E. 8th Street, Tempe, Arizona 85281). Their thermal jacketing is called Thermazip, alluding to the zip-on installation technique. Kits are available to insulate most fittings, as well as straight pipe runs.

We would recommend from four to six inches of foam or fiberglass insulation on the storage tank and at least one-half inch on pipe runs and fittings. Insulation can be reduced or completely eliminated as the piping enters the heated space, *provided* you have some way of controlling the heat addition to the room. Often it is best to insulate all circulation components except the room registers (the individual heating convectors for each room) so that unwanted heat addition can be avoided.

Proper placement of the storage tank can be especially useful. Losses from a unit which is located inside a living space can serve to warm your room (see Figure 4.33). Tanks located outside or underground do not share this advantage.

Circulation Loops: Detailed Considerations

We now have the background to examine the first and second circulation loops in detail. We'll look first at the collector/storage loop and then deal with the storage/interior loop.

Collector/Storage Loop

The function of the first loop is the collection and, often, the simultaneous storage of solar energy. In essence, we want to operate the first loop whenever there is the possibility of collecting solar energy. Clearly, then, the first loop should be turned on only when the collector surface is sufficiently warmer than the circulating medium in the storage unit. We say "sufficiently" because there is always heat loss associated with operating the first loop; the collector must be hot enough to effect a *net* heat transfer to the circulating medium, in spite of heat losses. In practice, collector temperatures should be about 15°F higher than the minimum storage temperature before the first loop is operated. Because we want to circulate the coldest storage fluid through the collector (to maximize heat transfer), the collector *feeder* should be located near the bottom of the storage unit, and the hot return leg of the loop should enter the top of the storage unit.

Knowing *when* to operate the first loop is part of the control function, and there are several means at your disposal. The most sophisticated method involves thermostatic control. Here, thermostats, judiciously located in the collector and storage units, are connected electrically so that a predetermined temperature differential between thermostats automatically turns on the first loop pump (or fan); an insufficient temperature differential stops the flow. Anyone with a little electrical know-how should be able to design such a system.

You also might consider the use of a simple photocell to initiate first-loop operation. The cell would sense the presence of sunshine, perhaps at a predetermined level, and trigger a simple relay to run the circulator. Finally, manual control is fine with us, if it's fine with *you*: when you see the sun is shining, simply flip a switch!

In both natural and forced circulation systems, it is important to prevent *reverse cycling* when the heating system is off. Reverse cycling operates on the same principle as thermosyphoning, except that it is a cooling cycle which occurs when the circulation fluid loses energy (usually through the collector, at night). A simple *check valve*, which permits flow only in one direction, will prevent reverse cycling in fluid systems.

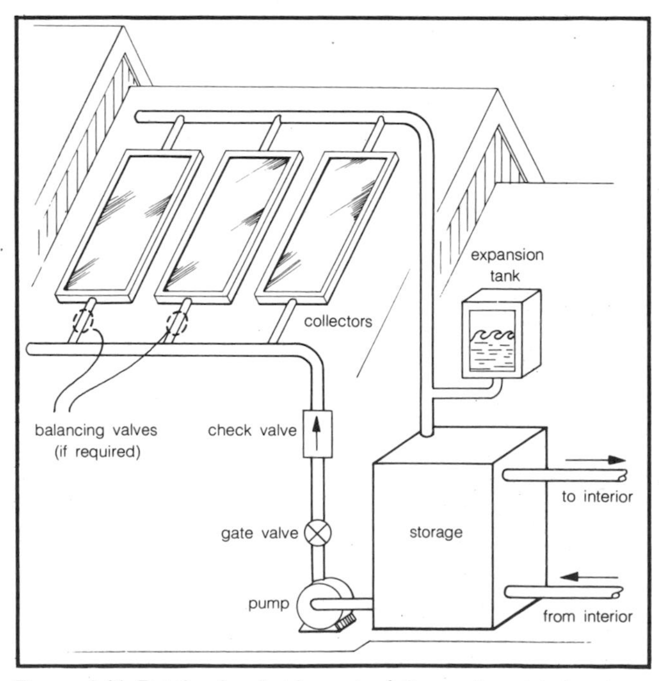

Figure 4.41 Details of a first-loop circulation system using water.

Figure 4.41 illustrates a typical collector/storage loop which might be used in a liquid system. We have included in the diagram an expansion tank, a check valve, and a pump. We can now ask some very pointed questions. For instance, what size pump do we need?

Our answer depends upon the number of collectors, the flow rate we choose, and the resulting frictional losses. Also, we might have to deal with the elevation change between the pump and the collector. Flow rates through collectors are usually on the order of 1 gpm. Manufacturers frequently recommend a range of flow rates. Slower flow rates expose the fluid to the sun for a longer time and higher temperatures are achieved at the outlet of the collector; faster flow rates result in lower exit temperatures. But don't be misled here. Often the faster flows collect energy more *efficiently*, since the temperatures are lower and heat losses from the collector are less. This is clearly illustrated by the efficiency curves in Figure 4.27. At higher average collector temperatures, we experience lower efficiencies; and high collector temperatures arise from low fluid flow rates.

Suppose each of the collectors in Figure 4.41 was designed for a flow rate of 1 gpm. Since there are three collectors, the total flow rate through the pump must be 3 gpm. (Often, it's a real problem to obtain an even division of flow between collectors plumbed in parallel. The manifolds must be designed so that the pressure drop along each branch is nearly equal. Fortunately, these effects are somewhat reduced at lower flow rates and/or when fluid flows "upward" in the collectors. Keep manifolding short and balanced.) To determine the required pump head, we would sum the frictional pressure drops along the flow path. The straight pipe, fittings, and valve can be handled using the methods in the preceding section on piping and ducting. The pressure drop in the collectors must be either measured or specified by the manufacturer. In our configuration, the collectors are parallel, so we include only the pressure drop along one collector. (If the collectors were connected in series, we would use three times the pressure drop across one panel, but then the overall system flow rate would be only 1 gpm.)

Example: Suppose the system in Figure 4.41 was comprised of 100 feet of three-quarter-inch copper tubing, with ten elbows and two valves. There is a 30-foot height difference between pump and collector. Further, let's assume that the pressure drop across each collector plate is 6 inches of H_2O at a flow of 1 gpm. What is the total head loss?

Solution: A rough calculation, using Tables 4.23 and 4.24, might go like this:

Element	*Equivalent Length (to nearest ft)*
100 ft of 0.75-inch pipe	100 ft
10 90-degree elbows (@ 25 diameters each)	15 ft
1 gate valve (18 diameters)	1 ft
1 check valve (est. 20 diameters)	1 ft
Storage tank (use "boiler"— 100 diameters)	6 ft
Misc. manifolding (est. 200 diameters)	13 ft
Total Equivalent Length:	136 ft
Pressure loss (2.3 ft-H_2O per 100 ft @ 3 gpm)	3.15 ft-H_2O
Collector (6 inches-H_2O @ 1 gpm)	0.50 ft-H_2O
Total Head Loss:	3.65 ft-H_2O

Thus the pump must handle a flow rate of 3 gpm with a frictional head of 3.65 ft-H_2O. But have we forgotten something? What about the 30-foot change in elevation as the fluid rises to the collector? This could easily be ignored in an air system, but water is much heavier.

Our question has a very subtle answer. If the circulation loop is completely closed and filled, the change in elevation can be ignored! The energy required to lift the fluid on one side will be returned when the fluid falls along the other. This effect is closely related to a syphon. We might say, then, that the pump need only handle the frictional pressure drop. In a way this is true, but we are faced with something of a philosophical dilemma. How do we fill the loop in the first place? Or suppose we wish to drain the collectors on cold evenings—how is the fluid flow started again in the morning?

These questions point to the advantages of a variable-speed pump. Referring to Figure 4.39, we see that pumps develop their maximum head at "zero" flow rate. A pump would approach zero flow rate as it attempted to lift the water to the highest point in the system (frictional losses would then be zero, due to zero flow). Once the highest point was reached, the downhill flow rate would quickly increase and the only head loss would be that due to friction. Unless you have an alternative way to charge the circulation loop, your pump must have the capacity to lift the water to the maximum elevation in the system. In our example, the pump must have a "zero flow" head capacity of at least 30 ft-H_2O, which is more than eight times the "steady flow" head loss. Once the flow is started, a speed controller can slow the pump as it takes on the much easier task of overcoming flow friction.

Much of our discussion applies equally well to air systems, except that changes in elevation can be neglected. The blower is sized in a way analogous to our preceding example; details can be found in any good text on heating and ventilating. A representative air system is sketched in Figure 4.42.

Storage/Interior Loop

Now we are interested in distributing energy to the interior living space in a controlled fashion. Once again, in this second loop, air and water are the prime candidates for circulation. Water systems follow well-developed design criteria established by the hydronics industry; similarly, years of practical experience are behind forced-air systems.

Slight modifications must be made to both types of systems in order to interface with a solar heating system. These changes are prompted by the temperature differences between conventional and solar systems. Storage units in solar systems operate in the range of 120 to 160°F. In contrast, standard hydronic and forced-air

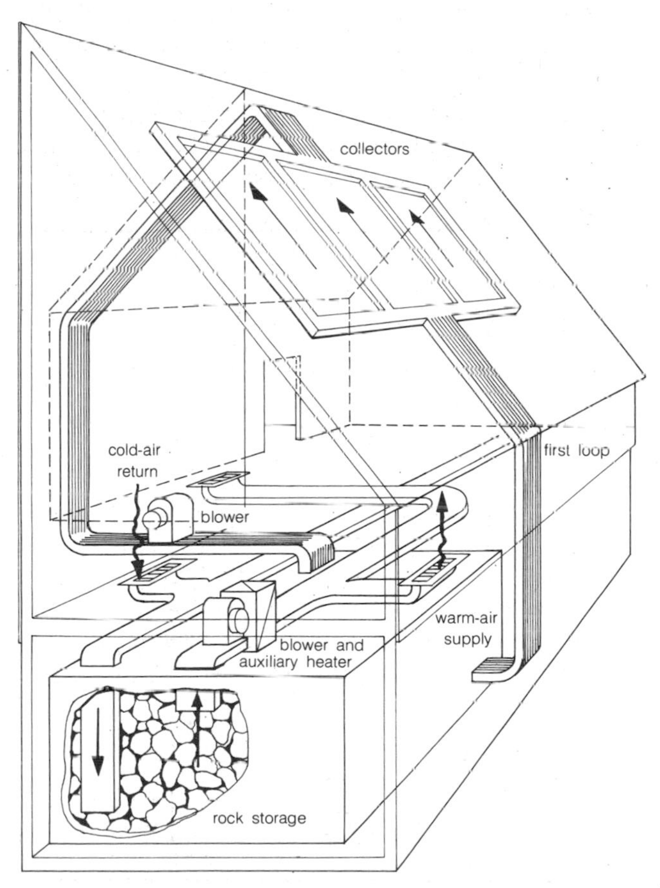

Figure 4.42 A schematic for a solar heating system utilizing air.

systems operate at higher temperatures. For instance, a gas-fired or electric water heater circulates water at temperatures in excess of 180°F.

To compensate for the lower temperatures in a solar system, most components must be oversized. Hydronic baseboard heaters (*convectors*) must have twice the thermal rating, and forced-air circulation rates must be correspondingly higher. Adaptation is otherwise quite straightforward.

A typical hydronic distribution system for a small dwelling is sketched in Figure 4.43. This configuration is known as the "one-pipe" scheme, probably the most popular for small systems. Hot water leaves the storage tank and traverses the circulation loop, returning anywhere from 10 to 30°F cooler than when it left. Each room in the house has one or more baseboard registers which perform the actual transfer of heat into the room air. An uninsulated length of pipe would do the job, but finned pipe is much more efficient. Baseboard convectors, like the one in Figure 4.44, are designed in this way. These heating units are available in a variety of lengths and cost \$2 to \$3 a foot (1974). Heat-transfer and pressure-drop characteristics are similar to those given in Table 4.25. Note that the ratings for these particular units are quoted at 1 and 4 gpm, with water temperatures from 170 to 240°F. For instance, a slant/fin model 10-75E baseboard operating at 1 gpm will release 680 Btu per hour per foot of length with water circulating at 200°F. This rating can easily be adjusted for the lower temperatures in a solar heating system. You will recall from our earlier discussion of heat loss that heat-transfer rates are proportional to the temperature difference. Since the table lists entering air at 65°F, the catalog rating we quoted corresponds to a 135°F temperature difference (200 − 65). If we were operating with 150°F water, the temperature difference (150 − 65) would be 85°F, so the rating would be reduced proportionally. We can use a general formula

E. 4.31 $$\frac{R_1}{R_2} = \frac{\Delta T_1}{\Delta T_2}$$

where R_1 equals the solar rating we are trying to determine; R_2 equals the catalog rating for the baseboard collector; ΔT_1 equals the temperature difference between water and inside catalog design temperature in our solar system; and ΔT_2 equals the temperature difference at the given catalog rate. Then we see that

$$Rating_{150^\circ} = Rating_{200^\circ} \frac{(150 - 65)°F}{(200 - 65)°F}$$

$$= (630\ Btu/hr\text{-}ft) \left(\frac{85}{135}\right)$$

$$= 428\ Btu/hr\text{-}ft$$

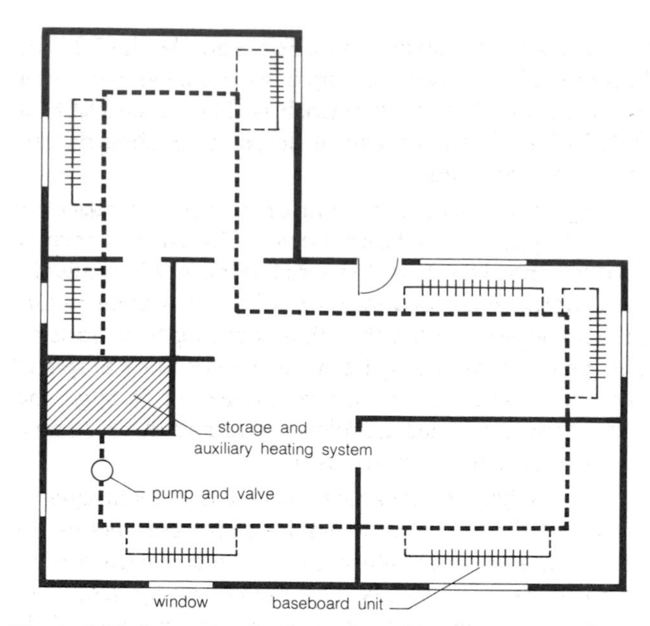

Figure 4.43 A one-pipe hydronics loop for a small dwelling.

Such proportioning can be done between any two temperatures.

This method is approximate, since the outside convective film coefficient is reduced as the system temperature is lowered. This tends to make the actual ratings lower than those computed using the method above. It might be wise to reduce the calculated values by 10 to 15 percent. On the other hand, some manufacturers may publish performance data at lower temperatures.

Since the room heaters normally operate at lower output ratings, the number required per room can be two to three times that of a conventional system. The attendant cost and aesthetic problem can be circumvented by using the solar heating system as a preheater and using the auxiliary system to boost the temperatures 30 to 40°F. This technique will be discussed more fully when we consider auxiliary systems.

Table 4.25 also lists the pressure drop per foot of baseboard. This information is useful in sizing the pump

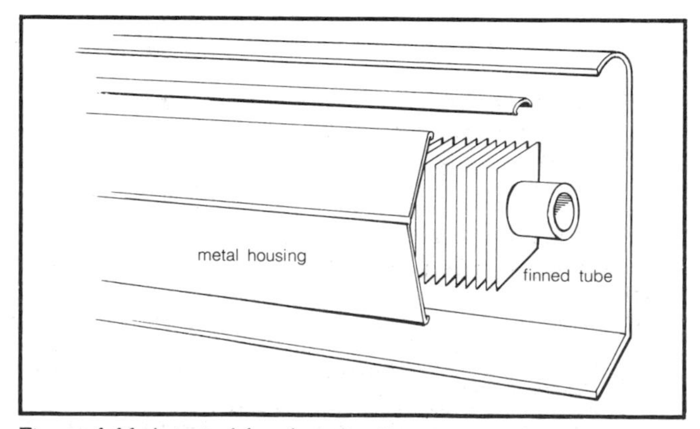

Figure 4.44 A typical baseboard unit.

for the interior circulation loop. For example, the 10-75E baseboard has a pressure drop of 47 *milinches* per foot at a flow rate of 1 gpm. A milinch is 0.001 inch-H_2O, or 0.0008 ft-H_2O; thus pressure drops in baseboards can often be neglected.

Figure 4.45 illustrates some of the piping details used with this type of baseboard heater. The valve shown is used to control the heat flow into the room. Alternatively, many units have a louvered control which restricts natural convection around the fins, thus reducing heat transfer. The pipe tee returning from the heater is of a special "venturi" design to encourage circulation through the unit. Finally, a small air bleed is provided to remove trapped air whenever necessary.

A sidelight to baseboard installation techniques is shown in Figure 4.46. These Fast/Flex connectors are flexible metal tubes which simplify installation around sharp corners and bends. The bellows design also compensates for thermal expansion and contraction of the piping system. This common problem often causes annoying clicking and popping sounds; strategically placed bellows units can quiet the system considerably. You also might use these connectors on the solar collector plates, where the expansion problem is particularly severe. The retail cost for a three-quarter-inch connector (diameter) is about $4 (1974).

Placement of the heating units is based on the heating requirements in specified parts of the dwelling. They are usually located along outside walls, often directly under windows to compensate for the higher local losses. The number and/or length of the baseboard convector units in any given room depends on the design heat loss from that room. Earlier, we computed the total design heat loss from an entire building. This figure can be subdivided into individual components reflecting different rooms or sections of the house. Heat loss in a particular room is determined by the losses through exterior walls (interior walls don't count), floor area,

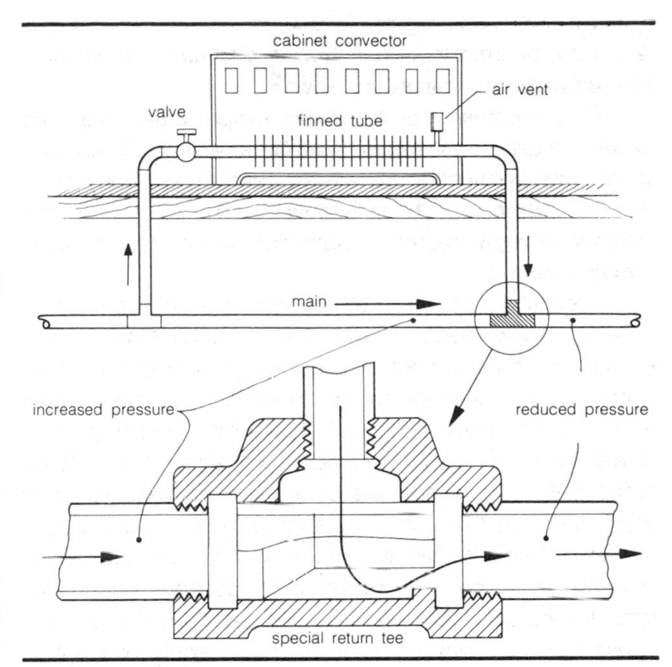

Figure 4.45 Details of plumbing to and from a baseboard convector.

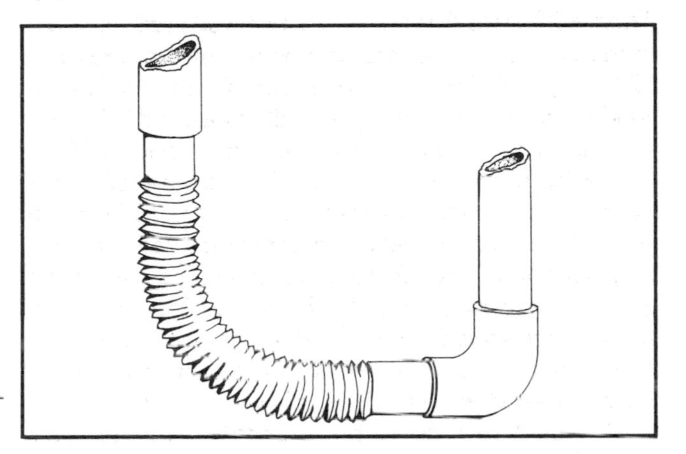

Figure 4.46 Flexible metal connectors for use with baseboard convectors and solar panels (Fast/Flex Connectors by Slant/Fin, Inc.).

Table 4.25 Typical Performance Data for a Baseboard Unit[a d]

Element	Water Flow	Pressure Drop[b]	170°	180°	190°	200°	210°	215°	220°	230°	240°
No. 10-75E Baseboard with E-75 element; ¾" nominal copper tubing, 2⅝" x 2⅛" x .009" aluminum fins bent to 2 19/64" x 2⅛", 55 fins/ft.[c]	1 gpm	47	480	550	620	680	750	780	820	880	950
	4 gpm	525	510	580	660	720	790	820	870	930	1000
No. 10-50 Baseboard with E-50 element; ½" nominal copper tubing, 2⅝" x 2⅛" x .009" aluminum fins bent to 2 5/16" x 2⅛", 55 fins/ft.[c]	1 gpm	260	490	550	610	680	740	770	800	860	920
	4 gpm	2880	520	580	640	720	780	810	850	910	970

Notes: a. Units of Btu/hr-ft with 65°F entering air; from Slant/Fin Corp., Publication 10-10.
b. Milinches per foot.
c. End fins are of tinned steel for extra ruggedness.
d. Ratings are for element installed with damper open, with expansion cradles. Ratings are based on active finned length (5" to 6" less than overall length) and include 15% heating effect factor. Use 4 gpm ratings only when flow is known to be equal to or greater than 4 gpm; otherwise 1 gpm ratings must be used.

Table 4.26 Room-by-Room Baseboard Requirements and Heat Loss of House in Figure 4.43

Room	Heat Loss @ Design (Btu/hr)	Required Baseboard (ft)[a]
Living Room	7000	17.5
Den	6000	15.0
Kitchen	6000	15.0
Bedroom	4000	10.0
Bathroom	2000	5.0
Total:	25,000	62.5

Notes: a. Based on Model 10-75E baseboard operated at 1 gpm and 150°F, with approximate rating of 400 Btu/hr-ft.

ceiling area, and the infiltration load. It is easy enough to divide the total heat loss figure into these component figures. Then using the published hot-water ratings for whatever heating unit we select (Table 4.25 is an example) the required number of feet of baseboard can be determined.

As an example, suppose the home depicted in Figure 4.43 had a total design heat loss of 25,000 Btu/hr. Assume the room-by-room heat load is divided as shown in Table 4.26. The storage tank contains water at 150°F, the piping is 200 feet of three-quarter-inch copper tubing, plus miscellaneous fittings, and we will design for a 20°F temperature drop—that is, we will select a flow rate such that the circulating water returns to the storage tank (150 − 20) at 130°F.

First, we compute the required flow rate. Using a related form of Equation 4.29, from our discussion of energy storage, we can equate the total heat load and flow rate in the following way:

E. 4.32 $$Q = 500\ (gpm)\ (\Delta T)$$

where Q equals the heat load; ΔT equals the temperature drop in the loop; *gpm* equals the flow rate; and 500 is a conversion factor accounting for water's density and specific heat. Rearranging, we have

E. 4.33 $$gpm = \frac{Q}{500(\Delta T)}$$

In our example, ΔT equals 20°F and Q equals 25,000 Btu/hr, so that gpm equals 2.5. This is the total flow rate we must have in the main supply branch.

The flow through the individual elements will be something less than 2.5 gpm, depending on the control-valve setting. Referring to Table 4.25, you can see that the hot-water ratings of the baseboard units differ by only a few percent between 1 and 4 gpm. We might use 1 gpm as a typical flow rate. The heat rating for a 10-75E unit has already been adjusted for 150°F (see Equation 4.31) and is on the order of 400 Btu/hr-ft. Naturally, convector units further down the line will operate with cooler water, but since the *total* temperature drop has been specified as 20°F, little error will be introduced if we assume a constant 150°F temperature for the water circulating in the convector. The required footage of baseboard for each room is summarized in Table 4.26. The figures were calculated by dividing the heat loss in each room by the heat rating of the unit (400 Btu/hr-ft).

The only remaining task is to size the pump. We already know the flow rate from Equation 4.33. We now need to find the system head loss. This can be done in an approximate fashion by computing the pressure loss experienced as the flow travels to and from the most distant convector unit. In a "one-pipe" system such as that shown in Figure 4.43, this is equivalent to the total circuit length, no matter which convector we choose. We use Tables 4.23 and 4.24: all elbows and fittings encountered along the way are included, as well as the pressure drop experienced in the convector itself (often negligible; we have estimated 15 feet of convector footage as representative for this system). Tees need only be counted if you are taking a right-angle turn in one, which usually occurs when you reach the convector of your choice. So even though there are fourteen tees in the circuit depicted in Figure 4.43, we only need account for two in our head-loss calculation. Table 4.27 summarizes the friction head losses. The total head loss is seen to be 4.16 ft-H_2O, with a total flow rate of 2.5 gpm. We now have enough information to select a circulation pump, and the job is complete!

Baseboard convectors aren't the only devices which

Table 4.27 Summary of Friction Losses in Sample Circuit of Figure 4.43[a]

Flow Element	Equivalent Length (to nearest foot)
200 ft of ¾″ copper pipe	200
8 90° elbows (@ 25 diameters ea.)	13
1 gate valve at pump (@ 18 diameters)	1
1 gate valve at convector (@ 18 diameters)	1
2 tees (@ 100 diameters ea.)	13
Storage tank (use "boiler"—100 diameters)	6
Subtotal: 234 ft	
	Pressure Loss
Circuit (@ about 1.75 ft-H_2O per 100 ft with 2.5 gpm)[b]	4.10 ft-H_2O
Convector (@ 47 milinches-H_2O per foot times 15 ft)[c]	0.06 ft-H_2O
Total: 4.16 ft-H_2O	

Notes: a. Data determined from Tables 4.23 and 4.24.
b. Head-loss value taken midway between 2 and 3 gpm.
c. From Table 4.26. We have figured a representative convector to be about 15 feet.

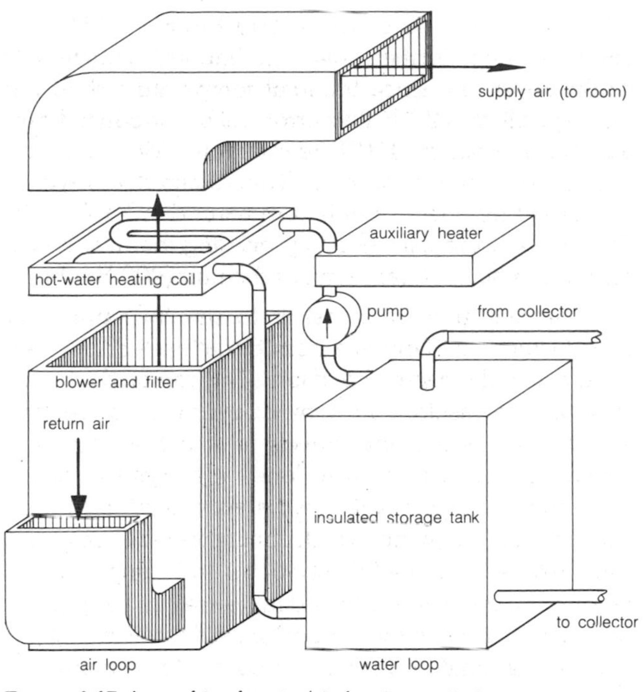

Figure 4.47 A combined water/air heating system.

can distribute energy throughout the dwelling. Hot water can be circulated beneath concrete floors, although the total heat-loss rate is somewhat higher when using this scheme. Systems such as these are used, however, to take the chill out of bare concrete floors. Sizing techniques are quite straightforward—not unlike those applicable to baseboard systems. Details can be found in any handbook on heating and ventilating.

Another possible second-loop scheme utilizes a combined water-air system, as depicted in Figure 4.47. Hot water is circulated from the storage tank into a coil located in an air-duct system. Air flowing across the coil is warmed before entering the living space. A typical coil, not unlike an automobile radiator, is shown in Figure 4.48.

This system is an attractive alternative for existing forced-air heating systems. The change to solar heating requires only the addition of a properly sized hot-water coil and the installation of a blower with approximately twice the capacity of the original. You will recall that higher flow rates for air are necessary because of the relatively low water temperatures in solar heating systems. Coil sizing is done in much the same way as for baseboard convector units; you should consult manufacturers' catalogs for detailed design data.

We close this section with a few remarks. First, let us emphasize that the material presented here has been necessarily sketchy. You should have enough information to design a simple, single-story second-loop system; but if your design becomes more complicated, we suggest that you begin to read some of the material listed in the Bibliography. In fact, it might be wise to have your proposed design checked by a competent heating, ventilating and air-conditioning engineer. You should be confident of your design before committing hard cash and construction time.

Second, you should recall that humidity plays an important part in a comfortable indoor environment. A good heating design accounts for the natural tendency to have too dry air in a heated room. House plants are remarkably effective in maintaining reasonable humidity levels; twenty-five plants can do the job of a small commercial humidifier!

Third, you needn't maintain high room temperatures at night. Night-setback of thermostats is an especially effective conservation technique in any heating system.

And finally, you might have noticed that this section was the very first to use the *design* heat-loss rate for any significant calculation. Uusally the heat loss over some time period (days or months) is used. However, in sizing the second loop, we must be prepared to handle the coldest day which can reasonably be expected. This, then, is the primary importance of the design heat-loss figure.

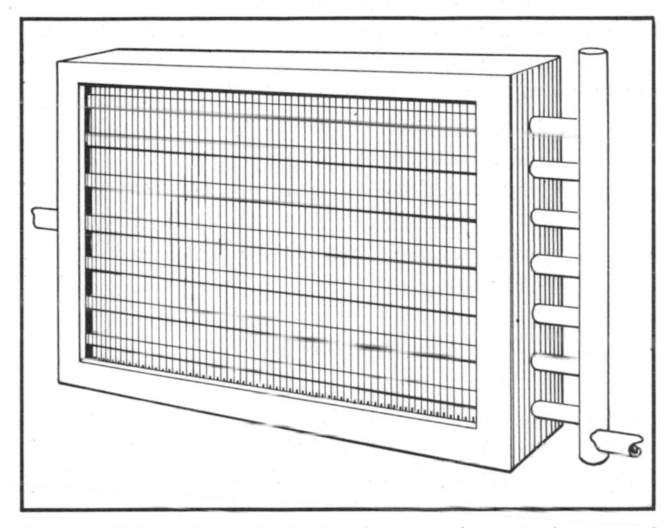

Figure 4.48 Typical water/air heat exchanger or hot-water heating coil (after Wave-Fin coils manufactured by Dunhan-Bush, Inc.).

Hot-Water Systems for Domestic Use

There are two methods for water heating which are compatible with component solar heating systems: the first is an independent collector and storage system; the second is an integral storage unit/heat exchanger. The former system is simply a solar heating system on a smaller scale, with a freshwater collector, potable storage tank, and perhaps a small pump. The latter system derives heat from the space-heating storage tank through a heat exchanger. Both schemes are shown in Figure 4.49.

The independent system might be advantageous in air-heated homes, since heat exchange from the storage unit may not be practical. Either system may be used for water-heated homes. The independent system, because it has a freshwater collector, risks possible freezing damage. On the other hand, it is a self-sufficient system, and it therefore may function even if the space-heating system is inoperable or its storage unit is depleted.

Hot water can be delivered to faucets using electrical pumps or the line pressure in the city water supply. An auxiliary heater may be added to increase the water temperature to any desired level. In this latter scheme, the solar systems serve as preheaters.

Energy requirements for water heating vary with the number of inhabitants, water use, and conservation habits. In general, the energy per day needed for water heating (QH) can be expressed as follows:

E. 4.34 $$QH = (gal/day\text{-}person) \times (no.\ people) \times (C_p) \times (\Delta T) \times \left(\frac{8.3\ lbm}{gal}\right)$$

where C_p equals the specific heat of water (1.0 Btu/lbm-°F); ΔT equals the average hot-water temperature minus the base cold-water temperature; and *gal/day-person* equals the volume of water used at the average hot-water temperature per day per person. In most regions, ΔT will be on the order of 60°F, the difference between normal hot-water temperature (about 110°F) and typical groundwater temperature (about 50°F). You should check these figures for your area before calculating hot-water heating requirements. Your local water utility will have this information. For illustrative purposes we'll use a ΔT of 60°F, so the equation becomes

E. 4.35 $$QH = (500\ Btu/gal) \times (gal/day\text{-}person) \times (no.\ of\ people)$$

Hot-water consumption varies considerably with lifestyle. The figure for domestic use in the United States is about 30 gallons/day per person. It is probably fair to say that typical American consumption rates are needlessly high; if you are willing to apply some conservation measures (such as foot-controlled valves and shorter showers), hot-water consumption can be drastically reduced. Let us assume that 15 gallons of hot water per day per person is sufficient for comfortable living. For 4 people, we would require

$$QH = (500\ Btu/gal) \times (15\ gal/day\text{-}person) \times (4\ people)$$

$$= 30{,}000\ Btu/day$$

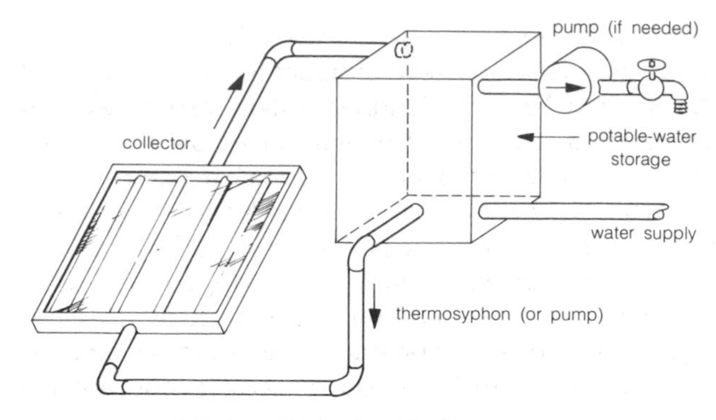

(a) Independent Hot-Water System

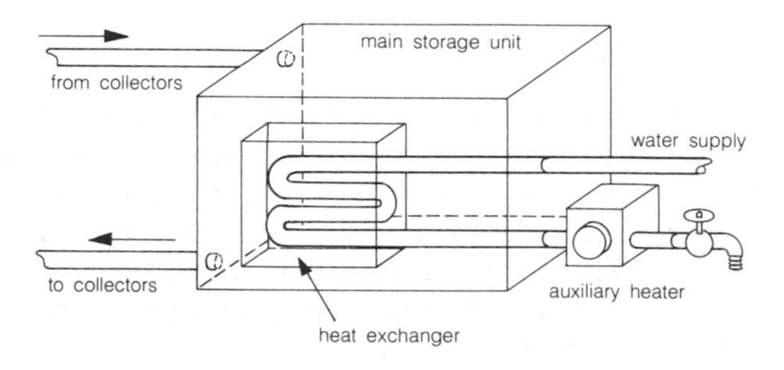

(b) System with Main Storage and Collectors

Figure 4.49 Schemes for domestic hot-water heating (note that pumps are not required if sufficient line pressure is present).

If we estimate (for lack of better information) that the system's heat loss will be about 10 percent of the above energy usage (30,000 + 3000), a revised QH gives us 33,000 Btu/day.

To size a separate collector system to handle this load, we simply use the methods discussed in sizing a flat plate collector. If you wish to draw heat from the heating storage system, both the storage unit and the collectors must reflect the revised thermal load. The monthly and three-day heat-loss figures we calculated earlier (for both units) must also be adjusted to include the new demand.

Integral Systems

Several existing designs combine the tasks of energy collection and storage. These most often involve a structural member of high mass and large frontal area—for instance, a thick south-facing wall or massive roof. The outside face is coated to increase its absorptive properties and often is covered with sheets of clear plastic or glass to take advantage of the greenhouse effect.

Unfortunately, the performance of these systems is difficult to predict without a fairly sophisticated computer analysis, based on specific design criteria. This complexity stems from the nonuniformity of temperature throughout the device. Yet this temperature nonuniformity (and the "thermal time delay" or heat lag associated with it) is the very reason such a system works. Energy absorbed during the day takes several hours to work its way into the dwelling. Properly designed systems capitalize on this fact, so that most of the energy enters the living space during the evening, when it is most needed.

Our treatment in this section is limited to a description of systems which are known to work. If you wish to utilize a similar heating concept, you will have to make an educated guess as to sizing and construction or else consult someone competent in heat-transfer analysis. The former alternative is quite viable if you are the experimental type, willing to refine your design as you gain operating experience.

Rooftop Waterbeds

The rooftop-waterbed solar house design, developed by Harold Hay, is an example of a complete or integrated solar heating system. Large bags of water, placed on a flat-roofed house, function as both collection and storage units. There is no need for a separate circulation system because the energy is transferred directly through a metal roof into the house interior.

The three-bedroom house of Hay's design is located in Atascadero, California. It is currently occupied by a five-person family and is under evaluation by a multidisciplinary group of professors at the California Polytechnic Institute. The one-story house has a slab roof composed of braced metal sheets which support large waterbeds underlaid by a double layer of plastic liner. Above the waterbeds are sliding insulated panels mounted on tracks. On sunny winter days, the panels are moved along the tracks and stacked over the car-port, uncovering the waterbeds to collect solar energy. Some of this energy is stored in the waterbeds; some is transferred by conduction through the metal roof and into the house interior, where the ceiling acts as a radiant heating device. During the cold nights, the panels are moved back to cover the waterbeds and prevent energy from reradiating into the night sky. Energy continues to be transferred to the living space, providing nocturnal heating.

During the summer, the positions and functions of the system are reversed. The panels are removed from the waterbeds at night to allow reradiation of heat, which cools the water. The lack of any glazing lets the radiation process proceed unhindered. During hot days, the waterbeds are covered to prevent solar heat collection and to allow the cooled water to absorb energy through the roof from warm air inside the dwelling. This heat is radiated to the atmosphere at night when the waterbeds are uncovered again.

In the Atascadero house, a one-quarter-horsepower motor is used for two minutes each morning and night to change the panel positions. This prototype positioning system is quite massive, and simpler and less expensive systems are under development.

The weight of the waterbed system requires special construction considerations for the roof support members. In addition, many plastics become brittle when exposed to direct sunlight for any length of time, so the waterbed material should be chosen with care.

South-wall Water Barrels

In this system, notably successful in the house of Steve and Holly Baer at Zomeworks, near Albuquerque, New Mexico, 55-gallon drums filled with water are stacked horizontally to form the entire south wall of the dwelling. The drums are positioned inside the house, behind large glass windows which provide a greenhouse effect. The south-facing surfaces of the drums are painted black, while those facing the interior are coated with ordinary white paint. Naturally, the drums become an integral part of the interior environment, but the Baers have carried this off quite well. The drum support structure is tastefully decorated with small book shelves and hanging plants.

Large honeycomb insulating panels are hinged outside the house, near the bottom of the glass windows. They can be raised to cover the barrels for insulation or lowered to lie on the ground in front (south) of the barrel wall; in the latter case, their aluminum coating reflects the sun's rays toward the barrels, increasing heat collection.

In winter, the insulating panels are open during the day and closed at night. During the summer, these positions are reversed, and windows and vents in the roof are also opened at night and closed during the day; the house is thus cooled at night and remains cooled during the day.

Steve Baer reports that the drum walls are extraordinarily efficient solar collectors; the collecting surface is never above 100°F, and collection is increased by reflection from the open aluminum-coated panels. On clear days during December, the drum walls collected nearly 1400 Btu per square foot of glass. The house has approximately 2000 square feet of floor area and only 400 square feet of solar collector.

Steve also notes that in winter the house usually stays between 63 and 70°F, although it has swung to extremes of 56 and 75°F. They use two wood stoves for auxiliary heating, but claim to have burned "less than a cord of wood, and in Albuquerque it was an exceptionally cold winter."

South-facing Concrete Walls

Designs utilizing a south-facing concrete wall are technically simple. Concrete is not only a structural asset, but is well suited to heat collection and storage, with a thermal capacity about 60 percent that of water, on a volumetric basis. Like most south-wall designs, the success of this heating system is closely tied to the greenhouse effect provided by a cover glass or plastic. Finally, it is important to provide for additional insulation on the outer portion of the glass or plastic, to prevent overheating in the summer and excessive heat loss during winter nights.

A system of this type was developed by Professor Trombe and built at the Solar Energy Lab of CNRS (architect, Jacques Michel). The prototype Trombe-Michel house is located in the French Pyrenees. It uses a south-facing concrete wall which is roughcast and painted black. The wall is sheathed in one or more layers of glass. Schematics are shown in Figure 4.50.

While a good deal of thermal energy passes through the wall and enters the room directly, the Trombe design also makes use of the energy leaving the outer face of the concrete slab. This energy management is accomplished by a natural circulation scheme using two air ducts, one placed at the top of the concrete wall and one below it. The heated air between the glass and the wall expands and rises, entering the house through an air duct near the ceiling. This causes the cooler air (which is more dense and therefore closer to the floor) to be pushed through a baseboard duct into the bottom of the air space, and circulated upward again as it is heated.

In the Trombe design, there is enough energy stored in any given day to last until the early morning hours. Natural circulation continues during the night, but at a reduced rate. A reverse cooling cycle is prevented in the early morning or on cloudy days by locating the bottom duct slightly above the bottom of the collector, as shown in the figure. In the summer, the natural circulation is also used for cooling. The duct at the top of the collector is closed to the inside rooms and opened to the outside, so that hot air is vented to the outside (instead of channeled back into the dwelling). This causes a chimney effect which draws cooler air through the opened vent of the wall opposite the collector. This effect is also depicted schematically in the figure.

Even though the house was not well insulated, the Trombe system reportedly supplied up to three quarters of the entire heating requirement during an average winter in the Pyrenees, where temperatures often fall below 0°F. For his particular climatological conditions, Trombe suggested a rule of thumb for wall sizing. He claims that a well-insulated house should have 1 square meter (10 square feet) of wall-collector area for every 10 cubic meters (35 cubic feet) of house volume. You should be warned that this is a very rough approximation and will vary substantially with location and specific design.

A similar "south wall" was modeled by computer techniques at Stanford University. Each square inch of a 12-inch-thick concrete wall was "exposed" to 1350 Btu of solar energy per day. The solar intensity varied with time to simulate the traverse of the sun during the day. The outside temperature varied from a maximum of 70°F at 2 P.M. to a minimum of 30°F at 2 A.M. The simulated design included a quarter-inch cover glass and a 6-inch air gap. A polystyrene cover was placed over the outer glass surface at "night."

The results of the analysis are shown in Figures 4.51 and 4.52. Figure 4.51 depicts the temperature distribution in the wall at various times over a 24-hour period. The numbers in parentheses indicate the outside air temperature at the times noted. The solutions reflect the conditions of the wall after five days of operation with full sun and the specified outside temperature conditions. Note that the average wall temperature is on the order of 110°F.

Figure 4.52 plots the energy flow into the room and also the energy lost from the outer wall face, both as functions of time. The units are Btu per hour per square foot of wall surface. Note that the maximum energy transfer into the room occurs around midnight and is on the order of 18 Btu/ft^2-hr. The energy lost from the front

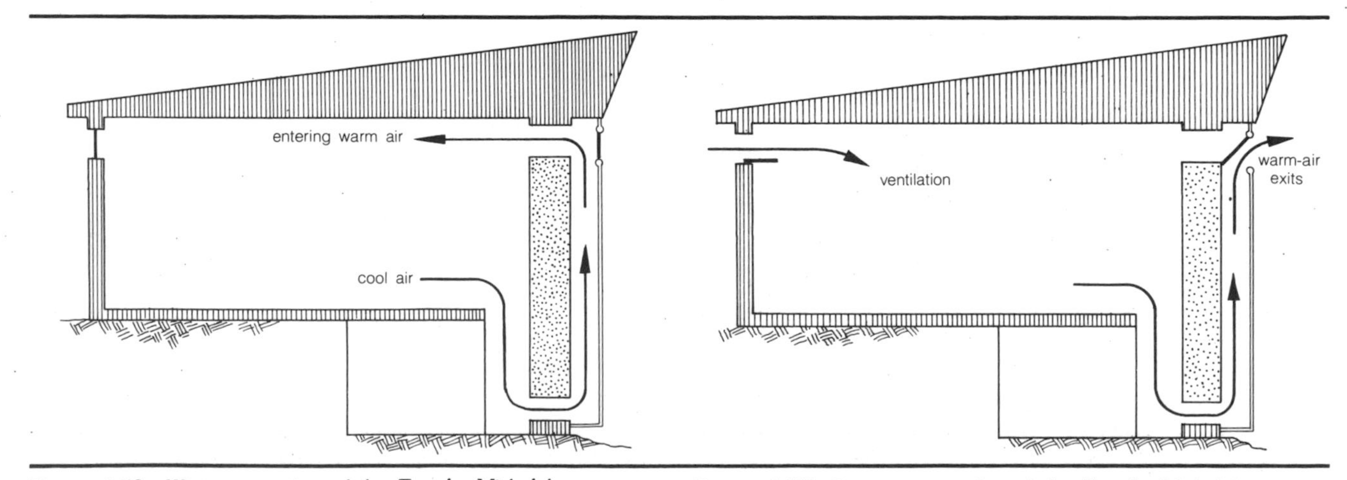

Figure 4.50a Winter operation of the Trombe-Michel house.

Figure 4.50b Summer operation of the Trombe-Michel house.

face is substantial during the daylight hours, but is dramatically reduced at night with the addition of the polystyrene insulator. Much of this energy could be directed into the dwelling with an appropriate natural- or forced-convection system, as in the Trombe design. Figure 4.52 also indicates the amount of thermal energy stored as a function of time, referenced to 70°F. For instance, a 400-square-foot wall, a foot thick, would contain nearly 400,000 Btu of thermal energy at 4 P.M.—a substantial cushion for inclement weather.

We should emphasize that the results of this modeling are crucially tied to the temperature and insolation modeling. It is difficult to extend these predictions to different design conditions without a computer analysis. Figures 4.51 and 4.52 are included merely to familiarize you with the underlying principles of south-wall systems.

The Biosphere

The aim of the biosphere concept is to integrate the heating and food-production functions (also, by the way, water and waste-disposal needs). This integration is accomplished by extending the south-wall scheme to incorporate a garden area between the wall and the glass or plastic cover system. In essence, a miniature greenhouse is attached to the south side of the dwelling.

Day Chahroudi, who conceived the biosphere, described it in the May 1973 issue of the *Tribal Messenger*. A sketch is shown in Figure 4.53. A plastic membrane is stretched over a plywood frame to enclose the garden area. Incident solar energy is absorbed in three ways: into the storage wall, by the plants, and into the ground. The heat storage capacity of the ground is greatly increased by laying plastic tubes that are filled with water in between the rows of the garden. The temperature inside the enclosure is held at about 100°F by venting whenever necessary. According to his scheme, if heat is needed inside the living area, the door is opened and air convection removes heat from the ground and distributes it throughout the living area. According to Chahroudi's calculations, three overcast days with an average temperature of 25°F or below are needed before the temperature inside the house drops below 70° (and then only at night).

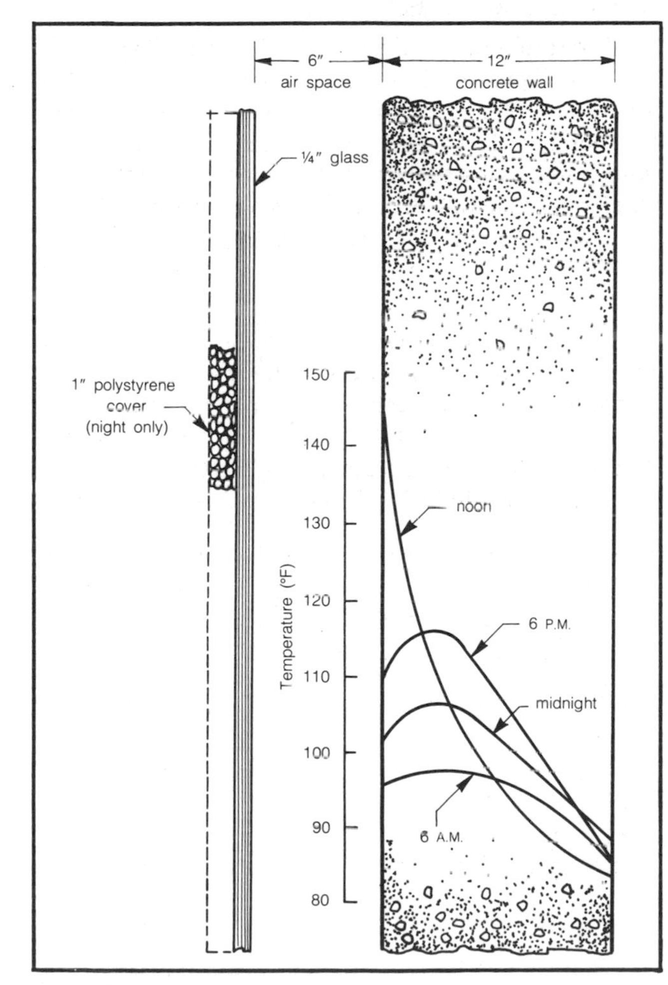

Figure 4.51 Results of a computer analysis for a south-facing concrete wall: temperature distribution as function of time.

Figure 4.52 Results of a computer analysis of a south-facing concrete wall: energy flows and storage capacity.

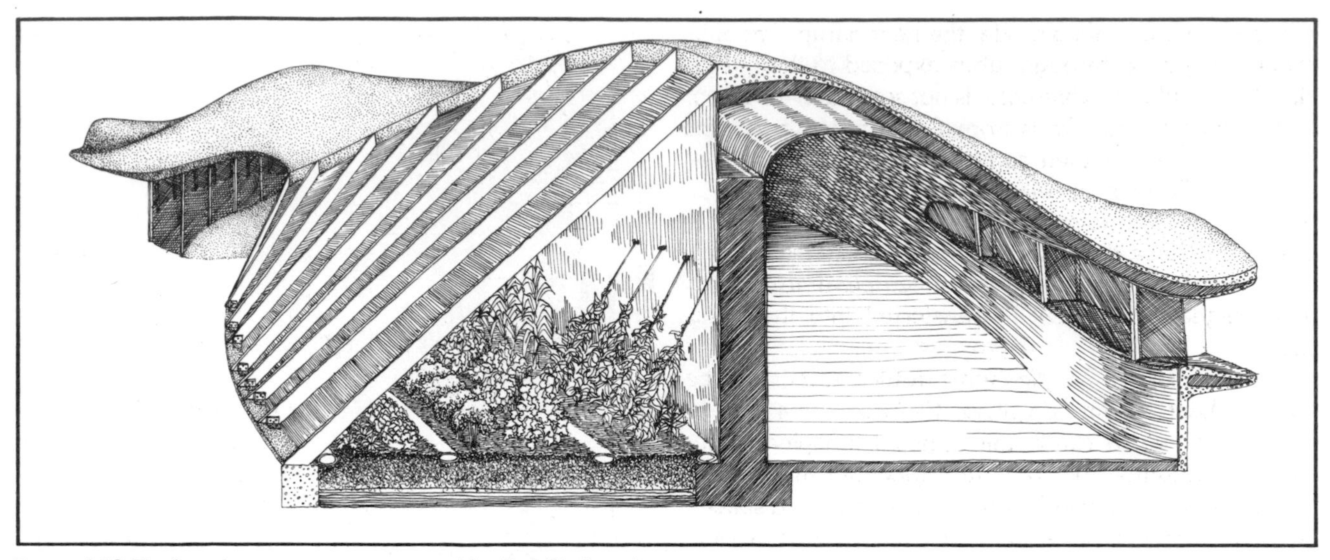

Figure 4.53 The biosphere concept as conceived by Day Chahroudi.

The usable garden area in this proposed design is 400 square feet—which could provide a welcome supplement of fresh fruit and vegetables for an average family. The garden is in a tropical environment, but since the system is closed, almost no water is consumed. The garden is self-sustaining and independent of the outside environment with respect to temperature and water. Only sunlight is required.

The membrane design consists of several layers. First, an outer skin of 16-mil clear vinyl serves as protection against wind and rain. The angle of the membrane allows it to shed snow and dust. Then thermal insulation is provided by a 3-inch layer of clear acrylic bubble-foam. The bubble size is half an inch and the walls are very thin, so that an estimated 70 percent of the sunlight passes through.

At this writing, the biosphere is only a concept. Many of the schemes are untried, but we believe the basic premise—that of self-sustaining (or nearly so) energy loops—has merit.

Chahroudi also has conceived an interesting insulation technique known as Beadwall. With this scheme, the air space between two or more layers of glass or plastic can be filled in seconds with plastic insulating beads. An air blower provides the driving mechanism. The air space can be emptied just as quickly. Beadwall can be applied to any collector design which features double glazing. At nightfall, the beads fill the insulating space with the flip of a switch, reducing unwanted heat loss and protecting the system from freezing. In the morning, the process is reversed and the collectors are readied to receive the sun's rays. As an added attraction, the dynamic patterns made by the beads are fascinating to watch. Design details can be purchased from the Zomeworks, Albuquerque, New Mexico.

Auxiliary Heating Systems

There will be times when the energy storage system is unable to provide sufficient energy to cope with long periods of sunlessness and/or cold. Or, due to economic or space constraints, you may not be able to install sufficient collector area to provide 100 percent solar heating for every month of the year. Consequently, an auxiliary heating system is strongly recommended. Suitable heaters include fireplaces, wood stoves, electric or gas heaters, coal or oil furnaces, and kerosene space heaters.

If your system is designed for extreme weather conditions, then a few electric heaters (expensive to run) or a central fireplace will probably suffice. But if frequent periods of auxiliary demand are anticipated, it's advisable to include a substantial back-up which utilizes an economical fuel source. Specifically, one might install a gas- or oil-fired water heater into the circulation scheme. A conventional water heater is probably adequate, since the duty cycle is significantly less than a conventional hydronic heater. Units with capacities of 75,000 Btu/hr can be purchased from Sears for under $150. Note that auxiliary water-heating units are included in Figures 4.43 and 4.47.

Auxiliary units can also be used to boost the temperature of the circulating fluid. The collectors then serve as preheaters, but still provide the bulk of the energy. This approach permits more conventional sizing of second-loop components, since they then operate in the temperature range for which they were originally designed. The boosting technique can be applied with equal success to air- and water-based systems.

The heat pump is a unique way to provide auxiliary heating. This device operates in much the same way as does a home refrigerator, except that the cycle can be reversed. In the heating mode, the heat pump runs cold, evaporating freon through tubes exposed to the outside air. The outside air, while cold, is not so cold as the freon, so the net heat transfer is from the outside air into the freon. The freon is then compressed to a higher pressure—usually with an electrically driven compressor—and allowed to condense in tubes exposed to the inside air. Since the condensation process occurs at a much higher pressure, the freon temperature is also much higher. Thus heat is transferred into the living space.

In principle, a simple home refrigerator could be used to heat a small dwelling. Perhaps some further explanation would clarify this concept. Imagine, if you will, removing the door of your refrigerator and moving the box to an open window or door. If the "cold space" faces the winter air, it can remove energy from this air as the freon is cycled through the walls of the unit. (This is the very same energy removal process that is going on in your refrigerator right now.) Energy captured in the "cold space" is run through a compressor and rejected from coils on the back or bottom of the refrigerator—presumably into your room. You can feel these warm coils on the back of any refrigerator. (Unfortunately, home refrigerators are not designed to handle the high heat loads in a typical home, so our scheme is impractical.)

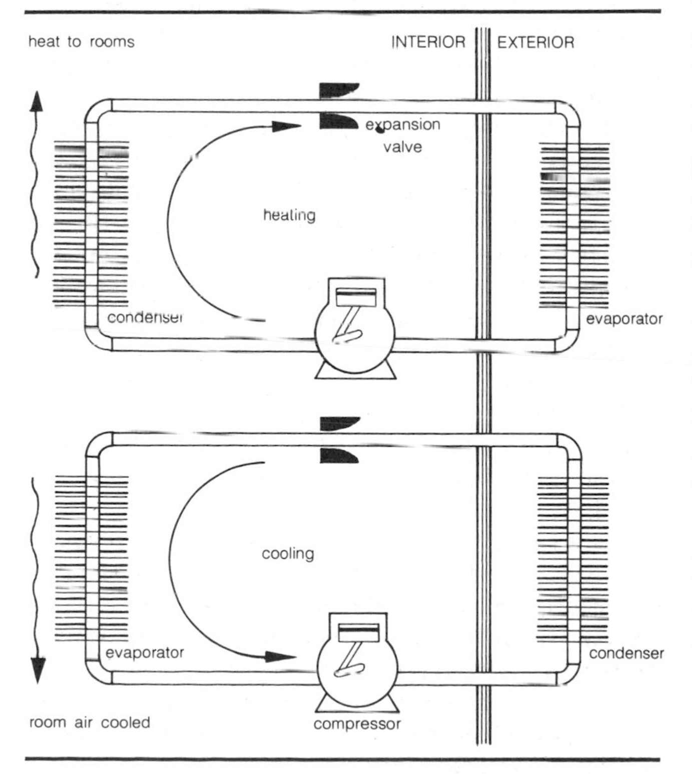

Figure 4.54a A heat pump in both heating and cooling modes. (Contrary to the flow diagram, the freon flow does *not* reverse. Instead, the functional roles of the condenser and evaporator coils are reversed. This is accomplished with internal valving.)

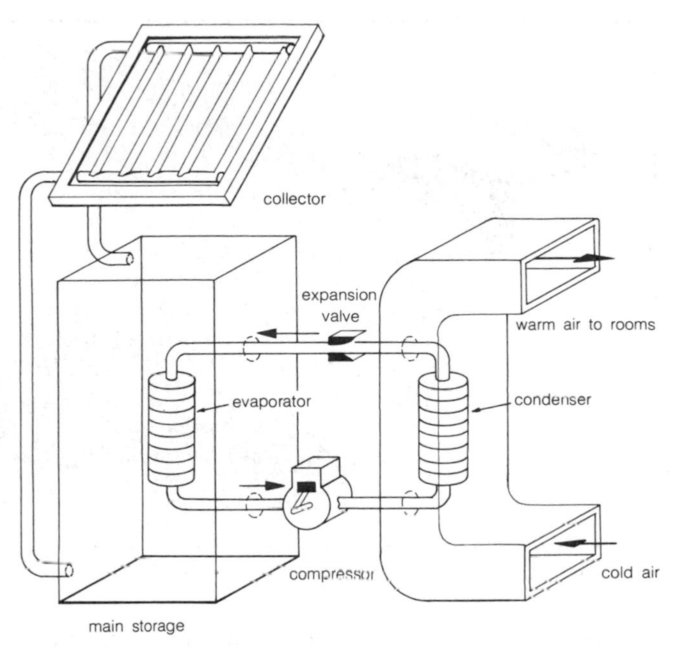

Figure 4.54b A heat pump used in conjunction with a low-temperature thermal-storage device.

As we have seen, the net effect of the heat pump is that energy is removed from a cooler region and "pumped" to a warmer one. Energy does not flow spontaneously from lower to higher temperatures, which is the reason for the compressor. The key question, of course, is how much energy is required to run the compressor? This depends on the operating temperatures involved, but usually we require one unit of compressor energy for every three or four units of energy delivered to the living space. This implies that we receive two or three units of energy "free" from the outside air! Sounds like magic, doesn't it?

Magic or not, these systems do work. Not only do they work, but the cycle can be reversed in the summer to cool the building! Your local General Electric or Westinghouse dealer can provide you with literature on heat pumps which have been available commercially for nearly two decades. A typical unit is shown in Figure 4.54a.

Heat pumps can be combined with a simple solar collection system for some striking results (see Figure 4.54b). The synergistic effects can be explained in terms of the heat pump's coefficient of performance (*COP* for short). The *COP* is defined in the following way:

$$\text{E. 4.36} \quad COP = \frac{\textit{Heat to Living Space}}{\textit{Energy Input to Compressor}}$$

In other words, the *COP* is the ratio of what you get out to what you put in. The *COP* is linked to the temperature difference between the heat supply (outside) and the heat sink (inside). The larger the temperature difference, or boost, the lower the *COP*. This is clearly shown in Figure 4.55. Observe that for a constant interior temperature of 70°F, the *COP* falls from 4.0 at an outside temperature of 50°F to 2.0 for 10°F. A *COP* of 1.0 implies that you are receiving only the energy you input to the compressor—much like an electrical resistance system!

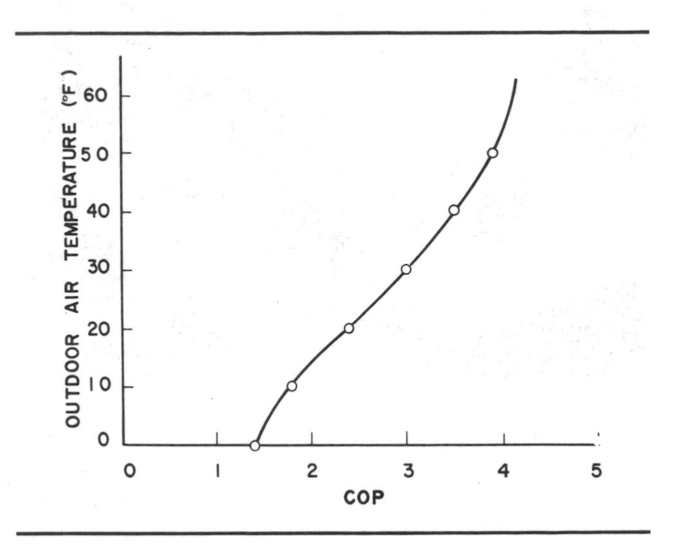

Figure 4.55 The *COP* of air-to-air heat pumps operating at various outdoor temperatures.

It has been proposed that swimming pools be covered with a plastic glazing material to collect energy during winter days. With proper nocturnal insulation, the pool temperature might be kept at 50°F, even though the average outside temperature is much lower. The 50°F water is useless to heat the house directly, but it can be used as a heat source for the heat pump. The *COP* would be significantly higher than if the unit were operated with the outside air. A possible pitfall in this scheme is the potential for localized freezing on the outside coils. Newer units do have automatic defrost cycles, but we suggest that you proceed with caution if you consider this approach.

Indeed, the final word on auxiliary systems should be one of warning. Many crude coal-, oil-, and wood-burning stoves are really dangerous. Modern space-conditioning systems have removed the specter of tragic home fires and asphyxiations from our lives; it would be nice to keep matters that way. Make *sure* you know how to install and operate these systems.

Cooling

In this chapter, we have considered only systems for house heating. There are several reasons for this emphasis. First, in many portions of the country, summer cooling is unnecessary. This is particularly true if simple architectural and landscaping techniques are employed to promote natural cooling. Second, if cooling *is* desired, many of the systems discussed can be modified to provide it easily. For instance, flat plate collectors can be run on clear nights (without glazing materials) to cool the fluid in the circulation and storage system. The water is cooled by radiant heat transfer from the circulating fluid to the night sky. (This same principle is used in many countries to freeze water—even though the surrounding air temperature is above 32°F!) During the day, the cool fluid can be circulated throughout the house, absorbing heat from the air along the way and thus reducing indoor temperatures.

Several absorption-type air conditioners are available which utilize a heat source instead of electrical energy. However, the driving temperature for these units must be at least 220 to 240°F. Flat plate collectors alone simply cannot achieve these temperatures without exotic designs.

Focused collectors have the temperature capabilities, but generally require expensive tracking mechanisms. One system, successfully used in a house built by Colorado State University, employs an auxiliary heater to boost the temperature of solar-preheated water. The resulting temperature is sufficient to run the absorption cycle.

Economics, Taxes, and Zoning

The solar heating industry is young and conventional heating prices are volatile, so it is truly difficult to assess the overall economic potential of solar heating for our society. In our presentation, we have kept the economics crude. We feel that it is safe to say that a solar heating system *can* compete economically, provided you are willing to supply significant amounts of labor and ingenuity. If you contract the work out or purchase prefabricated collectors, it's unlikely that you will realize any significant savings—at least over the short term.

An important exception involves the construction and financing of a new home. The cost of the collector system, to a great extent, can be absorbed in the first mortgage. Instead of paying monthly fuel bills, you pay the interest and principal on that part of the mortgage associated with the solar heating system. This scheme is not unlike buying your own utility company and using a loan to finance the transaction. There are two advantages here. First, interest payments are tax-deductible, whereas fuel bills are not. Second, your monthly payments reflect more of an investment than an operating cost. Essentially, you are capitalizing your utility costs.

Congressman Charles Vanik (D-Ohio) has introduced legislation to grant either a straight $1000 tax credit or a 25 percent deduction for equipment purchases up to $4000 to homeowners who install solar heating systems. Also, Senator Mike Gravel (D-Alaska) has been a major advocate of solar technology and, in the next few years, we can expect that legislative efforts will further encourage the economic viability of solar energy for home use.

You may run into some unique zoning and code regulations in your town. It has been our experience that code inspectors are eager to cooperate with new energy-conservative systems. This tone, however, may change after the first few poorly designed systems surface. Then, too, the aesthetics of solar heating may precipitate a confrontation with your local zoning board. The steep collector angles required for optimum solar collection require some rather novel building shapes at times. The results can be hideous unless careful thought is given to system integration. Haifa, Israel, should serve as a warning. Recently, rooftop solar water heaters were banned on grounds of "visual pollution." These units were mounted haphazardly on flat rooftops, and the City Powers finally lost their patience.

Existing Solar Homes

In the first rough version of this text, considerable space was devoted to a description of existing solar homes. This time, however, we have opted to explain the basic design criteria common to most of these homes. Those wishing a compilation of existing solar homes are directed to *Solar Heated Buildings, a Brief Survey,* edited by W.A. Shurcliff and distributed by the *Solar Energy Digest* (P.O. Box 17776, San Diego, California 92117). The sixth edition (October 1974) contains seventy-one one-page descriptions and is priced at $5. In addition, an excellent overview article entitled "Sun Power" appeared in the September 1974 issue of *Architecture Plus.* The photographs and sketches are quite striking. And, finally, the Bibliography contains many references to individual house designs.

So there it is: a cloudy day, the hidden sun ninety-odd million miles away, and you standing in front of your house or site with a speculative, pioneering grimace twisting your features. For decades there have been numerous practical examples of solar energy applied to space heating and the designers involved were truly pioneers, who had only their imagination and practical experience to guide them. Their work was admired, but seldom reproduced, since each system was unique to a specific dwelling and climate. It was easy enough to appreciate a solar-heated house in Arizona, but how do you apply the same technology to a town in Michigan?

All we can hope is that we've answered that question to your satisfaction and that we've provoked you to try a journey which will take you far beyond the examples we've shown here. Suddenly the words on this page seem sharper? Yes, it's true: the clouds are blowing away and there comes the sun. . . .

Appendix 4A Weather Design Data for the United States [abc]

State and Station	Lat. °	Winter: Median of Annual Extremes	Winter: 99%	Winter: 97½%	Winter: Coincident Wind Velocity
ALABAMA					
Alexander City	33	12	16	20	L
Anniston AP	33	12	17	19	L
Auburn	32	17	21	25	L
Birmingham AP	33	14	19	22	L
Decatur	34	10	15	19	L
Dothan AP	31	19	23	27	L
Florence AP	34	8	13	17	L
Gadsden	34	11	16	20	L
Huntsville AP	34	8	13	17	L
Mobile AP	30	21	26	29	M
Mobile CO	30	24	28	32	M
Montgomery AP	32	18	22	26	L
Selma-Craig AFB	32	18	23	27	L
Talladega	33	11	15	19	L
Tuscaloosa AP	33	14	19	23	L
ALASKA					
Anchorage AP	61	−29	−25	−20	VL
Barrow	71	−49	−45	−42	M
Fairbanks AP	64	−59	−53	−50	VL
Juneau AP	58	−11	− 7	− 4	L
Kodiak	57	4	8	12	M
Nome AP	64	−37	−32	−28	L
ARIZONA†					
Douglas AP	31	13	18	22	VL
Flagstaff AP	35	−10	0	5	VL
Fort Huachuca AP	31	18	25	28	VL
Kingman AP	35	18	25	29	VL
Nogales	31	15	20	24	VL
Phoenix AP	33	25	31	34	VL
Prescott AP	34	7	15	19	VL
Tucson AP	33	23	29	32	VL
Winslow AP	35	2	9	13	VL
Yuma AP	32	32	37	40	VL
ARKANSAS					
Blytheville AFB	36	6	12	17	L
Camden	33	13	19	23	L
El Dorado AP	33	13	19	23	L
Fayetteville AP	36	3	9	13	M
Fort Smith AP	35	9	15	19	M
Hot Springs Nat. Pk.	34	12	18	22	M
Jonesboro	35	8	14	18	M
Little Rock AP	34	13	19	23	M
Pine Bluff AP	34	14	20	24	L
Texarkana AP	33	16	22	26	M
CALIFORNIA					
Bakersfield AP	35	26	31	33	VL
Barstow AP	34	18	24	28	VL
Blythe AP	33	26	31	35	VL
Burbank AP	34	30	36	38	VL
Chico	39	23	29	33	VL
Concord	38	27	32	36	VL
Covina	34	32	38	41	VL
Crescent City AP	41	28	33	36	L
Downey	34	30	35	38	VL
El Cajon	32	26	31	34	VL
El Centro AP	32	26	31	35	VL
Escondido	33	28	33	36	VL
Eureka/Arcata AP	41	27	32	35	L
Fairfield-Travis AFB	38	26	32	34	VL
Fresno AP	36	25	28	31	VL
Hamilton AFB	38	28	33	35	VL
Laguna Beach	33	32	37	39	VL
CALIFORNIA (continued)					
Livermore	37	23	28	30	VL
Lompoc, Vandenburg AFB	34	32	36	38	VL
Long Beach AP	33	31	36	38	VL
Los Angeles AP	34	36	41	43	VL
Los Angeles CO	34	38	42	44	VL
Merced-Castle AFB	37	24	30	32	VL
Modesto	37	26	32	36	VL
Monterey	36	29	34	37	VL
Napa	38	26	31	34	VL
Needles AP	34	27	33	37	VL
Oakland AP	37	30	35	37	VL
Oceanside	33	33	38	40	VL
Ontario	34	26	32	34	VL
Oxnard AFB	34	32	35	37	VL
Palmdale AP	34	18	24	27	VL
Palm Springs	33	27	32	36	VL
Pasadena	34	31	36	39	VL
Petaluma	38	24	29	32	VL
Pomona CO	34	26	31	34	VL
Redding AP	40	25	31	35	VL
Redlands	34	28	34	37	VL
Richmond	38	28	35	38	VL
Riverside-March AFB	33	26	32	34	VL
Sacramento AP	38	24	30	32	VL
Salinas AP	36	27	32	35	VL
San Bernardino, Norton AFB	34	26	31	33	VL
San Diego AP	32	38	42	44	VL
San Fernando	34	29	34	37	VL
San Francisco AP	37	32	35	37	L
San Francisco CO	37	38	42	44	VL
San Jose AP	37	30	34	36	VL
San Luis Obispo	35	30	35	37	VL
Santa Ana AP	33	28	33	36	VL
Santa Barbara CO	34	30	34	36	VL
Santa Cruz	37	28	32	34	VL
Santa Maria AP	34	28	32	34	VL
Santa Monica CO	34	38	43	45	VL
Santa Paula	34	28	33	36	VL
Santa Rosa	38	24	29	32	VL
Stockton AP	37	25	30	34	VL
Ukiah	39	22	27	30	VL
Visalia	36	26	32	36	VL
Yreka	41	7	13	17	VL
Yuba City	39	24	30	34	VL
COLORADO					
Alamosa AP	37	−26	−17	−13	VL
Boulder	40	− 5	4	8	L
Colorado Springs AP	38	− 9	− 1	4	L
Denver AP	39	− 9	− 2	3	L
Durango	37	−10	0	4	VL
Fort Collins	40	−10	− 9	− 5	L
Grand Junction AP	39	− 2	8	11	VL
Greeley	40	−18	− 9	− 5	L
La Junta AP	38	−14	− 6	− 2	M
Leadville	39	−18	− 9	− 4	VL
Pueblo AP	38	−14	− 5	− 1	L
Sterling	40	−15	− 6	− 2	M
Trinidad AP	37	− 9	1	5	L

Notes: a. From ASHRAE *Handbook of Fundamentals.*
b. AP indicates airport; AFB indicates Air Force base; CO indicates cosmopolitan area; other may be taken to be semi-rural.
c. Wind velocity: VL = Very Light, 70% or more of cold extreme hours at less than 7 mph; L = Light, 50-69% cold extreme hours less than 7 mph; M = Moderate, 50-74% cold extreme hours more than 7 mph; H = High, 75% or more cold extreme hours more than 7 mph, 50% more than 12 mph.

State and Station	Lat. °	Winter: Median of Annual Extremes	Winter: 99%	Winter: 97½%	Winter: Coincident Wind Velocity
CONNECTICUT					
Bridgeport AP	41	−1	4	8	M
Hartford, Brainard Field	41	−4	1	5	M
New Haven AP	41	0	5	9	H
New London	41	0	4	8	H
Norwalk	41	−5	0	4	M
Norwich	41	−7	−2	2	M
Waterbury	41	−5	0	4	M
Windsor Locks, Bradley Field	42	−7	−2	2	M
DELAWARE					
Dover AFB	39	8	13	15	M
Wilmington AP	39	6	12	15	M
DISTRICT OF COLUMBIA					
Andrews AFB	38	9	13	16	M
Washington National AP	38	12	16	19	M
FLORIDA					
Belle Glade	26	31	35	39	M
Cape Kennedy AP	28	33	37	40	L
Daytona Beach AP	29	28	32	36	L
Fort Lauderdale	26	37	41	45	M
Fort Myers AP	26	34	38	42	M
Fort Pierce	27	33	37	41	M
Gainesville AP	29	24	28	32	L
Jacksonville AP	30	26	29	32	L
Key West AP	24	50	55	58	M
Lakeland CO	28	31	35	39	M
Miami AP	25	39	44	47	M
Miami Beach CO	25	40	45	48	M
Ocala	29	25	29	33	L
Orlando AP	28	29	33	37	L
Panama City, Tyndall AFB	30	28	32	35	M
Pensacola CO	30	25	29	32	M
St. Augustine	29	27	31	35	L
St. Petersburg	28	35	39	42	M
Sanford	28	29	33	37	L
Sarasota	27	31	35	39	M
Tallahassee AP	30	21	25	29	L
Tampa AP	28	32	36	39	M
West Palm Beach AP	26	36	40	44	M
GEORGIA					
Albany, Turner AFB	31	21	26	30	L
Americus	32	18	22	25	L
Athens	34	12	17	21	L
Atlanta AP	33	14	18	23	H
Augusta AP	33	17	20	23	L
Brunswick	31	24	27	31	L
Columbus, Lawson AFB	32	19	23	26	L
Dalton	34	10	15	19	L
Dublin	32	17	21	25	L
Gainesville	34	11	16	20	L
Griffin	33	13	17	22	L
La Grange	33	12	16	20	L
Macon AP	32	18	23	27	L
Marietta, Dobbins AFB	34	12	17	21	L
Moultrie	31	22	26	30	L
Rome AP	34	11	16	20	L
Savannah-Travis AP	32	21	24	27	L
Valdosta-Moody AFB	31	24	28	31	L
Waycross	31	20	24	28	L
HAWAII					
Hilo AP	19	56	59	61	L
Honolulu AP	21	58	60	62	L
Kaneohe	21	58	60	61	L
Wahiawa	21	57	59	61	L

State and Station	Lat. °	Winter: Median of Annual Extremes	Winter: 99%	Winter: 97½%	Winter: Coincident Wind Velocity
IDAHO					
Boise AP	43	0	4	10	L
Burley	42	−5	4	8	VL
Coeur d'Alene AP	47	−4	2	7	VL
Idaho Falls AP	43	−17	−12	−6	VL
Lewiston AP	46	1	6	12	VL
Moscow	46	−11	−3	1	VL
Mountain Home AFB	43	−3	2	9	L
Pocatello AP	43	−12	−8	−2	VL
Twin Falls AP	42	−5	4	8	L
ILLINOIS					
Aurora	41	−13	−7	−3	M
Belleville, Scott AFB	38	0	6	10	M
Bloomington	40	−7	−1	3	M
Carbondale	37	1	7	11	M
Champaign/Urbana	40	−6	0	4	M
Chicago, Midway AP	41	−7	−4	1	M
Chicago, O'Hare AP	42	−9	−4	0	M
Chicago, CO	41	−5	−3	1	M
Danville	40	−6	−1	4	M
Decatur	39	−6	0	4	M
Dixon	41	−13	−7	−3	M
Elgin	42	−14	−8	−4	M
Freeport	42	−16	−10	−6	M
Galesburg	41	−10	−4	0	M
Greenville	39	−3	3	7	M
Joliet AP	41	−11	−5	−1	M
Kankakee	41	−10	−4	1	M
La Salle/Peru	41	−9	−3	1	M
Macomb	40	−5	−3	1	M
Moline AP	41	−12	−7	−3	M
Mt. Vernon	38	0	6	10	M
Peoria AP	40	−8	−2	2	M
Quincy AP	40	−8	−2	2	M
Rantoul, Chanute AFB	40	−7	−1	3	M
Rockford	42	−13	−7	−3	M
Springfield AP	39	−7	−1	4	M
Waukegan	42	−11	−5	−1	M
INDIANA					
Anderson	40	−5	0	5	M
Bedford	38	−3	3	7	M
Bloomington	39	−3	3	7	M
Columbus, Bakalar AFB	39	−3	3	7	M
Crawfordsville	40	−8	−2	2	M
Evansville AP	38	1	6	10	M
Fort Wayne AP	41	−5	0	5	M
Goshen AP	41	−10	−4	0	M
Hobart	41	−10	−4	0	M
Huntington	40	−8	−2	2	M
Indianapolis AP	39	−5	0	4	M
Jeffersonville	38	3	9	13	M
Kokomo	40	−6	0	4	M
Lafayette	40	−7	−1	3	M
La Porte	41	−10	−4	0	M
Marion	40	−8	−2	2	M
Muncie	40	−8	−2	2	M
Peru, Bunker Hill AFB	40	−9	−3	1	M
Richmond AP	39	−7	−1	3	M
Shelbyville	39	−4	2	6	M
South Bend AP	41	−6	−2	3	M
Terre Haute AP	39	−3	3	7	M
Valparaiso	41	−12	−6	−2	M
Vincennes	38	−1	5	9	M

State and Station	Lat. °	Winter: Median of Annual Extremes	Winter: 99%	Winter: 97½%	Coincident Wind Velocity
IOWA					
Ames	42	−17	−11	−7	M
Burlington AP	40	−10	−4	0	M
Cedar Rapids AP	41	−14	−8	−4	M
Clinton	41	−13	−7	−3	M
Council Bluffs	41	−14	−7	−3	M
Des Moines AP	41	−13	−7	−3	M
Dubuque	42	−17	−11	−7	M
Fort Dodge	42	−18	−12	−8	M
Iowa City	41	−14	−8	−4	M
Keokuk	40	−9	−3	1	M
Marshalltown	42	−16	−10	−6	M
Mason City AP	43	−20	−13	−9	M
Newton	41	−15	−9	−5	M
Ottumwa AP	41	−12	−6	−2	M
Sioux City AP	42	−17	−10	−6	M
Waterloo	42	−18	−12	−8	M
KANSAS					
Atchison	39	−9	−2	2	M
Chanute AP	37	−3	3	7	H
Dodge City AP	37	−5	3	7	M
El Dorado	37	−3	4	8	H
Emporia	38	−4	3	7	H
Garden City AP	38	−10	−1	3	M
Goodland AP	39	−10	−2	4	M
Great Bend	38	−5	2	6	M
Hutchinson AP	38	−5	2	6	H
Liberal	37	−4	4	8	M
Manhattan, Fort Riley	39	−7	−1	4	H
Parsons	37	−2	5	9	H
Russell AP	38	−7	0	4	M
Salina	38	−4	3	7	H
Topeka AP	39	−4	3	6	M
Wichita AP	37	−1	5	9	H
KENTUCKY					
Ashland	38	1	6	10	L
Bowling Green AP	37	1	7	11	L
Corbin AP	37	0	5	9	L
Covington AP	39	−3	3	8	L
Hopkinsville, Campbell AFB	36	4	10	14	L
Lexington AP	38	0	6	10	M
Louisville AP	38	1	8	12	L
Madisonville	37	1	7	11	L
Owensboro	37	0	6	10	L
Paducah AP	37	4	10	14	L
LOUISIANA					
Alexandria AP	31	20	25	29	L
Baton Rouge AP	30	22	25	30	L
Bogalusa	30	20	24	28	L
Houma	29	25	29	33	L
Lafayette AP	30	23	28	32	L
Lake Charles AP	30	25	29	33	M
Minden	32	17	22	26	L
Monroe AP	32	18	23	27	L
Natchitoches	31	17	22	26	L
New Orleans AP	30	29	32	35	M
Shreveport AP	32	18	22	26	M
MAINE					
Augusta AP	44	−13	−7	−3	M
Bangor, Dow AFB	44	−14	−8	−4	M
Caribou AP	46	−24	−18	−14	L
Lewiston	44	−14	−8	−4	M
Millinocket AP	45	−22	−16	−12	L
Portland AP	43	−14	−5	0	L
Waterville	44	−15	−9	−5	M
MARYLAND					
Baltimore AP	39	8	12	15	M
Baltimore CO	39	12	16	20	M
Cumberland	39	0	5	9	L
Frederick AP	39	2	7	11	M
Hagerstown	39	1	6	10	L
Salisbury	38	10	14	18	M
MASSACHUSETTS					
Boston AP	42	−1	6	10	H
Clinton	42	−8	−2	2	M
Fall River	41	−1	5	9	H
Framingham	42	−7	−1	3	M
Gloucester	42	−4	2	6	H
Greenfield	42	−12	−6	−2	M
Lawrence	42	−9	−3	1	M
Lowell	42	−7	−1	3	M
New Bedford	41	3	9	13	H
Pittsfield AP	42	−11	−5	−1	M
Springfield, Westover AFB	42	−8	−3	2	M
Taunton	41	−9	−4	0	H
Worcester AP	42	−8	−3	1	M
MICHIGAN					
Adrian	41	−6	0	4	M
Alpena AP	45	−11	−5	−1	M
Battle Creek AP	42	−6	1	5	M
Benton Harbor AP	42	−7	−1	3	M
Detroit Met. CAP	42	0	4	8	M
Escanaba	45	−13	−7	−3	M
Flint AP	43	−7	−1	3	M
Grand Rapids AP	42	−3	2	6	M
Holland	42	−4	2	6	M
Jackson AP	42	−6	0	4	M
Kalamazoo	42	−5	1	5	M
Lansing AP	42	−4	2	6	M
Marquette CO	46	−14	−8	−4	L
Mt. Pleasant	43	−9	−3	1	M
Muskegon AP	43	−2	4	8	M
Pontiac	42	−6	0	4	M
Port Huron	43	−6	−1	3	M
Saginaw AP	43	−7	−1	3	M
Sault Ste. Marie AP	46	−18	−12	−8	L
Traverse City AP	44	−6	0	4	M
Ypsilanti	42	−3	−1	5	M
MINNESOTA					
Albert Lea	43	−20	−14	−10	M
Alexandria AP	45	−26	−19	−15	L
Bemidji AP	47	−38	−32	−28	L
Brainerd	46	−31	−24	−20	L
Duluth AP	46	−25	−19	−15	M
Faribault	44	−23	−16	−12	L
Fergus Falls	46	−28	−21	−17	L
International Falls AP	48	−35	−29	−24	L
Mankato	44	−23	−16	−12	L
Minneapolis/St. Paul AP	44	−19	−14	−10	L
Rochester AP	44	−23	−17	−13	M
St. Cloud AP	45	−26	−20	−16	L
Virginia	47	−32	−25	−21	L
Willmar	45	−25	−18	−14	L
Winona	44	−19	−12	−8	M
MISSISSIPPI					
Biloxi, Keesler AFB	30	26	30	32	M
Clarksdale	34	14	20	24	L
Columbus AFB	33	13	18	22	L
Greenville AFB	33	16	21	24	L

Appendix 4A—*Continued*

State and Station	Lat. °	Winter: Median of Annual Extremes	Winter: 99%	Winter: 97½%	Winter: Coincident Wind Velocity
MISSISSIPPI (continued)					
Greenwood	33	14	19	23	L
Hattiesburg	31	18	22	26	L
Jackson AP	32	17	21	24	L
Laurel	31	18	22	26	L
McComb AP	31	18	22	26	L
Meridian AP	32	15	20	24	L
Natchez	31	18	22	26	L
Tupelo	34	13	18	22	L
Vicksburg CO	32	18	23	26	L
MISSOURI					
Cape Girardeau	37	2	8	12	M
Columbia AP	39	−4	2	6	M
Farmington AP	37	−2	4	8	M
Hannibal	39	−7	−1	4	M
Jefferson City	38	−4	2	6	M
Joplin AP	37	1	7	11	M
Kansas City AP	39	−2	4	8	M
Kirksville AP	40	−13	−7	−3	M
Mexico	39	−7	−1	3	M
Moberly	39	−8	−2	2	M
Poplar Bluff	36	3	9	13	M
Rolla	38	−3	3	7	M
St. Joseph AP	39	−8	−1	3	M
St. Louis AP	38	−2	4	8	M
St. Louis CO	38	1	7	11	M
Sedalia, Whiteman AFB	38	−2	4	9	M
Sikeston	36	4	10	14	L
Springfield AP	37	0	5	10	M
MONTANA					
Billings AP	45	−19	−10	−6	L
Bozeman	45	−25	−15	−11	L
Butte AP	46	−34	−24	−16	VL
Cut Bank AP	48	−32	−23	−17	L
Glasgow AP	48	−33	−25	−20	L
Glendive	47	−28	−20	−16	L
Great Falls AP	47	−29	−20	−16	L
Havre	48	−32	−22	−15	M
Helena AP	46	−27	−17	−13	L
Kalispell AP	48	−17	−7	−3	VL
Lewiston AP	47	−27	−18	−14	L
Livingston AP	45	−26	−17	−13	L
Miles City AP	46	−27	−19	−15	L
Missoula AP	46	−16	−7	−3	VL
NEBRASKA					
Beatrice	40	−10	−3	1	M
Chadron AP	42	−21	−13	−9	M
Columbus	41	−14	−7	−3	M
Fremont	41	−14	−7	−3	M
Grand Island AP	41	−14	−6	−2	M
Hastings	40	−11	−3	1	M
Kearney	40	−14	−6	−2	M
Lincoln CO	40	−10	−4	0	M
McCook	40	−12	−4	0	M
Norfolk	42	−18	−11	−7	M
North Platte AP	41	−13	−6	−2	M
Omaha AP	41	−12	−5	−1	M
Scottsbluff AP	41	−16	−8	−4	M
Sidney AP	41	−15	−7	−2	M
NEVADA					
Carson City	39	−4	3	7	VL
Elko AP	40	−21	−13	−7	VL
Ely AP	39	−15	−6	−2	VL
Las Vegas AP	36	18	23	26	VL
Lovelock AP	40	0	7	11	VL
NEVADA (continued)					
Reno AP	39	−2	2	7	VL
Reno CO	39	8	12	17	VL
Tonopah AP	38	2	9	13	VL
Winnemucca AP	40	−8	1	5	VL
NEW HAMPSHIRE					
Berlin	44	−25	−19	−15	L
Claremont	43	−19	−13	−9	L
Concord AP	43	−17	−11	−7	M
Keene	43	−17	−12	−8	M
Laconia	43	−22	−16	−12	M
Manchester, Grenier AFB	43	−11	−5	1	M
Portsmouth, Pease AFB	43	−8	−2	3	M
NEW JERSEY					
Atlantic City CO	39	10	14	18	H
Long Branch	40	4	9	13	H
Newark AP	40	6	11	15	M
New Brunswick	40	3	8	12	M
Paterson	40	3	8	12	M
Phillipsburg	40	1	6	10	L
Trenton CO	40	7	12	16	M
Vineland	39	7	12	16	M
NEW MEXICO					
Alamagordo, Holloman AFB	32	12	18	22	L
Albuquerque AP	35	6	14	17	L
Artesia	32	9	16	19	L
Carlsbad AP	32	11	17	21	L
Clovis AP	34	2	14	17	L
Farmington AP	36	−3	6	9	VL
Gallup	35	−13	−5	−1	VL
Grants	35	−15	−7	−3	VL
Hobbs AP	32	9	15	19	L
Las Cruces	32	13	19	23	L
Los Alamos	35	−4	5	9	L
Raton AP	36	−11	−2	2	L
Roswell, Walker AFB	33	5	16	19	L
Santa Fe CO	35	−2	7	11	L
Silver City AP	32	8	14	18	VL
Socorro AP	34	6	13	17	L
Tucumcari AP	35	1	9	13	L
NEW YORK					
Albany AP	42	−14	−5	0	L
Albany CO	42	−5	1	5	L
Auburn	43	−10	−2	2	M
Batavia	43	−7	−1	3	M
Binghamton CO	42	−8	−2	2	L
Buffalo AP	43	−3	3	6	M
Cortland	42	−11	−5	−1	L
Dunkirk	42	−2	4	8	M
Elmira AP	42	−5	1	5	L
Geneva	42	−8	−2	2	M
Glens Falls	43	−17	−11	−7	L
Gloversville	43	−12	−6	−2	L
Hornell	42	−15	−9	−5	L
Ithaca	42	−10	−4	0	L
Jamestown	42	−5	1	5	M
Kingston	42	−8	−2	2	L
Lockport	43	−4	2	6	M
Massena AP	45	−22	−16	−12	M
Newburgh-Stewart AFB	41	−4	2	6	M
NYC-Central Park	40	6	11	15	H
NYC-Kennedy AP	40	12	17	21	H
NYC-LaGuardia AP	40	7	12	16	H
Niagara Falls AP	43	−2	4	7	M
Olean	42	−13	−8	−3	L

State and Station	Lat. °	Winter: Median of Annual Extremes	Winter: 99%	Winter: 97½%	Winter: Coincident Wind Velocity
NEW YORK (continued)					
Oneonta	42	−13	−7	−3	L
Oswego CO	43	−4	2	6	M
Plattsburg AFB	44	−16	−10	−6	L
Poughkeepsie	41	−6	−1	3	L
Rochester AP	43	−5	2	5	M
Rome-Griffiss AFB	43	−13	−7	−3	L
Schenectady	42	−11	−5	−1	L
Suffolk County AFB	40	4	9	13	H
Syracuse AP	43	−10	−2	2	M
Utica	43	−12	−6	−2	L
Watertown	44	−20	−14	−10	M
NORTH CAROLINA					
Asheville AP	35	8	13	17	L
Charlotte AP	35	13	18	22	L
Durham	36	11	15	19	L
Elizabeth City AP	36	14	18	22	M
Fayetteville, Pope AFB	35	13	17	20	L
Goldsboro, Seymour AFB	35	14	18	21	M
Greensboro AP	36	9	14	17	L
Greenville	35	14	18	22	M
Henderson	36	8	12	16	L
Hickory	35	9	14	18	L
Jacksonville	34	17	21	25	M
Lumberton	34	14	18	22	L
New Bern AP	35	14	18	22	L
Raleigh/Durham AP	35	13	16	20	L
Rocky Mount	36	12	16	20	L
Wilmington AP	34	19	23	27	L
Winston-Salem AP	36	9	14	17	L
NORTH DAKOTA					
Bismarck AP	46	−31	−24	−19	VL
Devil's Lake	48	−30	−23	−19	M
Dickinson AP	46	−31	−23	−19	L
Fargo AP	46	−28	−22	−17	L
Grand Forks AP	48	−30	−26	−23	L
Jamestown AP	47	−29	−22	−18	L
Minot AP	48	−31	−24	−20	M
Williston	48	−28	−21	−17	M
OHIO					
Akron/Canton AP	41	−5	1	6	M
Ashtabula	42	−3	3	7	M
Athens	39	−3	3	7	M
Bowling Green	41	−7	−1	3	M
Cambridge	40	−6	0	4	M
Chillicothe	39	−1	5	9	M
Cincinnati CO	39	2	8	12	L
Cleveland AP	41	−2	2	7	M
Columbus AP	40	−1	2	7	M
Dayton AP	39	−2	0	6	M
Defiance	41	−7	−1	1	M
Findlay AP	41	−6	0	4	M
Fremont	41	−7	−1	3	M
Hamilton	39	−2	4	8	M
Lancaster	39	−5	1	5	M
Lima	40	−6	0	4	M
Mansfield AP	40	−7	1	3	M
Marion	40	−5	1	6	M
Middletown	39	−3	3	7	M
Newark	40	−7	−1	3	M
Norwalk	41	−7	−1	3	M
Portsmouth	38	0	5	9	L
Sandusky CO	41	−2	4	8	M
Springfield	40	−3	3	7	M
OHIO (continued)					
Steubenville	40	−2	4	9	M
Toledo AP	41	−5	1	5	M
Warren	41	−6	0	4	M
Wooster	40	−7	−1	3	M
Youngstown AP	41	−5	1	6	M
Zanesville AP	40	−7	−1	3	M
OKLAHOMA					
Ada	34	6	12	16	H
Altus AFB	34	7	14	18	H
Ardmore	34	9	15	19	H
Bartlesville	36	−1	5	9	H
Chickasha	35	5	12	16	H
Enid-Vance AFB	36	3	10	14	H
Lawton AP	34	6	13	16	H
McAlester	34	7	13	17	H
Muskogee AP	35	6	12	16	M
Norman	35	5	11	15	H
Oklahoma City AP	35	4	11	15	H
Ponca City	36	1	8	12	H
Seminole	35	6	12	16	H
Stillwater	36	2	9	13	H
Tulsa AP	36	4	12	16	H
Woodward	36	−3	4	8	H
OREGON					
Albany	44	17	23	27	VL
Astoria AP	46	22	27	30	M
Baker AP	44	−10	−3	1	VL
Bend	44	−7	0	4	VL
Corvallis	44	17	23	27	VL
Eugene AP	44	16	22	26	VL
Grants Pass	42	16	22	26	VL
Klamath Falls AP	42	−5	1	5	VL
Medford AP	42	15	21	23	VL
Pendleton AP	45	−2	3	10	VL
Portland AP	45	17	21	24	L
Portland CO	45	21	26	29	L
Roseburg AP	43	19	25	29	VL
Salem AP	45	15	21	25	VL
The Dalles	45	7	13	17	VL
PENNSYLVANIA					
Allentown AP	40	−2	3	5	M
Altoona CO	40	−4	1	5	L
Butler	40	−8	−2	2	L
Chambersburg	40	0	5	9	L
Erie AP	42	1	7	11	M
Harrisburg AP	40	4	9	13	L
Johnstown	40	−4	1	5	L
Lancaster	40	−3	2	6	L
Meadville	41	−6	0	4	M
New Castle	41	−7	−1	4	M
Philadelphia AP	39	7	11	15	M
Pittsburgh AP	40	−1	5	9	M
Pittsburgh CO	40	1	7	11	M
Reading CO	40	1	6	9	M
Scranton/Wilkes-Barre	41	−3	2	6	L
State College	40	−3	2	6	L
Sunbury	40	−2	3	7	L
Uniontown	39	−1	4	8	L
Warren	41	−8	−3	1	L
West Chester	40	4	9	13	M
Williamsport AP	41	−5	1	5	L
York	40	−1	4	8	L
RHODE ISLAND					
Newport	41	1	5	11	H
Providence AP	41	0	6	10	M

State and Station	Lat. °	Winter			Coincident Wind Velocity
		Median of Annual Extremes	99%	97½%	
SOUTH CAROLINA					
Anderson	34	13	18	22	L
Charleston AFB	32	19	23	27	L
Charleston CO	32	23	26	30	L
Columbia AP	34	16	20	23	L
Florence AP	34	16	21	25	L
Georgetown	33	19	23	26	L
Greenville AP	34	14	19	23	L
Greenwood	34	15	19	23	L
Orangeburg	33	17	21	25	L
Rock Hill	35	13	17	21	L
Spartanburg AP	35	13	18	22	L
Sumter-Shaw AFB	34	18	23	26	L
SOUTH DAKOTA					
Aberdeen AP	45	−29	−22	−18	L
Brookings	44	−26	−19	−15	M
Huron AP	44	−24	−16	−12	L
Mitchell	43	−22	−15	−11	M
Pierre AP	44	−21	−13	− 9	M
Rapid City AP	44	−17	− 9	− 6	M
Sioux Falls AP	43	−21	−14	−10	M
Watertown AP	45	−27	−20	−16	L
Yankton	43	−18	−11	− 7	M
TENNESSEE					
Athens	33	10	14	18	L
Bristol-Tri City AP	36	6	11	16	L
Chattanooga AP	35	11	15	19	L
Clarksville	36	6	12	16	L
Columbia	35	8	13	17	L
Dyersburg	36	7	13	17	L
Greenville	35	5	10	14	L
Jackson AP	35	8	14	17	L
Knoxville AP	35	9	13	17	L
Memphis AP	35	11	17	21	L
Murfreesboro	35	7	13	17	L
Nashville AP	36	6	12	16	L
Tullahoma	35	7	13	17	L
TEXAS					
Abilene AP	32	12	17	21	M
Alice AP	27	26	30	34	M
Amarillo AP	35	2	8	12	M
Austin AP	30	19	25	29	M
Bay City	29	25	29	33	M
Beaumont	30	25	29	33	M
Beeville	28	24	28	32	M
Big Spring AP	32	12	18	22	M
Brownsville AP	25	32	36	40	M
Brownwood	31	15	20	25	M
Bryan AP	30	22	27	31	M
Corpus Christi AP	27	28	32	36	M
Corsicana	32	16	21	25	M
Dallas AP	32	14	19	24	H
Del Rio, Laughlin AFB	29	24	28	31	M
Denton	33	12	18	22	H
Eagle Pass	28	23	27	31	L
El Paso AP	31	16	21	25	L
Fort Worth AP	32	14	20	24	H
Galveston AP	29	28	32	36	M
Greenville	33	13	19	24	H
Harlingen	26	30	34	38	M
Houston AP	29	23	28	32	M
Houston CO	29	24	29	33	M
Huntsville	30	22	27	31	M
TEXAS (*continued*)					
Killeen-Gray AFB	31	17	22	26	M
Lamesa	32	7	14	18	M
Laredo AFB	27	29	32	36	L
Longview	32	16	21	25	M
Lubbock AP	33	4	11	15	M
Lufkin AP	31	19	24	28	M
McAllen	26	30	34	38	M
Midland AP	32	13	19	23	M
Mineral Wells AP	32	12	18	22	H
Palestine CO	31	16	21	25	M
Pampa	35	0	7	11	M
Pecos	31	10	15	19	L
Plainview	34	3	10	14	M
Port Arthur AP	30	25	29	33	M
San Angelo, Goodfellow AFB	31	15	20	25	M
San Antonio AP	29	22	25	30	L
Sherman-Perrin AFB	33	12	18	23	H
Snyder	32	9	15	19	M
Temple	31	18	23	27	M
Tyler AP	32	15	20	24	M
Vernon	34	7	14	18	H
Victoria AP	28	24	28	32	M
Waco AP	31	16	21	26	M
Wichita Falls AP	34	9	15	19	H
UTAH					
Cedar City AP	37	−10	− 1	6	VL
Logan	41	− 7	3	7	VL
Moab	38	2	12	16	VL
Ogden CO	41	− 3	7	11	VL
Price	39	− 7	3	7	L
Provo	40	− 6	2	6	L
Richfield	38	−10	− 1	3	L
St. George CO	37	13	22	26	VL
Salt Lake City AP	40	− 2	5	9	L
Vernal AP	40	−20	−10	− 6	VL
VERMONT					
Barre	44	−23	−17	−13	L
Burlington AP	44	−18	−12	− 7	M
Rutland	43	−18	−12	− 8	L
VIRGINIA					
Charlottsville	38	7	11	15	L
Danville AP	36	9	13	17	L
Fredericksburg	38	6	10	14	M
Harrisonburg	38	0	5	9	L
Lynchburg AP	37	10	15	19	L
Norfolk AP	36	18	20	23	M
Petersburg	37	10	15	18	L
Richmond AP	37	10	14	18	L
Roanoke AP	37	9	15	18	L
Staunton	38	3	8	12	L
Winchester	39	1	6	10	L
WASHINGTON					
Aberdeen	47	19	24	27	M
Bellingham AP	48	8	14	18	L
Bremerton	47	17	24	29	L
Ellensburg AP	47	− 5	2	6	VL
Everett-Paine AFB	47	13	19	24	L
Kennewick	46	4	11	15	VL
Longview	46	14	20	24	L
Moses Lake, Larson AFB	47	−14	− 7	− 1	VL
Olympia AP	47	15	21	25	L
Port Angeles	48	20	26	29	M
Seattle-Boeing Fld	47	17	23	27	L

State and Station	Lat. °	Winter: Median of Annual Extremes	99%	97½%	Coincident Wind Velocity
WASHINGTON (*continued*)					
Seattle CO	47	22	28	32	L
Seattle-Tacoma AP	47	14	20	24	L
Spokane AP	47	−5	−2	4	VL
Tacoma-McChord AFB	47	14	20	24	L
Walla Walla AP	46	5	12	16	VL
Wenatchee	47	−2	5	9	VL
Yakima AP	46	−1	6	10	VL
WEST VIRGINIA					
Beckley	37	−4	0	6	L
Bluefield AP	37	1	6	10	L
Charleston AP	38	1	9	14	L
Clarksburg	39	−2	3	7	L
Elkins AP	38	−4	1	5	L
Huntington CO	38	4	10	14	L
Martinsburg AP	39	1	6	10	L
Morgantown AP	39	−2	3	7	L
Parkersburg CO	39	2	8	12	L
Wheeling	40	0	5	9	L
WISCONSIN					
Appleton	44	−16	−10	−6	M
Ashland	46	−27	−21	−17	L
Beloit	42	−13	−7	−3	M
Eau Claire AP	44	−21	−15	−11	L
Fond du Lac	43	−17	−11	−7	M
Green Bay AP	44	−16	−12	−7	M
La Crosse AP	43	−18	−12	−8	M
Madison AP	43	−13	−9	−5	M
Manitowoc	44	−11	−5	−1	M
Marinette	45	−14	−8	−4	M
Milwaukee AP	43	−11	−6	−2	M
Racine	42	−10	−4	0	M
Sheboygan	43	−10	−4	0	M
Stevens Point	44	−22	−16	−12	M
Waukesha	43	−12	−6	−2	M
Wausau AP	44	−24	−18	−14	M
WYOMING					
Casper AP	42	−20	−11	−5	L
Cheyenne AP	41	−15	−6	−2	M
Cody AP	44	−23	−13	−9	L
Evanston	41	−22	−12	−8	VL
Lander AP	42	−26	−16	−12	VL
Laramie AP	41	−17	−6	−2	M
Newcastle	43	−18	−9	−5	M
Rawlins	41	−24	−15	−11	L
Rock Springs AP	41	−16	−6	−1	VL
Sheridan AP	44	−21	−12	−7	L
Torrington	42	−20	−11	−7	M

Appendix 4B Average Monthly and Yearly Degree-Days for the United States[abc]

State	Station	Avg. Winter Temp	July	Aug.	Sept.	Oct.	Nov.	Dec.	Jan.	Feb.	Mar.	Apr.	May	June	Yearly Total
Ala.	Birmingham.............A	54.2	0	0	6	93	363	555	592	462	363	108	9	0	2551
	Huntsville.............A	51.3	0	0	12	127	426	663	694	557	434	138	19	0	3070
	Mobile.................A	59.9	0	0	0	22	213	357	415	300	211	42	0	0	1560
	Montgomery.............A	55.4	0	0	0	68	330	527	543	417	316	90	0	0	2291
Alaska	Anchorage..............A	23.0	245	291	516	930	1284	1572	1631	1316	1293	879	592	315	10864
	Fairbanks..............A	6.7	171	332	642	1203	1833	2254	2359	1901	1739	1068	555	222	14279
	Juneau.................A	32.1	301	338	483	725	921	1135	1237	1070	1073	810	601	381	9075
	Nome...................A	13.1	481	496	693	1094	1455	1820	1879	1666	1770	1314	930	573	14171
Ariz.	Flagstaff..............A	35.6	46	68	201	558	867	1073	1169	991	911	651	437	180	7152
	Phoenix................A	58.5	0	0	0	22	234	415	474	328	217	75	0	0	1765
	Tucson.................A	58.1	0	0	0	25	231	406	471	344	242	75	6	0	1800
	Winslow................A	43.0	0	0	6	245	711	1008	1054	770	601	291	96	0	4782
	Yuma...................A	64.2	0	0	0	0	108	264	307	190	90	15	0	0	974
Ark.	Fort Smith.............A	50.3	0	0	12	127	450	704	781	596	456	144	22	0	3292
	Little Rock............A	50.5	0	0	9	127	465	716	756	577	434	126	9	0	3219
	Texarkana..............A	54.2	0	0	0	78	345	561	626	468	350	105	0	0	2533
Calif.	Bakersfield............A	55.4	0	0	0	37	282	502	546	364	267	105	19	0	2122
	Bishop.................A	46.0	0	0	48	260	576	797	874	680	555	306	143	36	4275
	Blue Canyon............A	42.2	28	37	108	347	594	781	896	795	806	597	412	195	5596
	Burbank................A	58.6	0	0	6	43	177	301	366	277	239	138	81	18	1646
	Eureka.................C	49.9	270	257	258	329	414	499	546	470	505	438	372	285	4643
	Fresno.................A	53.3	0	0	0	84	354	577	605	426	335	162	62	6	2611
	Long Beach.............A	57.8	0	0	9	47	171	316	397	311	264	171	93	24	1803
	Los Angeles............A	57.4	28	28	42	78	180	291	372	302	288	219	158	81	2061
	Los Angeles............C	60.3	0	0	6	31	132	229	310	230	202	123	68	18	1349
	Mt. Shasta.............C	41.2	25	34	123	406	696	902	983	784	738	525	347	159	5722
	Oakland................A	53.5	53	50	45	127	309	481	527	400	353	255	180	90	2870
	Red Bluff..............A	53.8	0	0	0	53	318	555	605	428	341	168	47	0	2515
	Sacramento.............A	53.9	0	0	0	56	321	546	583	414	332	178	72	0	2502
	Sacramento.............C	54.4	0	0	0	62	312	533	561	392	310	173	76	0	2419
	Sandberg...............C	46.8	0	0	30	202	480	691	778	661	620	426	264	57	4209
	San Diego..............A	59.5	9	0	21	43	135	236	298	235	214	135	90	42	1458
	San Francisco..........A	53.4	81	78	60	143	306	462	508	395	363	279	214	126	3015
	San Francisco..........C	55.1	192	174	102	118	231	388	443	336	319	279	239	180	3001
	Santa Maria............A	54.3	99	93	96	146	270	391	459	370	363	282	233	165	2967
Colo.	Alamosa................A	29.7	65	99	279	639	1065	1420	1476	1162	1020	696	440	168	8529
	Colorado Springs.......A	37.3	9	25	132	456	825	1032	1128	938	893	582	319	84	6423
	Denver.................A	37.6	6	9	117	428	819	1035	1132	938	887	558	288	66	6283
	Denver.................C	40.8	0	0	90	366	714	905	1004	851	800	492	254	48	5524
	Grand Junction.........A	39.3	0	0	30	313	786	1113	1209	907	729	387	146	21	5641
	Pueblo.................A	40.4	0	0	54	326	750	986	1085	871	772	429	174	15	5462
Conn.	Bridgeport.............A	39.9	0	0	66	307	615	986	1079	966	853	510	208	27	5617
	Hartford...............A	37.3	0	12	117	394	714	1101	1190	1042	908	510	205	33	6235
	New Haven..............A	39.0	0	12	87	347	648	1011	1097	991	871	543	245	45	5897
Del.	Wilmington.............A	42.5	0	0	51	270	588	927	980	874	735	387	112	6	4930
D. C.	Washington.............A	45.7	0	0	33	217	519	834	871	762	626	288	74	0	4224
Fla.	Apalachicola...........C	61.2	0	0	0	16	153	319	347	260	180	33	0	0	1308
	Daytona Beach..........A	64.5	0	0	0	0	75	211	248	190	140	15	0	0	879
	Fort Myers.............A	68.6	0	0	0	0	24	109	146	101	62	0	0	0	442
	Jacksonville...........A	61.9	0	0	0	12	144	310	332	246	174	21	0	0	1239
	Key West...............A	73.1	0	0	0	0	0	28	40	31	9	0	0	0	108
	Lakeland...............C	66.7	0	0	0	0	57	164	195	146	99	0	0	0	661
	Miami..................A	71.1	0	0	0	0	0	65	74	56	19	0	0	0	214

Notes: a. From ASHRAE *Guide and Data Book*; base temperature 65°F.
b. A indicates airport; C indicates city.
c. Average winter temperatures for October through April, inclusive.

Appendix 4B—*Continued*

State	Station	Avg. Winter Temp	July	Aug.	Sept.	Oct.	Nov.	Dec.	Jan.	Feb.	Mar.	Apr.	May	June	Yearly Total
Fla. (Cont'd)	Miami Beach............C	72.5	0	0	0	0	0	40	56	36	9	0	0	0	141
	Orlando..................A	65.7	0	0	0	0	72	198	220	165	105	6	0	0	766
	Pensacola...............A	60.4	0	0	0	19	195	353	400	277	183	36	0	0	1463
	Tallahassee..............A	60.1	0	0	0	28	198	360	375	286	202	36	0	0	1485
	Tampa.....................A	66.4	0	0	0	0	60	171	202	148	102	0	0	0	683
	West Palm Beach........A	68.4	0	0	0	0	6	65	87	64	31	0	0	0	253
Ga.	Athens....................A	51.8	0	0	12	115	405	632	642	529	431	141	22	0	2929
	Atlanta...................A	51.7	0	0	18	124	417	648	636	518	428	147	25	0	2961
	Augusta..................A	54.5	0	0	0	78	333	552	549	445	350	90	0	0	2397
	Columbus................A	54.8	0	0	0	87	333	543	552	434	338	96	0	0	2383
	Macon.....................A	56.2	0	0	0	71	297	502	505	403	295	63	0	0	2136
	Rome......................A	49.9	0	0	24	161	474	701	710	577	468	177	34	0	3326
	Savannah.................A	57.8	0	0	0	47	246	437	437	353	254	45	0	0	1819
	Thomasville..............C	60.0	0	0	0	25	198	366	394	305	208	33	0	0	1529
Hawaii	Lihue......................A	72.7	0	0	0	0	0	0	0	0	0	0	0	0	0
	Honolulu.................A	74.2	0	0	0	0	0	0	0	0	0	0	0	0	0
	Hilo........................A	71.9	0	0	0	0	0	0	0	0	0	0	0	0	0
Idaho	Boise......................A	39.7	0	0	132	415	792	1017	1113	854	722	438	245	81	5809
	Lewiston.................A	41.0	0	0	123	403	756	933	1063	815	694	426	239	90	5542
	Pocatello................A	34.8	0	0	172	493	900	1166	1324	1058	905	555	319	141	7033
Ill.	Cairo......................C	47.9	0	0	36	164	513	791	856	680	539	195	47	0	3821
	Chicago (O'Hare)........A	35.8	0	12	117	381	807	1166	1265	1086	939	534	260	72	6639
	Chicago (Midway).......A	37.5	0	0	81	326	753	1113	1209	1044	890	480	211	48	6155
	Chicago..................C	38.9	0	0	66	279	705	1051	1150	1000	868	489	226	48	5882
	Moline....................A	36.4	0	9	99	335	774	1181	1314	1100	918	450	189	39	6408
	Peoria.....................A.	38.1	0	6	87	326	759	1113	1218	1025	849	426	183	33	6025
	Rockford.................A	34.8	6	9	114	400	837	1221	1333	1137	961	516	236	60	6830
	Springfield..............A	40.6	0	0	72	291	696	1023	1135	935	769	354	136	18	5429
Ind.	Evansville...............A	45.0	0	0	66	220	606	896	955	767	620	237	68	0	4435
	Fort Wayne..............A	37.3	0	9	105	378	783	1135	1178	1028	890	471	189	39	6205
	Indianapolis.............A	39.6	0	0	90	316	723	1051	1113	949	809	432	177	39	5699
	South Bend..............A	36.6	0	6	111	372	777	1125	1221	1070	933	525	239	60	6439
Iowa	Burlington...............A	37.6	0	0	93	322	768	1135	1259	1042	859	426	177	33	6114
	Des Moines..............A	35.5	0	6	96	363	828	1225	1370	1137	915	438	180	30	6588
	Dubuque..................A	32.7	12	31	156	450	906	1287	1420	1204	1026	546	260	78	7376
	Sioux City...............A	34.0	0	9	108	369	867	1240	1435	1198	989	483	214	39	6951
	Waterloo.................A	32.6	12	19	138	428	909	1296	1460	1221	1023	531	229	54	7320
Kans.	Concordia................A	40.4	0	0	57	276	705	1023	1163	935	781	372	149	18	5479
	Dodge City...............A	42.5	0	0	33	251	666	939	1051	840	719	354	124	9	4986
	Goodland.................A	37.8	0	6	81	381	810	1073	1166	955	884	507	236	42	6141
	Topeka....................A	41.7	0	0	57	270	672	980	1122	893	722	330	124	12	5182
	Wichita...................A	44.2	0	0	33	229	618	905	1023	804	645	270	87	6	4620
Ky.	Covington................A	41.4	0	0	75	291	669	983	1035	893	756	390	149	24	5265
	Lexington................A	43.8	0	0	54	239	609	902	946	818	685	325	105	0	4683
	Louisville...............A	44.0	0	0	54	248	609	890	930	818	682	315	105	9	4660
La.	Alexandria...............A	57.5	0	0	0	56	273	431	471	361	260	69	0	0	1921
	Baton Rouge.............A	59.8	0	0	0	31	216	369	409	294	208	33	0	0	1560
	Lake Charles............A	60.5	0	0	0	19	210	341	381	274	195	39	0	0	1459
	New Orleans.............A	61.0	0	0	0	19	192	322	363	258	192	39	0	0	1385
	New Orleans.............C	61.8	0	0	0	12	165	291	344	241	177	24	0	0	1254
	Shreveport...............A	56.2	0	0	0	47	297	477	552	426	304	81	0	0	2184
Me.	Caribou...................A	24.4	78	115	336	682	1044	1535	1690	1470	1308	858	468	183	9767
	Portland.................A	33.0	12	53	195	508	807	1215	1339	1182	1042	675	372	111	7511
Md.	Baltimore................A	43.7	0	0	48	264	585	905	936	820	679	327	90	0	4654
	Baltimore................C	46.2	0	0	27	189	486	806	859	762	629	288	65	0	4111
	Frederich................A	42.0	0	0	66	307	624	955	995	876	741	384	127	12	5087
Mass.	Boston....................A	40.0	0	9	60	316	603	983	1088	972	846	513	208	36	5634
	Nantucket................A	40.2	12	22	93	332	573	896	992	941	896	621	384	129	5891
	Pittsfield................A	32.6	25	59	219	524	831	1231	1339	1196	1063	660	326	105	7578
	Worcester................A	34.7	6	34	147	450	774	1172	1271	1123	998	612	304	78	6969

Appendix 4B—*Continued*

State	Station	Avg. Winter Temp	July	Aug.	Sept.	Oct.	Nov.	Dec.	Jan.	Feb.	Mar.	Apr.	May	June	Yearly Total
Mich.	Alpena.................A	29.7	68	105	273	580	912	1268	1404	1299	1218	777	446	156	8506
	Detroit (City)...........A	37.2	0	0	87	360	738	1088	1181	1058	936	522	220	42	6232
	Detroit (Wayne).........A	37.1	0	0	96	353	738	1088	1194	1061	933	534	239	57	6293
	Detroit (Willow Run)....A	37.2	0	0	90	357	750	1104	1190	1053	921	519	229	45	6258
	Escanaba...............C	29.6	59	87	243	539	924	1293	1445	1296	1203	777	456	159	8481
	Flint..................A	33.1	16	40	159	465	843	1212	1330	1198	1066	639	319	90	7377
	Grand Rapids...........A	34.9	9	28	135	434	804	1147	1259	1134	1011	579	279	75	6894
	Lansing................A	34.8	6	22	138	431	813	1163	1262	1142	1011	579	273	69	6909
	Marquette..............C	30.2	59	81	240	527	936	1268	1411	1268	1187	771	468	177	8393
	Muskegon...............A	36.0	12	28	120	400	762	1088	1209	1100	995	594	310	78	6696
	Sault Ste. Marie.........A	27.7	96	105	279	580	951	1367	1525	1380	1277	810	477	201	9048
Minn.	Duluth.................A	23.4	71	109	330	632	1131	1581	1745	1518	1355	840	490	198	10000
	Minneapolis............A	28.3	22	31	189	505	1014	1454	1631	1380	1166	621	288	81	8382
	Rochester..............A	28.8	25	34	186	474	1005	1438	1593	1366	1150	630	301	93	8295
Miss.	Jackson................A	55.7	0	0	0	65	315	502	546	414	310	87	0	0	2239
	Meridian...............A	55.4	0	0	0	81	339	518	543	417	310	81	0	0	2289
	Vicksburg..............C	56.9	0	0	0	53	279	462	512	384	282	69	0	0	2041
Mo.	Columbia...............A	42.3	0	0	54	251	651	967	1076	874	716	324	121	12	5046
	Kansas City............A	43.9	0	0	39	220	612	905	1032	818	682	294	109	0	4711
	St. Joseph.............A	40.3	0	6	60	285	708	1039	1172	949	769	348	133	15	5484
	St. Louis...............A	43.1	0	0	60	251	627	936	1026	848	704	312	121	15	4900
	St. Louis...............C	44.8	0	0	36	202	576	884	977	801	651	270	87	0	4484
	Springfield.............A	44.5	0	0	45	223	600	877	973	781	660	291	105	6	4900
Mont.	Billings................A	34.5	6	15	186	487	897	1135	1296	1100	970	570	285	102	7049
	Glasgow................A	26.4	31	47	270	608	1104	1466	1711	1439	1187	648	335	150	8996
	Great Falls.............A	32.8	28	53	258	543	921	1169	1349	1154	1063	642	384	186	7750
	Havre..................A	28.1	28	53	306	595	1065	1367	1584	1364	1181	657	338	162	8700
	Havre..................C	29.8	19	37	252	539	1014	1321	1528	1305	1116	612	304	135	8182
	Helena.................A	31.1	31	59	294	601	1002	1265	1438	1170	1042	651	381	195	8129
	Kalispell...............A	31.4	50	99	321	654	1020	1240	1401	1134	1029	639	397	207	8191
	Miles City.............A	31.2	6	6	174	502	972	1296	1504	1252	1057	579	276	99	7723
	Missoula...............A	31.5	34	74	303	651	1035	1287	1420	1120	970	621	391	219	8125
Neb.	Grand Island...........A	36.0	0	6	108	381	834	1172	1314	1089	908	462	211	45	6530
	Lincoln.................C	38.8	0	6	75	301	726	1066	1237	1016	834	402	171	30	5864
	Norfolk.................A	34.0	9	0	111	397	873	1234	1414	1179	983	498	233	48	6979
	North Platte............A	35.5	0	6	123	440	885	1166	1271	1039	930	519	248	57	6684
	Omaha..................A	35.6	0	12	105	357	828	1175	1355	1126	939	465	208	42	6612
	Scottsbluff.............A	35.9	0	0	138	459	876	1128	1231	1008	921	552	285	75	6673
	Valentine...............A	32.6	9	12	165	493	942	1237	1395	1176	1045	579	288	84	7425
Nev.	Elko...................A	34.0	9	34	225	561	924	1197	1314	1036	911	621	409	192	7433
	Ely....................A	33.1	28	43	234	592	939	1184	1308	1075	977	672	456	225	7733
	Las Vegas..............A	53.5	0	0	0	78	387	617	688	487	335	111	6	0	2709
	Reno...................A	39.3	43	87	204	490	801	1026	1073	823	729	510	357	189	6332
	Winnemucca............A	36.7	0	34	210	536	876	1091	1172	916	837	573	363	153	6761
N. H.	Concord................A	33.0	6	50	177	505	822	1240	1358	1184	1032	636	298	75	7383
	Mt. Washington Obsv....	15.2	493	536	720	1057	1341	1742	1820	1663	1652	1260	930	603	13817
N. J.	Atlantic City............A	43.2	0	0	39	251	549	880	936	848	741	420	133	15	4812
	Newark.................A	42.8	0	0	30	248	573	921	983	876	729	381	118	0	4589
	Trenton................C	42.4	0	0	57	264	576	924	989	885	753	399	121	12	4980
N. M.	Albuquerque............A	45.0	0	0	12	229	642	868	930	703	595	288	81	0	4348
	Clayton.................A	42.0	0	6	66	310	699	899	986	812	747	429	183	21	5158
	Raton..................A	38.1	9	28	126	431	825	1048	1116	904	834	543	301	63	6228
	Roswell.................A	47.5	0	0	18	202	573	806	840	641	481	201	31	0	3793
	Silver City..............A	48.0	0	0	6	183	525	729	791	605	518	261	87	0	3705
N. Y.	Albany.................A	34.6	0	19	138	440	777	1194	1311	1156	992	564	239	45	6875
	Albany.................C	37.2	0	9	102	375	699	1104	1218	1072	908	498	186	30	6201
	Binghamton.............A	33.9	22	65	201	471	810	1184	1277	1154	1045	645	313	99	7286
	Binghamton.............C	36.6	0	28	141	406	732	1107	1190	1081	949	543	229	45	6451
	Buffalo.................A	34.5	19	37	141	440	777	1156	1256	1145	1039	645	329	78	7062
	New York (Cent. Park)..	42.8	0	0	30	233	540	902	986	885	760	408	118	9	4871
	New York (La Guardia)..A	43.1	0	0	27	223	528	887	973	879	750	414	124	6	4811

Appendix 4B—*Continued*

State	Station	Avg. Winter Temp	July	Aug.	Sept.	Oct.	Nov.	Dec.	Jan.	Feb.	Mar.	Apr.	May	June	Yearly Total
	New York (Kennedy)....A	41.4	0	0	36	248	564	933	1029	935	815	480	167	12	5219
	Rochester..............A	35.4	9	31	126	415	747	1125	1234	1123	1014	597	279	48	6748
	Schenectady............C	35.4	0	22	123	422	756	1159	1283	1131	970	543	211	30	6650
	Syracuse...............A	35.2	6	28	132	415	744	1153	1271	1140	1004	570	248	45	6756
N. C.	Asheville...............C	46.7	0	0	48	245	555	775	784	683	592	273	87	0	4042
	Cape Hatteras...........	53.3	0	0	0	78	273	521	580	518	440	177	25	0	2612
	Charlotte...............A	50.4	0	0	6	124	438	691	691	582	481	156	22	0	3191
	Greensboro.............A	47.5	0	0	33	192	513	778	784	672	552	234	47	0	3805
	Raleigh.................A	49.4	0	0	21	164	450	716	725	616	487	180	34	0	3393
	Wilmington.............A	54.6	0	0	0	74	291	521	546	462	357	96	0	0	2347
	Winston-Salem..........A	48.4	0	0	21	171	483	747	753	652	524	207	37	0	3595
N. D.	Bismarck...............A	26.6	34	28	222	577	1083	1463	1708	1442	1203	645	329	117	8851
	Devils Lake............C	22.4	40	53	273	642	1191	1634	1872	1579	1345	753	381	138	9901
	Fargo..................A	24.8	28	37	219	574	1107	1569	1789	1520	1262	690	332	99	9226
	Williston...............A	25.2	31	43	261	601	1122	1513	1758	1473	1262	681	357	141	9243
Ohio	Akron-Canton...........A	38.1	0	9	96	381	726	1070	1138	1016	871	489	202	39	6037
	Cincinnati..............C	45.1	0	0	39	208	558	862	915	790	642	294	96	6	4410
	Cleveland...............A	37.2	9	25	105	384	738	1088	1159	1047	918	552	260	66	6351
	Columbus...............A	39.7	0	6	84	347	714	1039	1088	949	809	426	171	27	5660
	Columbus...............C	41.5	0	0	57	285	651	977	1032	902	760	396	136	15	5211
	Dayton..................A	39.8	0	6	78	310	696	1045	1097	955	809	429	167	30	5622
	Mansfield...............A	36.9	9	22	114	397	768	1110	1169	1042	924	543	245	60	6403
	Sandusky...............C	39.1	0	6	66	313	684	1032	1107	991	868	495	198	36	5796
	Toledo..................A	36.4	0	16	117	406	792	1138	1200	1056	924	543	242	60	6494
	Youngstown.............A	36.8	6	19	120	412	771	1104	1169	1047	921	540	248	60	6417
Okla.	Oklahoma City..........A	48.3	0	0	15	164	498	766	868	664	527	189	34	0	3725
	Tulsa...................A	47.7	0	0	18	158	522	787	893	683	539	213	47	0	3860
Ore.	Astoria.................A	45.6	146	130	210	375	561	679	753	622	636	480	363	231	5186
	Burns...................C	35.9	12	37	210	515	867	1113	1246	988	856	570	366	177	6957
	Eugene..................A	45.6	34	34	129	366	585	719	803	627	589	426	279	135	4726
	Meacham................A	34.2	84	124	288	580	918	1091	1209	1005	983	726	527	339	7874
	Medford.................A	43.2	0	0	78	372	678	871	918	697	642	432	242	78	5008
	Pendleton...............A	42.6	0	0	111	350	711	884	1017	773	617	396	205	63	5127
	Portland................A	45.6	25	28	114	335	597	735	825	644	586	396	245	105	4635
	Portland................C	47.4	12	16	75	267	534	679	769	594	536	351	198	78	4109
	Roseburg...............A	46.3	22	16	105	329	567	713	766	608	570	405	267	123	4491
	Salem...................A	45.4	37	31	111	338	594	729	822	647	611	417	273	144	4754
Pa.	Allentown...............A	38.9	0	0	90	353	693	1045	1116	1002	849	471	167	24	5810
	Erie.....................A	36.8	0	25	102	391	714	1063	1169	1081	973	585	288	60	6451
	Harrisburg..............A	41.2	0	0	63	298	648	992	1045	907	766	396	124	12	5251
	Philadelphia.............A	41.8	0	0	60	297	620	965	1016	889	747	392	118	40	5144
	Philadelphia.............C	44.5	0	0	30	205	513	856	924	823	691	351	93	0	4486
	Pittsburgh..............A	38.4	0	9	105	375	726	1063	1119	1002	874	480	195	39	5987
	Pittsburgh..............C	42.2	0	0	60	291	615	930	983	885	763	390	124	12	5053
	Reading.................C	42.4	0	0	54	257	597	939	1001	885	735	372	105	0	4945
	Scranton................A	37.2	0	19	132	434	762	1104	1156	1028	893	498	195	33	6254
	Williamsport............A	38.5	0	9	111	375	717	1073	1122	1002	856	468	177	24	5934
R. I.	Block Island............A	40.1	0	16	78	307	594	902	1020	955	877	612	344	99	5804
	Providence..............A	38.8	0	16	96	372	660	1023	1110	988	868	534	236	51	5954
S. C.	Charleston..............A	56.4	0	0	0	59	282	471	487	389	291	54	0	0	2033
	Charleston..............C	57.9	0	0	0	34	210	425	443	367	273	42	0	0	1794
	Columbia................A	54.0	0	0	0	84	345	577	570	470	357	81	0	0	2484
	Florence................A	54.5	0	0	0	78	315	552	552	459	347	84	0	0	2387
	Greenville-Spartenburg...A	51.6	0	0	6	121	399	651	660	546	446	132	19	0	2980
S. D.	Huron...................A	28.8	9	12	165	508	1014	1432	1628	1355	1125	600	288	87	8223
	Rapid City..............A	33.4	22	12	165	481	897	1172	1333	1145	1051	615	326	126	7345
	Sioux Falls..............A	30.6	19	25	168	462	972	1361	1544	1285	1082	573	270	78	7839
Tenn.	Bristol..................A	46.2	0	0	51	236	573	828	828	700	598	261	68	0	4143
	Chattanooga.............A	50.3	0	0	18	143	468	698	722	577	453	150	25	0	3254
	Knoxville................A	49.2	0	0	30	171	489	725	732	613	493	198	43	0	3494
	Memphis.................A	50.5	0	0	18	130	447	698	729	585	456	147	22	0	3232

Appendix 4B—*Concluded*

State or Prov.	Station	Avg. Winter Temp	July	Aug.	Sept.	Oct.	Nov.	Dec.	Jan.	Feb.	Mar.	Apr.	May	June	Yearly Total
	Memphis C	51.6	0	0	12	102	396	648	710	568	434	129	16	0	3015
	Nashville . . . . A	48.9	0	0	30	158	495	732	778	644	512	189	40	0	3578
	Oak Ridge C	47.7	0	0	39	192	531	772	778	669	552	228	56	0	3817
Tex.	Abilene . . . . A	53.9	0	0	0	99	366	586	642	470	347	114	0	0	2624
	Amarillo . . . . A	47.0	0	0	18	205	570	797	877	664	546	252	56	0	3985
	Austin . . . . A	59.1	0	0	0	31	225	388	468	325	223	51	0	0	1711
	Brownsville . . . . A	67.7	0	0	0	0	66	149	205	106	74	0	0	0	600
	Corpus Christi . . . . A	64.6	0	0	0	0	120	220	291	174	109	0	0	0	914
	Dallas . . . . A	55.3	0	0	0	62	321	524	601	440	319	90	6	0	2363
	El Paso . . . . A	52.9	0	0	0	84	414	648	685	445	319	105	0	0	2700
	Fort Worth . . . . A	55.1	0	0	0	65	324	536	614	448	319	99	0	0	2405
	Galveston . . . . A	62.2	0	0	0	6	147	276	360	263	189	33	0	0	1274
	Galveston C	62.0	0	0	0	0	138	270	350	258	189	30	0	0	1235
	Houston . . . . A	61.0	0	0	0	6	183	307	384	288	192	36	0	0	1396
	Houston C	62.0	0	0	0	0	165	288	363	258	174	30	0	0	1278
	Laredo . . . . A	66.0	0	0	0	0	105	217	267	134	74	0	0	0	797
	Lubbock . . . . A	48.8	0	0	18	174	513	744	800	613	484	201	31	0	3578
	Midland . . . . A	53.8	0	0	0	87	381	592	651	468	322	90	0	0	2591
	Port Arthur . . . . A	60.5	0	0	0	22	207	329	384	274	192	39	0	0	1447
	San Angelo . . . . A	56.0	0	0	0	68	318	536	567	412	288	66	0	0	2255
	San Antonio . . . . A	60.1	0	0	0	31	204	363	428	286	195	39	0	0	1546
	Victoria . . . . A	62.7	0	0	0	6	150	270	344	230	152	21	0	0	1173
	Waco . . . . A	57.2	0	0	0	43	270	456	536	389	270	66	0	0	2030
	Wichita Falls . . . . A	53.0	0	0	0	99	381	632	698	518	378	120	6	0	2832
Utah	Milford . . . . A	36.5	0	0	99	443	867	1141	1252	988	822	519	279	87	6497
	Salt Lake City . . . . A	38.4	0	0	81	419	849	1082	1172	910	763	459	233	84	6052
	Wendover . . . . A	39.1	0	0	48	372	822	1091	1178	902	729	408	177	51	5778
Vt.	Burlington . . . . A	29.4	28	65	207	539	891	1349	1513	1333	1187	714	353	90	8269
Va.	Cape Henry C	50.0	0	0	0	112	360	645	694	633	536	246	53	0	3279
	Lynchburg . . . . A	46.0	0	0	51	223	540	822	849	731	605	267	78	0	4166
	Norfolk . . . . A	49.2	0	0	0	136	408	698	738	655	533	216	37	0	3421
	Richmond . . . . A	47.3	0	0	36	214	495	784	815	703	546	219	53	0	3865
	Roanoke . . . . A	46.1	0	0	51	229	549	825	834	722	614	261	65	0	4150
Wash.	Olympia . . . . A	44.2	68	71	198	422	636	753	834	675	645	450	307	177	5236
	Seattle-Tacoma . . . . A	44.2	56	62	162	391	633	750	828	678	657	474	295	159	5145
	Seattle . . . . U	46.9	50	47	129	329	543	657	738	599	577	396	242	117	4424
	Spokane . . . . A	36.5	9	25	168	493	879	1082	1231	980	834	531	288	135	6655
	Walla Walla C	43.8	0	0	87	310	681	843	986	745	589	342	177	45	4805
	Yakima . . . . A	39.1	0	12	144	450	828	1039	1163	868	713	435	220	69	5941
W. Va.	Charleston . . . . A	44.8	0	0	63	254	591	865	880	770	648	300	96	9	4476
	Elkins . . . . A	40.1	9	25	135	400	729	992	1008	896	791	444	198	48	5675
	Huntington . . . . A	45.0	0	0	63	257	585	856	880	764	636	294	99	12	4446
	Parkersburg C	43.5	0	0	60	264	606	905	942	826	691	339	115	6	4754
Wisc.	Green Bay . . . . A	30.3	28	50	174	484	924	1333	1494	1313	1141	654	335	99	8029
	La Crosse . . . . A	31.5	12	19	153	437	924	1339	1504	1277	1070	540	245	69	7589
	Madison . . . . A	30.9	25	40	174	474	930	1330	1473	1274	1113	618	310	102	7863
	Milwaukee . . . . A	32.6	43	47	174	471	876	1252	1376	1193	1054	642	372	135	7635
Wyo.	Casper . . . . A	33.4	6	16	192	524	942	1169	1290	1084	1020	657	381	129	7410
	Cheyenne . . . . A	34.2	28	37	219	543	909	1085	1212	1042	1026	702	428	150	7381
	Lander . . . . A	31.4	6	19	204	555	1020	1299	1417	1145	1017	654	381	153	7870
	Sheridan . . . . A	32.5	25	31	219	539	948	1200	1355	1154	1051	642	366	150	7680

Appendix 4C Solar Positions and Insolation Values for Various Latitudes[a]

Table I 24 Degrees North Latitude

DATE	SOLAR TIME AM	SOLAR TIME PM	SOLAR POSITION ALT	SOLAR POSITION AZM	BTUH/SQ. FT. TOTAL INSOLATION ON SURFACES NORMAL	HORIZ.	SOUTH FACING SURFACE ANGLE WITH HORIZ. 14	24	34	54	90
JAN 21	7	5	4.8	65.6	71	10	17	21	25	28	31
	8	4	16.9	58.3	239	83	110	126	137	145	127
	9	3	27.9	48.8	288	151	188	207	221	228	176
	10	2	37.2	36.1	308	204	246	268	282	287	207
	11	1	43.6	19.6	317	237	283	306	319	324	226
	12		46.0	0.0	320	249	296	319	332	336	232
	SURFACE DAILY TOTALS				2766	1622	1984	2174	2300	2360	1766
FEB 21	7	5	9.3	74.6	158	35	44	49	53	56	46
	8	4	22.3	67.2	263	116	135	145	150	151	102
	9	3	34.4	57.6	298	187	213	225	230	228	141
	10	2	45.1	44.2	314	241	273	286	291	287	168
	11	1	53.0	25.0	321	276	310	324	328	323	185
	12		56.0	0.0	324	288	323	337	341	335	191
	SURFACE DAILY TOTALS				3036	1998	2276	2396	2446	2424	1476
MAR 21	7	5	13.7	83.8	194	60	63	64	62	59	27
	8	4	27.2	76.8	267	141	150	152	149	142	64
	9	3	40.2	67.9	295	212	226	229	225	214	95
	10	2	52.3	54.8	309	266	285	288	283	270	120
	11	1	61.9	33.4	315	300	322	326	320	305	135
	12		66.0	0.0	317	312	334	339	333	317	140
	SURFACE DAILY TOTALS				3078	2270	2428	2456	2412	2298	1022
APR 21	6	6	4.7	100.6	40	7	5	4	4	3	2
	7	5	18.3	94.9	203	83	77	70	62	51	10
	8	4	32.0	89.0	256	160	157	149	137	122	16
	9	3	45.6	81.9	280	227	227	220	206	186	41
	10	2	59.0	71.8	292	278	282	275	259	237	61
	11	1	71.1	51.6	298	310	316	309	293	269	74
	12		77.6	0.0	299	321	328	321	305	280	79
	SURFACE DAILY TOTALS				3036	2454	2458	2374	2228	2016	488
MAY 21	6	6	8.0	108.4	86	22	15	10	9	9	5
	7	5	21.2	103.2	203	98	85	73	59	44	12
	8	4	34.6	98.5	248	171	159	145	127	106	15
	9	3	48.3	93.6	269	233	224	210	190	165	16
	10	2	62.0	87.7	280	281	275	261	239	211	22
	11	1	75.5	76.9	286	311	307	293	270	240	34
	12		86.0	0.0	288	322	317	304	281	250	37
	SURFACE DAILY TOTALS				3032	2556	2447	2286	2072	1800	246
JUN 21	6	6	9.3	111.6	97	29	20	12	12	11	7
	7	5	22.3	106.8	201	103	87	73	58	41	13
	8	4	35.5	102.6	242	173	158	142	122	99	16
	9	3	49.0	98.7	263	234	221	204	182	155	18
	10	2	62.6	95.0	274	280	269	253	229	199	18
	11	1	76.3	90.8	279	309	300	283	259	227	19
	12		89.4	0.0	281	319	310	294	269	236	22
	SURFACE DAILY TOTALS				2994	2574	2422	2230	1992	1700	204
JUL 21	6	6	8.2	109.0	81	23	16	11	10	9	6
	7	5	21.4	103.8	195	98	85	73	59	44	13
	8	4	34.8	99.2	239	169	157	143	125	104	16
	9	3	48.4	94.5	261	231	221	207	187	161	18
	10	2	62.1	89.0	272	278	270	256	235	206	21
	11	1	75.7	79.2	278	307	302	287	265	235	32
	12		86.6	0.0	280	317	312	298	275	245	36
	SURFACE DAILY TOTALS				2932	2526	2412	2250	2036	1766	246
AUG 21	6	6	5.0	101.3	35	7	5	4	4	4	2
	7	5	18.5	95.6	186	82	76	69	60	50	11
	8	4	32.2	89.7	241	158	154	146	134	118	16
	9	3	45.9	82.9	265	223	222	214	200	181	39
	10	2	59.3	73.0	278	273	275	268	252	230	58
	11	1	71.6	53.2	284	304	309	301	285	261	71
	12		78.3	0.0	286	315	320	313	296	272	75
	SURFACE DAILY TOTALS				2864	2408	2402	2316	2168	1958	470
SEP 21	7	5	13.7	83.8	173	57	60	60	59	56	26
	8	4	27.2	76.8	248	136	144	146	143	136	62
	9	3	40.2	67.9	278	205	218	221	217	206	93
	10	2	52.3	54.8	292	258	275	278	273	261	116
	11	1	61.9	33.4	299	291	311	315	309	295	131
	12		66.0	0.0	301	302	323	327	321	306	136
	SURFACE DAILY TOTALS				2878	2194	2342	2366	2322	2212	992
OCT 21	7	5	9.1	74.1	138	32	40	45	48	50	42
	8	4	22.0	66.7	247	111	129	139	144	145	99
	9	3	34.1	57.1	284	180	206	217	223	221	138
	10	2	44.7	43.8	301	234	265	277	282	279	165
	11	1	52.5	24.7	309	268	301	315	319	314	182
	12		55.5	0.0	311	279	314	328	332	327	188
	SURFACE DAILY TOTALS				2868	1928	2198	2314	2364	2346	1442
NOV 21	7	5	4.9	65.8	67	10	16	20	24	27	29
	8	4	17.0	58.4	232	82	108	123	135	142	124
	9	3	28.0	48.9	282	150	186	205	217	224	172
	10	2	37.3	36.3	303	203	244	265	278	283	204
	11	1	43.8	19.7	312	236	280	302	316	320	222
	12		46.2	0.0	315	247	293	315	328	332	228
	SURFACE DAILY TOTALS				2706	1610	1962	2146	2268	2324	1730
DEC 21	7	5	3.2	62.6	30	3	7	9	11	12	14
	8	4	14.9	55.3	225	71	99	116	129	139	130
	9	3	25.5	46.0	281	137	176	198	214	223	184
	10	2	34.3	33.7	304	189	234	258	275	283	217
	11	1	40.4	18.2	314	221	270	295	312	320	236
	12		42.6	0.0	317	232	282	308	325	332	243
	SURFACE DAILY TOTALS				2624	1474	1852	2058	2204	2286	1808

Notes: a. From ASHRAE *Transactions*; ground reflection not included.

Table II 32 Degrees North Latitude

DATE	SOLAR TIME AM	SOLAR TIME PM	SOLAR POSITION ALT	SOLAR POSITION AZM	BTUH/SQ. FT. TOTAL INSOLATION ON SURFACES NORMAL	HORIZ.	SOUTH FACING SURFACE ANGLE WITH HORIZ. 22	32	42	52	90
JAN 21	7	5	1.4	65.2	1	0	0	0	0	1	1
	8	4	12.5	56.5	203	56	93	106	115	123	115
	9	3	22.5	46.0	269	118	175	193	206	212	181
	10	2	30.6	33.1	295	167	235	256	269	274	221
	11	1	36.1	17.5	305	198	273	295	308	312	245
	12		38.0	0.0	310	209	235	308	321	324	253
	SURFACE DAILY TOTALS				2458	1288	1839	2008	2118	2166	1779
FEB 21	7	5	7.1	73.5	121	22	34	37	40	42	38
	8	4	19.0	64.4	247	95	127	136	140	141	108
	9	3	29.9	53.4	288	161	206	217	222	220	158
	10	2	39.1	39.4	306	212	266	278	283	279	193
	11	1	45.6	21.4	315	244	304	317	321	315	214
	12		48.0	0.0	317	255	316	330	334	328	222
	SURFACE DAILY TOTALS				2872	1724	2188	2300	2345	2322	1644
MAR 21	7	5	12.7	81.9	185	54	60	60	59	56	32
	8	4	25.1	73.0	260	129	146	147	144	137	78
	9	3	36.8	62.1	290	194	222	224	220	209	119
	10	2	47.3	47.5	304	245	280	283	278	265	150
	11	1	55.0	26.8	311	277	317	321	315	300	170
	12		58.0	0.0	313	287	329	333	327	312	177
	SURFACE DAILY TOTALS				3012	2084	2378	2403	2358	2246	1276
APR 21	6	6	6.1	99.9	66	14	9	6	6	5	3
	7	5	18.8	92.2	206	86	78	71	62	51	10
	8	4	31.5	84.0	255	158	156	148	136	120	35
	9	3	43.9	74.2	278	220	225	217	203	183	68
	10	2	55.7	60.3	290	267	279	272	256	234	95
	11	1	65.4	37.5	295	297	313	306	290	265	112
	12		69.6	0.0	297	307	325	318	301	276	118
	SURFACE DAILY TOTALS				3076	2390	2444	2356	2206	1994	764
MAY 21	6	6	10.4	107.2	119	36	21	13	13	12	7
	7	5	22.8	100.1	211	107	88	75	60	44	13
	8	4	35.4	92.9	250	175	159	145	127	105	15
	9	3	48.1	84.7	269	233	223	209	188	163	33
	10	2	60.6	73.3	280	277	273	259	237	208	56
	11	1	72.0	51.9	285	305	305	290	268	237	72
	12		78.0	0.0	286	315	315	301	278	247	77
	SURFACE DAILY TOTALS				3112	2582	2454	2284	2064	1788	469
JUN 21	6	6	12.2	110.2	131	45	26	16	15	14	9
	7	5	24.3	103.4	210	115	91	76	59	41	14
	8	4	36.9	96.8	245	180	159	143	122	99	16
	9	3	49.6	89.4	264	236	221	204	181	153	19
	10	2	62.2	79.7	274	279	268	251	227	197	41
	11	1	74.2	60.9	279	306	299	282	257	224	56
	12		81.5	0.0	280	315	309	292	267	234	60
	SURFACE DAILY TOTALS				3084	2634	2436	2234	1990	1690	370
JUL 21	6	6	10.7	107.7	113	37	22	14	13	12	8
	7	5	23.1	100.6	203	107	87	75	60	44	14
	8	4	35.7	93.6	241	174	158	143	125	104	16
	9	3	48.4	85.5	261	231	220	205	185	159	31
	10	2	60.9	74.3	271	274	269	254	232	204	54
	11	1	72.4	53.3	277	302	300	285	262	232	69
	12		78.6	0.0	279	311	310	296	273	242	74
	SURFACE DAILY TOTALS.				3012	2558	2422	2250	2030	1754	458
AUG 21	6	6	6.5	100.5	59	14	9	7	6	6	4
	7	5	19.1	92.8	190	85	77	69	60	50	12
	8	4	31.8	84.7	240	156	152	144	132	116	33
	9	3	44.3	75.0	263	216	220	212	197	178	65
	10	2	56.1	61.3	276	262	272	264	249	226	91
	11	1	66.0	38.4	282	292	305	298	281	257	107
	12		70.3	0.0	284	302	317	309	292	268	113
	SURFACE DAILY TOTALS				2902	2352	2388	2296	2144	1934	736
SEP 21	7	5	12.7	81.9	163	51	56	56	55	52	30
	8	4	25.1	73.0	240	124	140	141	138	131	75
	9	3	36.8	62.1	272	188	213	215	211	201	114
	10	2	47.3	47.5	287	237	270	273	268	255	145
	11	1	55.0	26.8	294	268	306	309	303	289	164
	12		58.0	0.0	296	278	318	321	315	300	171
	SURFACE DAILY TOTALS				2808	2014	2288	2308	2264	2154	1226
OCT 21	7	5	6.8	73.1	99	19	29	32	34	36	32
	8	4	18.7	64.0	229	90	120	128	133	134	104
	9	3	29.5	53.0	273	155	198	208	213	212	153
	10	2	38.7	39.1	293	204	257	269	273	270	188
	11	1	45.1	21.1	302	236	294	307	311	306	209
	12		47.5	0.0	304	247	306	320	324	318	217
	SURFACE DAILY TOTALS				2696	1654	2100	2208	2252	2232	1588
NOV 21	7	5	1.5	65.4	2	0	0	0	1	1	1
	8	4	12.7	56.6	196	55	91	104	113	119	111
	9	3	22.6	45.1	263	118	173	190	202	208	176
	10	2	30.8	33.2	289	166	233	252	265	270	217
	11	1	36.2	17.6	301	197	270	291	303	307	241
	12		38.2	0.0	304	207	282	304	316	320	249
	SURFACE DAILY TOTALS				2406	1280	1816	1980	2084	2130	1742
DEC 21	8	4	10.3	53.8	176	41	77	90	101	108	107
	9	3	19.8	43.6	257	102	161	180	195	204	183
	10	2	27.6	31.2	288	150	221	244	259	267	226
	11	1	32.7	16.4	301	180	258	282	298	305	251
	12		34.6	0.0	304	190	271	295	311	318	259
	SURFACE DAILY TOTALS				2348	1136	1704	1888	2016	2086	1794

Table III 40 Degrees North Latitude

DATE	SOLAR TIME		SOLAR POSITION		BTUH/SQ. FT. TOTAL INSOLATION ON SURFACES						
							SOUTH FACING SURFACE ANGLE WITH HORIZ.				
	AM	PM	ALT	AZM	NORMAL	HORIZ.	30	40	50	60	90
JAN 21	8	4	8.1	55.3	142	28	65	74	81	85	84
	9	3	16.8	44.0	239	83	155	171	182	187	171
	10	2	23.8	30.9	274	127	218	237	249	254	223
	11	1	28.4	16.0	289	154	257	277	290	293	253
	12		30.0	0.0	294	164	270	291	303	306	263
	SURFACE DAILY TOTALS				2182	948	1660	1810	1906	1944	1726
FEB 21	7	5	4.8	72.7	69	10	19	21	23	24	22
	8	4	15.4	62.2	224	73	114	122	126	127	107
	9	3	25.0	50.2	274	132	195	205	209	208	167
	10	2	32.8	35.9	295	178	256	267	271	267	210
	11	1	38.1	18.9	305	206	293	306	310	304	236
	12		40.0	0.0	308	216	306	319	323	317	245
	SURFACE DAILY TOTALS				2640	1414	2060	2162	2202	2176	1730
MAR 21	7	5	11.4	80.2	171	46	55	55	54	51	35
	8	4	22.5	69.6	250	114	140	141	138	131	89
	9	3	32.8	57.3	282	173	215	217	213	202	138
	10	2	41.6	41.9	297	218	273	276	271	258	176
	11	1	47.7	22.6	305	247	310	313	307	293	200
	12		50.0	0.0	307	257	322	326	320	305	208
	SURFACE DAILY TOTALS				2916	1852	2308	2330	2284	2174	1484
APR 21	6	6	7.4	98.9	89	20	11	8	7	7	4
	7	5	18.9	89.5	206	87	77	70	61	50	12
	8	4	30.3	79.3	252	152	153	145	133	117	53
	9	3	41.3	67.2	274	207	221	213	199	179	93
	10	2	51.2	51.4	286	250	275	267	252	229	126
	11	1	58.7	29.2	292	277	308	301	285	260	147
	12		61.6	0.0	293	287	320	313	296	271	154
	SURFACE DAILY TOTALS				3092	2274	2412	2320	2168	1956	1022
MAY 21	5	7	1.9	114.7	1	0	0	0	0	0	0
	6	6	12.7	105.6	144	49	25	15	14	13	9
	7	5	24.0	96.6	216	214	89	76	60	44	13
	8	4	35.4	87.2	250	175	158	144	125	104	25
	9	3	46.8	76.0	267	227	221	206	186	160	60
	10	2	57.5	60.9	277	267	270	255	233	205	89
	11	1	66.2	37.1	283	293	301	287	264	234	108
	12		70.0	0.0	284	301	312	297	274	243	114
	SURFACE DAILY TOTALS				3160	2552	2442	2264	2040	1760	724
JUN 21	5	7	4.2	117.3	22	4	3	3	2	2	1
	6	6	14.8	108.4	155	60	30	18	17	16	10
	7	5	26.0	99.7	216	123	92	77	59	41	14
	8	4	37.4	90.7	246	182	159	142	121	97	16
	9	3	48.8	80.2	263	233	219	202	179	151	47
	10	2	59.8	65.8	272	272	266	248	224	194	74
	11	1	69.2	41.9	277	296	296	278	253	221	92
	12		73.5	0.0	279	304	306	289	263	230	98
	SURFACE DAILY TOTALS				3180	2648	2434	2224	1974	1670	610
JUL 21	5	7	2.3	115.2	2	0	0	0	0	0	0
	6	6	13.1	106.1	138	50	26	17	15	14	9
	7	5	24.3	97.2	208	114	89	75	60	44	14
	8	4	35.8	87.8	241	174	157	142	124	102	24
	9	3	47.2	76.7	259	225	218	203	182	157	58
	10	2	57.9	61.7	269	265	266	251	229	200	86
	11	1	66.7	37.9	275	290	296	281	258	228	104
	12		70.6	0.0	276	298	307	292	269	238	111
	SURFACE DAILY TOTALS				3062	2534	2409	2230	2006	1728	702

DATE	SOLAR TIME		SOLAR POSITION		BTUH/SQ. FT. TOTAL INSOLATION ON SURFACES						
							SOUTH FACING SURFACE ANGLE WITH HORIZ.				
	AM	PM	ALT	AZM	NORMAL	HORIZ.	30	40	50	60	90
AUG 21	6	6	7.9	99.5	81	21	12	9	8	7	5
	7	5	19.3	90.0	191	87	76	69	60	49	12
	8	4	30.7	79.9	237	150	150	141	129	113	50
	9	3	41.8	67.9	260	205	216	207	193	173	89
	10	2	51.7	52.1	272	246	267	259	244	221	120
	11	1	59.3	29.7	278	273	300	292	276	252	140
	12		62.3	0.0	280	282	311	303	287	262	147
	SURFACE DAILY TOTALS				2916	2244	2354	2258	2104	1894	978
SEP 21	7	5	11.4	80.2	149	43	51	51	49	47	32
	8	4	22.5	69.6	230	109	133	134	131	124	84
	9	3	32.8	57.3	263	167	206	208	203	193	132
	10	2	41.6	41.9	280	211	262	265	260	247	168
	11	1	47.7	22.6	287	239	298	301	295	281	192
	12		50.0	0.0	290	249	310	313	307	292	200
	SURFACE DAILY TOTALS				2708	1788	2210	2228	2182	2074	1416
OCT 21	7	5	4.5	72.3	48	7	14	15	17	17	16
	8	4	15.0	61.9	204	68	106	113	117	118	100
	9	3	24.5	49.8	257	126	185	195	200	198	160
	10	2	32.4	35.6	280	170	245	257	261	257	203
	11	1	37.6	18.7	291	199	283	295	299	294	229
	12		39.5	0.0	294	208	295	308	312	306	238
	SURFACE DAILY TOTALS				2454	1348	1962	2060	2098	2074	1654
NOV 21	8	4	8.2	55.4	136	28	63	72	78	82	81
	9	3	17.0	44.1	232	82	152	167	178	183	167
	10	2	24.0	31.0	268	126	215	233	245	249	219
	11	1	28.6	16.1	283	153	254	273	285	288	248
	12		30.2	0.0	288	163	267	287	298	301	258
	SURFACE DAILY TOTALS				2128	942	1636	1778	1870	1908	1686
DEC	8	4	5.5	53.0	89	14	39	45	50	54	56
	9	3	14.0	41.9	217	65	135	152	164	171	163
	10	2	20.7	29.4	261	107	200	221	235	242	221
	11	1	25.0	15.2	280	134	239	262	276	283	252
	12		26.6	0.0	285	143	253	275	290	296	263
	SURFACE DAILY TOTALS				1978	782	1480	1634	1740	1796	1646

Table IV 48 Degrees North Latitude

DATE	SOLAR TIME		SOLAR POSITION		BTUH/SQ. FT. TOTAL INSOLATION ON SURFACES						
							SOUTH FACING SURFACE ANGLE WITH HORIZ.				
	AM	PM	ALT	AZM	NORMAL	HORIZ.	38	48	58	68	90
JAN 21	8	4	3.5	54.6	37	4	17	19	21	22	22
	9	3	11.0	42.6	185	46	120	132	140	145	139
	10	2	16.9	29.4	239	83	190	206	216	220	206
	11	1	20.7	15.1	261	107	231	249	260	263	243
	12		22.0	0.0	267	115	245	264	275	278	255
	SURFACE DAILY TOTALS				1710	596	1360	1478	1550	1578	1478
FEB 21	7	5	2.4	72.2	12	1	3	4	4	4	4
	8	4	11.6	60.5	188	49	95	102	105	106	96
	9	3	19.7	47.7	251	100	178	187	191	190	167
	10	2	26.2	33.3	278	139	240	251	255	251	217
	11	1	30.5	17.2	290	165	278	290	294	288	247
	12		32.0	0.0	293	173	291	304	307	301	258
	SURFACE DAILY TOTALS				2330	1080	1880	1972	2024	1978	1720
MAR 21	7	5	10.0	78.7	153	37	49	49	47	45	35
	8	4	19.5	66.8	236	96	131	132	129	122	96
	9	3	28.2	53.4	270	147	205	207	203	193	152
	10	2	35.4	37.8	287	187	263	266	261	248	195
	11	1	40.3	19.8	295	212	300	303	297	283	223
	12		42.0	0.0	298	220	312	315	309	294	232
	SURFACE DAILY TOTALS				2780	1578	2208	2228	2132	2074	1632
APR 21	6	6	8.6	97.8	108	27	13	9	8	7	5
	7	5	18.6	86.7	205	85	76	69	59	48	21
	8	4	28.5	74.9	247	142	149	141	129	113	69
	9	3	37.8	61.2	268	191	216	208	194	174	115
	10	2	45.8	44.6	280	228	268	260	245	223	152
	11	1	51.5	24.0	286	252	301	294	278	254	177
	12		53.6	0.0	288	260	313	305	289	264	185
	SURFACE DAILY TOTALS				3076	2106	2358	2266	2114	1902	1262
MAY 21	5	7	5.2	114.3	41	9	4	4	[illegible]	3	2
	6	6	14.7	103.7	162	61	27	16	15	13	10
	7	5	24.6	93.0	219	118	89	75	6[illegible]	43	13
	8	4	34.7	81.6	248	171	156	142	12[illegible]	101	45
	9	3	44.3	68.3	264	217	217	202	18[illegible]	156	86
	10	2	53.0	51.3	274	252	265	251	223	200	120
	11	1	59.5	28.6	279	274	296	281	252	228	141
	12		62.0	0.0	280	281	306	292	26[illegible]	238	149
	SURFACE DAILY TOTALS				3254	2482	2418	2234	201[illegible]	1728	982
JUN 21	5	7	7.9	116.5	77	21	9	9	[illegible]	7	5
	6	6	17.2	106.2	172	74	33	19	1[illegible]	16	12
	7	5	27.0	95.8	220	129	93	77	[illegible]	39	15
	8	4	37.1	84.6	246	181	157	140	1[illegible]9	95	35
	9	3	46.9	71.6	261	225	216	198	1[illegible]5	147	74
	10	2	55.8	54.8	269	259	262	244	2[illegible]	189	105
	11	1	62.7	31.2	274	280	291	273	248	216	126
	12		65.5	0.0	275	287	301	283	2[illegible]8	225	133
	SURFACE DAILY TOTALS				3312	2626	2420	2204	19[illegible]	1644	874
JUL 21	5	7	5.7	114.7	43	10	5	5	4	4	3
	6	6	15.2	104.1	156	62	28	18	[illegible]6	15	11
	7	5	25.1	93.5	211	118	89	75	[illegible]9	42	14
	8	4	35.1	82.1	240	171	154	140	1[illegible]1	99	43
	9	3	44.8	68.8	256	215	214	199	1[illegible]8	153	33
	10	2	53.5	51.9	266	250	261	246	2[illegible]4	195	116
	11	1	60.1	29.0	271	272	291	276	25[illegible]	223	137
	12		62.6	0.0	272	279	301	286	25[illegible]	232	144
	SURFACE DAILY TOTALS				3158	2474	2386	2200	19[illegible]4	1694	956
AUG 21	6	6	9.1	98.3	99	28	14	10	9	8	6
	7	5	19.1	87.2	190	85	75	67	58	47	20
	8	4	29.0	75.4	232	141	145	137	125	109	65
	9	3	38.4	61.8	254	189	210	201	187	168	110
	10	2	46.4	45.1	266	225	260	252	237	214	146
	11	1	52.2	24.3	272	248	293	285	268	244	169
	12		54.3	0.0	274	256	304	296	279	255	177
	SURFACE DAILY TOTALS				2898	2086	2300	2200	2046	1836	1208
SEP 21	7	5	10.0	78.7	131	35	44	44	43	40	31
	8	4	19.5	66.8	215	92	124	124	121	115	90
	9	3	28.2	53.4	251	142	196	197	193	183	143
	10	2	35.4	37.8	269	181	251	254	248	236	185
	11	1	40.3	19.8	278	205	287	289	284	269	212
	12		42.0	0.0	280	213	299	302	296	281	221
	SURFACE DAILY TOTALS				2568	1522	2102	2118	2070	1966	1546
OCT 21	7	5	2.0	71.9	4	0	1	1	1	1	1
	8	4	11.2	60.2	165	44	86	91	95	95	87
	9	3	19.3	47.4	233	94	167	176	180	178	157
	10	2	25.7	33.1	262	133	228	239	242	239	207
	11	1	30.0	17.1	274	157	266	277	281	276	237
	12		31.5	0.0	278	166	279	291	294	288	247
	SURFACE DAILY TOTALS				2154	1022	1774	1860	1890	1866	1626
NOV 21	8	4	3.6	54.7	36	5	17	19	21	22	22
	9	3	11.2	42.7	179	46	117	129	137	141	135
	10	2	17.1	29.5	233	83	186	202	212	215	201
	11	1	20.9	15.1	255	107	227	245	255	258	238
	12		22.2	0.0	261	115	241	259	270	272	250
	SURFACE DAILY TOTALS				1668	596	1336	1448	1518	1544	1442
DEC 21	9	3	8.0	40.9	140	27	87	98	105	110	109
	10	2	13.6	28.2	214	63	164	180	192	197	190
	11	1	17.3	14.4	242	86	207	226	239	244	231
	12		18.6	0.0	250	94	222	241	254	260	244
	SURFACE DAILY TOTALS				1444	446	1136	1250	1326	1364	1304

Table V 56 Degrees North Latitude

DATE	SOLAR TIME AM	SOLAR TIME PM	SOLAR POSITION ALT	SOLAR POSITION AZM	BTUH/SQ. FT. TOTAL INSOLATION ON SURFACES NORMAL	HORIZ.	SOUTH FACING SURFACE ANGLE WITH HORIZ. 46	56	66	76	90
JAN 21	9	3	5.0	41.8	78	11	50	55	59	60	60
	10	2	9.9	28.5	170	39	135	146	154	156	153
	11	1	12.9	14.5	207	58	183	197	206	208	201
	12		14.0	0.0	217	65	198	214	222	225	217
	SURFACE DAILY TOTALS				1126	282	934	1010	1058	1074	1044
FEB 21	8	4	7.6	59.4	129	25	65	69	72	72	69
	9	3	14.2	45.9	214	65	151	159	162	161	151
	10	2	19.4	31.5	250	98	215	225	228	224	208
	11	1	22.8	16.1	266	119	254	265	268	263	243
	12		24.0	0.0	270	126	268	279	282	276	255
	SURFACE DAILY TOTALS				1986	740	1640	1716	1742	1716	1598
MAR 21	7	5	8.3	77.5	128	28	40	40	39	37	32
	8	4	16.2	64.4	215	75	119	120	117	111	97
	9	3	23.3	50.3	253	118	192	193	189	180	154
	10	2	29.0	34.9	272	151	249	251	246	234	205
	11	1	32.7	17.9	282	172	285	288	282	268	236
	12		34.0	0.0	284	179	297	300	294	280	246
	SURFACE DAILY TOTALS				2586	1268	2066	2084	2040	1938	1700
APR 21	5	7	1.4	108.8	0	0	0	0	0	0	0
	6	6	9.6	96.5	122	32	14	9	8	7	6
	7	5	18.0	84.1	201	81	74	66	57	46	29
	8	4	26.1	70.9	239	129	143	135	123	108	82
	9	3	33.6	56.3	260	169	208	200	186	167	133
	10	2	39.9	39.7	272	201	259	251	236	214	174
	11	1	44.1	20.7	278	220	292	284	268	245	200
	12		45.6	0.0	280	227	303	295	279	255	209
	SURFACE DAILY TOTALS				3024	1892	2282	2186	2038	1830	1458
MAY 21	4	8	1.2	125.5	0	0	0	0	0	0	0
	5	7	8.5	113.4	93	25	10	9	8	7	6
	6	6	16.5	101.5	175	71	28	17	15	13	11
	7	5	24.8	89.3	219	119	88	74	58	41	16
	8	4	33.1	76.3	244	163	153	138	119	98	63
	9	3	40.9	61.6	259	201	212	197	176	151	109
	10	2	47.6	44.2	268	231	259	244	222	194	146
	11	1	52.3	23.4	273	249	288	274	251	222	170
	12		54.0	0.0	275	255	299	284	261	231	178
	SURFACE DAILY TOTALS				3340	2374	2374	2188	1962	1682	1218
JUN 21	4	8	4.2	127.2	21	4	2	2	2	2	1
	5	7	11.4	115.3	122	40	14	13	11	10	8
	6	6	19.3	103.6	185	86	34	19	17	15	12
	7	5	27.6	91.7	222	132	92	76	57	38	15
	8	4	35.9	78.8	243	175	154	137	116	92	55
	9	3	43.8	64.1	257	212	211	193	170	143	98
	10	2	50.7	46.4	265	240	255	238	214	184	133
	11	1	55.6	24.9	269	258	284	267	242	210	156
	12		57.5	0.0	271	264	294	276	251	219	164
	SURFACE DAILY TOTALS				3438	2562	2388	2166	1910	1606	1120
JUL 21	4	8	1.7	125.8	0	0	0	0	0	0	0
	5	7	9.0	113.7	91	27	11	10	9	8	6
	6	6	17.0	101.9	169	72	30	18	16	14	12
	7	5	25.3	89.7	212	119	88	74	58	41	15
	8	4	33.6	76.7	237	163	151	136	117	96	61
	9	3	41.4	62.0	252	201	208	193	173	147	106
	10	2	48.2	44.6	261	230	254	239	217	189	142
	11	1	52.9	23.7	265	248	283	268	245	216	165
	12		54.6	0.0	267	254	293	278	255	225	173
	SURFACE DAILY TOTALS				3240	2372	2342	2152	1926	1646	1186
AUG 21	5	7	2.0	109.2	1	0	0	0	0	0	0
	6	6	10.2	97.0	112	34	16	11	10	9	7
	7	5	18.5	84.5	187	82	73	65	56	45	28
	8	4	26.7	71.3	225	128	140	131	119	104	78
	9	3	34.3	56.7	246	168	202	193	179	160	126
	10	2	40.5	40.0	258	199	251	242	227	206	166
	11	1	44.8	20.9	264	218	282	274	258	235	191
	12		46.3	0.0	266	225	293	285	269	245	200
	SURFACE DAILY TOTALS				2850	1884	2218	2118	1966	1760	1392
SEP 21	7	5	8.3	77.5	107	25	36	36	34	32	28
	8	4	16.2	64.4	194	72	111	111	108	102	89
	9	3	23.3	50.3	233	114	181	182	178	168	147
	10	2	29.0	34.9	253	146	236	237	232	221	193
	11	1	32.7	17.9	263	166	271	273	267	254	223
	12		34.0	0.0	266	173	283	285	279	265	233
	SURFACE DAILY TOTALS				2368	1220	1950	1962	1918	1820	1594
OCT 21	8	4	7.1	59.1	104	20	53	57	59	59	57
	9	3	13.8	45.7	193	60	138	145	148	147	138
	10	2	19.0	31.3	231	92	201	210	213	210	195
	11	1	22.3	16.0	248	112	240	250	253	248	230
	12		23.5	0.0	253	119	253	263	266	261	241
	SURFACE DAILY TOTALS				1804	688	1516	1586	1612	1588	1480
NOV 21	9	3	5.2	41.9	76	12	49	54	57	59	58
	10	2	10.0	28.5	165	39	132	143	149	152	148
	11	1	13.1	14.5	201	58	179	193	201	203	196
	12		14.2	0.0	211	65	194	209	217	219	211
	SURFACE DAILY TOTALS				1094	284	914	986	1032	1046	1016
DEC 21	9	3	1.9	40.5	5	0	3	4	4	4	4
	10	2	6.6	27.5	113	19	86	95	101	104	103
	11	1	9.5	13.9	166	37	141	154	163	167	164
	12		10.6	0.0	180	43	159	173	182	186	182
	SURFACE DAILY TOTALS				748	156	620	678	716	734	722

Table VI 64 Degrees North Latitude

DATE	SOLAR TIME AM	SOLAR TIME PM	SOLAR POSITION ALT	SOLAR POSITION AZM	BTUH/SQ. FT. TOTAL INSOLATION ON SURFACES NORMAL	HORIZ.	SOUTH FACING SURFACE ANGLE WITH HORIZ. 54	64	74	84	90
JAN 21	10	2	2.8	28.1	22	2	17	19	20	20	20
	11	1	5.2	14.1	81	12	72	77	80	81	81
	12		6.0	0.0	100	16	91	98	102	103	103
	SURFACE DAILY TOTALS				306	45	268	290	302	306	304
FEB 21	8	4	3.4	58.7	35	4	17	19	19	19	19
	9	3	8.6	44.8	147	31	103	108	111	110	107
	10	2	12.6	30.3	199	55	170	178	181	178	173
	11	1	15.1	15.3	222	71	212	220	223	219	213
	12		16.0	0.0	228	77	225	235	237	232	226
	SURFACE DAILY TOTALS				1432	400	1230	1286	1302	1282	1252
MAR 21	7	5	6.5	76.5	95	18	30	29	29	27	25
	8	4	20.7	62.6	185	54	101	102	99	94	89
	9	3	18.1	48.1	227	87	171	172	169	160	153
	10	2	22.3	32.7	249	112	227	229	224	213	203
	11	1	25.1	16.6	260	129	262	265	259	246	235
	12		26.0	0.0	263	134	274	277	271	258	246
	SURFACE DAILY TOTALS				2296	932	1856	1870	1830	1736	1556
APR 21	5	7	4.0	108.5	27	5	2	2	2	1	1
	6	6	10.4	95.1	133	37	15	9	8	7	6
	7	5	17.0	81.6	194	76	70	63	54	43	37
	8	4	23.3	67.5	228	112	136	128	116	102	91
	9	3	29.0	52.3	248	144	197	189	176	158	145
	10	2	33.5	36.0	260	169	246	239	224	203	188
	11	1	36.5	18.4	265	184	278	270	255	233	216
	12		97.6	0.0	268	190	289	281	266	243	225
	SURFACE DAILY TOTALS				2982	1644	2176	2082	1936	1736	1594
MAY 21	4	8	5.8	125.1	51	11	5	4	4	3	3
	5	7	11.6	112.1	132	42	13	11	10	9	8
	6	6	17.9	99.1	185	79	29	16	14	12	11
	7	5	24.5	85.7	218	117	86	72	56	39	28
	8	4	30.9	71.5	239	152	143	133	115	94	80
	9	3	36.8	56.1	252	182	204	190	170	145	128
	10	2	41.6	38.9	261	205	249	235	213	186	167
	11	1	44.9	20.1	265	219	278	264	242	213	193
	12		46.0	0.0	267	224	288	274	251	222	201
	SURFACE DAILY TOTALS				3470	2236	2312	2124	1898	1624	1436
JUN 21	3	9	4.2	139.4	21	4	2	2	2	2	1
	4	8	9.0	126.4	93	27	10	9	8	7	6
	5	7	14.7	113.6	154	60	16	15	13	11	10
	6	6	21.0	100.8	194	96	34	19	17	14	13
	7	5	27.5	87.5	221	132	91	74	55	36	23
	8	4	34.0	73.3	239	166	150	133	112	88	73
	9	3	39.9	57.8	251	195	204	187	164	137	119
	10	2	44.9	40.4	258	217	247	230	206	177	157
	11	1	48.3	20.9	262	231	275	258	233	202	181
	12		49.5	0.0	263	235	284	267	242	211	189
	SURFACE DAILY TOTALS				3650	2488	2342	2118	1862	1558	1356
JUL 21	4	8	6.4	125.3	53	13	6	5	5	4	4
	5	7	12.1	112.4	128	44	14	13	11	10	9
	6	6	18.4	99.4	179	81	30	17	16	13	12
	7	5	25.0	86.0	211	118	86	72	56	38	28
	8	4	31.4	71.8	231	152	146	131	113	91	77
	9	3	37.3	56.3	245	182	201	186	166	141	124
	10	2	42.2	39.2	253	204	245	230	208	181	162
	11	1	45.4	20.2	257	218	273	258	236	207	187
	12		46.6	0.0	259	223	282	267	245	216	195
	SURFACE DAILY TOTALS				3372	2248	2280	2090	1864	1588	1400
AUG 21	5	7	4.6	108.8	29	6	3	3	2	2	2
	6	6	11.0	95.5	123	39	16	11	10	8	7
	7	5	17.6	81.9	181	77	69	61	52	42	35
	8	4	23.9	67.8	214	113	132	123	112	97	87
	9	3	29.6	52.6	234	144	190	182	169	150	138
	10	2	34.2	36.2	246	168	237	229	215	194	179
	11	1	37.2	18.5	252	183	268	260	244	222	205
	12		38.3	0.0	254	188	278	270	255	232	215
	SURFACE DAILY TOTALS				2808	1646	2108	1008	1860	1662	1522
SEP 21	7	5	6.5	76.5	77	16	25	25	24	23	21
	8	4	12.7	72.6	163	51	92	92	90	85	81
	9	3	18.1	48.1	206	83	159	159	156	147	141
	10	2	22.3	32.7	229	108	212	213	209	198	189
	11	1	25.1	16.6	240	124	246	248	243	230	220
	12		26.0	0.0	244	129	258	260	254	241	230
	SURFACE DAILY TOTALS				2074	892	1726	1736	1696	1608	1532
OCT 21	8	4	3.0	58.5	17	2	9	9	10	10	10
	9	3	8.1	44.6	122	26	86	91	93	92	90
	10	2	12.1	30.2	176	50	152	159	161	159	155
	11	1	14.6	15.2	201	65	193	201	203	200	295
	12		15.5	0.0	208	71	207	215	217	213	208
	SURFACE DAILY TOTALS				1238	358	1088	1136	1152	1134	1106
NOV 21	10	2	3.0	28.1	23	3	18	20	21	21	21
	11	1	5.4	14.2	79	12	70	76	79	80	79
	12		6.2	0.0	97	17	89	96	100	101	100
	SURFACE DAILY TOTALS				302	46	266	286	298	302	300
DEC 21	11	1	1.8	13.7	4	0	3	4	4	4	4
	12		2.6	0.0	16	2	14	15	16	17	17
	SURFACE DAILY TOTALS				24	2	20	22	24	24	24

Appendix 4D Design Values of Various Building and Insulation Materials[a]

Table I Conductivities (k), Conductances (C), and Resistances (R) of Various Construction Materials

Material	Description	Density (Lb per Cu Ft)	Mean Temp F	Conductivity (k)	Conductance (C)	Resistance (R) Per inch thickness (1/k)	Resistance (R) For thickness listed (1/C)	Specific Heat, Btu per (lb) (F deg)
BUILDING BOARD Boards, Panels, Subflooring, Sheathing, Woodbased Panel Products	Asbestos-cement board	120	75	4.0	—	0.25	—	
	Asbestos-cement board ⅛ in.	120	75	—	33.00	—	0.033	
	Asbestos-cement board ¼ in.	120	75	—	16.50	—	0.07	
	Gypsum or plaster board ⅜ in.	50	75	—	3.10	—	0.32	
	Gypsum or plaster board ½ in.	50	75	—	2.25	—	0.45	
	Plywood	34	75	0.80	—	1.25	—	0.29
	Plywood ¼ in.	34	75	—	3.20	—	0.31	0.29
	Plywood ⅜ in.	34	75	—	2.13	—	0.47	0.29
	Plywood ½ in.	34	75	—	1.60	—	0.62	0.29
	Plywood or wood panels ¾ in.	34	75	—	1.07	—	0.93	0.29
	Insulating board							
	Sheathing, regular density ½ in.	18	75	—	0.76	—	1.32	0.31
	25/32 in.	18	75	—	0.49	—	2.06	0.31
	Sheathing intermediate density ½ in.	22	75	—	0.82	—	1.22	0.31
	Nail-base sheathing ½ in.	25	75	—	0.88	—	1.14	0.31
	Shingle backer ⅜ in.	18	75	—	1.06	—	0.94	0.31
	Shingle backer 5/16 in.	18	75	—	1.28	—	0.78	0.31
	Sound deadening board ½ in.	15		—	0.74	—	1.35	0.30
	Tile and lay-in panels, plain or acoustic	18	75	0.40	—	2.50	—	0.32
	½ in.	18	75	—	0.80	—	1.25	0.32
	¾ in.	18	75	—	0.53	—	1.89	0.32
	Laminated paperboard	30	75	0.50	—	2.00	—	
	Homogeneous board from repulped paper	30	75	0.50	—	2.00	—	0.28
	Hardboard							
	Medium density siding 7/16 in.	40	75	—	1.49	—	0.67	0.28
	Other medium density	50	75	0.73	—	1.37	—	0.31
	High density, service temp. service, underlay	55	75	0.82	—	1.22	—	0.33
	High density, std. tempered	63	75	1.00	—	1.00	—	0.33
	Particleboard							
	Low density	37	75	0.54	—	1.85	—	0.31
	Medium density	50	75	0.94	—	1.06	—	0.31
	High density	62.5	75	1.18	—	0.85	—	0.31
	Underlayment ⅝ in.	40	75	—	1.22	—	0.82	0.29
	Wood subfloor ¾ in.		75	—	1.06	—	0.94	0.34
BUILDING PAPER	Vapor—permeable felt	—	75	—	16.70	—	0.06	
	Vapor—seal, 2 layers of mopped 15 lb felt	—	75	—	8.35	—	0.12	
	Vapor—seal, plastic film	—	75	—	—	—	Negl.	
FINISH FLOORING MATERIALS	Carpet and fibrous pad	—	75	—	0.48	—	2.08	
	Carpet and rubber pad	—	75	—	0.81	—	1.23	0.34
	Cork tile ⅛ in.	—	75	—	3.60	—	0.28	
	Terrazzo 1 in.	—	75	—	12.50	—	0.08	
	Tile—asphalt, linoleum, vinyl, rubber	—	75	—	20.00	—	0.05	0.30
INSULATING MATERIALS Blanket and Katt	Mineral Fiber, fibrous form processed from rock, slag, or glass							
	approx. 2–2¾ in.	—	75	—	—	—	7	0.18
	approx. 3–3½ in.	—	75	—	—	—	11	0.18
	approx. 5¼–6½ in.	—	75	—	—	—	19	0.18
Board and Slabs	Cellular glass	9	75	0.40	—	2.50	—	0.24
	Glass fiber, organic bonded	4–9	75	0.25	—	4.00	—	0.19
	Expanded rubber (rigid)	4.5	75	0.22	—	4.55	—	
	Expanded polystyrene extruded, plain	1.8	75	0.25	—	4.00	—	0.29
	Expanded polystyrene extruded, (R-12 exp.)	2.2	75	0.20	—	5.00	—	0.29
	Expanded polystyrene extruded, (R-12 exp.) (Thickness 1 in. and greater)	3.5	75	0.19	—	5.26	—	0.29
	Expanded polystyrene, molded beads	1.0	75	0.28	—	3.57	—	0.29
	Expanded polyurethane (R-11 exp.) (Thickness 1 in. or greater)	1.5 2.5	75	0.16	—	6.25	—	0.38 0.38
	Mineral fiber with resin binder	15	75	0.29	—	3.45	—	0.17
	Mineral fiberboard, wet felted Core or roof insulation	16–17	75	0.34	—	2.94	—	

Notes: a. From ASHRAE *Handbook of Fundamentals.*

Table I—*Continued*

Material	Description	Density (Lb per Cu Ft)	Mean Temp F	Conductivity (k)	Conductance (C)	Resistance (R) Per inch thickness (1/k)	Resistance (R) For thickness listed (1/C)	Specific Heat, Btu per (lb)(F deg)
BOARD AND SLABS (*Continued*)	Acoustical tile	18	75	0.35	—	*2.86*	—	
	Acoustical tile	21	75	0.37	—	*2.73*	—	
	Mineral fiberboard, wet molded Acoustical tile	23	75	0.42	—	*2.38*	—	
	Wood or cane fiberboard Acoustical tile ½ in.	—	75	—	0.80	—	*1.25*	0.30
	Acoustical tile ¾ in.	—	75	—	0.53	—	*1.89*	0.30
	Interior finish (plank, tile)	15	75	0.35	—	*2.86*	—	0.32
	Insulating roof deck Approximately 1½ in.	—	75	—	0.24	—	*4.17*	
	Approximately 2 in.	—	75	—	0.18	—	*5.56*	
	Approximately 3 in.	—	75	—	0.12	—	*8.33*	
	Wood shredded (cemented in preformed slabs)	22	75	0.60	—	*1.67*	—	0.38
LOOSE FILL	Cellulose insulation (milled paper or wood pulp)	2.5–3	75	0.27	—	*3.70*	—	0.33
	Sawdust or shavings	0.8–1.5	75	0.45	—	*2.22*	—	0.33
	Wood fiber, softwoods	2.0–3.5	75	0.30	—	*3.33*	—	0.33
	Perlite, expanded	5.0–8.0	75	0.37	—	*2.70*	—	
	Mineral fiber (rock, slag or glass) approx. 3 in.	—	75	—	—	*9*	—	0.18
	approx. 4½ in.	—	75	—	—	*13*	—	0.18
	approx. 6¼ in.	—	75	—	—	*19*	—	0.18
	approx. 7¼ in.	—	75	—	—	*24*	—	0.18
	Silica aerogel	7.6	75	0.17	—	*5.88*	—	
	Vermiculite (expanded)	7.0–8.2	75	0.47	—	*2.13*	—	
		4.0–6.0	75	0.44	—	*2.27*	—	
ROOF INSULATION	Preformed, for use above deck Approximately ½ in.	—	75	—	0.72	—	*1.39*	
	Approximately 1 in.	—	75	—	0.36	—	*2.78*	
	Approximately 1½ in.	—	75	—	0.24	—	*4.17*	
	Approximately 2 in.	—	75	—	0.19	—	*5.56*	
	Approximately 2½ in.	—	75	—	0.15	—	*6.67*	
	Approximately 3 in.	—	75	—	0.12	—	*8.33*	
	Cellular glass	9	75	0.40	—	*2.50*	—	0.24
MASONRY MATERIALS CONCRETES	Cement mortar	116		5.0	—	*0.20*	—	
	Gypsum-fiber concrete 87½% gypsum, 12½% wood chips	51		1.66	—	*0.60*	—	
	Lightweight aggregates including expanded shale, clay or slate; expanded slags; cinders; pumice; vermiculite; also cellular concretes	120		5.2	—	*0.19*	—	
		100		3.6	—	*0.28*	—	
		80		2.5	—	*0.40*	—	
		60		1.7	—	*0.59*	—	
		40		1.15	—	*0.86*	—	
		30		0.90	—	*1.11*	—	
		20		0.70		*1.43*		
	Sand and gravel or stone aggregate (oven dried)	140		9.0	—	*0.11*	—	
	Sand and gravel or stone aggregate (not dried)	140		12.0	—	*0.08*	—	
	Stucco	116		5.0	—	*0.20*	—	
MASONRY UNITS	Brick, common	120	75	5.0	—	*0.20*	—	
	Brick, face	130	75	9.0	—	*0.11*	—	
	Clay tile, hollow: 1 cell deep 3 in.	—	75	—	1.25	—	*0.80*	
	1 cell deep 4 in.	—	75	—	0.90	—	*1.11*	
	2 cells deep 6 in.	—	75	—	0.66	—	*1.52*	
	2 cells deep 8 in.	—	75	—	0.54	—	*1.85*	
	2 cells deep 10 in.	—	75	—	0.45	—	*2.22*	
	3 cells deep 12 in.	—	75	—	0.40	—	*2.50*	
	Concrete blocks, three oval core: Sand and gravel aggregate 4 in.	—	75	—	1.40	—	*0.71*	
	8 in.	—	75	—	0.90	—	*1.11*	
	12 in.	—	75	—	0.78	—	*1.28*	
	Cinder aggregate 3 in.	—	75	—	1.16	—	*0.86*	
	4 in.	—	75	—	0.90	—	*1.11*	
	8 in.	—	75	—	0.58	—	*1.72*	
	12 in.	—	75	—	0.53	—	*1.89*	
	Lightweight aggregate (expanded shale, clay, slate or slag; pumice) 3 in.	—	75	—	0.79	—	*1.27*	
	4 in.	—	75	—	0.67	—	*1.50*	
	8 in.	—	75	—	0.50	—	*2.00*	
	12 in.	—	75	—	0.44	—	*2.27*	

Table I—*Concluded*

Material	Description	Density (Lb per Cu Ft)	Mean Temp F	Conductivity (k)	Conductance (C)	Resistance (R) Per inch thickness (1/k)	Resistance (R) For thickness listed (1/C)	Specific Heat Btu per (lb) (F deg)
	Concrete blocks, rectangular core.							
	Sand and gravel aggregate							
	2 core, 8 in. 36 lb.	—	45	—	0.96	—	*1.04*	
	Same with filled cores	—	45	—	0.52	—	*1.93*	
	Lightweight aggregate (expanded shale, clay, slate or slag, pumice):							
	3 core, 6 in. 19 lb.	—	45	—	0.61	—	*1.65*	
	Same with filled cores	—	45	—	0.33	—	*2.99*	
	2 core, 8 in. 24 lb.	—	45	—	0.46	—	*2.18*	
	Same with filled cores	—	45	—	0.20	—	*5.03*	
	3 core, 12 in. 38 lb.	—	45	—	0.40	—	*2.48*	
	Same with filled cores	—	45	—	0.17	—	*5.82*	
	Stone, lime or sand	—	75	12.50	—	*0.08*	—	
	Gypsum partition tile:							
	3 × 12 × 30 in. solid	—	75	—	0.79	—	*1.26*	
	3 × 12 × 30 in. 4-cell	—	75	—	0.74	—	*1.35*	
	4 × 12 × 30 in. 3-cell	—	75	—	0.60	—	*1.67*	
METALS	(See Chapter 30, Table 3)							
PLASTERING MATERIALS	Cement plaster, sand aggregate	116	75	5.0	—	*0.20*	—	
	Sand aggregate ... 3/8 in.	—	75	—	13.3	—	*0.08*	
	Sand aggregate ... 3/4 in.	—	75	—	6.66	—	*0.15*	
	Gypsum plaster:							
	Lightweight aggregate ... 1/2 in.	45	75	—	3.12	—	*0.32*	
	Lightweight aggregate ... 5/8 in.	45	75	—	2.67	—	*0.39*	
	Lightweight agg. on metal lath. 3/4 in.	—	75	—	2.13	—	*0.47*	
	Perlite aggregate	45	75	1.5	—	*0.67*	—	
	Sand aggregate	105	75	5.6	—	*0.18*	—	
	Sand aggregate ... 1/2 in.	105	75	—	11.10	—	*0.09*	
	Sand aggregate ... 5/8 in.	105	75	—	9.10	—	*0.11*	
	Sand aggregate on metal lath .. 3/4 in.	—	75	—	7.70	—	*0.1*	
	Vermiculite aggregate	45	75	1.7	—	*0.59*	—	
ROOFING	Asbestos-cement shingles	120	75	—	4.76	—	*0.21*	
	Asphalt roll roofing	70	75	—	6.50	—	*0.15*	
	Asphalt shingles	70	75	—	2.27	—	*0.44*	
	Built-up roofing ... 3/8 in.	70	75	—	3.00	—	*0.33*	0.35
	Slate ... 1/2 in.	—	75	—	20.00	—	*0.05*	
	Wood shingles, plain a plastic film faced	—	75	—	1.06	—	*0.94*	0.31
SIDING MATERIALS (ON FLAT SURFACE)	Shingles							
	Asbestos-cement	120	75	—	4.76	—	*0.21*	
	Wood, 16 in., 7½ exposure	—	75	—	1.15	—	*0.87*	0.31
	Wood, double, 16-in., 12-in. exposure	—	75	—	0.84	—	*1.19*	0.31
	Wood, plus insul. backer board. 5/16 in.	—	75	—	0.71	—	*1.40*	0.31
	Siding							
	Asbestos-cement, 1/4 in., lapped	—	75	—	4.76	—	*0.21*	
	Asphalt roll siding	—	75	—	6.50	—	*0.15*	
	Asphalt insulating siding (1/2 in. bd.)	—	75	—	0.69	—	*1.46*	
	Wood, drop, 1 × 8 in.	—	75	—	1.27	—	*0.79*	0.31
	Wood, bevel, 1/2 × 8 in., lapped	—	75	—	1.23	—	*0.81*	0.31
	Wood, bevel, 3/4 × 10 in., lapped	—	75	—	0.95	—	*1.05*	0.31
	Wood, plywood, 3/8 in., lapped	—	75	—	1.59	—	*0.59*	0.29
	Aluminum or Steel -, over sheathing							
	Hollow-backed	—	—	—	1.61	—	*0.61*	
	Insulating-board backed nominal 3/8 in.	—	—	—	0.55	—	*1.82*	
	Insulating-board backed nominal 3/8 in. foil backed	—	—	—	0.34	—	*2.96*	
	Architectural glass	—	75	—	10.00	—	*0.10*	
WOODS	Maple, oak, and similar hardwoods	45	75	1.10	—	*0.91*	—	0.30
	Fir, pine, and similar softwoods	32	75	0.80	—	*1.25*	—	0.33
	Fir, pine, and similar softwoods ... 3/4 in.	32	75	—	1.06	—	*0.94*	0.33
	.. 1½ in.	32	75	—	0.53	—	*1.89*	0.33
	.. 2½ in.	32	75	—	0.32	—	*3.12*	0.33
	.. 3½ in.	32	75	—	0.23	—	*4.35*	0.33

Table II Thermal Conductivity (*k*) of Various Construction Materials[a]

Form	Material (Composition)	Accepted Max Temp for Use, F	Typical Density (lb/cu ft)	Typical Conductivity k at Mean Temp F													
				−100	−75	−50	−25	0	25	50	75	100	200	300	500	700	900
BLANKETS & FELTS	MINERAL FIBER (Rock, Slag, or Glass)																
	Blanket, Metal Reinforced	1200	6–12									0.26	0.32	0.39	0.54		
		1000	2.5–6									0.24	0.31	0.40	0.61		
	Mineral Fiber, Glass Blanket, Flexible, Fine-Fiber Organic Bonded	350	0.65				0.25	0.26	0.28	0.30	0.33	0.36	0.53				
			0.75				0.24	0.25	0.27	0.29	0.32	0.34	0.48				
			1.0				0.23	0.24	0.25	0.27	0.29	0.32	0.43				
			1.5				0.21	0.22	0.23	0.25	0.27	0.28	0.37				
			2.0				0.20	0.21	0.22	0.23	0.25	0.26	0.33				
			3.0				0.19	0.20	0.21	0.22	0.23	0.24	0.31				
	Blanket, Flexible, Textile-Fiber Organic Bonded	350	0.65				0.27	0.28	0.29	0.30	0.31	0.32	0.50	0.68			
			0.75				0.26	0.27	0.28	0.29	0.31	0.32	0.48	0.66			
			1.0				0.24	0.25	0.26	0.27	0.29	0.31	0.45	0.60			
			1.5				0.22	0.23	0.24	0.25	0.27	0.29	0.39	0.51			
			3.0				0.20	0.21	0.22	0.23	0.24	0.25	0.32	0.41			
	Felt, Semi-Rigid Organic Bonded	400	3–8						0.24	0.25	0.26	0.27	0.35	0.44			
		850	3	0.16	0.17	0.18	0.19	0.20	0.21	0.22	0.23	0.24	0.35	0.55			
	Laminated & Felted Without Binder	1200	7.5											0.35	0.45	0.60	
	VEGETABLE & ANIMAL FIBER																
	Hair Felt or Hair Felt plus Jute	180	10						0.26	0.28	0.29	0.30					
BLOCKS, BOARDS & PIPE INSULATION	ASBESTOS																
	Laminated Asbestos Paper	700	30									0.40	0.45	0.50	0.60		
	Corrugated & Laminated Asbestos Paper																
	4-ply	300	11–13								0.54	0.57	0.68				
	6-ply	300	15–17								0.49	0.51	0.59				
	8-ply	300	18–20								0.47	0.49	0.57				
	MOLDED AMOSITE & BINDER	1500	15–18									0.32	0.37	0.42	0.52	0.62	0.72
	85% MAGNESIA	600	11–12									0.35	0.38	0.42			
	CALCIUM SILICATE	1200	11–13									0.38	0.41	0.44	0.52	0.62	0.72
		1800	12–15												0.63	0.74	0.95
	CELLULAR GLASS	800	9			0.32	0.33	0.35	0.36	0.38	0.40	0.42	0.48	0.55			
	DIATOMACEOUS SILICA	1600	21–22												0.64	0.68	0.72
		1900	23–25												0.70	0.75	0.80
	MINERAL FIBER																
	Glass, Organic Bonded, Block and Boards	400	3–10	0.16	0.17	0.18	0.19	0.20	0.22	0.24	0.25	0.26	0.33	0.40			
	Non-Punking Binder	1000	3–10									0.26	0.31	0.38	0.52		
	Pipe Insulation, slag or glass	350	3–4					0.20	0.21	0.22	0.23	0.24	0.29				
		500	3–10					0.20	0.22	0.24	0.25	0.26	0.33	0.40			
	Inorganic Bonded-Block	1000	10–15									0.33	0.38	0.45	0.55		
		1800	15–24									0.32	0.37	0.42	0.52	0.62	0.74
	Pipe Insulation slag or glass	1000	10–15									0.33	0.38	0.45	0.55		
	MINERAL FIBER Resin Binder		15			0.23	0.24	0.25	0.26	0.28	0.29						
	Rigid Polystryrene																
	Extruded, R-12 exp	*170	3.5	0.16	0.16	0.15	0.16	0.16	0.17	0.18	0.19	0.20					
	Extruded, R-12 exp	170	2.2	0.16	0.16	0.17	0.16	0.17	0.18	0.19	0.20						
	Extruded	170	1.8	0.17	0.18	0.19	0.20	0.21	0.23	0.24	0.25	0.27					
	Molded Beads	170	1	0.18	0.20	0.21	0.23	0.24	0.25	0.26	0.28						
	Polyurethane R-11 exp	210	1.5–2.5	0.16	0.17	0.18	0.18	0.18	0.17	0.16	0.16	0.17					
	RUBBER, Rigid Foamed	150	4.5						0.20	0.21	0.22	0.23					
	VEGETABLE & ANIMAL FIBER																
	Wool Felt (Pipe Insulation)	180	20						0.28	0.30	0.31	0.33					
INSULATING CEMENTS	MINERAL FIBER (Rock, Slag, or Glass)																
	With Colloidal Clay Binder	1800	24–30									0.49	0.55	0.61	0.73	0.85	
	With Hydraulic Setting Binder	1200	30–40									0.75	0.80	0.85	0.95		
LOOSE FILL	Cellulose insulation (Milled pulverized paper or wood pulp)		2.5–3							0.26	0.27	0.29					
	Mineral fiber, slag, rock or glass		2–5			0.19	0.21	0.23	0.25	0.26	0.28	0.31					
	Perlite (expanded)		5–8	0.25	0.27	0.29	0.30	0.32	0.34	0.35	0.37	0.39					
	Silica aerogel		7.6			0.13	0.14	0.15	0.15	0.16	0.17	0.18					
	Vermiculite (expanded)		7–8.2			0.39	0.40	0.42	0.44	0.45	0.47	0.49					
			4–6			0.34	0.35	0.38	0.40	0.42	0.44	0.46					

Appendix 4E Coefficients of Transmission (U) for Various Structural Elements[a,b]

Table I Coefficients of Transmission (U) of Frame Construction Ceilings and Floors[c]

Example—Floor E 5

Heated room below unheated space

Construction (heat flow up)	Resistance (R)
1. Top surface (still air)	*0.61*
2. Linoleum or tile (avg R)	*0.05*
3. Felt	*0.06*
4. Plywood ($\frac{5}{8}$ in.)	*0.78*
5. Wood subfloor ($\frac{25}{32}$ in.)	*0.98*
6. Air space ($7\frac{1}{2}$ in.)	*0.85*
7. Metal lath and $\frac{3}{4}$ in. plas. (lt. wt. agg.)	*0.47*
8. Bottom surface (still air)	*0.61*
Total resistance	*4.41*
$U = 1/R = 1/4.41 =$	0.23

See value 0.23 in boldface type in table below.

Example of Substitution

Assume heated room is above unheated space so heat flow is down

Total resistance		*4.41*
Deduct 1. Top surface (heat flow up)	*0.61*	
6. Air space (heat flow up)	*0.85*	
8. Bottom surface (heat flow up)	*0.61*	*2.07*
Difference		*2.34*
Add 1. Top surface (heat flow down)	*0.92*	
6. Air space (heat flow down)	*1.25*	
8. Bottom surface (heat flow down)	*0.92*	*3.09*
Total resistance		*5.43*
$U = 1/R = 1/5.43 =$		0.18

Direction of Heat →		Heat Flow Upward						Heat Flow Downward						
		Type of Floor												
				Wood subfloor ($^{25}/_{32}$ in.), felt, and—						Wood subfloor ($^{25}/_{32}$ in.), felt, and—				
		None	Wood subfloor ($^{25}/_{32}$ in.)	Cement (1½ in.) and ceramic tile (½ in.)	Hardwood floor (¾ in.)	Plywood (⅝ in.) and floor tile or linoleum (⅛ in.)	Insul. bd. (⅜ in.) and hard bd. (¼ in.) and floor tile or linoleum (⅛ in.)	None	Wood subfloor ($^{25}/_{32}$ in.)	Cement (1½ in.) and ceramic tile (½ in.)	Hardwood floor (¾ in.)	Plywood (⅝ in.) and floor tile or linoleum (⅛ in.)	Insul. bd. (⅜ in.) and hard bd. (¼ in.) and floor tile or linoleum (⅛ in.)	Number
Type of Ceiling	Resistance ↓ →	—	0.98	1.38	1.72	1.87	2.26	—	0.98	1.38	1.72	1.87	2.26	
Material	R	U	U	U	U	U	U	U	U	U	U	U	U	
		A	B	C	D	E	F	G	H	I	J	K	L	
None	—	—	0.45	0.38	0.34	0.32	0.29	—	0.35	0.31	0.28	0.26	0.24	**1**
Gypsum bd. (⅜ in.)	*0.32*	0.65	0.30	0.27	0.24	0.23	0.22	0.46	0.24	0.22	0.21	0.20	0.19	**2**
Gypsum lath (⅜ in.) and ½ in. plas. (lt. wt. agg.)	*0.64*	0.54	0.27	0.24	0.23	0.22	0.20	0.40	0.22	0.21	0.19	0.19	0.17	**3**
Gypsum lath (⅜ in.) and ½ in. plas. (sand agg.)	*0.41*	0.61	0.29	0.26	0.24	0.23	0.21	0.44	0.24	0.22	0.20	0.20	0.18	**4**
Metal lath and ⅜ in. plas. (lt. wt. agg.)	*0.47*	0.59	0.28	0.26	0.23	**0.23**	0.21	0.43	0.23	0.21	0.20	0.19	0.18	**5**
Metal lath and ⅜ in. plas. (sand agg.)	*0.13*	0.74	0.31	0.28	0.26	0.25	0.22	0.51	0.25	0.23	0.21	0.21	0.19	**6**
Insul. bd. (½ in.)	*1.43*	0.38	0.22	0.20	0.19	0.19	0.17	0.31	0.19	0.18	0.17	0.16	0.15	**7**
Insul. bd. lath (½ in.) and ½ in. plas. (sand agg.)	*1.52*	0.36	0.22	0.20	0.19	0.18	0.17	0.30	0.19	0.17	0.17	0.16	0.15	**8**
Acoustical tile (½ in.) on gypsum bd. (⅜ in.)	*1.51*	0.37	0.22	0.20	0.19	0.18	0.17	0.30	0.19	0.17	0.17	0.16	0.15	**9**
(½ in.) on furring	*1.19*	0.41	0.24	0.22	0.20	0.19	0.18	0.33	0.20	0.19	0.17	0.17	0.16	**10**
(¾ in.) on gypsum bd. (⅜ in.)	*2.10*	0.30	0.19	0.18	0.17	0.17	0.15	0.25	0.17	0.16	0.15	0.15	0.14	**11**
(¾ in.) on furring	*1.78*	0.33	0.21	0.19	0.18	0.17	0.16	0.28	0.18	0.17	0.16	0.15	0.15	**12**
Wood lath and ½ in. plas. (sand agg.)	*0.40*	0.62	0.29	0.26	0.24	0.23	0.21	0.45	0.24	0.22	0.20	0.20	0.18	**13**

Notes: a. From ASHRAE *Handbook of Fundamentals*.
b. Coefficients are expressed in Btu/hr-ft²-°F where °F is temperature difference between air on two sides.
c. Still air conditions on both sides.
d. Outside wind velocity of 15 mph.
e. Winter conditions, upward flow.
f. Outside wind velocity of 15 mph for upward flow, and 7.5 mph wind velocity for heat flow downward.

Appendix 4E—*Continued*

Table II Coefficients of Transmission (U) of Concrete Floor-Ceiling Construction[e]

Example—Floor J 4

Heated room below unheated space

Construction (heat flow up)	Resistance (R)
1. Top surface (still air)	*0.61*
2. Asphalt tile and felt	*0.11*
3. Plywood (⅝ in.)	*0.78*
4. Air space	*0.85*
5. Concrete slab 4 in. (avg R)	*0.40*
6. Air space (8 in.)	*0.85*
7. Metal lath and ¾ plas. (sand agg.)	*0.13*
8. Bottom surface (still air)	*0.61*
Total resistance	*4.34*
$U = 1/R = 1/4.34 =$	0.23

See value 0.23 in boldface type in table below.

Example of Substitution

Replace items 2, 3, and 4 with hardwood block (13/16 in.) on slab

Total resistance		*4.34*
Deduct 2. Asphalt tile and felt	*0.11*	
3. Plywood (⅝ in.)	*0.78*	
4. Air space	*0.85*	*1.74*
Difference		*2.60*
Add 2. Wood block (13/16 in.)		*0.74*
Total resistance		*3.34*
$U = 1/R = 1/3.34 =$		0.30

Type of Deck: Material	R	avg R	Type of Finish Floor: Material	avg R	Type of Ceiling: None	Ceiling Applied Directly to Slab: Plas. (Lt. wt. agg.) ⅛ in.	Ceiling Applied Directly to Slab: Plas. (Sand agg.) ⅛ in.	Ceiling Applied Directly to Slab: Acoustical tile—glued ½ in.	Ceiling Applied Directly to Slab: Acoustical tile—glued ¾ in.	Suspended Ceiling: Gypsum bd. (⅜ in.) and Plas.: No plas.	Suspended Ceiling: Gypsum bd. (⅜ in.) and Plas.: (Lt. wt. agg.) ½ in.	Suspended Ceiling: Gypsum bd. (⅜ in.) and Plas.: (Sand agg.) ¾ in.	Suspended Ceiling: Metal lath and plas.: (Lt. wt. agg.) ½ in.	Suspended Ceiling: Metal lath and plas.: (Sand agg.) ¾ in.	Suspended Ceiling: Acoustical tile: On furring or channels ½ in.	Suspended Ceiling: Acoustical tile: On furring or channels ¾ in.	Suspended Ceiling: Acoustical tile: On gypsum bd. (⅜ in.) ½ in.	Suspended Ceiling: Acoustical tile: On gypsum bd. (⅜ in.) ¾ in.	Number
			Resistance		—	**0.08**	**0.02**	***1.19***	***1.78***	**0.32**	**0.64**	**0.41**	**0.47**	**0.13**	***1.19***	***1.78***	***1.51***	**2.10**	
					U A	*U* B	*U* C	*U* D	*U* E	*U* F	*U* G	*U* H	*U* I	*U* J	*U* K	*U* L	*U* M	*U* N	
Concrete (sand agg.) (4 in.) (6 in.)	*0.32* *0.48*	*0.40*	None	—	0.62	0.59	0.61	0.36	0.29	0.36	0.32	0.35	0.34	0.38	0.27	0.24	0.25	0.22	**1**
			Floor tile or linoleum (⅛ in.)	*0.05*	0.60	0.57	0.59	0.35	0.29	0.35	0.32	0.34	0.33	0.38	0.27	0.23	0.25	0.22	**2**
			Wood block (13/16 in.) on slab	*0.74*	0.42	0.41	0.42	0.28	0.24	0.28	0.26	0.28	0.27	0.30	0.23	0.20	0.21	0.19	**3**
			Floor on sleepers Plywood subfloor (⅝ in.), felt and floor tile or linoleum (⅛ in.)	*0.89*	0.30	0.29	0.30	0.22	0.19	0.22	0.21	0.22	0.21	**0.23**	0.19	0.17	0.17	0.16	**4**
			Wood subfloor (25/32 in.), felt and hardwood (¾ in.)	*1.72*	0.24	0.24	0.24	0.19	0.17	0.19	0.18	0.18	0.18	0.19	0.16	0.15	0.15	0.14	**5**
Concrete (sand agg.) (8 in.) (10 in.)	*0.64* *0.80*	*0.72*	None	—	0.52	0.50	0.51	0.32	0.27	0.32	0.29	0.31	0.31	0.34	0.25	0.22	0.23	0.20	**6**
			Floor tile or linoleum (⅛ in.)	*0.05*	0.50	0.48	0.50	0.31	0.26	0.32	0.29	0.31	0.30	0.34	0.25	0.22	0.23	0.20	**7**
			Wood block (13/16 in.) on slab	*0.74*	0.37	0.36	0.37	0.26	0.22	0.26	0.24	0.25	0.25	0.27	0.21	0.19	0.20	0.18	**8**
			Floor on sleepers Plywood subfloor (⅝ in.), felt and floor tile or linoleum (⅛ in.)	*0.89*	0.27	0.27	0.27	0.21	0.18	0.21	0.19	0.20	0.20	0.21	0.17	0.16	0.17	0.15	**9**
			Wood subfloor (25/32 in.), felt and hardwood (¾ in.)	*1.72*	0.22	0.22	0.22	0.18	0.16	0.18	0.17	0.17	0.17	0.18	0.16	0.14	0.15	0.13	**10**

Table III Coefficients of Transmission (U) of Flat Masonry Roofs with Built-up Roofing and with and without Suspended Ceilings[d,e]

Example—Roof J 2

Construction (heat flow up)	Resistance (R)
1. Outside surface (15 mph wind)	0.17
2. Built-up roofing—3/8 in.	0.33
3. Roof insulation (none)	—
4. Concrete slab (lt. wt. agg.) (2 in.)	2.22
5. Corrugated metal	0
6. Air space	0.85
7. Metal lath and 3/4 in. plas. (lt. wt. agg.)	0.47
8. Inside surface (still air)	0.61
Total resistance	4.65
$U = 1/R = 1/4.65 =$	0.22

See value 0.22 in boldface type in table below.

Example of Substitution

Replace item 4 with 4 in. concrete slab (gravel agg.) and roof insulation (C = 0.36) on top of slab.

Total resistance		4.65
Deduct 4. Concrete slab (lt. wt. agg.) (2 in.)		2.22
Difference		2.43
Add 3. Roof insulation (C = 0.36)	2.78	
4. Concrete slab (gravel agg.) 4 in.	0.44	3.22
Total resistance		5.65
$U = 1/R = 1/5.65 =$		0.18

Type of Deck: Material	R	Type of Form: Material	R	Type of Ceiling: Roof Insulation No Ceiling — C value of roof insulation: None	0.72	0.36	0.24	0.19	0.15	0.12	Suspended Ceiling — Gypsum bd. (3/8 in.) and plas.: No plas.	Lt. wt. agg. 1/2 in.	Sand agg. 1/2 in.	Metal lath and plas.: Lt. wt. agg. 3/4 in.	Sand agg. 3/4 in.	Acoustical tile — On furring or channels: 1/2 in.	3/4 in.	On gypsum bd. (3/8 in.): 1/2 in.	3/4 in.	Number
		Resistance →		—	1.39	2.78	4.17	5.26	6.67	8.33	0.32	0.64	0.41	0.47	0.13	1.19	1.75	1.51	2.10	
				U	U	U	U	U	U	U	U	U	U	U	U	U	U	U	U	
				A	B	C	D	E	F	G	H	I	J	K	L	M	N	O	P	
Concrete slab Gravel agg. (4 in.)	0.32	Temporary	—	0.70	0.35	0.24	0.18	0.15	0.12	0.10	0.38	0.34	0.37	0.36	0.41	0.29	0.25	0.26	0.23	1
(6 in.)	0.48	Temporary	—	0.63	0.34	0.23	0.17	0.15	0.12	0.10	0.36	0.32	0.35	0.34	0.39	0.27	0.24	0.25	0.22	2
(8 in.)	0.64	Temporary	—	0.57	0.32	0.22	0.17	0.14	0.12	0.10	0.34	0.31	0.33	0.33	0.37	0.26	0.23	0.24	0.21	3
Lt. wt. agg. (2 in.)	2.22	Corrugated metal	0	0.30	0.21	0.16	0.13	0.12	0.10	0.09	0.22	0.21	0.22	**0.22**	0.23	0.19	0.17	0.18	0.16	4
		Insulation bd. (1 in.)	2.78	0.16	0.13	0.11	0.10	0.09	0.08	0.07	0.14	0.13	0.14	0.13	0.14	0.12	0.11	0.12	0.11	5
		Insulation bd. (1½ in.)	4.17	0.13	0.11	0.10	0.09	0.08	0.07	0.06	0.12	0.11	0.11	0.11	0.12	0.10	0.10	0.10	0.10	6
		Glass fiber bd. (1 in.)	4.00	0.14	0.11	0.10	0.09	0.08	0.07	0.06	0.12	0.11	0.12	0.12	0.12	0.11	0.10	0.10	0.10	7
(3 in.)	3.33	Corrugated metal	0	0.23	0.17	0.14	0.12	0.10	0.09	0.08	0.18	0.17	0.18	0.17	0.18	0.15	0.14	0.15	0.14	8
		Insulation bd. (1 in.)	2.78	0.14	0.12	0.10	0.09	0.08	0.07	0.06	0.12	0.11	0.12	0.12	0.12	0.11	0.10	0.10	0.10	9
		Insulation bd. (1½ in.)	4.17	0.12	0.10	0.09	0.08	0.07	0.07	0.06	0.10	0.10	0.10	0.10	0.10	0.09	0.09	0.09	0.09	10
		Glass fiber bd. (1 in.)	4.00	0.12	0.10	0.09	0.08	0.07	0.07	0.06	0.10	0.10	0.10	0.10	0.11	0.10	0.09	0.09	0.09	11
(4 in.)	4.44	Corrugated metal	0	0.18	0.14	0.12	0.10	0.09	0.08	0.07	0.15	0.14	0.15	0.15	0.15	0.13	0.12	0.13	0.12	12
		Insulation bd. (1 in.)	2.78	0.12	0.10	0.09	0.08	0.07	0.07	0.06	0.11	0.10	0.10	0.10	0.11	0.10	0.09	0.09	0.09	13
		Insulation bd. (1½ in.)	4.17	0.10	0.09	0.08	0.07	0.07	0.06	0.06	0.09	0.09	0.09	0.09	0.09	0.09	0.08	0.08	0.08	14
		Glass fiber bd. (1 in.)	4.00	0.10	0.09	0.08	0.07	0.07	0.06	0.06	0.09	0.09	0.09	0.09	0.09	0.09	0.08	0.08	0.08	15
Gypsum slab (2 in.)	1.20	Gypsum bd. (½ in.)	0.45	0.36	0.24	0.18	0.14	0.12	0.11	0.09	0.25	0.24	0.25	0.25	0.27	0.21	0.19	0.20	0.18	16
		Insulation bd. (1 in.)	2.78	0.20	0.15	0.13	0.11	0.10	0.09	0.07	0.16	0.15	0.16	0.16	0.16	0.14	0.13	0.13	0.12	17
		Insulation bd. (1½ in.)	4.17	0.15	0.13	0.11	0.09	0.09	0.08	0.07	0.13	0.13	0.13	0.13	0.13	0.12	0.11	0.11	0.11	18
		Asbestos-cement bd. (¼ in.)	0.06	0.40	0.26	0.19	0.15	0.13	0.11	0.09	0.27	0.25	0.26	0.26	0.29	0.22	0.19	0.20	0.18	19
		Glass fiber bd. (1 in.)	4.00	0.16	0.13	0.11	0.10	0.09	0.08	0.07	0.13	0.13	0.13	0.13	0.14	0.12	0.11	0.12	0.11	20
(3 in.)	1.80	Gypsum bd. (½ in.)	0.45	0.30	0.21	0.16	0.13	0.12	0.10	0.09	0.22	0.21	0.22	0.21	0.23	0.19	0.17	0.17	0.16	21
		Insulation bd. (1 in.)	2.78	0.18	0.14	0.12	0.10	0.09	0.08	0.07	0.15	0.14	0.14	0.14	0.15	0.13	0.12	0.12	0.12	22
		Insulation bd. (1½ in.)	4.17	0.14	0.12	0.10	0.09	0.08	0.07	0.06	0.12	0.12	0.12	0.12	0.12	0.11	0.10	0.11	0.10	23
		Asbestos-cement bd. (¼ in.)	0.06	0.34	0.23	0.17	0.14	0.12	0.10	0.09	0.24	0.22	0.24	0.23	0.25	0.20	0.18	0.19	0.17	24
		Glass fiber bd. (1 in.)	4.00	0.14	0.12	0.10	0.09	0.08	0.07	0.07	0.12	0.12	0.12	0.12	0.13	0.11	0.10	0.11	0.10	25
(4 in.)	2.40	Gypsum bd. (½ in.)	0.45	0.25	0.19	0.15	0.12	0.11	0.09	0.08	0.19	0.18	0.19	0.19	0.20	0.17	0.15	0.16	0.14	26
		Insulation bd. (1 in.)	2.78	0.16	0.13	0.11	0.10	0.09	0.08	0.07	0.13	0.13	0.13	0.13	0.14	0.12	0.11	0.12	0.11	27
		Insulation bd. (1½ in.)	4.17	0.13	0.11	0.10	0.08	0.08	0.07	0.06	0.11	0.11	0.11	0.11	0.12	0.10	0.10	0.10	0.09	28
		Asbestos-cement bd. (¼ in.)	0.06	0.28	0.20	0.16	0.13	0.11	0.10	0.08	0.21	0.20	0.21	0.20	0.22	0.18	0.16	0.17	0.15	29
		Glass fiber bd. (1 in.)	4.00	0.13	0.11	0.10	0.09	0.08	0.07	0.06	0.12	0.11	0.11	0.11	0.12	0.10	0.10	0.10	0.10	30

Table IV Coefficients of Transmission (*U*) of Wood or Metal Construction Flat Roofs and Ceilings[de]

Example—K 4

Construction (Heat flow up)	Resistance (*R*)
1. Outside surface (15 mph wind)	*0.17*
2. Built-up roofing, ⅜ in.	*0.33*
3. Roof insulation (*C* = 0.72)	*1.39*
4. Wood deck (1 in.)	*0.98*
5. Air space	*0.85*
6. Gypsum wall board (⅜ in.)	*0.32*
7. Acoustical tile (½ in.)—glued	*1.19*
8. Inside surface (still air)	*0.61*
Total resistance	*5.84*
U = 1/*R* = 1/*5.84*	0.17

See value 0.17 in boldface type in table below.

Example of Substitution

Replace item 4 with 2 in. wood deck (exposed to inside) and omit items 5, 6, and 7.

Total resistance		*5.84*
Deduct 4. Wood deck (1 in.)	*0.98*	
5. Air space	*0.85*	
6. Gypsum wall board (⅜ in.)	*0.32*	
7. Acoustical tile (½ in.) glued	*1.19*	*3.34*
Difference		*2.50*
Add 4. Wood deck (2 in.)		*2.03*
Total resistance		*4.53*
U = 1/*R* = 1/*4.53* =		0.22

Type of Deck (Built-up Roof in All Cases)		Insulation Added on Top of Deck		Type of Ceiling											Number
				None	Gypsum Bd. (⅜ in. and Plas.)			Metal Lath and Plas.		Insul. Bd. (½ in.)	Acoustical Tile				
		Con-duct-ance of insul.	Resis-tance		None	Lt. wt. agg. ½ in.	Sand agg. ½ in.	Lt. wt. agg. ¾ in.	Sand agg. ¾ in.	Plain (1.43) or ½ in. plas. sand agg. (1.52)	On Furring ½ in.	On Furring ¾ in.	On Gypsum Bd. (⅜ in.) ½ in.	On Gypsum Bd. (⅜ in.) ¾ in.	
Resistance →			→		0.32	0.64	0.41	0.47	0.13	1.47	1.19	1.78	1.51	2.10	
Material	R	C	R	U A	U B	U C	U D	U E	U F	U G	U H	U I	U J	U K	
Wood 1 in.	*0.98*	None	—	0.48	0.31	0.28	0.30	0.29	0.33	0.23	0.24	0.21	0.22	0.20	**1**
		0.72	*1.39*	0.29	0.22	0.20	0.21	0.21	0.22	0.17	0.18	0.16	**0.17**	0.16	**2**
		0.36	*2.78*	0.21	0.17	0.16	0.16	0.16	0.17	0.14	0.14	0.13	0.14	0.13	**3**
		0.24	*4.17*	0.16	0.13	0.13	0.13	0.13	0.14	0.12	0.12	0.11	0.12	0.11	**4**
		0.19	*5.26*	0.14	0.12	0.11	0.12	0.12	0.12	0.10	0.11	0.10	0.10	0.10	**5**
		0.15	*6.67*	0.11	0.10	0.10	0.10	0.10	0.10	0.09	0.09	0.09	0.09	0.09	**6**
		0.12	*8.33*	0.10	0.09	0.08	0.09	0.09	0.09	0.08	0.08	0.08	0.08	0.07	**7**
Wood 2 in.	*2.03*	None	—	0.32	0.23	0.22	0.23	0.22	0.24	0.18	0.19	0.17	0.18	0.16	**8**
		0.72	*1.39*	0.22	0.18	0.17	0.17	0.17	0.18	0.15	0.15	0.14	0.15	0.13	**9**
		0.36	*2.78*	0.17	0.14	0.13	0.14	0.14	0.14	0.12	0.13	0.12	0.12	0.11	**10**
		0.24	*4.17*	0.14	0.12	0.11	0.12	0.12	0.12	0.10	0.11	0.10	0.10	0.10	**11**
		0.19	*5.26*	0.12	0.10	0.10	0.10	0.10	0.11	0.09	0.10	0.09	0.09	0.09	**12**
		0.15	*6.67*	0.10	0.09	0.09	0.09	0.09	0.09	0.08	0.08	0.08	0.08	0.08	**13**
		0.12	*8.33*	0.09	0.08	0.08	0.08	0.08	0.08	0.07	0.07	0.07	0.07	0.07	**14**
Wood 3 in.	*3.28*	None	—	0.23	0.18	0.17	0.18	0.18	0.19	0.15	0.16	0.14	0.15	0.14	**15**
		0.72	*1.39*	0.17	0.14	0.14	0.14	0.14	0.15	0.12	0.13	0.12	0.12	0.11	**16**
		0.36	*2.78*	0.14	0.12	0.12	0.12	0.12	0.12	0.11	0.11	0.10	0.10	0.10	**17**
		0.24	*4.17*	0.12	0.10	0.10	0.10	0.10	0.10	0.09	0.09	0.09	0.09	0.09	**18**
		0.19	*5.26*	0.10	0.09	0.09	0.09	0.09	0.09	0.08	0.09	0.08	0.08	0.08	**19**
		0.15	*6.67*	0.09	0.08	0.08	0.08	0.08	0.08	0.07	0.08	0.07	0.07	0.07	**20**
		0.12	*8.33*	0.08	0.07	0.07	0.07	0.07	0.07	0.07	0.07	0.07	0.07	0.06	**21**
Preformed slabs—wood fiber and cement binder 2 in.	*3.60*	None	—	0.21	0.17	0.16	0.17	0.17	0.18	0.14	0.15	0.14	0.14	0.13	**22**
3 in.	*5.40*	None	—	0.15	0.13	0.13	0.13	0.13	0.13	0.11	0.12	0.11	0.11	0.11	**23**
Flat metal roof deck	*0*	None	—	0.90	0.44	0.38	0.42	0.41	0.48	0.29	0.32	0.27	0.29	0.25	**24**
		0.72	*1.39*	0.40	0.27	0.25	0.27	0.26	0.29	0.21	0.22	0.19	0.21	0.18	**25**
		0.36	*2.78*	0.26	0.20	0.19	0.19	0.19	0.21	0.16	0.17	0.15	0.16	0.15	**26**
		0.24	*4.17*	0.19	0.16	0.15	0.15	0.15	0.16	0.13	0.14	0.13	0.13	0.12	**27**
		0.19	*5.26*	0.16	0.13	0.13	0.13	0.13	0.14	0.12	0.12	0.11	0.11	0.11	**28**
		0.15	*6.67*	0.13	0.11	0.11	0.11	0.11	0.11	0.10	0.10	0.10	0.10	0.09	**29**
		0.12	*8.33*	0.11	0.09	0.09	0.09	0.09	0.10	0.09	0.09	0.08	0.08	0.08	**30**

Table V Coefficients of Transmission (U) of Frame Partitions of Interior Walls[c]

Example—Wall B 1		*Example of Substitution*	
Construction	Resistance (R)	Replace item 2 with wood fiber hardboard (¼ in.).	
1. Surface (still air)	*0.68*	Total resistance	*2.97*
2. Gypsum bd. (⅜ in.)	*0.32*	Deduct 2. Gypsum wall board (⅜ in.)	*0.32*
3. Air space	*0.97*		
4. Gypsum wall board (⅜ in.)	*0.32*	Difference	*2.65*
5. Surface (still air)	*0.68*	Add 2. Hardboard (¼ in.)	*0.18*
Total resistance	*2.97*	Total resistance	*2.83*
$U = 1/R = 1/2.97$	0.34	$U = 1/R = 1/2.83 =$	0.35
See value 0.34 in boldface type in table below.			

Type of Interior Finish		*Single Partition (Finish on Only One Side of Studs)*	*Double Partition (Finish on Both Sides of Studs)*	*Number*
Material	*R*	*U*	*U*	
		A	B	
Gypsum bd. (⅜ in.)	*0.32*	0.60	**0.34**	1
Gypsum lath (⅜ in.) and ½ in. plas. (lt. wt. agg.)	*0.64*	0.50	0.28	2
Gypsum lath (⅜ in.) and ½ in. plas. (sand agg.)	*0.41*	0.56	0.32	3
Metal lath and ¾ in. plas. (lt. wt. agg.)	*0.47*	0.55	0.31	4
Metal lath and ¾ in. plas. (sand agg.)	*0.13*	0.67	0.39	5
Insul. bd. (½ in.)	*1.43*	0.36	0.19	6
Insul. bd. lath (½ in.) and ½ plas. (sand agg.)	*1.52*	0.35	0.19	7
Plywood: (¼ in.)	*0.31*	0.60	0.34	8
(⅜ in.)	*0.47*	0.55	0.31	9
(½ in.)	*0.63*	0.50	0.28	10
Wood panels (¾ in.)	*0.94*	0.43	0.24	11
Wood-lath and ½ in. plas. (sand agg.)	*0.40*	0.57	0.32	12
Sheet-metal panels adhered to wood (framing)	*0*	0.74	0.43	13

Table VI Coefficients of Transmission (*U*) of Masonry Partitions[c]

Example—Wall C 2

Construction	Resistance (*R*)
1. Inside surface (still air)	*0.68*
2. Plas. (lt. wt. agg.) 5/8 in.	*0.39*
3. Cement block (cinder agg.) (4 in.)	*1.11*
4. Plas. (lt. wt. agg.) 5/8 in.	*0.39*
5. Inside surface (still air)	*0.68*
Total resistance	*3.25*
$U = 1/R = 1/3.25 =$	0.31

See value 0.31 in boldface type in table below.

Example of Substitution

Replace item 3 with gypsum tile (4 in.)	
Total resistance	*3.25*
Deduct 3. Cement block (cinder agg.) (4 in.)	*1.11*
Difference	*2.14*
Add 3. Gypsum tile (4 in.)	*1.67*
Total resistance	*3.81*
$U = 1/R = 1/3.81 =$	0.26

Type of Partition	Resistance	Surface Finish: None	Plas. (lt. wt. agg.) 5/8 in. One side 0.39	Plas. (lt. wt. agg.) 5/8 in. Two sides 0.78	Plas. (sand agg.) 5/8 in. One side 0.11	Plas. (sand agg.) 5/8 in. Two sides 0.22	Number
Material	*R*	*U*	*U*	*U*	*U*	*U*	
		A	B	C	D	E	
Hollow concrete block							
(Cinder agg.)							
(3 in.)	*0.86*	0.45	0.38	0.33	0.43	0.41	**1**
(4 in.)	*1.11*	0.40	0.35	**0.31**	0.39	0.37	**2**
(8 in.)	*1.72*	0.32	0.29	0.26	0.31	0.30	**3**
(12 in.)	*1.89*	0.31	0.27	0.25	0.30	0.29	**4**
(Lt. wt. agg.)							
(3 in.)	*1.27*	0.38	0.33	0.30	0.36	0.35	**5**
(4 in.)	*1.50*	0.35	0.31	0.27	0.34	0.32	**6**
(8 in.)	*2.00*	0.30	0.27	0.24	0.29	0.28	**7**
(12 in.)	*2.27*	0.28	0.25	0.23	0.27	0.26	**8**
(Gravel agg.)							
(8 in.)	*1.11*	0.40	0.35	0.31	0.39	0.37	**9**
(12 in.)	*1.28*	0.38	0.33	0.29	0.36	0.35	**10**
Hollow clay tile							
(3 in.)	*0.80*	0.46	0.39	0.34	0.44	0.42	**11**
(4 in.)	*1.11*	0.40	0.35	0.31	0.39	0.37	**12**
(6 in.)	*1.52*	0.35	0.31	0.27	0.33	0.32	**13**
(8 in.)	*1.85*	0.31	0.28	0.25	0.30	0.29	**14**
Hollow gypsum tile							
(3 in.)	*1.35*	0.37	0.32	0.29	0.35	0.34	**15**
(4 in.)	*1.67*	0.33	0.29	0.26	0.32	0.31	**16**
Solid plaster walls							
Gypsum lath (1/2 in.) and plas.							
3/4 in. each side							
(Lt. wt. agg.)	*1.39*	0.36	—	—	—	—	**17**
(Sand agg.)	*0.71*	0.48	—	—	—	—	**18**
1 in. each side							
(Lt. wt. agg.)	*1.73*	0.32	—	—	—	—	**19**
(Sand agg.)	*0.81*	0.46	—	—	—	—	**20**
Metal lath and plas.							
2 in. total thickness							
(Lt. wt. agg.)	*1.28*	0.38	—	—	—	—	**21**
(Sand agg.)	*0.36*	0.58	—	—	—	—	**22**
2 1/2 in. total thickness							
(Lt. wt. agg.)	*1.60*	0.34	—	—	—	—	**23**
(Sand agg.)	*0.45*	0.55	—	—	—	—	**24**

Table VII Coefficients of Transmission (*U*) of Masonry Walls[d]

Example—Wall G 1

Construction	Resistance (*R*)
1. Outside surface (15 mph wind)	*0.17*
2. Face brick (4 in.) (avg *R*)	*0.39*
3. Cement mortar (½ in.)	*0.10*
4. Concrete block (cinder agg.) (4 in.)	*1.11*
5. Air space	*0.97*
6. Gypsum lath (⅜ in.)	*0.32*
7. Plas. (sand agg.) (½ in.)	*0.09*
8. Inside surface (still air)	*0.68*
Total resistance	*3.83*
$U = 1/R = 1/3.83 =$	0.26

See value 0.26 in boldface type in table below.

Example of Substitution

Replace items 6 and 7 with wood panels (¾ in.) and vapor barrier applied over furring strips

Total resistance		*3.83*
Deduct 6. Gypsum lath (⅜ in.)	*0.32*	
7. Plas. (sand agg.) (½ in.)	*0.09*	*0.41*
Difference		*3.42*
Add 6. Vapor barrier	*0.06*	
7. Wood panel (¾ in.)	*0.94*	*1.00*
Total resistance		*4.42*
$U = 1/R = 1/4.42 =$		0.23

Exterior Facing			*Backing*		*Interior Finish*											Number
					None	Plas. 5/8 in. on Wall		Metal Lath and 3/4 in. Plas. on Furring		Gypsum Lath (3/8 in.) and 1/2 in. Plas. on Furring			Insul. Bd. Lath (1/2 in.) and 1/2 in. Plas. on Furring		Wood Lath 1/2 in. Plas.	
						(Sand agg.)	(Lt. wt. agg.)	(Sand agg.)	(Lt. wt. agg.)	No plas.	(Sand agg.)	(Lt. wt. agg.)	No plas.	(Sand agg.)	(Sand agg.)	
			Resistance ↓→			*0.11*	*0.39*	*0.13*	*0.47*	*0.32*	*0.41*	*0.64*	*1.43*	*1.52*	*0.40*	
Material	*R*	avg *R*	Material	*R*	*U*	*U*	*U*	*U*	*U*	*U*	*U*	*U*	*U*	*U*	*U*	
					A	B	C	D	E	F	G	H	I	J	K	
			Concrete block													
			(Cinder agg.)													
			(4 in.)	*1.11*	0.41	0.39	0.35	0.28	0.26	0.27	**0.26**	0.25	0.21	0.20	0.26	**1**
			(8 in.)	*1.72*	0.33	0.32	0.29	0.24	0.22	0.23	0.23	0.21	0.18	0.18	0.23	**2**
			(12 in.)	*1.89*	0.31	0.30	0.28	0.23	0.21	0.22	0.22	0.21	0.18	0.17	0.22	**3**
			(Lt. wt. agg.)													
			(4 in.)	*1.50*	0.35	0.34	0.31	0.25	0.23	0.24	0.24	0.22	0.19	0.19	0.24	**4**
Face brick			(8 in.)	*2.00*	0.30	0.29	0.27	0.23	0.21	0.22	0.21	0.20	0.17	0.17	0.21	**5**
4 in.	*0.44*		(12 in.)	*2.27*	0.28	0.27	0.25	0.21	0.20	0.20	0.20	0.19	0.17	0.16	0.20	**6**
Stone			(Sand agg.)													
4 in.	*0.32*		(4 in.)	*0.71*	0.49	0.46	0.41	0.32	0.29	0.30	0.29	0.27	0.22	0.22	0.29	**7**
Precast concrete		*0.39*	(8 in.)	*1.11*	0.41	0.39	0.35	0.28	0.26	0.27	0.26	0.25	0.21	0.20	0.26	**8**
(sand agg.)			(12 in.)	*1.28*	0.38	0.37	0.33	0.27	0.25	0.26	0.25	0.24	0.20	0.20	0.25	**9**
4 in.	*0.32*															
6 in.	*0.48*		Hollow clay tile													
			(4 in.)	*1.11*	0.41	0.39	0.35	0.28	0.26	0.27	0.26	0.25	0.21	0.20	0.26	**10**
			(8 in.)	*1.85*	0.31	0.30	0.28	0.23	0.22	0.22	0.22	0.21	0.18	0.18	0.22	**11**
			(12 in.)	*2.50*	0.26	0.25	0.24	0.20	0.19	0.19	0.19	0.18	0.16	0.16	0.19	**12**
			Concrete													
			(Sand agg.)													
			(4 in.)	*0.32*	0.60	0.56	0.49	0.36	0.32	0.34	0.33	0.31	0.25	0.24	0.33	**13**
			(6 in.)	*0.48*	0.55	0.52	0.45	0.34	0.31	0.32	0.31	0.29	0.24	0.23	0.31	**14**
			(8 in.)	*0.64*	0.51	0.48	0.42	0.32	0.29	0.31	0.30	0.28	0.23	0.22	0.30	**15**
			Concrete block													
			(Cinder agg.)													
			(4 in.)	*1.11*	0.36	0.35	0.32	0.26	0.24	0.25	0.24	0.23	0.19	0.19	0.24	**16**
			(8 in.)	*1.72*	0.29	0.29	0.26	0.22	0.21	0.21	0.21	0.20	0.17	0.17	0.21	**17**
			(12 in.)	*1.89*	0.28	0.27	0.25	0.21	0.20	0.21	0.20	0.19	0.17	0.17	0.20	**18**
			(Lt. wt. agg.)													
			(4 in.)	*1.50*	0.32	0.30	0.28	0.23	0.22	0.22	0.22	0.21	0.18	0.18	0.22	**19**
			(8 in.)	*2.00*	0.27	0.26	0.25	0.21	0.20	0.20	0.20	0.19	0.16	0.16	0.20	**20**
			(12 in.)	*2.27*	0.25	0.25	0.23	0.20	0.19	0.19	0.19	0.18	0.16	0.16	0.19	**21**
Common brick			(Sand agg.)													
4 in.	*0.80*		(4 in.)	*0.71*	0.42	0.40	0.36	0.29	0.26	0.27	0.27	0.25	0.21	0.21	0.27	**22**
Precast concrete		*0.72*	(8 in.)	*1.11*	0.36	0.35	0.32	0.26	0.24	0.25	0.24	0.23	0.19	0.19	0.24	**23**
(sand agg.)			(12 in.)	*1.28*	0.34	0.33	0.30	0.25	0.23	0.24	0.23	0.22	0.19	0.18	0.23	**24**
8 in.	*0.64*															
			Hollow clay tile													
			(4 in.)	*1.11*	0.36	0.35	0.32	0.26	0.24	0.25	0.24	0.23	0.19	0.19	0.24	**25**
			(8 in.)	*1.85*	0.28	0.28	0.26	0.22	0.20	0.21	0.20	0.19	0.17	0.17	0.20	**26**
			(12 in.)	*2.50*	0.24	0.23	0.22	0.19	0.18	0.18	0.18	0.17	0.15	0.15	0.18	**27**
			Concrete													
			(Sand agg.)													
			(4 in.)	*0.32*	0.50	0.48	0.42	0.32	0.29	0.30	0.30	0.28	0.23	0.22	0.30	**28**
			(6 in.)	*0.48*	0.47	0.44	0.39	0.31	0.28	0.29	0.28	0.27	0.22	0.22	0.28	**29**
			(8 in.)	*0.64*	0.43	0.41	0.37	0.29	0.27	0.28	0.27	0.26	0.21	0.21	0.27	**30**

Table VIII Coefficients of Transmission (*U*) of Masonry Cavity Walls[d]

Example—Wall H 6

Construction	Resistance (*R*)
1. Outside surface (15 mph wind)	*0.17*
2. Common brick (4 in.) (avg *R*)	*0.76*
3. Air space	*0.97*
4. Concrete block (gravel agg.) (4 in.)	*0.71*
5. Air space	*0.97*
6. Gypsum lath (⅜ in.)	*0.32*
7. Plas. (lt. wt. agg.) (½ in.)	*0.32*
8. Inside surface (still air)	*0.68*
Total resistance	*4.90*
$U = 1/R = 1/4.90 =$	0.20

See value 0.20 in boldface type in table below.

Example of Substitution

Replace item 4 with 8 in. concrete block and items 6 and 7 with ⅝ in. plas. (sand agg.) applied directly to concrete block

Total resistance		*4.90*
Deduct 4. Concrete block (gravel agg.) 4 in.	*0.71*	
5. Air space	*0.97*	
6. Gypsum lath (⅜ in.)	*0.32*	
7. Plas. (lt. wt. agg.) (½ in.)	*0.32*	*2.32*
Difference		*2.58*
Add 4. Concrete block (gravel agg.) 8 in.	*1.11*	
7. Plas. (sand agg.) (⅝ in.)	*0.11*	*1.22*
Total resistance		*3.80*
$U = 1/R = 1/3.80 =$		0.26

Exterior Construction: Material	R	avg R	Inner Section: Material	R	Interior Finish: None	Plas. ⅝ in. on Wall (Sand agg.)	Plas. ⅝ in. on Wall (Lt. wt. agg.)	Metal Lath and ¾ in. Plas. on Furring (Sand agg.)	Metal Lath and ¾ in. Plas. on Furring (Lt. wt. agg.)	Gypsum Lath (⅜ in.) and ½ in. Plas. on Furring (No plas.)	Gypsum Lath (⅜ in.) and ½ in. Plas. on Furring (Sand agg.)	Gypsum Lath (⅜ in.) and ½ in. Plas. on Furring (Lt. wt. agg.)	Insul. Bd. Lath and ½ in. Plas. on Furring (No plas.)	Insul. Bd. Lath and ½ in. Plas. on Furring (Sand agg.)	Wood Lath and ½ in. Plas. (Sand agg.)	Number
			Resistance →			*0.11*	*0.39*	*0.13*	*0.47*	*0.32*	*0.41*	*0.64*	*1.43*	*1.52*	*0.40*	
					U	U	U	U	U	U	U	U	U	U	U	
					A	B	C	D	E	F	G	H	I	J	K	
Face brick (4 in.)		*0.44*	Concrete block (4 in.) (Gravel agg.)	*0.71*	0.34	0.32	0.30	0.25	0.23	0.23	0.23	0.22	0.19	0.18	0.23	1
			(Cinder agg.)	*1.11*	0.30	0.29	0.27	0.22	0.21	0.21	0.21	0.20	0.17	0.17	0.21	2
			(Lt. wt. agg.)	*1.50*	0.27	0.26	0.24	0.21	0.19	0.20	0.19	0.19	0.16	0.16	0.19	3
			Common brick (4 in.)	*0.80*	0.33	0.32	0.29	0.24	0.22	0.23	0.23	0.21	0.18	0.18	0.23	4
			Clay tile (4 in.)	*1.11*	0.30	0.29	0.27	0.22	0.21	0.21	0.21	0.20	0.17	0.17	0.21	5
Common brick (4 in.)	*0.80*	*0.76*	Concrete block (4 in.) (Gravel agg.)	*0.71*	0.30	0.29	0.27	0.23	0.21	0.22	0.21	**0.20**	0.18	0.17	0.21	6
Concrete block (gravel agg.) (4 in.)	*0.71*		(Cinder agg.)	*1.11*	0.27	0.26	0.25	0.21	0.19	0.20	0.20	0.19	0.16	0.16	0.20	7
			(Lt. wt. agg.)	*1.50*	0.25	0.24	0.22	0.19	0.18	0.19	0.18	0.18	0.15	0.15	0.18	8
			Common brick (4 in.)	*0.80*	0.30	0.29	0.27	0.22	0.21	0.21	0.21	0.20	0.17	0.17	0.21	9
			Clay tile (4 in.)	*1.11*	0.27	0.26	0.25	0.21	0.19	0.20	0.20	0.19	0.16	0.16	0.20	10
Concrete block (cinder agg.) (4 in.)		*1.11*	Concrete block (4 in.) (Gravel agg.)	*0.71*	0.27	0.27	0.25	0.21	0.20	0.20	0.20	0.19	0.17	0.16	0.20	11
			(Cinder agg.)	*1.11*	0.25	0.24	0.23	0.19	0.18	0.19	0.18	0.18	0.16	0.15	0.18	12
			(Lt. wt. agg.)	*1.50*	0.23	0.22	0.21	0.18	0.17	0.17	0.17	0.17	0.15	0.14	0.17	13
			Common brick (4 in.)	*0.80*	0.27	0.26	0.24	0.21	0.19	0.20	0.20	0.19	0.16	0.16	0.20	14
			Clay tile (4 in.)	*1.11*	0.25	0.24	0.23	0.19	0.18	0.19	0.18	0.18	0.16	0.15	0.18	15

Table IX Coefficients of Transmission (*U*) of Frame Walls[d]

Example—Wall D 4	
Construction	Resistance (*R*)
1. Outside surface (15 mph wind)	*0.17*
2. Siding, wood, ½ in. × 8 in. lapped (avg *R*)	*0.85*
3. Building paper	*0.06*
4. Wood sheathing ($\frac{25}{32}$ in.)	*0.98*
5. Air space	*0.97*
6. Gypsum lath (⅜ in.)	*0.32*
7. Plaster (sand agg.) (½ in.)	*0.09*
8. Inside surface (still air)	*0.68*
Total resistance	*4.12*
$U = 1/R = 1/4.12 =$	0.24

See value 0.24 in boldface type in table below.

Example of Substitution		
Replace items 3 and 4 with insul. bd. sheathing ($\frac{25}{32}$ in.) and items 6 and 7 with gypsum wall board ½ in.)		
Total resistance		*4.12*
Deduct 3. Building paper	*0.06*	
4. Wood sheathing ($\frac{25}{32}$ in.)	*0.98*	
6. Gypsum lath (⅜ in.)	*0.32*	
7. Plaster (sand agg.) (½ in.)	*0.09*	*1.45*
Difference		*2.67*
Add 4. Insul. bd. sheathing ($\frac{25}{32}$ in.)	*2.06*	
6., 7. Gypsum bd. (½ in.)	*0.45*	*2.51*
Total resistance		*5.18*
$U = 1/R = 1/5.18 =$		0.19

Exterior			*Interior Finish*		*Type of Sheathing*						Number
					None, Building Paper	Gypsum Board ½ in.	Plywood $\frac{5}{16}$ in.	Wood, $\frac{25}{32}$ in. and Building Paper	Insulation Board Sheathing ½ in.	Insulation Board Sheathing $\frac{25}{32}$ in.	
				Resistance ↓→	0.06	0.45	0.39	1.04	1.32	2.06	
Material	*R*	avg *R*	Material	*R*	*U* A	*U* B	*U* C	*U* D	*U* E	*U* F	
			None	—	0.57	0.47	0.48	0.36	0.33	0.27	1
			Gypsum bd. (⅜ in.)	*0.32*	0.33	0.29	0.30	0.25	0.23	0.20	2
Wood siding			Gypsum lath (⅜ in.) and ½ in. plas. (lt. wt. agg.)	*0.64*	0.30	0.27	0.27	0.23	0.22	0.19	3
Drop—(1 in. × 8 in.)	*0.79*		Gypsum lath (⅜ in.) and ½ in. plas. (sand agg.)	*0.41*	0.32	0.28	0.29	**0.24**	0.23	0.19	4
Bevel (½ in. × 8 in.)			Metal lath and ¾ in. plas. (lt. wt. agg.)	*0.47*	0.31	0.28	0.28	0.24	0.22	0.19	5
	0.81	*0.85*	Metal lath and ¾ in. plas. (sand agg.)	*0.13*	0.35	0.31	0.31	0.26	0.24	0.21	6
Wood shingles 7½ in. exposure	*0.87*		Insul. bd. (½ in.)	*1.43*	0.24	0.22	0.22	0.19	0.18	0.16	7
Wood panels (¾ in.)	*0.94*		Insul. bd. lath (½ in.) and ½ in. plas. (sand agg.)	*1.52*	0.24	0.22	0.22	0.19	0.18	0.16	8
			Plywood (¼ in.)	*0.31*	0.33	0.29	0.30	0.25	0.23	0.20	9
			Wood panels (¾ in.)	*0.94*	0.27	0.25	0.25	0.22	0.20	0.18	10
			Wood lath and ½ in. plas. (sand agg.)	*0.40*	0.32	0.28	0.29	0.24	0.23	0.19	11
			None	—	0.73	0.56	0.58	0.42	0.38	0.30	12
			Gypsum bd. (⅜ in.)	*0.32*	0.37	0.33	0.33	0.27	0.25	0.21	13
			Gypsum lath (⅜ in.) and ½ in. plas. (lt. wt. agg.)	*0.64*	0.33	0.30	0.30	0.25	0.24	0.20	14
			Gypsum lath (⅜ in.) and ½ in. plas. (sand agg.)	*0.41*	0.36	0.32	0.32	0.27	0.25	0.21	15
Face-brick veneer	*0.44*	*0.45*	Metal lath and ¾ in. plas. (lt. wt. agg.)	*0.47*	0.35	0.31	0.32	0.26	0.25	0.21	16
Plywood (⅜ in.)	*0.47*		Metal lath and ¾ in. plas. (sand agg.)	*0.13*	0.40	0.35	0.36	0.29	0.27	0.22	17
			Insul. bd. (½ in.)	*1.43*	0.26	0.24	0.24	0.21	0.20	0.17	18
			Insul. bd. lath (½ in.) and ½ in. plas. (sand agg.)	*1.52*	0.26	0.23	0.24	0.21	0.19	0.17	19
			Plywood (¼ in.)	*0.31*	0.38	0.33	0.33	0.27	0.26	0.21	20
			Wood panels (¾ in.)	*0.94*	0.30	0.27	0.28	0.23	0.22	0.19	21
			Wood lath and ½ in. plas. (sand agg.)	*0.40*	0.36	0.32	0.32	0.27	0.25	0.21	22

Table IX—*Concluded*

Exterior			*Interior Finish*		*Type of Sheathing*						
					None, Building Paper	Gypsum Board 1/2 in.	Plywood 5/16 in.	Wood, 25/32 in. and Building Paper	Insulation Board Sheathing 1/2 in.	Insulation Board Sheathing 25/32 in.	*Number*
			Resistance ↓→		0.06	0.45	0.39	1.04	1.32	2.06	
					U	*U*	*U*	*U*	*U*	*U*	
Material	*R*	*avg R*	*Material*	*R*	A	B	C	D	E	F	
			None	—	0.43	0.37	0.38	0.30	0.28	0.23	23
			Gypsum bd. (3/8 in.)	*0.32*	0.28	0.25	0.25	0.22	0.20	0.18	24
			Gypsum lath (3/8 in.) and 1/2 in. plas. (lt. wt. agg.)	*0.64*	0.25	0.23	0.23	0.20	0.19	0.17	25
			Gypsum lath (3/8 in.) and 1/2 in. plas. (sand agg.)	*0.41*	0.27	0.24	0.25	0.21	0.20	0.18	26
Wood shingles over insul.: backer bd. (5/16 in.)	*1.40*	*1.42*	Metal lath and 3/4 in. plas. (lt. wt. agg.)	*0.47*	0.27	0.24	0.24	0.21	0.20	0.17	27
Asphalt insul. siding	*1.45*		Metal lath and 3/4 in. plas. (sand agg.)	*0.13*	0.29	0.26	0.27	0.23	0.21	0.18	28
			Insul. bd. (1/2 in.)	*1.43*	0.21	0.20	0.20	0.18	0.17	0.15	29
			Insul. bd. lath (1/2 in.) and 1/2 in. plas. (sand agg.)	*1.52*	0.21	0.19	0.19	0.17	0.16	0.15	30
			Plywood (1/4 in.)	*0.31*	0.28	0.25	0.25	0.22	0.20	0.18	31
			Wood panels (3/4 in.)	*0.94*	0.24	0.22	0.22	0.19	0.18	0.16	32
			Wood lath and 1/2 in. plas. (sand agg.)	*0.40*	0.27	0.24	0.25	0.21	0.20	0.18	33
			None	—	0.91	0.67	0.70	0.48	0.42	0.32	34
			Gypsum bd. (3/8 in.)	*0.32*	0.42	0.36	0.37	0.30	0.27	0.23	35
			Gypsum lath (3/8 in.) and 1/2 in. plas. (lt. wt. agg.)	*0.64*	0.37	0.32	0.33	0.27	0.25	0.21	36
			Gypsum lath (3/8 in.) and 1/2 in. plas. (sand agg.)	*0.41*	0.40	0.35	0.36	0.29	0.27	0.22	37
Asbestos-cement siding	*0.21*		Metal lath and 3/4 in. plas. (lt. wt. agg.)	*0.47*	0.39	0.34	0.35	0.28	0.26	0.22	38
Stucco 1 in.	*0.20*	*0.19*	Metal lath and 3/4 in. plas. (sand agg.)	*0.13*	0.45	0.39	0.40	0.31	0.29	0.24	39
Asphalt roll siding	*0.15*		Insul. bd. (1/2 in.)	*1.43*	0.29	0.26	0.26	0.22	0.21	0.18	40
			Insul. bd. lath (1/2 in.) and 1/2 in. plas. (sand agg.)	*1.52*	0.28	0.25	0.26	0.22	0.21	0.18	41
			Plywood (1/4 in.)	*0.31*	0.42	0.36	0.37	0.30	0.27	0.23	42
			Wood panels (3/4 in.)	*0.94*	0.33	0.29	0.30	0.25	0.23	0.20	43
			Wood lath and 1/2 in. plas. (sand agg.)	*0.40*	0.40	0.35	0.36	0.29	0.27	0.22	44
			None	—	0.37	0.32	0.33	0.27	0.25	0.21	45
			Gypsum bd. (3/8 in.)	*0.32*	0.25	0.23	0.23	0.20	0.19	0.17	46
			Gypsum lath (3/8 in.) and 1/2 in. plas. (lt. wt. agg.)	*0.64*	0.23	0.21	0.22	0.19	0.18	0.16	47
Aluminum or steel siding over sheathing			Gypsum lath (3/8 in.) and 1/2 in. plas. (sand agg.)	*0.41*	0.24	0.22	0.23	0.20	0.19	0.16	48
		1.8	Metal lath and 3/4 in. plas. (lt. wt. agg.)	*0.47*	0.24	0.22	0.22	0.20	0.18	0.16	49
Hollow backed	*0.61*		Metal lath and 3/4 in. plas. (sand agg.)	*0.13*	0.26	0.24	0.24	0.21	0.20	0.17	50
			Insul. bd. (1/2 in.)	*1.43*	0.20	0.18	0.18	0.16	0.16	0.14	51
Ins. board backed, nominal 3/8 in.	*1.8*		Insul. bd lath (1/2 in.) and 1/2 in. plas. (sand agg.)	*1.52*	0.19	0.18	0.18	0.16	0.15	0.13	52
			Plywood (1/4 in.)	*0.31*	0.25	0.23	0.23	0.20	0.19	0.17	53
Reflective foil ins. board backed	*2.96*		Wood panels (3/4 in.)	*0.94*	0.22	0.20	0.20	0.18	0.17	0.15	54
			Wood lath and 1/2 in. plas. (sand agg.)	*0.40*	0.24	0.22	0.23	0.20	0.19	0.16	55

Table X Coefficients of Transmission (*U*) of Solid Masonry Walls[d]

Example—Wall G 2

Construction	Resistance (R)
1. Outside surface (15 mph wind)	*0.17*
2. Face brick (4 in.)	*0.44*
3. Common brick (4 in.)	*0.80*
4. Air space	*0.97*
5. Gypsum lath (3/8 in.)	*0.32*
6. Plas. (sand agg.) (1/2 in.)	*0.09*
7. Inside surface (still air)	*0.68*
Total resistance	*3.47*
$U = 1/R = 1/3.47 =$	0.29

See value 0.29 in boldface type in table below.

Example of Substitution

Assume plain wall—no furring or plaster.

Total resistance		*3.47*
Deduct 4. Air space	*0.97*	
5. Gypsum lath (3/8 in.)	*0.32*	
6. Plas. (sand agg.) (1/2 in)	*0.09*	*1.38*
Total resistance		*2.09*
$U = 1/R = 1/2.09 =$		0.48

Exterior Construction		*Interior Finish*											Number
		None	Plas. 5/8 in. on Wall		Metal Lath and 3/4 in. Plas. on Furring		Gypsum Lath (3/8 in.) and 1/2 in. Plas. on Furring			Insul. Bd. Lath (1/2 in.) and 1/2 in. Plas. on Furring		Wood Lath and 1/2 in. Plas.	
			(Sand agg.)	(Lt. wt. agg.)	(Sand agg.)	(Lt. wt. agg.)	No plas.	(Sand agg.)	(Lt. wt. agg.)	No plas.	(Sand agg.)	(Sand agg.)	
Resistance →			*0.11*	*0.39*	*0.13*	*0.47*	*0.32*	*0.41*	*0.64*	*1.43*	*1.52*	*0.40*	
Material	R	U	U	U	U	U	U	U	U	U	U	U	
		A	B	C	D	E	F	G	H	I	J	K	
Brick (face and common)													
(6 in.)	*0.61*	0.68	0.64	0.54	0.39	0.34	0.36	0.35	0.33	0.26	0.25	0.35	**1**
(8 in.)	*1.24*	0.48	0.45	0.41	0.31	0.28	0.30	**0.29**	0.27	0.22	0.22	0.29	**2**
(12 in.)	*2.04*	0.35	0.33	0.30	0.25	0.23	0.24	0.23	0.22	0.19	0.19	0.23	**3**
(16 in.)	*2.84*	0.27	0.26	0.25	0.21	0.19	0.20	0.20	0.19	0.16	0.16	0.20	**4**
Brick (common only)													
(8 in.)	*1.60*	0.41	0.39	0.35	0.28	0.26	0.27	0.26	0.25	0.21	0.20	0.26	**5**
(12 in.)	*2.40*	0.31	0.30	0.27	0.23	0.21	0.22	0.22	0.21	0.18	0.17	0.22	**6**
(16 in.)	*3.20*	0.25	0.24	0.23	0.19	0.18	0.19	0.18	0.18	0.16	0.15	0.18	**7**
Stone (lime and sand)													
(8 in.)	*0.64*	0.67	0.63	0.53	0.39	0.34	0.36	0.35	0.32	0.26	0.25	0.35	**8**
(12 in.)	*0.96*	0.55	0.52	0.45	0.34	0.31	0.32	0.31	0.29	0.24	0.23	0.31	**9**
(16 in.)	*1.28*	0.47	0.45	0.40	0.31	0.28	0.29	0.28	0.27	0.22	0.22	0.29	**10**
(24) in.	*1.92*	0.36	0.35	0.32	0.26	0.24	0.25	0.24	0.23	0.19	0.19	0.24	**11**
Hollow clay tile													
(8 in.)	*1.85*	0.36	0.36	0.32	0.26	0.24	0.25	0.25	0.23	0.20	0.19	0.25	**12**
(10 in.)	*2.22*	0.33	0.31	0.29	0.24	0.22	0.23	0.22	0.21	0.18	0.18	0.23	**13**
(12 in.)	*2.50*	0.30	0.29	0.27	0.22	0.21	0.22	0.21	0.20	0.17	0.17	0.21	**14**
Poured concrete													
30 lb per cu ft													
(4 in.)	*4.44*	0.19	0.19	0.18	0.16	0.15	0.15	0.15	0.14	0.13	0.13	0.15	**15**
(6 in.)	*6.66*	0.13	0.13	0.13	0.12	0.11	0.11	0.11	0.11	0.10	0.10	0.11	**16**
(8 in.)	*8.88*	0.10	0.10	0.10	0.09	0.09	0.09	0.09	0.09	0.08	0.08	0.09	**17**
(10 in.)	*11.10*	0.08	0.08	0.08	0.08	0.07	0.08	0.08	0.07	0.07	0.07	0.08	**18**
80 lb per cu ft													
(6 in.)	*2.40*	0.31	0.30	0.27	0.23	0.21	0.22	0.22	0.21	0.18	0.17	0.22	**19**
(8 in.)	*3.20*	0.25	0.24	0.23	0.19	0.18	0.19	0.18	0.18	0.16	0.15	0.18	**20**
(10 in.)	*4.00*	0.21	0.20	0.19	0.17	0.16	0.16	0.16	0.15	0.14	0.14	0.16	**21**
(12 in.)	*4.80*	0.18	0.17	0.17	0.15	0.14	0.14	0.14	0.14	0.12	0.12	0.14	**22**
140 lb per cu ft													
(6 in.)	*0.48*	0.75	0.69	0.58	0.41	0.36	0.38	0.37	0.34	0.27	0.26	0.37	**23**
(8 in.)	*0.64*	0.67	0.63	0.53	0.39	0.34	0.36	0.35	0.32	0.26	0.25	0.35	**24**
(10 in.)	*0.80*	0.61	0.57	0.49	0.36	0.32	0.34	0.33	0.31	0.25	0.24	0.33	**25**
(12 in.)	*0.96*	0.55	0.52	0.45	0.34	0.31	0.32	0.31	0.29	0.24	0.23	0.31	**26**
Concrete block													
(Gravel agg.) (8 in.)	*1.11*	0.52	0.48	0.43	0.33	0.29	0.31	0.30	0.28	0.23	0.22	0.30	**27**
(12 in.)	*1.28*	0.47	0.45	0.40	0.31	0.28	0.29	0.28	0.27	0.22	0.22	0.29	**28**
(Cinder agg.) (8 in.)	*1.72*	0.39	0.37	0.34	0.27	0.25	0.26	0.25	0.24	0.20	0.20	0.25	**29**
(12 in.)	*1.89*	0.36	0.35	0.32	0.26	0.24	0.25	0.24	0.23	0.19	0.19	0.24	**30**
(Lt. wt. agg.) (8 in.)	*2.00*	0.35	0.34	0.31	0.26	0.23	0.24	0.24	0.22	0.19	0.19	0.24	**31**
(12 in.)	*2.27*	0.32	0.31	0.28	0.24	0.22	0.23	0.22	0.21	0.18	0.18	0.22	**32**

Table XI Coefficients of Transmission (U) of Pitched Roofs[f]

Example—Roof C 4

Construction (Heat flow up)	Resistance (R)
1. Outside surface (15 mph wind)	*0.17*
2. Slate shingles (½ in.)	*0.05*
3. Building paper	*0.06*
4. Wood sheathing (25⁄32 in.)	*0.98*
5. Air space	*0.90*
6. Gypsum lath (⅜ in.)	*0.32*
7. Plas. (sand agg.) (½ in.)	*0.09*
8. Inside surface (still air)	*0.62*
Total resistance	*3.19*
$U = 1/R = 1/3.19$	0.31

See value 0.31 in boldface type in table below.

Example of Substitution

Find U value for same construction with heat flow down (summer conditions)

Total resistance		*3.19*
Deduct 1. Outside surface (15 mph wind)	*0.17*	
5. Air space	*0.90*	
8. Inside surface (still air)	*0.62*	*1.69*
Difference		*1.50*
Add 1. Outside surface (7.5 mph wind)	*0.25*	
5. Air space	*0.89*	
8. Inside surface (still air)	*0.76*	*1.90*
Total resistance		*3.40*
$U = 1/R = 1/3.40$		0.29

Direction of Heat Flow →		*Upward Flow Winter Conditions*					*Downward Flow Summer Conditions*					
		Rafter Space					*Rafter Space*					
		Unventilated, Not to be Further Insulated				Insulated	Unventilated, Not to be Further Insulated				Insulated	
Type of Ceiling (Applied Directly to Roof Rafters)		Asphalt shingles building paper		Asbestos-cement slate or tile shingles, building paper on wood sheathing (25⁄32 in.)	Wood shingles on 1 x 4 in. wood strips on 6-in. centers	Roof and sheathing disregarded	Asphalt shingles building paper		Asbestos-cement slate, or tile shingles, building paper on wood sheathing (25⁄32 in.)	Wood shingles on 1 x 4 in. wood strips on 6 in. centers	Roof and sheathing disregarded	*Number*
		On plywood sheathing (5⁄16 in.)	On wood sheathing (25⁄32 in.)				On plywood sheathing (5⁄16 in.)	On wood sheathing (25⁄32 in.)				
Resistance →		0.95	1.48	1.09	0.87	—	0.95	1.48	1.09	0.87	—	
Material	R	U	U	U	U	U	U	U	U	U	U	
		A	B	C	D	E	F	G	H	I	J	
None	—	0.57	0.44	0.53	0.60	0.66	0.51	0.40	0.48	0.53	0.56	1
Gypsum bd. (⅜ in.)	*0.32*	0.34	0.29	0.32	0.35	0.54	0.30	0.26	0.29	0.31	0.47	2
Gypsum lath (⅜ in.) & ½ in. plas. (lt. wt. agg.)	*0.64*	0.30	0.26	0.29	0.31	0.46	0.28	0.24	0.27	0.28	0.41	3
Gypsum lath (⅜ in.) and ½ in. plas. (sand agg.)	*0.41*	0.33	0.28	**0.31**	0.34	0.52	0.29	0.25	0.28	0.30	0.45	4
Metal lath and ¾ in. plas. (lt. wt. agg.)	*0.47*	0.32	0.27	0.31	0.33	0.50	0.29	0.25	0.28	0.30	0.44	5
Metal lath and ¾ in. plas. (sand agg.)	*0.13*	0.36	0.30	0.34	0.37	0.61	0.32	0.27	0.31	0.33	0.52	6
Insul. bd. (½ in.)	*1.43*	0.25	0.22	0.24	0.25	0.34	0.23	0.20	0.22	0.23	0.31	7
Insul. bd. lath and ½ in. plas. (sand agg.)	*1.52*	0.24	0.21	0.23	0.25	0.33	0.22	0.20	0.22	0.23	0.30	8
Acoustical tile												
(½ in.) on gypsum bd. (⅜ in.)	*1.51*	0.24	0.21	0.23	0.25	0.33	0.22	0.20	0.22	0.23	0.30	9
(½ in.) on furring	*1.19*	0.26	0.23	0.25	0.27	0.37	0.24	0.21	0.23	0.24	0.34	10
(¾ in.) on gypsum bd. (⅜ in.)	*2.10*	0.21	0.19	0.20	0.21	0.26	0.20	0.18	0.19	0.20	0.26	11
(¾ in.) on furring	*1.78*	0.23	0.20	0.22	0.23	0.30	0.21	0.19	0.20	0.21	0.28	12
Wood lath and ½ in. plas. (sand agg.)	*0.40*	0.33	0.28	0.31	0.34	0.52	0.29	0.26	0.28	0.30	0.46	13

This section is designed to help you locate manufacturers of collectors and components. We have divided the material into several classifications: collector plates, collector modules (plate, insulation glass, etc.), low-temperature hot-water heaters, and miscellaneous hardware. Descriptions are given when possible, but many manufacturers have entered the field only recently and we can supply only names and addresses. We have tested few of these units, so no recommendations are offered. When writing for further information, we would suggest that you include a self-addressed, stamped, 9-by-12-inch envelope. (Nowadays, *everyone* is requesting information and your preaddressed envelope will be appreciated.)

A more detailed listing of solar energy equipment can be found in the Whole Earth *Energy Primer*, available for $4.50 prepaid (plus 6 percent sales tax for California residents) through the Whole Earth Truck Store, 558 Santa Cruz Ave., Menlo Park, California 94025. For further information, see our Bibliography.

COLLECTORS

ROLL-BOND

Olin Brass
ROLL-BOND Products
East Alton, Illinois 62024

Olin uses a silk-screen process and pressure weld to make tube-type heat exchangers to almost any specification. Recently, Olin has begun to market solar panels made of aluminum or copper. Reductions in the price of copper as well as the quality of manufacture make these panels attractive as the basis for a solar collector. Performance data and prices (1/3/75) are listed below. Cost will vary directly with the size produced. In addition, cost estimates will vary with current material and labor costs and the amount of fabrication required by the customer.

Copper Plate Collector
Size—22″ × 96″
Gauge—0.040″
Material—Alloy 122 Copper
Weight—27.29 lbs
Area—14.67 ft²
Price:
100 pieces—$65.85 each
250 pieces—$62.00 each
500 pieces—$60.10 each

Single units and small quantities available for a limited time at $70 each plus a $25 packing charge per shipment. Prices are based on Olin's 1/3/75 price on sheet with copper at $.68/lb.

Aluminum Plate Collector
Material—Alloy 1100 Aluminum
Gauge—0.060″

Article SP–7610
Size—17″ × 50″
Area—5.90 ft²
Price—$20 each plus $15 packing charge per shipment

Article FS–7727
Size—33.75″ × 96″
Area—22.5 ft²
Price:
100 pieces—$37.07 each
250 pieces—$32.17 each
500 pieces—$29.96 each
1000 pieces—$28.85 each

Tailor-made samples are also available with the following price schedule applied. These are alloy 1100 Aluminum panels with a gauge thickness of 0.060″:

Engineering and setup—$300
Manufacture setup—$375
Panels (0.847 lbs/ft²)—$1 per pound
Connectors (two, heliarc welded)—$3 per panel

ECONOCOIL

Tranter Manufacturing, Inc.
c/o Cal Pacific Equipment Corporation
960 Harrison
San Francisco, California 94107
(415) 391–1330

These people have been making heat exchangers for a long time and have recently started to paint them black and sell them as solar collectors. Price is negotiable with order size. The panels are made of carbon or stainless steel. Maximum pressure is 75 psig.

ENERGEX

Energex Corporation
481 Tropicana
Las Vegas, Nevada 89109
(702) 736–2994

Cost:
$3.70/ft² (for a basic flat plate collector) to $15.00/ft² (for a glazed finned-tube collector module)

Unit Size:
Collector Plate, Roll-Bond Design— 36″ × 104″ (smaller sizes available)

Collector Plate, Finned-Tube Design— 16″ × 144″

Module, Roll-Bond Design— 36″ × 104″

Module, Finned-Tube Design— 36″ × 150″

Construction:
Initially, Energex produced only aluminum Roll-Bond collectors. Recently they decided that copper was a more corrosion-resistant material and two constructions are offered. The standard Roll-Bond flat plate collectors can be purchased in module form. Modules are adaptable to single, double, or triple glazing. Energex also offers an aluminum, finned-tube collector consisting of a series of copper tubes with pressed aluminum fins. The entire unit is painted flat-black.

PPG

PPG Industries, Inc.
c/o Mr. N.M. Barker
Manager—Solar System Sales
One Gateway Center
Pittsburgh, Pennsylvania 15222

Cost:
$5.80/ft² (cost decreases as order size increases)

Unit Size:
34-3/16″ × 76-3/16″ × 1-5/16″

Construction:
Two layers of ⅛″ Herculite tempered glass covering a flat-black Roll-Bond aluminum collector surface, backed by 3″ of fiberglass insulation. Cost of this unit is commensurate with its quality (high).

Design Characteristics:
Maximum Temperature—450–500°F (long-term component durability)

Recommended Circulation Fluid — water/antifreeze mixture

REVERE

Revere Copper and Brass, Inc.
P.O. Box 151
Rome, New York 13440
(315) 338–2022

Cost:
$7/ft² (for standard 2′ × 8′ panel with accessories for three tubes not including glass or freight)

Unit Size:
Variable with application

Construction:
A modification of the Revere copper composite building panel for application to solar energy collection. To convert this plywood-copper laminate to a solar collector, a copper batten is added to the standard panel. The batten supports one or two layers of glazing for the panel. Rectangular copper tubes, which carry the circulating fluid, are fastened to the copper sheet at varying intervals.

Design Characteristics:
Maximum Temperature—210°F

Typical Flow Rate—no velocity greater than 4 ft/sec in tubes

SOLERGY

Solergy
150 Green Street
San Francisco, California 94111
(415) 398–6813

Cost:
$3/ft² (estimated)

Unit Size:
2′ × 4′ module

Construction:
This unit represents an interesting idea which retains the benefits of a focusing collector (i.e., higher temperatures), while avoiding the necessity of expensive tracking equipment. By placing a copper tube at the focus of a spiraling curve, radiation from a variety of incidence angles is directed onto the tube surface. Production details are not known at this time.

Design Characteristics:
Maximum Temperature—300°F

SUNWORKS

Sunworks, Inc.
669 Boston Post Road
Guilford, Connecticut 06437
(203) 453–6191

Cost:
Based on order size (two modules offered) $10–$12/ft²

Unit Size:
Flush mounted—23.625″ × 94″ × 5.5″

Surface mounted—36″ × 84″ × 4″

Construction:
Both modules are of the same construction. The flush-mounted module is for use in integral house construction, and its dimensions are based on standard-roof joist widths. A single 3/16″ tempered-glass cover overlays a flat-black copper collector surface with a gridwork of ¼″ copper tubing for coolant transfer. The surface-mounted module is designed for exterior mounting. Specifics of construction can be altered to suit the needs of the purchaser.

Design Characteristics:
Maximum Pressure—15 atmospheres (tested)

Typical Flow Rate—1 gal/hr-ft² of collector

Recommended Circulation Fluid—water/antifreeze mixture

OTHER MANUFACTURERS OF SOLAR COLLECTORS

When requesting information from manufacturers, include a self-addressed, stamped envelope along with a description of intended use and performance requirements.

Reynolds Aluminum
Richmond, Virginia 23261

Produces an aluminum collector designated Torex-14.

Physical Industries Corporation
P.O. Box 357
Lakeside, California 92040

Offers a number of solar products including an aluminum and copper plate collector, a steam generator, and an engine to go with it. Research work being done on heat engines using heat from eutectic salts for power.

Sunpower Inc.
Route 4, P.O. Box 275
Athens, Ohio 45701

Marketing a solar water pump.

Solar Energy Development Incorporated
Nicholas S. Macron, President
1437 Alameda Avenue
Lakewood, Ohio 44107

Marketing a flat plate collector for use in the Midwest costing approximately $9.85/ft².

Environmental Designs
A Division of Steelcraft Corporation
Gary Ford, Manager
P.O. Box 12408
Memphis, Tennessee 38112

Flat plate collector at an anticipated cost $10/ft².

Solarsystems, Inc.
1515 W.S.W. Loop 323
Tyler, Texas 75701

Flat plate collector with vacuum-space insulation.

Powell Brothers, Inc.
5903 Firestone Boulevard
South Gate, California 90280

Inexpensive module collectors. In batches of 8–15 panels, cost is $7.30 each (introductory price) or $1.36/ft². Connecting hose available at $.25/ft.

Sunwater Company
1112 Pioneer Way
El Cajon, California 92020

Makers of solar pool heater, air heaters, and domestic water heaters.

Edwards Engineering Corporation
Pompton Plains, New Jersey 07444

Panels designed to replace normal roofing material consisting of one inch of aluminum-covered fiberglass, heat absorbing aluminum plates, copper or aluminum tubes, and one or two layers of plastic glazing.

AAI Corporation
P.O. Box 6767
Baltimore, Maryland 21204

Complete installation services. These people did a series of school conversions to solar heating for the National Science Foundation early in 1974.

Inter Technology Corporation
P.O. Box 340
Warrenton, Virginia 22186

Also did solar conversion work for the National Science Foundation in 1974. Designed the collector for the Fauquier High School in Virginia.

Brown Manufacturing
P.O. Box 14546
Oklahoma City, Oklahoma 73114

Corning Glass Works
Corning, New York 14830

A vacuum-insulated cylindrical collector made from glass.

Emerson Electric Company
8100 W. Florissant
St. Louis, Missouri 63136

Energy Systems, Inc.
634 Crest Drive
El Cajon, California 92021

General Industries
2238 Moffett Drive
Fort Collins, Colorado 80521

Itek Corporation
Optical Systems Division
10 Maguire Road
Lexington, Massachusetts 02117

Unitspan, Inc.
6606 Variel Avenue
Canoga Park, California 91303

E.K. Swanson Engineering
5615 South Jolly Roger Rd.
Tempe, Arizona 85283

Obelitz Industries, Inc.
P.O. Box 2788
Seal Beach, California 90740

GLAZING

SUN-LITE

Kalwall Corporation
1111 Condia Road
Manchester, New Hampshire 03105
(603) 627–3861

A company with 20 years experience in the field of manufacturing light-transmitting and insulating panels, Kalwall is offering two polyester-reinforced fiberglass sheets called Sun-Lite and Long-Life Sun-Lite. Sun-Lite panels have a 95 percent transmittance and the manufacturer says that a 10 percent loss in transmittance can be expected over a 7-year period. Long-Life Sun-Lite has an 86 percent transmittance, but the Kalwall people say that it will not show any appreciable transmittancy loss in its 20-year lifetime. Maximum operating temperature for both panels is 140°F. In small orders (less than 3000 ft²), prices vary with thickness:

Thickness	*Sun-Lite Price*	*Long-Life Price*
0.030″	$.28/ft²	$.35/ft²
0.037″	$.30/ft²	$.375/ft²
0.045″	$.32/ft²	$.40/ft²

TEDLAR

E.I. DuPont deNemours & Co.
Film Department
Specialty Markets Division
Wilmington, Delaware 19898

Tedlar is a high-transmittancy (92–94 percent) plastic used in lieu of glass for glazing. Cost, impact resistance, durability, and ease of installation make this a useful alternative to glass.

HOT-WATER HEATERS

When discussing hot-water heaters, we are referring to those units that are limited to providing water at temperatures below 140°F. This limit may be imposed by the nature of the materials used in construction, as in the case of plastic panels, or simply by the inability of the system to generate higher temperatures. For this reason, solar hot-water heaters are not generally considered useful for space heating and cooling, since these uses require transfer media with significantly higher temperatures.

FAFCO

FAFCO, Inc.
138 Jefferson Drive
Menlo Park, California 94025
(415) 321-6311

Cost:
$1.75/ft²

Unit Size:
10′ × 4′4″
8′ × 4′4″
(or any length on special order)

Construction:
Fafco collectors are designed primarily to heat swimming pools, but would easily adapt to a hot-water supply system. The panel is of a simple and reliable construct and is now made entirely of polypropylene. Manufacture is a three-step process, beginning with the thermoforming of extruded polypropylene board to generate a flange on each end. Holes are then punched in the flange to access flow channels that run the length of the board. Finally, a manifold pipe of polypropylene is butt-welded to the flange, completing the module.

Design Characteristics:
Maximum Temperature—160°F

Maximum Pressure—5 psig (from manufacturer; it's probably higher)

Typical Flow Rate—8 gpm/panel (maximum)

Circulation Fluid—water

SAV CYLINDRICAL HEATER

Fred Rice Productions
6313 Peach Avenue
Van Nuys, California 91401

Cost:
$185.95/unit

Unit Size:
Capacity—12 U.S. gallons

Construction:
This unit is composed of a series of cylindrical elements around a common axis. Two exterior air spaces provide insulation and a greenhouse effect. The core is made up of a central storage tank and a cylindrical guide. When the sun heats water in the cylindrical guide, a vigorous thermosyphon action is generated, cycling hot water into the storage area and drawing cooler water into the guide from below.

SOLARATOR

Fun & Frolic, Inc.
P.O. Box 277
Madison Heights, Michigan 48071

Cost:
$2.50/ft²

Unit Size:
32″ × 73″ (operational)

Construction:
Two flexible PVC sheets thermowelded to form flow channels. We tend to think that this design is more apt for providing hot water in your Winnebago than for a long-term application such as swimming-pool heating.

Design Characteristics:
Maximum Temperature—manufacturer states 212°F, a number not in line with the known qualities of flexible PVC sheets.

Maximum Pressure—0–4 psi

Typical Flow Rate—1.5 gpm (maximum)

Circulation Fluid—water

SUNSOURCE

Daylin Sunsource
9606 Santa Monica Boulevard
Beverly Hills, California 90210
(213) 878-3211

Cost:
?

Unit Size:
1 meter × 2 meters × 10 cm (2 panels in system); 40-gallon tank

Construction:
A complete hot-water supply unit (collectors, storage tank, boost heater, controls). Made from galvanized-steel panels with Tabor "Selective Black" coating, covered by a single glass layer and insulated with 2″ rockwool.

Design Characteristics:
Maximum Temperature—140°F (nominal)
Maximum Pressure—tank head pressure

OTHER HOT-WATER HEATERS

Sol-Therm Corporation
7 West 14th Street
New York, New York 10011

Included in this hot-water heating unit are two flat collectors, a 32-gallon storage tank, a 1.5-KW thermostatically controlled electric heating element, mounting frame, and in-system plumbing. Unit cost of $495 (FOB New York).

Sunburst, Inc.
70 N.W. 94th Street
Miami Shores, Florida 33150

Will be marketing a one-piece unit for heating and storage of hot water. The unit will come with an electric booster element for supplemental heating. Projected unit cost of $500.

Basic Designs, Inc.
3000 Bridgeway
Sausalito, California 94965

Sun Shower—fill it with water, and in a few hours it will deliver a 2.5-gallon hot shower. $6.95.

PUMPS

JABSCO Impeller Pumps
1485 Dale Way
Costa Mesa, California 92626

Little Giant Pump Co.
3810 N. Tulsa Street
Oklahoma City, Oklahoma 73112

Oberdorfer Pump Division
Syracuse, New York 13201

Bibliography

General Reading

Alternative Sources of Energy. *ASE Newsletter*. Alternative Sources of Energy Publications, Route 2, P.O. Box 90A, Milaca, Minn. 56353.

American Society of Heating, Refrigerating and Air-Conditioning Engineers. *ASHRAE Journal*. ASHRAE, 345 E. 47th Street, New York, N.Y.

Clark, P. 1974. *The Natural Energy Workbook*, vol. 1. Berkeley, Calif.: Village Design.

Daniels, F. 1964. *Direct Use of the Sun's Energy*. New Haven, Conn.: Yale University Press.

Duffie, J. and Beckman, W. 1974. *Solar Energy Thermal Processes*. New York: John Wiley and Sons.

Federal Energy Administration. *Energy Reporter*. FEA, Washington, D.C. Published monthly.

Gravel, M. *Newsletter*. Washington, D.C.: U.S. Senate. Published quarterly by the senator from Alaska.

International Solar Energy Society. *Solar Energy*. New York: Pergamon Press Ltd.

Kahn, L., ed. 1973. *Shelter*. Bolinas, Calif.: Shelter Publications.

Merrill, C., et al. 1974. *Energy Primer*. Menlo Park, Calif.: Portola Institute.

Mother Earth News. *Mother Earth News*. Available from Mother Earth News Ink, P.O. Box 70, Hendersonville, N.C.

Reinhold Publishing. *Heating/Piping/Air Conditioning*. New York: Industrial Press.

Reynolds, W.C. 1974. *Energy From Nature to Man*. New York: McGraw-Hill.

Villecco, M. 1974. "Sunpower." *Architecture Plus*. September—October 1974.

Yellott, J.I. 1973. "Utilization of Sun and Sky Radiation for Heating and Cooling of Buildings." *ASHRAE Journal*, December 1973.

Zarem, E. and Erway, D. 1963. *Introduction to the Utilization of Solar Energy*. New York: McGraw-Hill. Now available through University Microfilms, Ann Arbor, Michigan.

Thermal Comfort

Akerman, J. "Indoor Climate." Seminar on Human Biometeriology, January 1974. Washington, D.C.: U.S. Department of Health, Education and Welfare.

Angus, T.C. 1968. *Control of Indoor Climate*. Oxford: Pergamon Press.

Folk, G.E. 1966. *Introduction to Environmental Physiology*. Philadelphia: Lea and Febiger.

McNall, P., and Nevins, R. 1968. "A Critique of ASHRAE Comfort Standard 55–66." *ASHRAE Journal*, June 1968.

Nevins, R.G. 1965. *Criteria for Thermal Comfort*. Manhattan, Kansas: Institute for Environmental Research, Kansas State University.

Climatology

Kendrewix, W.G. 1959. *Climates of the Continents*. Oxford: Clarendon Press.

National Oceanic and Atmospheric Administration. 1968. *Climatic Atlas of the United States*. Washington, D.C.: National Climate Center, National Oceanic and Atmospheric Administration, U.S. Department of Commerce.

———. 1969. *Selective Guide to Climatic Data Sources*. Ibid.

———. *Local Climatological Data*. Ibid. Published by state.

Heat Loss

American Society of Heating, Refrigerating and Air-Conditioning Engineers. 1967 or 1972. *Handbook of Fundamentals*. New York: ASHRAE.

———. 1970. *Guide and Data Book*. New York: ASHRAE.

Baumeister, T., ed. 1967. *Standard Handbook for Mechanical Engineers*. 7th ed. New York: McGraw-Hill.

Harris, N. and Conde, D. 1974. *Modern Air Conditioning Practice*. New York: McGraw-Hill.

McGuinness, W.J., and Stein, B. 1971. *Mechanical and Electrical Equipment for Buildings*. 5th ed. New York: John Wiley and Sons.

Olivieri, J.B. 1973. *How to Design a Heating-Cooling Comfort System*. Birmingham, Mich.: Business News Publishing.

Perry, J. 1967. *Engineering Manual*. New York: McGraw-Hill.

Strock, C., and Konal, R. 1965. *Handbook of Air Conditioning, Heating and Ventilation*. New York: Industrial Press.

Solar Heating

American Society of Heating, Refrigerating and Air-Conditioning Engineers. 1967. *Low-Temperature Engineering Application of Solar Energy*. New York: ASHRAE. $4.

Baer, S. 1973. "A Suspension Medium for Heat Storage Materials." *Proceedings of the Solar Heating and Cooling for Buildings Workshop, March 21–23, 1973*, vol. 1. Washington, D.C.: National Science Foundation.

Barkman, H. 1973. "Arroyo Barranca Project." *Proceedings of the Solar Heating and Cooling for Buildings Workshop, March 21–23, 1973*, vol. 1. Washington, D.C.: National Science Foundation.

Bell and Gossett. 1967. *Engineering Design Manual; Centrifugal Pump Selection*. Manual no. 3100. Morton Grove, Ill.: Bell and Gosset. $2.

Chahroudi, D. 1973. "What is a Biosphere?" *The Tribal Messenger* (Albuquerque), April 30 May 14, 1973.

Chemical Rubber Company. *Handbook of Chemistry and Physics*. Cleveland, Ohio: Chemical Rubber Company. Published yearly.

———. *Handbook of Engineering Science*. Cleveland, Ohio: Chemical Rubber Comany. Published yearly.

Committee on Science and Astronautics. 1973. *Solar Energy for Heating and Cooling: Hearings*. Washington, D.C: U.S. Government Printing Office.

———. 1974. *Solar Heating and Cooling Demonstration: Background and History, Act of 1974 (HR 11864)*. Washington, D.C.: U.S. Government Printing Office.

Copper Development Association. 1973. *Copper Tube Handbook*. New York: Copper Development Association.

de Winter, F. 1973. *How to Design and Build a Solar Swimming-Pool Heater*. New York: Copper Development Association.

Dotter, B. 1974. "Top Inventor in Sun's Dream Department." *Our Sun* (Philadelphia), Summer 1974. Publication of Sun Oil Company.

Eibling, J.A. 1973. "A Study of Energy Storage." *Proceedings of the Solar Heating and Cooling for Buildings Workshop, March 21–23, 1973*, vol. 1. Washington, D.C.: National Science Foundation.

Hottel, H.C., and Woertz, B.B. 1942. "The Performance of Flat-Plate Solar-Heat Collectors." *Transactions of ASME*, February 1942.

Löf, G. 1963. "Heating with Solar Heated Air: Colorado House." *ASHRAE Journal*, October 1963.

Löf, G., and Tybout, R. 1970. "Solar House Heating." *Natural Resources Journal*, April 1970.

———. 1972. "A Model for Optimizing Solar Heating Design." Paper presented at winter annual meeting of the American Society of Mechanical Engineers, November 26–30, 1972 at New York.

McGraw-Hill Information Systems. 1973. *Sweet's Architectural Catalog*. New York: McGraw-Hill.

McLaughlin, R.H. 1974. "Solar Energy Research: Can it Survive the Congressional Hot Air Treatment?" *New Engineer*, July–August 1974.

Moorcraft, C. 1973. "Solar Energy in Housing." *Architectural Design*, October 1973.

National Bureau of Standards. 1974. *Method of Testing for Rating Thermal Storage Based on Thermal Performance*. Washington, D.C.: Center for Building Technology, National Bureau of Standards.

Shondiff, W.A., ed. 1974. *Solar Heated Buildings: A Brief Survey*. 6th ed. San Diego: Solar Energy Digest.

Swet, C.J. 1973. "A Prototype Solar Kitchen." Paper 73-WA/SOL-4, presented at winter meeting, November 1973, of American Society of Mechanical Engineers, at Detroit.

Telkes, M. 1974. "Storage of Solar Energy." *ASHRAE Transactions* 80b.

Thomason, H.E. 1972. *Solar Houses and Solar House Models*. Barrington, N.J.: Edmund Scientific Company.

Walton, Jr., J.D. 1973. "Space Heating with Solar Energy at the C.N.R.S. Laboratory, Odeillo, France." *Proceedings of the Solar Heating and Cooling for Buildings Workshop, March 21–23, 1973*, vol. 1. Washington, D.C.: National Science Foundation.

Whillier, A. 1964. "Thermal Resistance of the Tube-Plate Bond in Solar Heat Collectors." *Solar Energy*, July–September 1964.

Collector Design Information

ERDA (Energy Research and Development Authority). "List of Current Solar Collector Manufacturers." Technical Information Center, P.O. Box 62, Oak Ridge Tennessee 37830.

Knauer, Virgina. "Guidelines for Solar Collector Purchase." Office of Consumer Affairs, Washington, D.C.

Solar Energy Industry Association. "List of Collector Manufacturers." Suite 632, 1001 Connecticut Avenue, Washington, D.C. 20036.

WASTE

Wastes as resources

The methane digester:
why, how it works, its products
and what to do with them

The Clivus Multrum: what and how

Outhouses and septic tanks

Oxidation ponds: why, when, and how

Deciding which method is best for you

WASTE-HANDLING SYSTEMS

INTRODUCTION

We cannot escape our wastes. In the United States, we have developed this illusion of escape, but "out of sight, out of mind" simply does not work when we are talking about the day-to-day production of wastes. In controlled municipal settings, the disposal problem is an ever-growing headache. And in more rural environments, inadequate waste-handling techniques often have led to poisoned water and grave sicknesses.

We are talking, as you may gather, about more than litter and junk, which are avoidable through conservative use and recycling. We are talking about animal manures, food scraps, harvest dross, about the water that is lost each time we flush the toilet and the human excreta that go with it. No matter how conscientiously conservative we are, there still will be wastewater and solid wastes to dispose of. To say the least, it is a historical problem.

There are many ways of handling the disposal of human and animal wastes. What we would like to show you is how to deal with your wastes in a manner which can allow you to gain some positive benefit from them—to turn your wastes into resources. We have taken the approach here of providing basic information in a context and format which will allow you to make responsible, intelligent choices for yourself. We have gathered together information and data in a manner designed to allow computations, for example, on any possible combination of circumstances in the case of a methane digester. We consider the type of raw material available to the digester, appropriate nutrient balance, and expected gas production. In addition to the methane digester, such waste-handling techniques as the Clivus Multrum, outhouses, septic tanks, and oxidation ponds are considered. Composting, another excellent technique, is dealt with in Chapter 7.

Obviously, you cannot develop your own waste-handling facilities without paying a price. The cost is a combination of dollars and physical and mental effort, and there are certainly going to be trade-offs between them. You may not have had any previous experience with methane digesters, pumps, and oxidation ponds, and it may not be among the easiest of your experiences—but you can do it! Once you master the concepts, the calculations are simple and relatively straightforward.

METHANE DIGESTERS

Methane digestion has been used by sanitary engineers in the treatment of domestic sewage sludge and organic wastes for decades, but as a process found in nature, it is far older than man. Recently, its usefulness for waste reduction has received new attention among farmers here and abroad. Human excreta, animal manures, garbage, and even refuse—all previously thought of as undesirable, troublesome "wastes" of raw materials—can be digested under suitable conditions, resulting in the production of valuable bio-gas and fertilizer. So magic survives in the twentieth century: complete the cycle, close the loop, and "wastes" are transformed into new raw materials.

Why a Digester?

What is a methane digester anyway? A methane digester is nothing more than a container which holds our organic wastes in a manner which allows natural bacterial degradation of the organic matter to occur in the absence of oxygen. By causing this anaerobic (oxygen-free) process to occur in a container, we control the conditions inside and outside the container to promote the efficiency of the process and capture the bio-gas product. The fact is that the methane digestion process will occur quite naturally without our interference—all we do is try to improve the process somewhat. What else can a digester do for us besides provide some bio-gas? First, it rids us of our organic wastes—horse manure, human excreta, vegetable wastes, and so forth; and second, it converts these wastes into resources. These resources are (in addition to the bio-gas) sludge and effluent, both excellent fertilizing materials.

But bio-gas and sludge are only resources if you both need and are in a position to utilize them. What this means is that digesters basically are only suitable for a rural or semi-rural setting. If this is your situation and you are interested in building a digester, you will want to consider the costs and the potential returns before embarking on a digester trip.

Bio-gas consists of methane mixed with carbon dioxide, approximately two parts methane to one part carbon dioxide by volume, and with very small additional amounts of oxygen, nitrogen, hydrogen, carbon monoxide, and hydrogen sulfide. Any appliance that runs on natural gas, which is primarily methane, runs well on bio-gas pressurized in the proper range (2 to 8 inches of water or 0.07 to 0.3 pounds per square inch [psi]), including gas stoves, refrigerators, hot-water heaters, lamps, incubators, and space heaters. Butane and propane appliances also have been run on bio-gas, and it can be used to operate steam and internal-combustion engines, both of which can operate electrical generators. In *Mother Earth News,* Keith Gilbert reports: "There are currently available from Japan several models of steam engines which can be used for any number of things on the farm or in a small factory. They are quite inexpensive (the starting cost is about $100 for a small one), and will operate a wide variety of equipment including such things as: Electric generators, hammer mills, shredders, pumps, power saws for producing lumber, compressors, irrigation pumps, combines for threshing grain and beans, and other power machinery. One Japanese steam plant I observed was being used to operate a small saw mill and it did an effective job. It was a wood burner and cost only $60. Contact the Japanese Trade Legation for further information."

A small-scale digester will produce relatively small amounts of energy. We must maintain a perspective when we plan how to make use of the methane produced. As a point of reference, per capita consumption of natural gas in the United States is about 350 cubic feet per day. This represents about 30 percent of the total consumption of energy in this country for all purposes. It is clearly unreasonable, at normal consumption levels, to expect to drive a car or to take care of space heating with the output of a small digester, as you can see from the rates at which various appliances use methane (Table 5.1).

If we want to use a methane digester to provide all or some of our gas-related energy needs, we need to know what our present consumption is or is likely to be in the near future. An average family of five, using natural gas for cooking, heating water, drying clothes, and space heating, will consume on the order of 8000 to 10,000 cubic feet per month during the winter in California (mild winters). If space and water heating are handled by solar energy or by a wood-burning stove, and clothes are dried

Table 5.1 Rates of Use of Methane[a]

Use	Rate (ft^3)
Lighting, Methane	2.5 per mantle per hour
Cooking, Methane	8–16 per hour per 2- to 4-inch burner
	12–15 per person per day
Incubator, Methane	0.5–0.7 per hour per cubic foot of incubator
Gas Refrigerator, Methane	1.2 per hour per cubic foot of refrigerator
Gasoline Engine (25 percent efficiency)	
Methane	11 per brake horsepower per hour
Bio-Gas	16 per brake horsepower per hour
As Gasoline Alternative	
Methane	135–160 per gallon
Bio-Gas	180–250 per gallon
As Diesel Oil Alternative	
Methane	150–188 per gallon
Bio-Gas	200–278 per gallon

Notes: a. Adapted from *Methane Digester for Fuel Gas and Fertilizer*, by Merrill and Fry.

in the sun, the winter monthly consumption can be cut to about 2000 to 4000 cubic feet, easily within the range of a digester of moderate size.

If you presently use natural gas, you can estimate your requirements by looking at your utility bills for the last year (if you can't locate your bills, the utility company will usually provide you with copies). Then you can estimate cubic feet of natural gas used per month over an annual cycle, as well as the maximum monthly use (usually during winter). Once you have this estimate and when you are able to calculate the quantity of methane available from your waste materials, then you can evaluate the level of self-sufficiency you can achieve. Even if you are able only to satisfy half of your estimated needs, you may still consider it worth your while to build a digester as a waste-handling unit and energy supplement.

Natural gas in the United States now costs on the order of 1 to 2 cents per 10 cubic feet (in 1975). If we compare raw gas from a methane digester to natural gas, we find that natural gas has about 1000 Btu/ft^3 while bio-gas has about 600 Btu/ft^3; thus, it takes roughly one-third more bio-gas to give the same heat value as natural gas. If we consider a small digester producing on the order of 50 cubic feet of bio-gas per day (two-thirds methane), we save about $12 to $24 per year at current prices. It becomes obvious that small-scale digesters are not built primarily for savings on gas alone. If we consider other aspects of digesters—the value of the sludge and supernatant as fertilizer—our benefits are somewhat more apparent. If you are located in a remote area, the value of the gas as well as the fertilizer would be significantly increased because of transportation costs or the absence of a steady gas supply.

In quantity, commercial dried-sludge fertilizer costs close to $1 per 100 pounds. This is equivalent to about 100 gallons of wet sludge effluent from a digester, since a gallon weighs about 10 pounds and the effluent is almost 10 percent solids. Besides this modest return, you then have the added comfort of knowing that sludge from methane digesters operating on concentrated slurries is known to have a superior fertilizer value compared to sludges from municipal sewage-treatment plants (which operate on a very dilute input). Another factor you may consider is that, for many locations, the cost and labor involved in installing a septic tank and leaching field are avoided.

It is very difficult to estimate the cost of building a digester, as expenses vary with the ingenuity of the builder, the good fortune of finding used or surplus equipment, and the amount of labor hired out. Ram Bux Singh suggests a design for a digester capable of producing 100 cubic feet of gas per day, and calculates that it would cost $400 to build, as the construction is largely of concrete and masonry (in 1971). William Olkowski built a 300-gallon digester for $150, out of surplus tanks, valves, and hoses. John Fry was able to produce about 8000 cubic feet of methane per day with a digester initially costing $10,000. With gas at a value of 10 cents per 100 cubic feet, the value of Fry's daily gas production is about $8, giving an annual value of $2920. At this rate, he was able to pay for the cost of his digester within three years, giving him "free" gas from then on.

Even though our main concern in this book is not economics, the costs and returns of each potential energy source cannot be ignored and, in fact, must be weighed very carefully before you embark on a given project. If you are thinking seriously about designing a methane digester for a small-scale application, you must ask yourself one basic question: Is a small-scale digester practical? Will the methane digester provide you with enough gas and fertilizer to make it worth your time and money to build and maintain it? The answer to this question depends upon your available wastes and money and your willingness to alter your lifestyle to make the time to maintain the digester properly. The larger the scale of its operation, the more practical and feasible a methane digester becomes. Thus, what might not be practical for a single family may well be practical for a large farm or a small group of families. The ideal, of course, is an integrated energy-resource system. Figure 5.1 depicts such a system, where the methane digester constitutes a focal point for the other energy sources as a waste-to-resource converter.

The present state-of-the-art in methane digesters for small-scale operations is limited to homebuilt and owner-operated units. There are no commercially available units at the present time—you will have to design

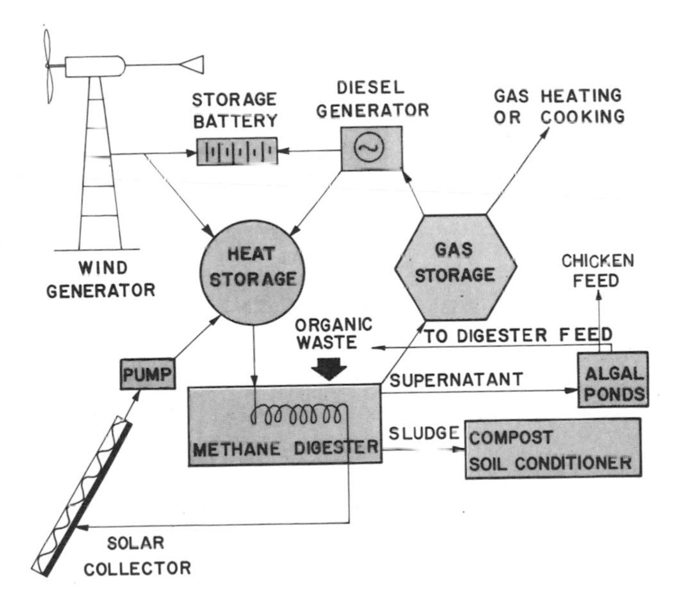

Figure 5.1 An integrated energy and resource system utilizing a digester.

and build your own methane digester if you want one. Your decision to build or not build a digester rests with an evaluation of your basic needs—waste-disposal and energy needs, fertilizer requirements, and any other appropriate considerations related to your particular situation. We can't give you a ready-made guide for evaluating your needs and requirements—you must do that. What we can say, however, is that if you do not own one or two horses or cows, or several hundred chickens, a methane digester is not going to be worth your investment. Beyond this lower limit, you must do some careful estimating. The design considerations in this chapter should help you out and there are several excellent publications on the construction and operation of small digesters for home or farm use. One of the most recent, *Methane Digesters for Fuel Gas and Fertilizer,* was produced in 1973 by the members of the New Alchemy Institute. The NAI work is based in part on an earlier, extremely helpful publication, *Bio-Gas Plant* (1971), produced under the direction of the Indian investigator Ram Bux Singh. A third very practical resource is *Practical Building of Methane Power Plants* by L. John Fry. All three publications are strongly recommended source material for anyone interested in evaluating, building, and operating a digester.

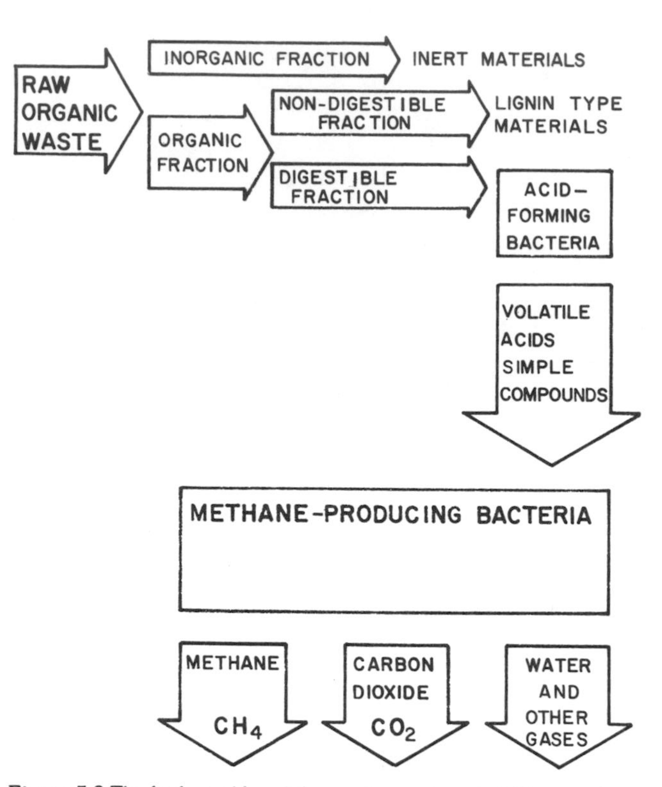

Figure 5.3 The biological breakdown of organic material in a methane digester.

The Digestion Process

Methane digestion is an *anaerobic* process; it occurs in the absence of oxygen. In anaerobic digestion, organic waste is mixed with large populations of microorganisms under conditions where air is excluded. Under these conditions, bacteria grow which are capable of converting the organic waste to carbon dioxide (CO_2) and methane gas (CH_4). The anaerobic conversion to methane gas yields relatively little energy to the microorganisms themselves. Thus, their rate of growth is slow and only a small portion of the degradable waste is converted to new microorganisms; most is converted to methane gas (for animal manures, about 50 percent is converted to methane). Since this gas is insoluble, it escapes from the digester fluid where it can then be collected and burned to carbon dioxide and water for heat. It turns out that as much as 80 to 90 percent of the degradable organic portion of a waste can be stabilized in this manner, even in highly loaded systems.

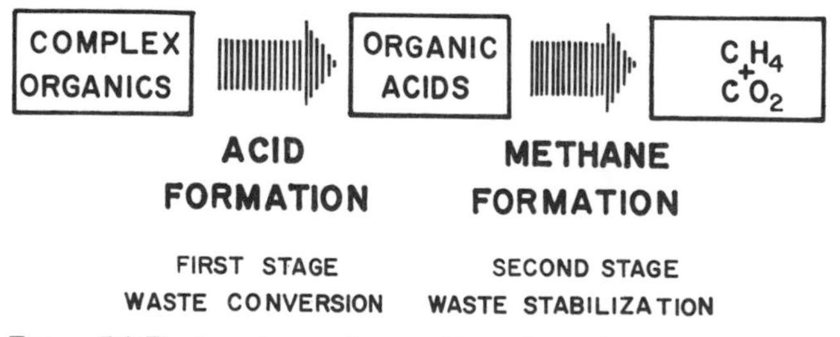

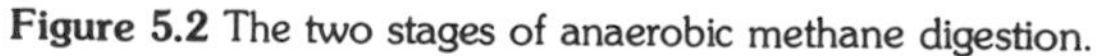

Figure 5.2 The two stages of anaerobic methane digestion.

Anaerobic treatment of complex organic materials is normally considered to be a two-stage process, as indicated in Figure 5.2. In the first stage, there is no methane production. Instead the complex organics are changed in form by a group of bacteria commonly called the "acid formers." Such complex materials as fats, proteins, and carbohydrates are biologically converted to more simple organic materials—for the most part, organic fatty acids. Acid-forming bacteria bring about these initial transformations to obtain small amounts of energy for growth and reproduction. This first phase is required to transform the organic matter to a form suitable for the second stage of the process.

It is in the second stage of anaerobic digestion that methane is produced. During this phase, the organic acids are converted by a special group of bacteria called the "methane formers" into gaseous carbon dioxide and methane. The methane-forming bacteria are strictly anaerobic and even small amounts of oxygen are harmful to them. There are several different types of these bacteria, and each type is characterized by its ability to convert a relatively limited number of organic compounds into methane. Consequently, for complete digestion of the complex organic materials, several different types are needed. The most important variety, which makes its living on acetic and propionic acids, grows quite slowly and hence must be retained in the digester for four days or longer; its slow rate of growth (and low rate of acid utilization) normally represents one of the limiting steps around which the anaerobic process must be designed.

The methane-forming bacteria have proven to be very difficult to isolate and study, and relatively little is known of their basic biochemistry. The conversion of organic matter into methane no doubt proceeds through a long sequence of complex biochemical steps. These complexities, however, need not concern us here; we can represent the overall process schematically (Figure 5.3) and derive the level of understanding necessary for our purposes. The two major volatile acids formed during the anaerobic treatment, as we implied a moment ago, are acetic acid and propionic acid. The importance of these two acids is indicated in Figure 5.4, which shows the pathways by which mixed complex organic materials are converted to methane gas.

Digester Design Process

There are two basic designs for the anaerobic process. One is the "conventional" process most widely used for the digestion of such concentrated wastes as animal manures and primary and secondary sludges at municipal treatment plants. The other process is one designed to handle more dilute wastes and has been termed the "anaerobic contact" process. These two process designs are depicted schematically in Figure 5.5. We will concentrate our discussion on the conventional process since most of our waste materials will come in concentrated form.

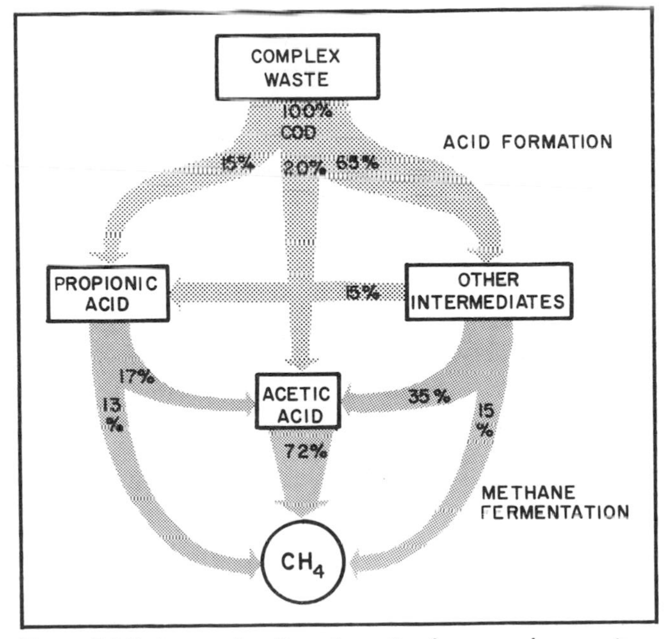

Figure 5.4 Pathways of methane formation from complex organic wastes.

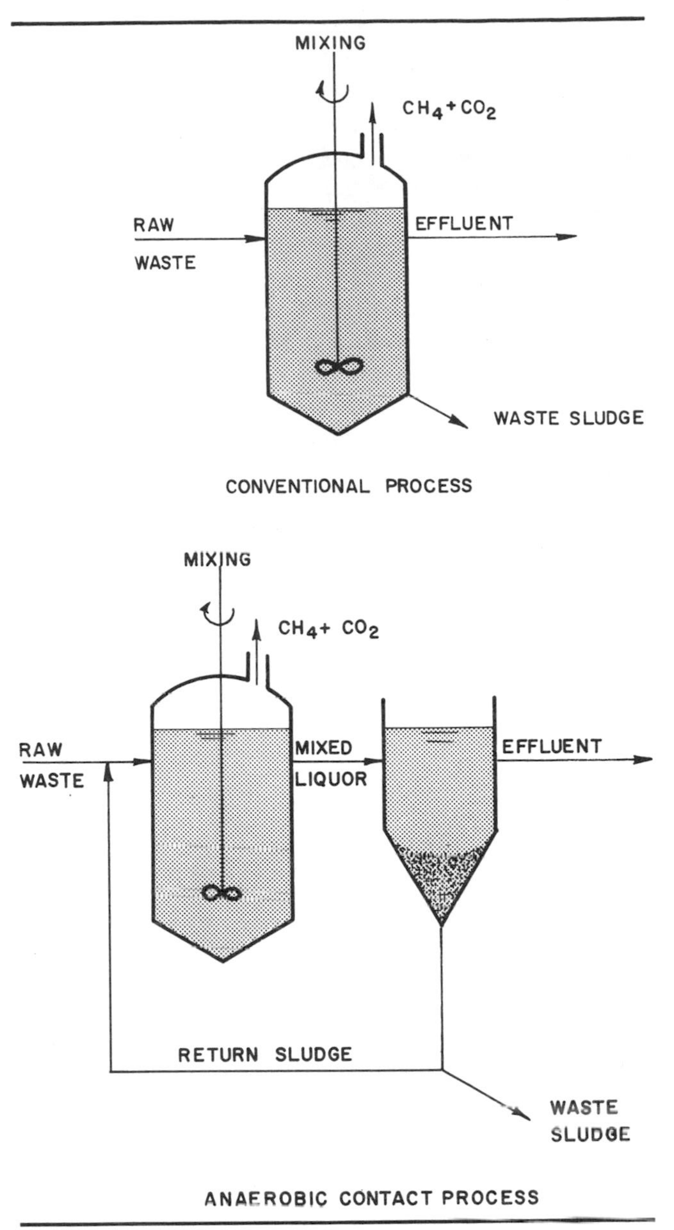

Figure 5.5 The two basic methane digester designs.

The conventional anaerobic treatment setup consists of a digestion tank containing waste and the bacteria responsible for the anaerobic process. Raw waste is introduced either periodically or continuously and is preferably mixed with the digester contents. The treated waste and microorganisms are usually removed together as treated sludge. Sometimes this mixture is introduced into a second tank where the suspended material is allowed to settle and concentrate before the sludge is removed.

There are many variations on a theme possible for methane digesters. A digester can be fed either on a batch basis (the more simple) or on a continuous basis, depending upon the trade-offs between initial cost, sophistication of design, and the cost of maintenance and operation. Any decision as to feed type also will depend on the projected scale of the operation. Another factor that you will need to consider is heat for the digester (to optimize the rate of methane production) and insulation of the digester tank (to reduce heat loss). These are cost considerations and will require you to do some estimating on your own specific design details once you have arrived at a tank size for your needs. A design decision which you will need to make early in the design process is whether you want a single-stage or a multiple-stage digestion system. This is really a question of how many tanks you want in series and is related to the residence time of the materials in the digestion system—details we will cover. Mixing of the digester contents is desirable since this agitation helps to increase the rate of methane production. The decision to mix mechanically or to allow natural mixing (much slower) depends again on a cost/benefit analysis which you must perform for yourself. A final consideration has to do with the manner in which you add your feed and remove your supernatant effluent (liquid by-product). There are basically only two possibilities: digesters with fixed covers and ones with floating covers.

Like any other design process, there are many interrelated factors involved in the design of a methane digester. Let us assume for purposes of discussion that we want to evaluate the design of a methane digester for a small five-acre farm—say, several cows, horses, and goats, and maybe fifty chickens. What kind of information must we have available before we can sit down and make our calculations? The first and most obvious consideration is the type and quantity of organic waste which can be used as feed for the digester. There will be a number of wastes—cow manure, chicken droppings, goat turds, and maybe the green trimmings from the vegetable garden (in late summer, early fall)—and we will need to know something about the composition of the various waste materials to insure that our friendly bacteria have a well-balanced diet (with special attention paid to nutrients such as nitrogen). Knowing the quantity and quality (composition) of the waste materials available as feed for the digester, we can calculate the mixture and size of our actual input into the digester (slurry feed). Then we can estimate the required size of the digester tank for specified conditions of temperature and residence time of waste in the tank—the average time that the waste is in the digester before leaving as sludge or supernatant liquid.

Once we know the quantity and quality of our organic wastes and the temperature and residence time of wastes in the digester tank, we will be in a position to estimate the amount of gas which will be produced—thus allowing us to pick a size for the gas collection tank. Your gas tank should be of sufficient size (several days' use) to insure that you will have gas available for occasional high-consumption use.

Notice that our attitude toward the methane digester is slightly different from the design procedures involved with deriving power from wind, water, or sun. We do not start out with design considerations to develop a methane digester to cover *all* your power needs; instead, we concentrate on your available resources in the way of waste products and investigate how much benefit you can derive from them. This is a reasonable attitude not only because here we are concerned with waste-handling, but also because of the economics involved. As we will see later, two cows and a horse can provide around 10 percent of an average individual's methane needs; if we then multiplied by a factor of 10 to accommodate 100 percent of our needs, we have a small ranch! So, all in all, we attempt to provide you with data to analyze what you have available right now; this same data, of course, *can* be used to calculate a 100 percent self-sufficient household or small community, if you have the requisite time, money, and waste already at hand.

Raw Materials

As we mentioned earlier, anaerobic organisms—bacteria that grow in the complete absence of oxygen—are responsible for converting the various organic raw materials into useful methane gas (CH_4), with carbon dioxide (CO_2) and water (H_2O) as by-products. Chemical analyses of anaerobic bacteria show the presence of carbon, oxygen, hydrogen, nitrogen, phosphorus, potassium, sodium, magnesium, calcium, and sulfur. This formidable list of elements, along with a number of organic and inorganic trace materials usually present in most raw materials used in methane fermentation, is essential for the growth of anaerobic bacteria; the hard-working bacteria must have a well-balanced diet if you expect them to perform at their best. Here, a well-balanced diet means adequate quantities of such nutrients as nitrogen and phosphorus. Nitrogen is generally the most important because it is the nutrient most likely to limit bacterial growth and, therefore, the rate and efficiency of methane production. A well-balanced diet for anaerobic bacteria also requires about thirty times more carbon than nitrogen, so you will need to know something about the carbon and nitrogen composition of the waste materials you are feeding your digester.

Phosphorus is third in importance only because smaller amounts are commonly needed. Although phosphorus and other elements that are found in even smaller percentages are necessary for growth, digesters are rarely inhibited by a lack of any of them because normal waste materials contain sufficient amounts to satisfy the bacteria's needs. (Interestingly enough, most detergents are "polluting" because they contain relatively large amounts of phosphorus compounds and other growth nutrients which stimulate growth of microorganisms and algae in

Table 5.2 Production of Raw Materials [a]

Average Adult Animal	Urine Portion [b]	Fecal Portion [b]	Livestock Units
Bovine (1000 lbs)	20.0	52.0	
Bulls			130.0–150.0
Dairy cow			120.0
Under 2 years			50.0
Calves			10.0
Horses (850 lbs)	8.0	36.0	
Heavy			130.0–150.0
Medium			100.0
Pony			50.0– 70.0
Swine (160 lbs)	4.0	7.5	
Boar, sow			25.0
Pig, over 160 lbs			20.0
Pig, under 160 lbs			10.0
Weaners			2.0
Sheep (67 lbs)	1.5	3.0	
Ewes, rams			8.0
Lambs			4.0
Humans (150 lbs)			5.0
Urine	2.2		
Feces		0.5	
Poultry			
Geese, turkey (15 lbs)		0.5 [c]	2.0
Ducks (6 lbs)		—	1.5
Layer chicken (3.5 lbs)		0.3 [c]	1.0

Notes: a. Adapted from *Methane Digesters for Fuel Gas and Fertilizer* (Merrill and Fry) and "Anaerobic Digestion of Solid Wastes" (Klein).
b. Pounds of wet manure per animal per day.
c. Total production.

receiving waters, leading to overproduction and stagnation.)

Assuming sufficient food is available in the proper form, other environmental conditions must also be satisfied for bacterial growth. The size and amount of the solid particles feeding the digester, the amount of water present in the feed slurry, how well the contents of the digester are mixed, the temperature range, and the acidity or alkalinity of the digesting mixture must all be favorable. Also, there are several kinds of waste that you must *not* use or your digester will cease to function properly. Let's start our discussion by looking into the composition of various organic wastes, and then we can talk about how to prepare a nourishing diet for your bacteria.

General Composition of Wastes

Table 5.2 is intended to give you a *general* idea of what to expect from animals as sources of energy. At least three variables (size of animal, degree of livestock confinement, and portion of manure collected) make this data nothing more than a series of rough approximations, but nevertheless it can be useful. Table 5.3 shows more specifically how the production of wet manure can vary (proportionally, in this case) with the weight of the animal. Manure deposited in open fields is obviously hard to collect and transport to the digester. NAI figures that open grazing during the day with confinement at night will yield about one-half of the total output as collectable raw material.

The *portion* of manure collected is important because urination is an animal's way of disposing of excess nitrogen and the nitrogen content of the raw material is a primary design consideration. Fecal material collected from cattle, horses, swine, and sheep in confinement may contain some or all of the urine output. You must make some judgment as to about how much urine is contained in the collected raw material. Human wastes are easily combined or separated, while poultry waste is produced in one combined load.

It is difficult to estimate the output of a garbage grinder in the kitchen. When sanitary engineers design treatment facilities for a city in which most people use garbage disposals, they increase the projected per capita output of sewage solids by 60 percent. If you have a garbage grinder, it might be practical to increase your daily output by about one-half (instead of 0.6) of the total sewage output of the household.

NAI uses a useful term called the *Livestock Unit* as a means of comparison of outputs between different kinds of animals (see Table 5.2). Taking the smallest tabulated output (that of a standard 3.5-pound layer chicken) as the unit of comparison, all other outputs are calculated as multiples of the chicken standard. Under this system, a common dairy cow is seen to be as valuable as 120 chickens from the standpoint of manure production. The Livestock Unit system is based on output of *digestible solids,* a term we must now define.

Wet manure or raw material is composed of both

Table 5.3 Hog Manure Production vs. Weight [a]

Hog Weight (lbs)	Feces [b]	Urine [b]	Total Manure [b]
40–80	2.7	2.9	5.6
80–120	5.4	6.1	11.5
120–160	6.5	8.1	14.6
160–200	8.5	9.1	17.6

Notes: a. From "Properties of Farm Animal Excreta" (Taiganides and Hazen).
b. Pounds per day.

water and solids. We define the solid portion *(total solids)* as the "dry weight" of the raw material, or the portion which would remain if the wet material were dried at a temperature of 212°F until no more weight was lost by drying (sun-dried manure still contains up to 30 percent water). It turns out that, of the total solids (*TS*), the fraction which would be digested by bacteria in a normal digester is proportional to the portion which would be *burned off* if it (total solids) were again heated, this time to about 1100°F, and kept at that temperature until once again there was no more weight loss. This digestible portion is called *volatile solids* (or *VS*). The remaining portion (after the water is evaporated and the volatile solids are burned off) is called the *fixed solids*. Figure 5.6 demonstrates these relationships for horse manure.

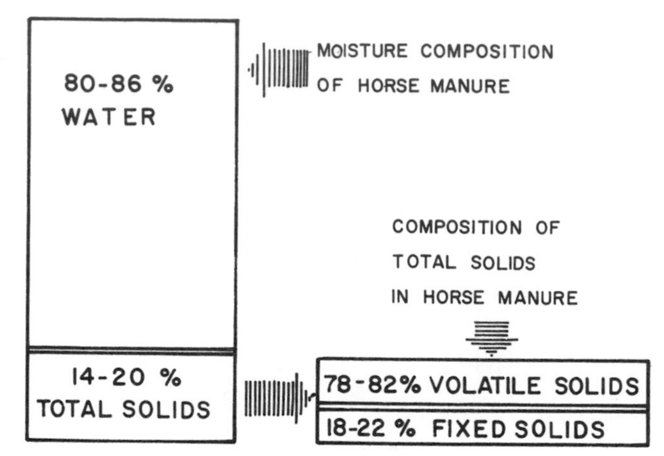

Figure 5.6 Moisture and solids content of horse manure.

Table 5.4 gives approximate values for the digestible portions of a variety of raw materials. These figures must be used realizing that the values have been arrived at under specific experimental conditions which may differ from those present in any other case.

The first part of the table (green garbage, kraft paper, newspaper, garden debris, white fir, average refuse, chicken manure, and steer-manure fertilizer) is drawn from an article by S.A. Klein (see Bibliography). Notice that his values for percent moisture are *extremely* low (green garbage has only 1 percent moisture). Except for the steer-manure fertilizer, all of his raw materials were freeze-dried before examination. His percentages of volatile solids are usable directly because volatile solids are taken as a percentage of dry *total solids*, not wet raw material. However, unless you can duplicate Klein's freeze-dried initial conditions (unlikely), his values for total solids should not be applied directly to your wet raw material.

There is a further complication with calculation of total solids of raw materials. If we assume that the raw manure is collected from animal enclosures, it is reasonable to assume that fresh fecal material will contain some urine as liquid. This urine content will raise the percentage of moisture in the manure above the "normal" level if the urine does not evaporate before the manure is used in the digester. A bit later we will detail the chemical significance of this addition of urine. The values in the rest of Table 5.4 are calculated with the assumption that the manure is used as a raw material for the digester *before* the urine component has evaporated significantly. If a rough separation of fecal material and urine is assumed, 20 percent total solids can be used as a good approximation for the fecal material for steers (fresh), horses, swine, and sheep. Separate values are given for human urine and fecal material because separation is more practical, and chicken manure will always contain the animal's urine excreta because, as we mentioned, the two components are deposited in the same load.

Table 5.4 Composition of Raw Materials [a]

Material	%Moisture	%Total Solids (*TS*)	Volatile Solids (% of *TS*)	%C	%N	*C/N* Ratio
Green garbage	1.0	99.0	77.8	54.7	3.04	18
Kraft paper	6.0	94.0	99.6	40.6	0	∞
Newspaper	7.0	93.0	97.1	40.6	0.05	813
Garden debris	24.8	75.2	87.0	–	–	–
White fir	9.3	90.7	99.5	46.0	0.06	767
Average refuse	7.3	92.7	63.6	33.4	0.74	45
Chicken manure	9.8	90.2	56.2	23.4	3.2	7
Steer manure (prepared fertilizer)	45.7	54.3	68.5	34.1	1.35	25
Steer manure (fresh) [b]	86.0	14.0	80.0	30.8	1.7	18
Horse manure [b]	84.0	16.0	80.0	57.5	2.3	25
Swine manure [b]	87.0	13.0	85.0	–	3.8	–
Sheep manure [b]	89.0	11.0	80.0	–	3.8	–
Human urine	94.0	6.0	75.0	14.4	18.0	0.8
Human feces	73.0	27.0	92.0	36.0–60.0	6.0	6–10
Chicken manure (fresh)	65.0	35.0	65.0	–	–	–

Notes: a. From "Anaerobic Digestion of Solid Wastes" (Klein) and *Methane Digesters for Fuel Gas and Fertilizer* (Merrill and Fry).
b. Includes urine.

Table 5.5 Gas Production as a Function of Volatile Solids[a]

Material	Proportion (%)	Cubic Feet of Gas[b]	Methane Content of Gas
Chicken manure	100	5.0	59.8
Chicken manure	31	7.8	60.0
& paper pulp	69		
Chicken manure	50	4.1	66.1
& newspaper	50		
Chicken manure	50	5.9	68.1
& grass clippings	50		
Steer manure	100	1.4	65.2
Steer manure	50	4.3	51.1
& grass clippings	50		
Steer manure	50	3.4	61.9
& chicken manure	50		
Steer manure	50	5.0	63.9
& sewage sludge	50		
Grass clippings	50	7.8	69.5
& sewage sludge	50		
White fir (wood)	10	9.3	68.9
& sewage sludge	90		
White fir (wood)	60	4.3	69.7
& sewage sludge	40		
Newspaper	10	9.9	67.1
& sewage sludge	90		
Newspaper	20	8.8	69.0
& sewage sludge	80		
Newspaper	30	7.5	69.5
& sewage sludge	70		

Notes: a. From "Anaerobic Digestion of Solid Wastes," by S.A. Klein.
b. Per pound of volatile solids (*VS*) added.

If you are an adventuresome experimenter, you can perform your own laboratory analysis for total solids by drying a known weight of wet manure in a kitchen oven at 212°F and measuring the weight lost (evaporated water). This weight loss is divided by the original wet weight and multiplied by 100 to give the percentage of moisture content in the original wet manure:

E. 5.1 $\% \text{ moisture} + \% \text{ total solids} = 100\%$

Figure 5.6 demonstrates this point.

We also should note that Klein's value for percentage of volatile solids for steer fertilizer is lower than the value for fresh steer manure (from NAI) because he is using manure which has been partially digested before examination, during the composting/preparation process involved in fertilizer production. His chicken manure (% total solids) is also high in comparison to the fresh manure value (NAI) because his extremely dry starting material will naturally have a higher content of total solids.

Tables 5.5 and 5.6 can be used to get a rough idea of the potential production of bio-gases (60 percent methane, 40 percent carbon dioxide) from typical raw materials if the volatile-solids content of the raw material is known. Equations presented in a later section can be used to determine the production of methane from other materials, if you know or can determine their volatile-solids content.

Table 5.6 Gas Production as a Function of Total Solids[a]

Material	Bio-gas (ft^3)[b]
Pig manure	6.0–8.0
Cow manure (India)	3.1–4.7
Chicken manure	6.0–13.2
Conventional sewage	6.0–9.0

Notes: a. From *Methane Digesters for Fuel Gas and Fertilizer* (Merrill and Fry). Note that *total solids* rather than *volatile solids* is used as the determinant of produced gas volume.
b. Per pound of total solids (*TS*) added.

Example: Let's calculate the cubic feet of methane per pound of raw chicken manure we can reasonably expect on the basis of information given in Table 5.5.

Solution: First we need information on percent *TS* and percent *VS* for fresh chicken manure given in Table 5.4. The cubic feet of methane per pound of fresh chicken manure is given by

$$\frac{\text{Cubic Feet of Methane}}{\text{Pound of Raw Material}} = 1 \times \%TS \times \%VS \times \frac{ft^3 \text{ Gas}}{lb} \times \frac{\% \text{ Methane}}{\text{Gas}}$$

E. 5.2
$$= 1 \times 0.35 \times 0.65 \times \frac{5\ ft^3}{lb} \times 0.598$$
$$= 0.68\ ft^3/lb$$

This gives us a rough estimate of the quantity of gas we can expect from a pound of fresh chicken manure, but what we eventually want is to be able to calculate the gas production for a mixture of raw wastes in a general way. Knowing specific information on gas production per pound of waste is helpful, but limited to the specific waste material.

Ram Bux Singh has recorded outputs of gas from vegetable matter that were seven times the output from common manures. NAI also found great increases in gas production brought about by the addition of fluids pressed from succulents (cacti). However, because the carbon content of plant material (compared to nitrogen content) is very high, the digestion of plant material releases a far greater percentage of carbon dioxide than the digestion of manures. Since carbon dioxide does not burn, we must view it as an impurity, and plant-generated bio-gas is therefore qualitatively inferior to manure-generated bio-gas.

Substances Inhibiting Digester Operation

Plant material also contains a high percentage of fixed solids, thus leaving a lower fraction of volatile solids. If the material itself is used instead of the pressed juices, scum formation in the digester will be greatly accelerated and can cause problems. McNary and Walford found that citrus peels ruined their digesters, due to the presence in the peels of the chemical inhibitor, d-limonene. Other substances that are known to inhibit the digestion process include heavy metals (zinc, lead, mercury, copper), high amounts of ammonia (above 1500 parts per million, or ppm), and the alkali elements (sodium, potassium, calcium, and magnesium). Another serious problem is the presence of sulfides, observed in many sewage-treatment digesters. Sulfides are easily detectable by their distinctive rotten-egg odor. It is very unlikely that any toxic materials will be present in small digester units utilizing manures and household wastes. If, however, toxicity is suspected, further study in the matter is required. P. McCarty has presented problems of digester toxicity in simple and concise terms and he is recommended for further reading on this aspect (see Bibliography). Later in the chapter, we also consider the digestibility of algae.

Carbon/Nitrogen Ratios

Our bacteria use carbon and nitrogen in the production of new cells and methane, and most people agree that, since carbon and nitrogen are used in the cells in the approximate ratio (*C/N*) of 30:1, the optimum ratio in the feed slurry also should be 30:1. By altering the composition of the inflow slurry, the digester can be "tuned" for efficient output. And, since the percentage of methane is in some ways determined by the carbon/nitrogen ratio of the feed, the quality of the gas also can be regulated.

The *C/N* ratio is difficult to establish for a general category of raw material because of the difficulty in testing for the *available* quantities and because the actual content of a material can vary with the maturity of the plants involved or the storage time of the manure. For these reasons, the figures given in our tables must be taken as approximations and guides rather than exact design parameters. It is the best we can do at this time.

Nitrogen is present in waste in many chemical forms, not all of which are equally available to anaerobic bacteria. The nitrogen in ammonia, for example, is more readily available (see Figure 5.7). Also, since urination is an animal's method of eliminating excess nitrogen, the amount of urine present in the manure will strongly affect the *C/N* ratio. Poultry waste is high in available nitrogen because urine and feces are excreted in the same load. Cattle and other ruminants (cud chewers) produce manure with an especially low nitrogen content since the bacteria essential to their digestion process live in one of their two stomachs and consume much of the nitrogen contained in the animal's diet. Vegetable waste is typically quite low in nitrogen content, while algae is quite high. Stable manure will usually be higher in nitrogen because it contains more urine than pasture manure (however, the straw included in stable manure can act to offset this increase because of its *low* nitrogen content).

Tables 5.4 and 5.7 give values for percent of *dry weight* (total solids) for nitrogen, since both volatile-solids and fixed-solids sources of nitrogen are available to the bacteria. The carbon percentages are for volatile or nonlignin portions whenever possible, again using dry weight. Using the weights of various raw materials and their *C/N* ratios, a recipe for a total-inflow *C/N* ratio of 30:1 can be derived from Tables 5.4 and 5.7. Further qualitative information about the importance of the *C/N* ratio in determining the quality of the bio-gas produced is provided in Table 5.8. Be sure that you are working with the *digestible* portion of the raw materials in your calculations. Singh recommends that the *C/N* ratio never exceed 35:1, but NAI notes that a level of 46:1 would be acceptable if it were unavoidable.

Although the addition of plant waste raises the carbon content of the inflow slurry significantly and aggravates the scum problem, it also tends to buffer the system at an alkaline level, protecting against a dangerous drop in pH to the acid level. Since overloading the digester with too much raw material lowers the pH and stops digestion if allowed to continue, the presence of plant material in the digester helps to protect the system from failure due to overloading.

Example: If we have 50 pounds of cow manure and 20 pounds of horse manure, what is the *C/N* ratio of the mixture?

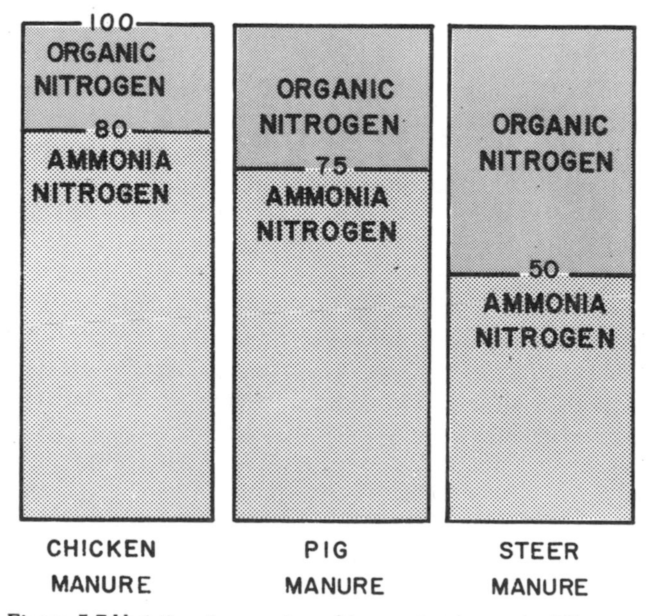

Figure 5.7 Variations in organic and inorganic nitrogen in different manures. The nitrogen of ammonia is more readily available.

Table 5.7 Nitrogen Content and C/N Ratio[a]

Material	Total Nitrogen (% dry weight)	C/N Ratio
Animal wastes		
Urine	16.0	0.8
Blood	12.0	3.5
Bone meal	—	3.5
Animal tankage	—	4.1[b]
Dry fish scraps	—	5.1[b]
Manure		
Human feces	6.0	6.0–10.0
Human urine	18.0	—
Chicken	6.3	15.0
Sheep	3.8	
Pig	3.8	
Horse	2.3	25.0[b]
Cow	1.7	18.0[b]
Steer	1.35	25.3
Sludge		
Milorganite	—	5.4[b]
Activated sludge	5.0	6.0
Fresh sewage	—	11.0[b]
Plant meals		
Soybean	—	5.0
Cottonseed	—	5.0[b]
Peanut hull	—	36.0[b]
Plant wastes		
Green garbage	3.0	18.0
Hay, young grass	4.0	12.0
Hay, alfalfa	2.8	17.0[b]
Hay, blue grass	2.5	19.0
Seaweed	1.9	19.0
Nonleguminous vegetables	2.5–4.0	11.0–19.0
Red clover	1.8	27.0
Straw, oat	1.1	48.0
Straw, wheat	0.5	150.0
Sawdust	0.1	200.0–500.0
White fir wood	0.06	767.0
Other wastes		
Newspaper	0.05	812.0
Refuse	0.74	45.0

Notes: a. From "Anaerobic Digestion of Solid Wastes" (Klein) and *Methane Digesters for Fuel Gas and Fertilizer* (Merrill and Fry).
b. Nitrogen is the percentage of total dry weight while carbon is calculated from either the total carbon percentage of dry weight or the percentage of dry weight of nonlignin carbon.

Table 5.8 C/N Ratio and Composition of Bio-gas[a]

Material	Gas: Methane	CO_2	Hydrogen	Nitrogen
C/N low (high nitrogen)	little	much	little	much
Blood				
Urine				
C/N high (low nitrogen)	little	much	much	little
Sawdust				
Straw				
Sugar and starch				
potatoes				
corn				
sugar beets				
C/N balanced (near 30:1)	much	some	little	little
Manures				
Garbage				

Notes: a. Adapted from *Methane Digesters for Fuel Gas and Fertilizer* (Merrill and Fry).

Solution: Cow manure is 86 percent water and 14 percent solids (Table 5.4). So its dry weight (50 × 0.14) equals 7 pounds of total solids. These dry solids are 1.7 percent nitrogen (also from Table 5.4) and so the weight of nitrogen (7 × 0.017) is *0.119 pounds*. Cow manure's C/N ratio is 18:1; so, since it contains 18 times as much carbon as nitrogen, it has (0.119 × 18) *2.14 pounds of carbon,* dry weight.

Using the same procedure for horse manure (84 percent water, 16 percent solids, 2.3 percent nitrogen by dry weight, and C/N ratio of 25:1—Tables 5.4 and 5.7), we find that horse manure has a dry weight of 3.2 pounds, and contains 0.074 pounds of nitrogen and 1.85 pounds of carbon, dry weight.

The total carbon content, then, is 2.14 pounds plus 1.85 pounds, or 3.99 pounds, dry weight. The total nitrogen content of the mixture is 0.193 pounds, dry weight. Thus, the total C/N ratio of the mixture is:

$$\frac{3.99 \text{ lbs}}{0.193 \text{ lbs}} = \frac{21}{1}$$

A C/N ratio of 21:1 isn't all that bad, but it's lower than the 30:1 we would like. A little bit of thinking will reveal the fact that we can never do better than a C/N ratio of 25:1, that of horse manure, because each time raw material is added, both the carbon and nitrogen components will be added proportionately. Obviously, the less cow manure we use (18:1), the better our C/N total will be; but we would lose total volume unless we found some horses to replace the cows.

Our hope for improving the C/N total beyond 25:1 lies in adding some raw material with a C/N of *greater than* 25:1 (or better yet, greater than 30:1)—wheat straw, for example (C/N = 150:1).

Assuming that we want to use all of our cow and horse manure and that we have some wheat straw to spare, take the total dry solids (10.2 pounds) with a total C/N ratio of 21:1 and combine it with some amount of wheat straw (you can guess at a figure for percent moisture; how about 10%?). Then use the above procedures to get the C/N ratio for your new mixture. Run through the calculations more than once to get a feel for the principles and an idea of how much wheat straw would be required to bring the total C/N up to around 30:1. (Try 1.5 pounds of wheat straw; you should get a C/N ratio of just about 30:1.)

Feed Slurry

Moisture Content, Volume, and Weight

For proper digestion, the raw materials must contain a certain amount of water, and experience with operating digesters has shown that a feed slurry containing 7 to 9 percent solids is optimum for digestion. To calculate the amount of water that must be combined with our raw materials to give this consistency, the moisture content of the raw material must be known.

Moisture values for various feed materials often used in digestion appear in Table 5.4. If there is any doubt as to the correctness of these values, or if the raw materials under consideration do not appear in this table, a direct determination of the moisture is usually feasible: we require only a small balance and an oven set at 212°F.

First, a small pan or plate is weighed and filled with a sample of the raw material. Care should be taken that the portion of the material used in this determination is representative of the entire batch to be used in the digestion process. Pan and raw material are then weighed together. After this step, pan and material are placed in the oven and allowed to dry until all the water has evaporated. When drying is completed, the pan and contents are allowed to cool in a dry place and are weighed once more. Percent moisture is given by

E. 5.3 $$\%M = \frac{W_i - W_f}{W_i - W_p} \times 100$$

where $\%M$ equals the percent moisture of the raw material; W_p equals the weight of the pan or plate (lbs); W_i equals the initial weight of the pan and sample, before drying (lbs); and W_f equals the final weight of the pan and sample, after drying (lbs).

Say the weight of your pan is 0.1 pounds, the initial weight is 2.1 pounds, and the final weight is 0.7 pounds. Then the percent moisture is

$$\%M = \frac{W_i - W_f}{W_i - W_p} \times 100$$

$$= \frac{2.1 - 0.7}{2.1 - 0.1} \times 100$$

$$= 70\%$$

Once the moisture content of the feed material(s) is known, the amount of water that must be added in order to give an 8 percent slurry (midway between 7 and 9 percent) can be calculated.

Let's first deal with the situation when the total weight of raw material we intend to put in the digester is known (by either weighing or estimation). We are going to introduce a long list of definitions and a very formidable-looking series of equations, but don't worry—we arrive at the other end with a few fairly simple formulas. Here are the definitions:

$\%M$ = moisture content of raw material in percent
M = moisture content of raw material as decimal fraction
$\%TS$ = percent of total solids in raw material
TS = total solids as decimal fraction of weight of raw material
W_r = total weight of raw material
W_s = weight of total solids in raw material
W_w' = weight of water already in raw material
W_w'' = total weight of water in 8 percent slurry
W_w = weight of water to be added to make 8 percent slurry
V_w = volume of water to be added to make 8 percent slurry
W_{sl} = total weight of slurry
V_{sl} = total volume of slurry
V_r = volume of raw material
D_r = apparent density of raw material

The moisture content of the raw material can be equated with the percentage of total solids in the raw material, either as percentages or as decimal fractions:

E. 5.4 $$\%TS = 100 - \%M$$

E. 5.5 $$M = \frac{\%M}{100}$$

E. 5.6 $$TS = 1 - M$$

The weight of solids in the raw material is given by the following two relationships:

E. 5.7 $$W_s = W_r TS$$

$$= W_r \times (1 - M)$$

To produce an 8 percent slurry, 8 pounds of solids should be mixed with 92 pounds of water, including, of course, the water already present in the raw materials:

E. 5.8 $$W_w' = W_r M$$

The total amount of water contained in an 8% slurry is

E. 5.9 $$W_w'' = 92\,\frac{W_s}{8}$$

$$= 11.5W_s$$

and the amount of water which you will need to add is the difference between the total moisture needed and the moisture already present:

E. 5.10 $$W_w = W''_w - W'_w$$

$$= 11.5\ W_s - W_r M$$

$$= 11.5\ W_r \times (1 - M) - W_r M$$

$$= 11.5\ W_r - 12.5\ W_r M$$

The volume of water to be added is

E. 5.11 $$V_w = \frac{W_w}{62.3}$$

$$= 0.1845\ W_r - 0.2\ W_r M$$

since water weighs 62.3 pounds per cubic foot at 60 to 70°F.

The weight of the 8 percent slurry can be calculated by adding the weight of water added to the original weight of raw material:

E. 5.12 $$W_{sl} = W_w + W_r$$

$$= 12.5\ W_r \times (1 - M)$$

and this must be equal to the sum of total water and total solids:

E. 5.13 $$W_{sl} = W''_w + W_s$$

By using 65 pounds per cubic foot as an average density of the 8 percent slurry, the volume of the slurry is

E. 5.14 $$V_{sl} = \frac{W_{sl}}{65}$$

$$= 0.192\ W_r \times (1 - M)$$

where V_{sl} is in cubic feet (1 gal = 7.5 ft³).

So, using Equations 5.10, 5.11, 5.12, and 5.14, we can figure out the weight and volume of the water to be added and the weight and volume of our 8 percent slurry, once we know the total weight of raw material and the moisture content (don't mix up M with $\%M$!):

$$W_w = 11.5\ W_r - 12.5\ W_r M$$

$$V_w = 0.1845\ W_r - 0.2\ W_r M = \frac{W_w}{62.3}$$

$$W_{sl} = 12.5\ W_r \times (1 - M)$$

$$V_{sl} = 0.192\ W_r \times (1 - M)$$

When the volume, density, and moisture content of a raw material are known (instead of the total weight), calculation of the volume of 8 percent slurry can be done in the following manner. The weight of raw material can be estimated by multiplying volume times density:

E. 5.15 $$W_r = V_r D_r$$

where D_r is in pounds per cubic foot. The volume of slurry then becomes

E. 5.16 $$V_{sl} = 0.192\ V_r D_r \times (1 - M)$$

The apparent density of the raw material can be estimated by weighing a known volume of material (without compacting the raw material) and dividing the weight by the volume.

When a mixture of, say, three materials ($\%A + \%B + \%C = 100\%$) is to be used in a digester, the moisture content of the mixture can be computed by

E. 5.17 $$M_{mix} = \frac{\%A \times \%M_a}{100} + \frac{\%B \times \%M_b}{100} + \frac{\%C \times \%M_c}{100}$$

where the subscripts denote the individual components of the mixture, A, B, and C. The weight of the mixture is

E. 5.18 $$W_{mix} - W_a + W_b + W_c$$

and the volume of the mixture is computed in a similar manner:

E. 5.19 $$V_{mix} = V_a + V_b + V_c$$

This same procedure can be followed for any number of mixture components.

Example: From two cows and a horse, you have 50 pounds of cow manure ($\%M = 86\%$) and 20 pounds of horse manure ($\%M = 84\%$). What is the volume of the slurry?

Solution: We must first compute the moisture content of the mixture. Total weight of mixture is

$$W_a + W_b = 50 + 20 = 70 \text{ lbs}$$

Percent cow manure in the mixture is

$$\%A = \frac{50}{70} \times 100 = 71.5\%$$

Percent horse manure in the mixture is

$$\%B = \frac{20}{70} \times 100 = 28.5\%$$

The percent moisture of the mixture is

$$\%M_{mix} = \frac{71.5 \times 86}{100} + \frac{28.5 \times 84}{100}$$

$$= 85.5\%$$

The volume of water to be added (from Equation 5.11) is

$$V_w = 0.1845 \times 70 - (0.2 \times 70 \times 0.855)$$

$$= 0.95\ ft^3$$

The weight of water to be added (also from Equation 5.11) is the volume times the density of water:

$$W_w = V_w \times 62.3$$

$$= 59.2\ lbs$$

The volume of the 8 percent slurry (from Equation 5.14) is

$$V_{sl} = 0.192 \times 70 \times (1.0 - 0.855)$$

$$= 1.95\ ft^3$$

Now we have come to a potentially confusing point. The raw material contains a lot of void spaces and when water is added to make up the slurry, it is "soaked" up by the raw manure—thus, the volume of added water and the volume of raw material *cannot* simply be added to find the volume of slurry.

Particle Size

The solids in your slurry should be in small particles so that bacterial action can proceed at a maximum rate (the previous calculations of the water needed to produce an 8 percent slurry assume that the solids are of sufficiently small size so that a slurry *is* produced!). Reducing the particle size also will facilitate transport of the slurry in pumps and pipes if these are used.

Manure does not require much reduction in the size of its solids—thorough mixing with water is sufficient in most cases. But when garbage, garden debris, or other kinds of refuse are to be digested, they should be shredded or chopped up by hand if a shredder is not available.

A good way to judge the proper particle size in a slurry is to observe how fast the solids settle out or if there are many solids floating on the surface after the water has been added. If there is a fast settling, the solids will accumulate on the bottom of the digester too quickly and make it difficult for the bacteria to do their work. In the case of floating material, the bacteria may never reach the solids to degrade them. With the raw materials normally used in digestion operations, flotation might be more of a problem than rapid settling. If the mixture is viscous enough, flotation can be avoided by proper mixing to entrap the particles. If flotation takes place, scum problems will appear in the working digester. L.J. Fry has studied the scum problem and has offered some useful information on handling the scum in digesters (see Bibliography).

Acid/Base Considerations (pH)

The term pH refers to the amount of acid or base present in solution. Too much of either can kill the methane-producing bacteria. As we described earlier, methane production is a two-stage process. In the first stage, one group of bacteria (the acid formers) utilize the organic matter of the feed solution (slurry) as a food source and produce organic acids. These acids are utilized in the second stage of digestion by another group of bacteria called methane formers. The methane formers utilize organic acids as food and produce methane. A balance of these two groups of bacteria must be maintained inside the digester at all times.

The methane formers multiply much more slowly than the acid formers, and this fact can result in an acidic environment that inhibits the growth of the methane formers. When you first start up your digester, such an imbalance is very likely. To help the situation, artificial means for raising the pH (to make the solution neutral—pH 7.0) of the feed have been successfully employed. Bicarbonate of soda can be used for this purpose at about 0.003 to 0.006 pounds per cubic foot of feed solution. This should be added to the slurry routinely during start-up and only when necessary while the digester is in full operation. Lime also can be used, but it is not as safe as bicarbonate and should be avoided if at all possible.

The pH of the feed slurry or the supernatant liquor can be determined in a number of ways. The most inexpensive ways are *pH paper* and the indicator *bromthymol blue.* Both may be obtained from a chemical supply house. The pH paper is easier to use, but it does not work well in the presence of sulfides, a common substance in anaerobic digesters.

Methane digestion will proceed quite well when the pH lies in the range 6.6–7.6. The optimum range is 7.0–7.2. In this range, a drop of bromthymol blue indicator will be dark blue-green in color (about one drop of indicator to ten drops of solution). If the mixture becomes more green, an acidic environment exists and bicarbonate of soda should be added. A deep blue color indicates a basic solution. In this case the cure is patience; in time, the digester will return to normal by itself.

Calculating Detention Time

Now that we have calculated the components of our feed slurry, we must find out *how long* the feed will need to remain in the digester to be processed. Temperature considerations play a large factor, so we will need to discuss them a bit. Then we can add in a safety factor and find out exactly how large a tank we will require. In order to find out the digestion time, we must return to a discussion of volatile solids and do a few preliminary calculations.

Chemical Oxygen Demand (COD)

Volatile solids, you will remember, are that portion of the total solids which burn off at a temperature of about 1100°F, and they represent the organic fraction of the total solids. Organics which can be decomposed by bacteria are called biodegradable. This portion is what the bacteria will use as a food source. It is also the portion responsible for methane production during digestion.

To calculate the minimum time required for digestion of certain raw materials, the *chemical oxygen demand* (the *COD*) of the feed slurry is required. This quantity represents the amount of oxygen required to oxidize—that is, to degrade or destroy—the organics by chemical means. In order to make the best use of available data and formulas, it is more convenient to express the *COD* in parts per million (ppm) than in pounds per cubic foot (where 1 lb/ft³ equals 16,000 ppm). We also require the moisture content and percentage of volatile solids of the raw material (both appear in Table 5.4). For mixtures of raw materials, the amount of volatile solids in pounds for each component *A*, is given by

E. 5.20 $$VS_a = W_a \times (1 - M_a) \times \frac{\%VS_a}{100}$$

where VS_a equals the weight of volatile solids of component *A* (lbs); W_a equals the weight of raw *A* used (lbs); M_a equals the moisture content of *A* as a decimal fraction; and $\%VS_a$ equals the percent of total solids that are volatile (from Table 5.4). The total amount of volatile solids then can be found by adding up the amounts of volatile solids of each component.

The concentration of volatile solids in the feed slurry (VS_{con}) is equal to the total amount of volatile solids present in the feed (VS_{total}) divided by the volume of the feed (which we calculated in the previous section):

E. 5.21 $$VS_{con} = \frac{VS_{total}}{V_{sl}}$$

The *COD* concentration of most materials can be approximated as equal to 1.5 times the volatile-solids concentration (or, of any specific sample, the total *COD* equals 1.5 times the volatile-solids total). But we must also account for the fact that only about 50 percent of the volatile solids are biodegradable:

E 5.22 $$COD = 0.5 \times 1.5 \times VS_{con}$$

If we want to express the *COD* in parts per million (after measuring the VS_{con} in pounds per cubic foot), the conversion equation is

E. 5.23 $$COD = 0.5 \times 1.5 \times 16{,}000 \times VS_{con}$$

$$= 12{,}000 \times VS_{con}$$

These last equations can also give us the *total COD* of a particular batch of feed, which we will need in later calculations.

Example: If 50 pounds of cow manure with 86 percent moisture and 80 percent *VS* is mixed with 20 pounds of horse manure having 84 percent moisture and 80 percent *VS* to produce, upon addition of water, an 8 percent slurry, calculate the biodegradable *COD* (in ppm).

Solution: As we calculated previously, the volume of this particular slurry is 1.95 cubic feet. Using Equation 5.20, the volatile solids of cow manure is

$$VS_{cm} = W_{cm} \times (1 - M_{cm}) \times \frac{\%VS_{cm}}{100}$$

$$= 50 \times (1 - 0.86) \times \left(\frac{80}{100}\right)$$

$$= 50 \times (0.14) \times (0.80)$$

$$= 5.6 \text{ lbs}$$

The volatile solids of horse manure, by the same equation, is

$$VS_{hm} = 20 \times (1 - 0.84) \times \left(\frac{80}{100}\right)$$

$$= 2.56 \text{ lbs}$$

so, the total amount of volatile solids in the mixture is

$$VS_{total} = 5.6 + 2.56 = 8.16 \text{ lbs}$$

The concentration of volatile solids in pounds per cubic foot (from Equation 5.21) is

$$VS_{con} = \frac{VS_{total}}{V_{sl}}$$

$$= \frac{8.16}{1.95} = 4.18 \text{ lbs/ft}^3$$

Then the biodegradable *COD* of the feed slurry in ppm is

$$COD = 12{,}000 \times VS_{con}$$

$$= 12{,}000 \times 4.18$$

$$= 50{,}300 \text{ ppm}$$

Solids Retention Time

Now that we have the *COD* of our slurry mixture, we are in a position to figure out how long it must remain in the digester. The solids retention time (*SRT*) is the average time that the incoming solids stay in the tank. We assume here that the mechanical design of our digester—its inflow and outflow schemes—is such that when new raw materials are introduced into the digester, they will replace old and digested material (a bottom inflow and top outflow arrangement, for example).

The *SRT* relates the digestion operation to the age and quantity of microorganisms in the system, and it is a sound parameter for design. An *SRT* of at least 10 days is a good rule-of-thumb value for the conventional digestion process. There also is a *minimum SRT* which reflects the ability of the microorganisms to consume the food source and reproduce themselves. If the *SRT* is less than the minimum *SRT*, you will literally wash out the bacteria faster than they can reproduce themselves and the digester will begin to lose efficiency. If the *SRT* is not increased, eventually the digestion process will stop. The minimum *SRT* is given by the following equation:

E. 5.24 $$\frac{1}{SRT_m} = \left[a \times k \times \left[1 - \left(\frac{K_c}{K_c + COD}\right)^{1/2}\right] \right] - b$$

where SRT_m equals the minimum solids retention time (days); *COD* equals the biodegradable chemical oxygen demand (ppm); *a* equals a constant showing how many bacteria can be produced per amount of food (*COD*) available (equal to about 0.04); *b* equals a constant showing how fast the bacteria die (equal to about 0.015); *k* is a factor for how fast the bacteria will consume food (depending on the temperature of digestion); and K_c equals the minimum amount of food required before bacteria can start multiplying (also dependent upon the temperature of digestion).

Values of *k* and K_c for the temperature range 59 to 95°F appear below:

Temperature	***k***	K_c
59°F	3.37	18,500
68°F	3.97	10,400
77°F	4.73	6,450
86°F	5.60	3,800
95°F	6.67	2,235

You can see from this table that favorable conditions for digestion increase with increasing temperature. If the digestion time is well above the minimum retention time, it has been found experimentally that the efficiency of the process (how much methane is produced) is about the same for digester temperatures ranging from 77 to 86°F. At 68°F it is a little lower and at 59°F it is about one-fourth the efficiency of the range from 77 to 86°F. Below 59°F very little if any methane appears to be produced. For the raw materials in our last example (50 pounds cow manure and 20 pounds horse manure) the *COD* was calculated as 50,300 ppm. Then, at 68°F for instance, the SRT_m of these raw materials is

$$\frac{1}{SRT_m} = \left[0.04 \times 3.97 \times \left[1 - \left(\frac{10{,}400}{10{,}400 + 50{,}300}\right)^{1/2}\right] \right] - 0.015$$

$$= 0.108$$

and the SRT_m is the inverse of this quantity:

$$SRT_m = \frac{1}{0.108}$$

$$= 9.25 \text{ days}$$

Therefore it will take about 9 days at a minimum to have the digestion of 70 pounds of combined manure going full blast.

Temperature Considerations

We just saw that temperature plays a very significant role in the digestion process. Anaerobic bacteria can operate either in a low or in a high temperature range. Bacteria that grow well in the range of from 77 to 95°F are called *mesophilic,* while bacteria that grow at higher temperatures (120 to 140°F) are called *thermophilic*. The higher temperatures required by thermophilic bacteria make them economically prohibitive for the small digesters we are considering here. Moreover, digestion within the thermophilic range produces a supernatant effluent much higher in colloidal (hard to settle out) solids, as shown below:

Characteristics of Supernatant from Laboratory Digesters	Thermophilic Range (130°F ± 10)	Normal Range (90°F ± 10)
Total solids (ppm)	0.309	0.231
Volatile solids (%)	67.1	58.5
Settleable solids (cc/1)	17.2	12.0
Suspended solids (ppm)	1490.0	773.0
Nonsettleable solids (ppm)	451.0	107.0

Therefore, only mesophilic bacteria are considered, and you should take the time to get to know your bacteria.

Mesophilic bacteria are very sensitive to temperature and temperature variations. This sensitivity, in turn, has a very noticeable effect on digester design and operation. As an example, consider the minimum solids retention time we just calculated, based on a temperature of digester operation of 68°F. At that temperature, the minimum solids retention time was calculated as approximately 9 days. By keeping all conditions of that example constant except for the temperature, you can calculate that the minimum solids retention time at 95°F is approximately 3 days! In other words, a 27°F increase in the digester operating temperature results in a reduction of the minimum solids retention time by two-thirds. This kind of impact in the minimum solids retention time will have an obvious impact on how fast the process can go and therefore on the size of the digestion and gas storage tanks needed. The higher temperatures result in a decreased detention time while more methane also is produced. So, the digestion tank will be smaller while the gas storage tank is larger. These considerations must be included in the selection of a safety factor, which we will discuss in a moment.

Any increase in temperature increases the rate of gas ebullition (bubbling) and so increases the solids bubbling about in the supernatant. There is not an appreciable change in this solids content over the temperature range of 70 to 95°F, but an unheated tank warming in the summer months may have rapid enough temperature changes and consequent changes in tank activity to show a high overflow of solids and even a complete overturn of the tank! Uniformity of tank temperature through controlled heating and insulation can reduce the possibility of any such unpleasantness.

Remember that, unless the digester is to be located in a tropical climate, temperatures of 95°F (optimum gas production) require artificial heating and/or insulation of the digester tank. The most common method of heating the tank uses methane-burning heaters, and this drain of methane production should be included in your calculations of total gas output. The heat losses and heat requirements of your digester must be calculated; if, for example, all methane produced has to be used for heating your digester, heating should not be employed. (For detailed calculations, see listings under Metcalf and Eddy, Eckenfelder and O'Connor, and Perry in the Bibliography.) An alternate method might involve the use of solar heating, although we've never seen it done.

Insulating the digester in some fashion will help to reduce extreme temperature variations. A number of digester designs place the digester totally or partially underground, so that the surrounding soil provides some insulation. Housing the digester inside a building also provides some protection against extreme temperature variations. This approach is, however, handicapped by the fact that feed slurry and sludge and supernatant products then have to be transported in and out of some structure, which might interfere with its other uses. Another pitfall of this approach is that the storage tank for the bio-gas should be located *outside* the building, to minimize explosion hazards inside.

Safety Factor

As in most design operations, a minimum is never taken as the basis for final calculations—it is multiplied by a safety factor before it is used to calculate other design parameters. Because it is difficult and time-consuming to get a methane digester operating properly, but it is easy to "kill" the digester operation by overloading, our safety factor is used to prevent accidental overloading.

The magnitude of a safety factor for the digestion process lies in the range of 5 to 100 and depends on the following considerations: 1) expected variations in temperature—the greater the variation, the greater the safety factor; 2) expected variations in raw materials (flow concentration and type of material)—dealt with as temperature variations; 3) other raw-material characteristics (*C/N* ratio, presence of phosphorus and other nutrients) —increase the safety factor as the *C/N* ratio moves away from 30 or the nutrient content drops; 4) competence of operators and attendance of the process; and 5) confidence in SRT_m value and the numbers used in calculating it.

We can express the use of the safety factor in the following equation:

E. 5.25 $$DT = SF \times SRT_m$$

where DT equals the digester detention time (days); SF equals the safety factor; and SRT_m equals the minimum solids retention time (days).

Unfortunately, no precise value can be given for the safety factor. Its usefulness rests on the fact that, as the degree of variation and uncertainty increases in factors important to the design and operation of any digester, the greater the potential for digester failure. We can only make some qualitative recommendations here and suggest that you do some experimentation and observe the results of your own digester. A reasonable rule of thumb for the safety factor is a minimum of 5 for a well-controlled digester with constant feed rate and composition, and a stable temperature of about 90°F. A safety factor of 10 would be advisable for a system with fluctuations in feed rate and composition, but a rather stable temperature at about 90°F. And finally, a safety factor of 20 or more for systems poorly controlled as to feed rate, composition, *C/N* ratio, or temperature.

Example: For the example considered thus far (50 pounds of cow manure and 20 pounds of horse manure), let us say that it is known that temperature variations will occur, but that the raw materials will not change appreciably as to nature and amounts. Further, let us suppose that methane production is employed only when these raw materials are available and that the digester is not attended extensively.

Solution: Based on these considerations, a reasonable choice of a safety factor will be about 20. Then

$$DT = SF \times SRT_m$$

$$= 20 \times 9.25 = 185 \text{ days}$$

Digester Characteristics

Calculating Tank Volume

Once the digester detention time is determined for our available raw materials, the volume of the digester tank can be calculated. In the case of a batch-system operation, the volume of the feed slurry (which we calculated previously) is equal to the volume of the digester. But in the case of a continuous or semi-continuous feeding system, the amount of raw materials processed per unit time must be known before we can determine the volume of the digester required. In either case, the following calculations *do not include* the volume of the methane gas produced. As we speak of it here, the digester volume is the volume that the slurry occupies excluding the gas that is produced during the process. Provisions obviously must be made to take care of this gas volume—a separate gas-collection tank or a floating-top digestion tank, for example—and we will consider these systems a bit later.

We should add that a reduction in liquid volume occurs in the digester once methane is being produced. This reduction will create volume variations inside the digester, but their effect is insignificant.

The equation we use to determine the volume of a digester tank (slurry feed, supernatant effluent, and sludge) for a continuous-feed operation is

E. 5.26 $$V_t = V_{sl} \times DT$$

where V_t equals the volume of the tank; V_{sl} equals the daily volume of the slurry concocted from the raw materials (from Equation 5.14); and *DT* equals the detention time (Equation 5.25).

Example: For the case of two cows and a horse we have been using as a sample, it is known that 70 pounds of combined manure is produced *daily* and that the volume of the 8 percent slurry is 1.95 cubic feet. The required digestion (detention) time also has been estimated as 185 days (with a safety factor of 20). What is the volume of the required digester tank?

Solution: The volume of the digester is

$$V_t = V_{sl} \times DT$$

$$= 1.95 \times 185$$

$$= 360 \text{ ft}^3$$

At 7.5 gallons per cubic foot, the tank should be able to hold 2700 gallons.

Now let's change our safety factor from 20 to 10 (that is, assume we have a much better controlled system). The new detention time is then 92.5 days, and the digester tank volume is then

$$V_t = 1.95 \times 92.5$$

$$= 180 \text{ft}^3 = 1352 \text{ gal}$$

Thus, for the same feed rate, the tank volume is a simple function of the detention time—double the detention time, double the tank volume.

Operation and Types of Digesters

First of all, digesters should be located to minimize the distances for transporting manure and wastes, for piping the gas, and for transporting the sludge and supernatant effluents. Secondly, since the digestion process can continue only under anaerobic conditions and since combinations of 5 to 15 percent methane in air are highly explosive, it is of the utmost importance that no air enter your digester with the incoming slurry. If the outside opening of the inflow pipe is well above the *highest possible* level of the liquid in the tank *and* the inside opening of the pipe to the tank is well below the *lowest possible* level of liquid in the tank *and* the inflow pipe is open to the atmosphere *only* during periods of slurry addition, the chances for air/methane mixing are minimized. Indian experimenters also have found that an 8 percent slurry is more dense than the digesting sludge in the tank. If the slurry is added at the bottom of the tank, its density will keep it below the older sludge until it, in turn, begins to digest and is pushed upward by new additions. This mode of addition affords the digestion process a natural mixing which supplements any mechanical techniques.

A pipe 3 inches in diameter should be sufficient for the inflow of the slurry. The pipe should be straight and without bends or the slurry is apt to cake on the inside and clog it, requiring periodic reaming to allow free flow.

An 8 percent slurry is very similar in consistency to cream. The raw material must be reduced to an adequate particle size and then must be well dispersed in the water medium before it is added to the tank. This mixing process may require a sturdy basin or trough to allow vigorous stirring.

Certain design characteristics determine your control over the quality of the supernatant in your digester. Single-stage (one tank) digesters can incorporate either fixed covers, floating covers, or multiple outflow valves (see Figures 5.8 and 5.9).

Fixed-cover digesters require that supernatant, or perhaps sludge, be removed for the introduction of fresh slurry. If the supernatant is removed simply by overflow through an outflow pipe, then the rate and time of excess liquid removal is obviously identical to the fresh slurry introduced (if no sludge is withdrawn at the same time). And, by the way, since continuous agitation or agitation due to the addition of raw slurry causes an increase in solids in the supernatant (due merely to mechanical action), slurry input should be slow and constant for the highest quality supernatant.

Supernatant preferably should be removed using an outflow valve at a more convenient and advantageous time—for example, before feed addition. Removal flow rates have to be high to prevent clogging of the valve. Floating-cover tanks offer the best timing control of supernatant removal: the supernatant need not be removed each time slurry is fed into the digester.

Two-stage digesters provide a far superior quality of supernatant. Here, the first tank is agitated and heated and the consequent overflow, full of suspended solids, runs into an unagitated tank for settling. The second tank need not be very large since it only serves the purpose of separating the solids from the mixture. Once the solids have settled, they can be recycled into the first tank to promote biological activity. Obviously, the two-stage digester system is preferred if your resources allow for construction of the second tank. However, problems can arise due to the septic properties of the supernatant. These properties can cause particle suspension in the settling tank if the settled solids are not removed periodically.

Overloading of a digester can cause serious impairment of sludge-supernatant separation. Rudolfs and Fontenelli (see Bibliography) studied the problem of overloading and we can summarize some of their findings here. They found an optimum loading rate for a two-stage digester operating at 82 to 84°F to be about 0.1 pounds of volatile solids per cubic foot of primary (the first tank) digester capacity per day. Doubling this loading rate increased the solids content in the supernatant to a point where sludge-supernatant separation was not easily achieved. For single-stage digesters, it was determined that even at loading rates of only 0.042 pounds of volatile solids per cubic foot per day, the supernatant contained about 3 percent solids. In sanitary-engineering practice, loading rates of 0.03 to 0.1 pounds of volatile solids per cubic foot per day are used for single-stage digesters with detention times on the order of 90 days. These figures will give you an idea of the range of loading rates which have been found experimentally to give reliable perform-

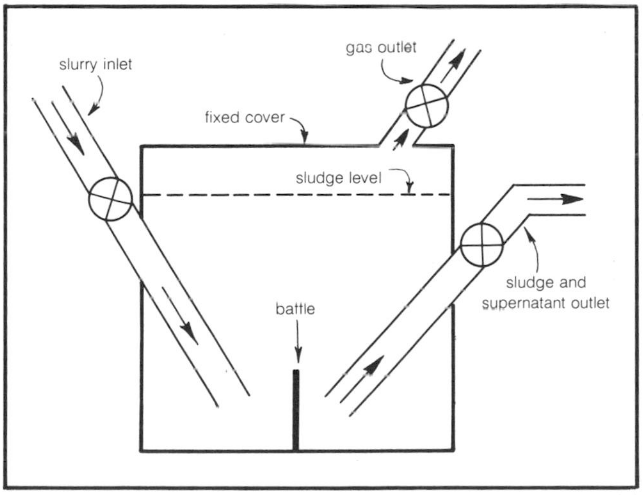

Figure 5.8 A fixed-cover digester with outlet placed at the desired sludge level in the digester.

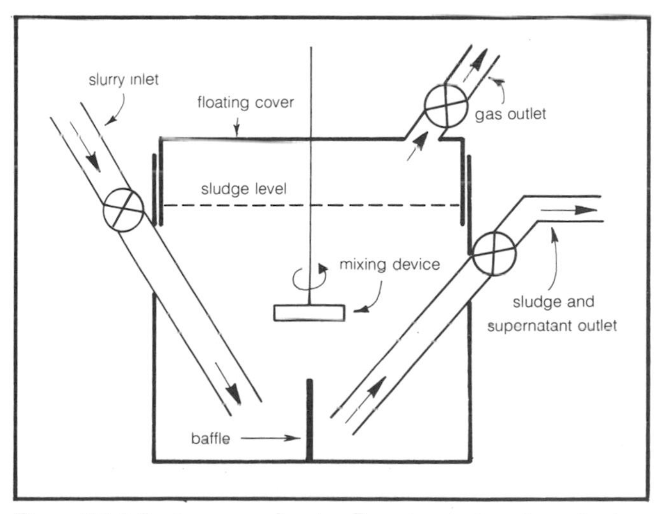

Figure 5.9 A floating-cover digester. For a two-tank system, simply connect two tanks in series. Mixing devices are useful, but optional.

ance from a digester. The engineering book by Metcalf and Eddy has a lot of useful detailed information if you are keen to go into greater depth and L.J. Fry has collected the best set of practical notes and experiences presently available concerning operation and maintenance of small-scale digesters (see Bibliography).

The importance of good sludge-supernatant separation is manifested first in terms of the effective capacity of the digester. With poor separation, solids—particularly fresh undigested solids—are much more likely to escape by overflow in the effluent. Because these solids are then no longer available as a fuel source, there is a decrease in digester efficiency and gas production. These factors, too, should be considered when selecting a safety factor for the digester. Poor separation also results in thinner and larger volumes of sludge. A sludge of 5 percent solids will contain twice as much water as a sludge of 10 percent solids and this excess water can create a handling problem, particularly in drying the digested sludge.

The sludge can be pumped out or, more simply, an outlet pipe 2 to 4 inches in diameter can be fitted as near to the bottom as possible, to allow for the periodic removal of the digested sludge into a dolly or wheelbarrow. Or, given the proper initial elevation, a large-diameter pipe can carry sludge directly into your garden or fields. A sludge outlet is not essential in batch digesters, though it simplifies unloading.

Start-up Considerations

A very important aspect of digester operation is the initial development of a good gas-producing sludge. You cannot overestimate the importance of this phase of digester operation.

It is advantageous to start the digester with a "seed" containing anaerobic bacteria. A sample of digester sludge from a properly operating municipal sewage-treatment facility or sludge from another methane digester would be ideal. Anaerobic muds from swamps or lake bottoms also can serve the purpose. If none of these is available, it would be wise to prepare a tightly sealed container of soil, water, and organic matter. This should sit in a warm place, about 95°F, for a few weeks. When the digester is ready to begin operation, everything but the gritty particles of soil should be decanted into the digester with as little exposure to air as possible.

We also must remind you about pH considerations. During start-up, the digester is often too acidic for the methane-forming bacteria. Review the section on pH for testing and rectifying techniques.

Products of a Digester

A normal unmixed digester will separate into layers as shown in Figure 5.10. Each of these layers can be used as a resource if the proper opportunities are available.

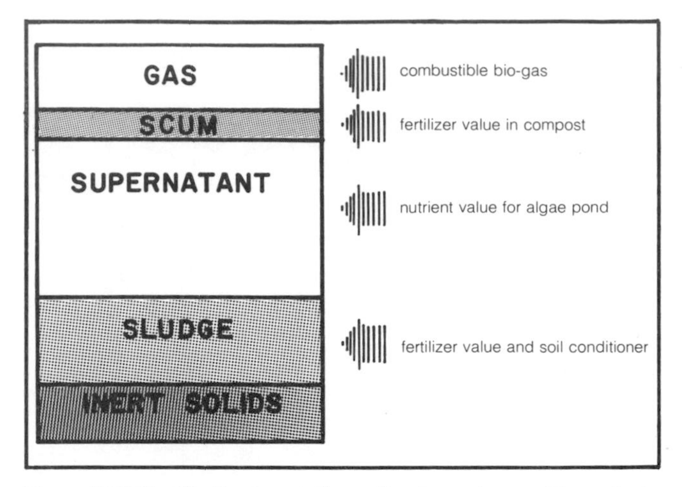

Figure 5.10 Stratification in a methane digester and uses of its products.

Methane gas, supernatant liquor or effluent (the liquid product of the digestion process), and sludge are withdrawn in a continuous or semi-continuous digester setup, at a volume rate equal to that of feed after an initial detention time period has elapsed.

Bio-gas and Gas Storage

Given the *COD* (assuming proper pH, temperature, and *C/N* ratio), an estimate of gas production can be made. We use the formula

E. 5.27 $$C = 5.62 \times [(e \times COD) - 1.42W_2]$$

where *C* equals the cubic feet of methane produced at 32°F and 14.7 psi (lbs/in^2) pressure; *e* equals the efficiency of raw-material utilization—how efficient the bacteria are in converting raw material to methane—with 6.0 being a recommended value; *COD* equals the total biodegradable chemical oxygen demand of the raw materials (this time we have a total, not a concentration: in pounds); and W_2 equals the weight of solids produced due to bacteria (lbs). This last factor we have not spoken of yet and, indeed, we *still* don't have to (see "Sludge"). The reason? For anaerobic decomposition the second term in the equation ($1.42W_2$) is small compared to the first term; and because of this comparative smallness, we can approximate our production using the shortened equation

E. 5.28 $$C = 5.62 \times e \times COD$$

You may recall from Equation 5.22 that the *COD* equals 0.75 VS_{total}; therefore

E. 5.29 $$C = 5.62 \times 0.6 \times 0.75\ VS_{total}$$

$$= 2.5VS_{total}$$

A nice, neat way to estimate the volume of methane gas!

For our 50 pounds of cow manure and 20 pounds of horse manure, the VS_{total} was 8.16 pounds. The methane produced will be

$$C = 2.5 \times 8.16$$

$$= 20.4\ ft^3$$

Remember, this is *methane* produced and bio-gas is only about two-thirds methane. The volume of bio-gas can be obtained by multiplying *C* by a factor of 1.5. Certain materials and their gas productions were listed in Table 5.6; also remember that the quality of the bio-gas will vary (see Table 5.8).

So, now we know that when the digester continuously is fed 70 pounds of combined manure daily (of a specified 50/20 mix), it will produce about 20 cubic feet of methane per day, after an initial period of time equal to the detention time. (At current prices for natural gas, this total is equivalent to 4 cents a day and would satisfy maybe 10 percent of the needs of an average individual.) The inherent assumption here is that all previously mentioned conditions—size solids, water content of feed slurry, etc.—are met. We also assume that the *C/N* ratio is favorable and that there are enough nutrients present. If the digester is not producing approximately the above calculated volume of methane, you should try to eliminate any possible malfunctions and insure that all assumed conditions are met. A bit later we will present a trouble-shooting summary. But now let's speak of how to collect and store the gas.

As methane is insoluble in water, it bubbles to the top of the digester tank. In order to maximize methane production and help eliminate oxygen from the system, the digester tank should be kept fairly full. Depending on your digester design, an additional tank for gas storage may be needed. Singh also suggests a few possible designs for gas storage tanks (see Bibliography).

A balance will have to be struck between the daily production of methane and the rate at which the methane is used. Since methane is a very dilute fuel, a compressor will have to be employed if periodic production of methane in considerable excess of the daily capacity to use it is anticipated. However, this seldom is the case. Ideally, the storage tank should be sufficiently large to hold at least several days' worth of optimal daily gas production. This capacity will allow some leeway in your rate of consumption as well as provide sufficient bio-gas for any short-term task that requires a high rate of energy input. Storage capacity certainly should be in excess of the anticipated peak daily demand. In the case of batch digestion, a few digesters with staggered digestion periods should be operated in order to maintain a relatively uniform rate of gas production.

For purposes of keeping oxygen out of the system and for maintaining a slight positive gas pressure, a floating-tank setup is best (see Figure 5.11). This is a concrete or steel tank filled with water on which the gas-holding tank floats. If concrete is used, the tank should be sunk in the ground to be able to withstand the pressure of the liquid inside. In any case, a below-ground design is desirable in colder climates to prevent freezing of the water; a thin oil layer on top of the water also helps to prevent freezing. The water-holding tank should be taller than the gas-holding tank, so that all the air can be flushed out of the gas-holding tank (by opening the top valve and pushing down on the floating tank until the top of the tank reaches the water level) before methane production begins. If you want, a few cups of lime can be added to the water in order to increase the fuel value of the gas by removing inert carbon dioxide—alkaline water will dissolve a larger quantity of carbon dioxide than water at neutral pH.

A simple apparatus can be set up to monitor the gas pressure in the gas storage tank (see Merrill and Fry). Or, a standard pressure gauge can be mounted on the gas line. The pressure can be controlled within a broad range by adding weight to the cover of the float tank or by using counterweights on pulleys. A compressor may be needed if large quantities of gas are needed at distances too far for piping (perhaps 100 feet, when we use weights on the cover to increase pressure) or to fuel a moving machine such as a car or rototiller. Small quantities of gas also can be made mobile by storing them in inner tubes.

Supernatant

For all practical purposes, the supernatant produced is approximately equal to the amount of water added to produce the 8 percent slurry that feeds the digester (see Equation 5.10). For removal techniques, see "Operation and Types of Digesters."

This supernatant can vary in color from clear,

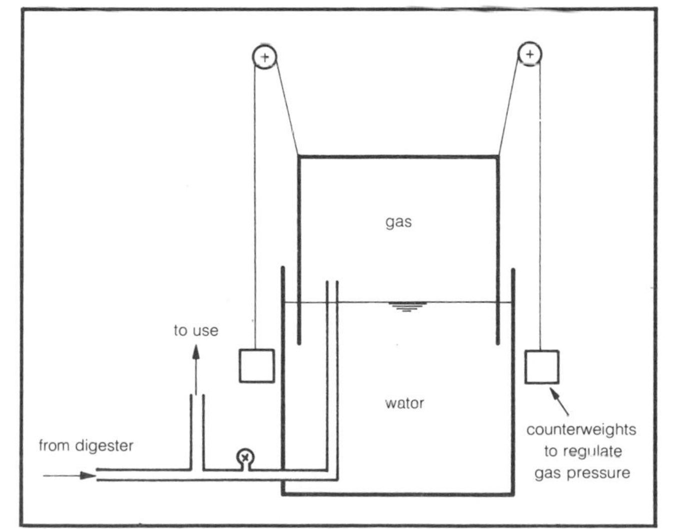

Figure 5.11 Floating-cover gas collection tank.

through various shades of yellow, to a very unsightly black. The odor may be unnoticeable or extremely offensive and nauseating. A properly operating digester has a clear (to slightly yellow) supernatant with few suspended solids and with no offensive odor.

As we discussed earlier in digester operations, various parameters can affect the quality of the supernatant resulting from digestion. As you would suspect, the characteristics of the raw slurry affect the supernatant: solids increase in the supernatant with the fineness of division of the slurry, and they also increase as the volatile-matter content of the slurry increases. This is due to greater activity and agitation in the digestion of these solids. Neither of these two factors is easily controllable. The amount of water in the slurry is, however, a controllable factor. Excessive volumes of slurry water produce excessive volumes of supernatant.

As we also have mentioned, two-stage digesters provide a far superior quality of supernatant. Single-stage, fixed-cover digesters are less expensive, but two-stage digesters offer vast improvements for a minimal extra cost. There are also temperature parameters, which we mentioned in the section "Temperature Considerations."

Supernatant has its uses—fertilizer for your fields, material for your compost pile, feed for your algae (see "Oxidation Ponds" in this chapter and also Chapter 7). But if you have an excess quantity, you must know what to do with it. In the past, supernatant was viewed primarily as a disposal problem rather than as a utilizable resource, and several methods were evolved for the liquor's disposal. For example, the supernatant was disposed of on sand beds. This method is unsatisfactory because of prohibitive costs due to clogging and odor problems. Centrifugation also has been used, but, on a small-scale basis, this is neither practical nor economical.

In large-scale digesters, the predominant method of disposal is to return the supernatant to the input of the digester. A.J. Fischer reports that this can cause problems in these digesters if the total-solids content exceeds 0.30 to 0.50 percent (see Bibliography), but this still is probably the most practical method for supernatant disposal in a small-scale, manual-fed digester: without perfect separation of the supernatant, the liquors can be recycled into the digester as solvent for making dry waste into a slurry.

Sludge

The sludge produced in a digester is a combination of the nonbiodegradable portion of the solids introduced into the digester and the amount of bacteria produced during the digestion process. After operation has started and the digester is running continuously, the amount of sludge solids produced by nonbiodegradable solids (W_1) is given by

E. 5.30 $$W_1 = W \times \left[1 - (0.5)\left(\frac{\%VS}{100}\right)\right]$$

$$= W_s \times (1 - M) \times \left[1 - (0.5)\left(\frac{\%VS}{100}\right)\right]$$

where W_s equals the weight of sludge solids produced due to initial solids (lbs); W_r equals the weight of raw material added (lbs); M equals moisture of raw material as a decimal fraction; and $\%VS$ equals the percent of volatile solids of the raw material. In case mixtures of raw materials are used, both M and $\%VS$ must be those of the mixture, as calculated previously.

The amount of sludge solids produced in the form of bacteria is

E. 5.31 $$W_2 = \frac{0.04 \times COD}{1 + (0.015 \times DT)}$$

where W_2 equals the weight of sludge solids of bacterial origin (lbs); COD equals the total biodegradable chemical oxygen demand (lbs); and DT equals the detention time of the digester (days).

The total amount of sludge solids produced is then

E. 5.32 $$W_{ss} = W_1 + W_2$$

Since the digested sludge is approximately 10 percent solids by weight, the amount of total sludge (10 percent solids + 90 percent water) is

E. 5.33 $$W_{sludge} = 10W_{ss}$$

If we assume an average density of 65 pounds per cubic foot for the digested sludge, its volume is

E. 5.34 $$V_{sludge} = \frac{W_{sludge}}{65}$$

Example: In the example considered thus far (50 pounds of cow manure mixed with 20 pounds of horse manure, with an average moisture of 85.5 percent, a $\%VS$ of 80 percent, and a VS_{total} of 8.16 pounds, digested for 185 days), how much sludge will be produced?

Solution: The amount of sludge solids produced due to incoming solids is

$$W_1 = W_r \times (1 - M) \times \left[1 - (0.5)\left(\frac{\%VS}{100}\right)\right]$$

$$= 70 \times (1 - 0.855) \times \left[1 - (0.5)\left(\frac{80}{100}\right)\right]$$

$$= 70 \times 0.145 \times [1 - (0.5 \times 0.8)]$$

$$= 70 \times 0.145 \times 0.6 = 6.09 \text{ lbs}$$

The biodegradable COD is given by

$$COD = 0.75VS_{total}$$

$$= 0.75 \times 8.16$$

$$= 6.12 \text{ lbs}$$

The sludge solids due to bacteria are

$$W_2 = \frac{(0.04 \times COD)}{[1 + (0.015 \times DT)]}$$

$$= \frac{(0.04 \times 6.12)}{[1 + (0.015 \times 185)]}$$

$$= \frac{0.245}{3.775}$$

$$= 0.065 \text{ lbs}$$

The total weight of sludge solids is

$$W_{ss} = W_1 + W_2$$

$$= 6.09 + 0.065$$

$$= 6.155 \text{ lbs}$$

The total weight of sludge is

$$W_{sludge} = 10W_{ss}$$

$$= 10 \times 6.155$$

$$= 61.55 \text{ lbs}$$

The total volume of the sludge is

$$V_{sludge} = \frac{W_{sludge}}{65}$$

$$= \frac{61.55}{65}$$

$$= 0.947 \text{ ft}^3 = 7.1 \text{ gal}$$

The above figures are for a single load in one day. If the digester is run continuously with 70 pounds of combined manure a day, the production figures represent production per day (for removal techniques, see "Operation and Types of Digesters"). And depending on the degree of agitation, the digestion period, and other factors, poor separation of the sludge and supernatant may occur. This larger volume of dilute sludge will require more labor to produce usable fertilizer.

Containing nitrogen (principally as ammonium ion NH_4), phosphorus, potassium, and trace elements, the digested sludge is an excellent fertilizer—it has a higher quality than the digested sludge from sewage plants, which use very dilute waste. It is also a very good soil conditioner. Recently, ammonium has been found to be superior to nitrate (an oxidized form of nitrogen and a standard nitrogen fertilizer) since it adsorbs well to soil particles and is therefore not as easily leached away, a serious problem with nitrate fertilizers. When exposed to air, the nitrogen in sludge is lost by the evaporation of ammonia (NH_3). Adsorption to soil particles can prevent this evaporative loss, and so the fresh sludge should be blended or mixed into the soil by shovel, fork, or tilling. If not used promptly, it should be stored in a covered container, or else stored temporarily in a hole in the ground and covered with a thick layer of straw.

The capacity of soils to take up sludge varies considerably. For example, sludge has to be spread more thinly on clay soils than on loamy soils, at least until the soil structure is improved. In any case, the soil should not be allowed to become waterlogged; waterlogging prevents aerobic microorganisms and processes from eliminating any disease-producing organisms which might not have been destroyed in the digester (if human waste was used).

Digested sludges produced from human waste should be used with some caution. Though the area is not well studied and no firm information is available, it is suggested by K. Gilbert (see Bibliography) that when batch digestion is carried out, a digestion period of at least three months is desirable so that adequate destruction of pathogenic organisms and parasites occurs. However, the minimum digestion period required for adequate sanitation has not yet been definitely demonstrated. And since mixing occurs in a continuous-flow digester, there is no way to insure that all introduced material will undergo a lengthy digestion period.

Incorporation of the sludge into the soil provides a set of conditions unfavorable for pathogenic organisms that thrive in the human body, but common sense dictates that sludge should not be used on soil growing food (excluding orchards) to be eaten raw. The fertilization of these soils with sludge should be done several months before planting and preferably used on land not to be cultivated for at least a year to insure complete exposure to aerobic conditions.

Digestibility of Algae

Because of a growing interest in the subject, we include a few separate remarks about algae. Difficulty has been encountered in the anaerobic digestion of algae. This is due to several factors. Firstly, algal material is highly proteinaceous. As a result, high ammonia concentrations arise in the culture media, pH increases, and bacterial activity decreases. Other problems arise due to the resistance of living algal cells to bacterial attacks. However, if algae is not the only feed source (mix it with manure, for example) for the digester, these problems virtually disappear. Algae also is a good source of carbon for balancing the nutrients of your slurry.

Further promising aspects of algae digestion are that alum-flocculated algae (see "Harvesting and Processing of Algae" in this chapter) digest just as well as algae that do not contain the 4 percent inorganic aluminum. Detention times as short as 11 days are possible, and variation of the detention time from 11 to 30 days has little effect on gas production. Loading rates can be as high as 0.18 pounds of volatile solids per cubic foot of digester capacity per day without deleterious effects. Digesters using algae also are much less affected by variations in loading rates.

For raw sewage sludge, there are 9.2 to 9.9 cubic feet of gas produced per pound of volatile matter introduced. For algae at mesophilic temperatures, only 6.1 to 7.0 cubic feet of gas is produced per pound of volatile solids.

The sludge produced by algae has undesirable characteristics due to the fact that it is not completely digested. There is an odor problem not encountered in sewage sludges. The algal sludge is highly colloidal and gelatinous. As a result, it dewaters poorly and disposal becomes a problem.

The use of algae to capture energy from the digester supernatant and the sun has some future possibilities (see "Oxidation Ponds"). Major problems now involve the conversion of energy stored in algae to usable forms. Digestion seems marginally applicable to such conversions.

Summary of Methodology

So far we have considered the various aspects of digester design separately. Now if we are to pull everything together, we can summarize the design process in a series of steps as follows:

1. Knowing the daily weight of available wastes to be used as digester feed, calculate the characteristics of the waste mixture, W_{mix}, $\%M_{mix}$, VS_{mix}, using Tables 5.4 through 5.7.
2. Calculate the *C/N* ratio for the mixed waste and make any adjustments which are necessary to achieve a reasonably balanced diet (*C/N* = 30:1 optimal).
3. Compute the *COD,* choose a projected digestion temperature, and then compute a minimum solids retention time.
4. Pick a safety factor appropriate to the situation and compute the detention time and then the volume of the digester.
5. Make your design decisions about the nature of your digester (one-tank or two-tank, fixed-cover or floating-cover, etc.).
6. Estimate the daily rate of gas production from VS_{total}, subtract gas necessary to heat tank (if applicable), and size collection or storage tank.

Indications of Poor Performance and How to Avoid It

A good indication of poor performance is the amount and quality of bio-gas your digester produces. If gas production is well below the value calculated using VS_{total}, the digestion is not proceeding at the optimum rate. When the carbon dioxide (CO_2) content of the bio-gas exceeds 50 percent, the digester is performing poorly. In both cases, corrections can be made to improve the digester operation.

The percentage of CO_2 in your bio-gas can be found either by devising a homemade gas analysis unit or by purchasing a commercial kit. A rather simple, easy-to-use manual gas analyzer is made by the Brenton Equipment Company (P.O. Box 34300, San Francisco, California 94134), called the Bacharach Duplex Kit. This analyzer is available in a form for measuring CO_2 in the 0 to 60 percent range, perfect for digester analysis.

If you prefer to devise a system for yourself, you will need some way of measuring the gas volume. Displacement of water inside a container of known volume can be employed if the container is marked at different volume capacities. If a known volume of bio-gas is bubbled through a lime solution, the carbon dioxide of the bio-gas will react with the lime and thus will be removed from the bio-gas mixture. If the volume of the gas remaining after bubbling through the lime solution is measured, the percentage of carbon dioxide in the bio-gas mixture is given by

E. 5.35 $$\%CO_2 = \frac{(V_{bb} - V_{ab})}{V_{bb}} \times 100$$

where V_{bb} equals the volume of bio-gas before bubbling through lime and V_{ab} equals the volume after bubbling.

Factors that cause poor digester performance or even complete failure include:

1. Sudden change in temperature (either due to climatic changes or failure of the heating system if one is used).
2. Sudden change in the rate of loading (how fast raw materials are introduced into the digester).
3. Sudden change in the nature of raw materials (materials or mixtures of raw materials other than what is routinely added).
4. Presence of toxic materials.
5. Extreme drop in pH (the digester has become acidic).
6. Slow bacterial growth during the start-up (especially important at initial stages of operation).

In case of poor performance or failure, the following steps should be followed:

1. Provide pH control.
2. Determine the cause of the upset: improper environmental conditions (pH, temperature); nutrient insufficiencies (*C/N* ratio, phosphorus); or toxic materials present (limonene, heavy metals, sulfides).
3. Correct the cause of the imbalance.

Poor operation can cause foul odors in the bio-gas, sludge, and supernatant. With proper and careful operation, such problems are minimized. The case of poor performance and failure is a whole study in itself and hardly enough material can be presented here on the subject. P.L. McCarty (see Bibliography) gives a good review that is concerned mainly with sewage-treatment digesters, but also is applicable in every case.

Safety Considerations

Methane/air mixtures are explosive when methane is present in 5 to 15 percent by volume. In an atmosphere of an inert gas (such as the carbon dioxide in bio-gas), oxygen must be present at least to the extent of about 13 percent before an explosion can occur. Obviously, you must take precautions to prevent explosive mixtures from occurring. Although no accidents have been reported in the literature for small digestion units (they rarely are. . .), it is highly advisable that the entire gas-handling system—piping, valves, storage tank, and so on—be designed with the utmost care. You should give special consideration to any possible leaks that might develop at any point of gas transport or storage. Needless to say, methane is merely another name for the natural gas commonly used for home cooking and heating. It therefore should be handled and used with the same caution. There are numerous examples of asphyxiation and death due to gas leakage from stoves and other household devices.

How supernatant and sludge are used becomes critical when human excreta are used as raw materials. Very little is known about the fate of pathogenic organisms during an anaerobic digestion process. The direct application of sludge and supernatant as fertilizer material is not recommended on vegetables or any other plants which are consumed by humans. Using them in orchards is quite safe, however. If no human excreta are used, both sludge and supernatant are safe for use anywhere as fertilizer material. In the case of pig manure, the precautions stated for human excreta apply, since certain pathogens common to pigs are transmittable to humans.

In case the water table is near the surface of application, sludge and supernatant should not be used as fertilizer but rather transported where there is no possibility of the sludge and/or supernatant leaching through soils into the groundwater. If this is not possible, provisions should be made for drying these products in impermeable basins and using the resulting dry solids in a manner that avoids groundwater contamination.

Final Thoughts

Your decision to build or not to build a methane digester ultimately will be based upon an analysis of the costs and benefits to you and your willingness to alter your lifestyle sufficiently to be compatible with the day-to-day operation of a digester. If you construct a digester, it should be planned and designed in a manner that takes into consideration the potential impacts—visual, physical, and chemical—on the environment. A digester potentially can free you from total dependence on your local utility company, but it surely will tie you to the routine maintenance and operation of the digester itself. Digesters are not for everyone, but if they fit into your lifestyle, we wish you well and hope that you produce the best gas around!

OTHER WASTE-HANDLING TECHNIQUES

In the first section we considered the use of a methane digester as both a waste-handling technique and as a potential waste-to-resource converter. But a digester does not fit into everyone's plans and alternative approaches to waste handling must be considered. Several alternatives are explored in this section, including outhouses, septics tanks, oxidation ponds, and a device called the Clivus Multrum. Composting is also an excellent technique; it is discussed at some length in Chapter 7.

The types of wastes considered are those commonly produced by any normal household: human excreta, food wastes, etc. Solid wastes such as paper, plastics, and wood are not considered here. The type of setting where these alternative waste-handling techniques might be used is more rural than urban; we assume that sewer lines are generally unavailable, but that land availability is not a problem.

Before moving on to a discussion of specifics, it is worth taking a few moments to look at the current state of affairs regarding household wastes. Tables 5.9 and 5.10 summarize data on the average waste concentration and volume generated by present methods of disposal using a water-carried sewage system. Note that a family of four requires on the order of 50 gallons of water per person per day, most of which ends up as wastewater; if we somehow can utilize the toilet in a manner which does not require water, we will have saved about 40 percent of our normal water requirements. Moreover, since it is easier to treat concentrated wastes, a good portion of our waste-handling problems are then also solved.

Clivus Multrum

The Clivus Multrum was introduced in Sweden some twenty years ago by Richard Lindstrom, but, although the unit is relatively cheap, simple, and fairly easy to install, the Clivus has not become popular in the United States and is, in fact, generally unknown in this country. Perhaps the affluent American, being accustomed to water-flushed toilets, is turned off by the prospect of having his wastes decompose directly beneath him in the cellar.

The Clivus Multrum is primarily intended for use in single-family houses—each unit is capable of handling the wastes of about four or five people. The unit consists of a container with a sloping bottom, and center and top sections (see Figure 5.12). The length of the Clivus container is divided by vertical partitions into three chambers which are interconnected at the bottom. The excrement and refuse chambers are equipped with vertical tubes, one leading to the toilet (specially designed) and the other to a garbage chute. The decomposition products eventually find their way to the third (lowest) chamber, where they accumulate.

The Clivus Multrum container has openings which admit air and the air moves through the channels into the waste mass. The air inlet is situated in the end wall of the storage chamber and is usually provided with a damper to regulate air flow. The air outlet is at the highest point in

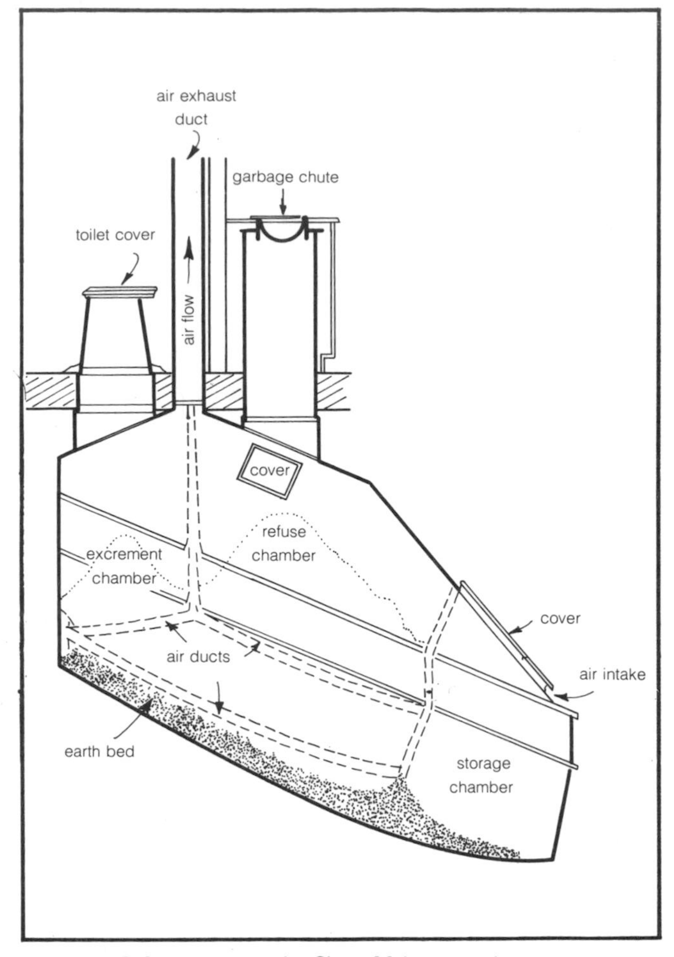

Figure 5.12 Cross section of a Clivus Multrum in place.

the container and is connected to an exhaust duct which extends above the roof. The exhaust duct opening should be larger than the air inlet to assure continuous ventilation.

During the decomposition process, the waste material generates heat which, in turn, warms the passing air, causing the air to rise through the exhaust duct. For thorough ventilation and to prevent water vapor from condensing on the duct walls, it is essential that the heat

Table 5.9 Average Waste Loads and Wastewater Volume from a Domestic Household with Four Members[a]

Wastewater Event	Number per day	Water Vol. per use (gal)	Total Water Use (gal)	*BOD* (lb/day)	Suspended Solids (lb/day)
Toilet	16	5	80	0.208	0.272
Bath/Shower	2	25	50	0.078	0.050
Laundry	1	40	40	0.085	0.065
Dishwashing	2	7	14	0.052	0.026
Garbage disposal	3	2	6	0.272	0.384
Total			190	0.695	0.797

Notes: a. Adapted from "Household Wastewater Characterization," by K. Ligman, et al.

be conserved by insulating the Clivus container as well as the entire exhaust duct. The exhaust duct must not be exposed to temperatures lower than those inside the duct; that is, it should not run through an unheated attic or above the roof of your building. In colder regions, you can maintain a warm temperature in the exhaust duct by locating it next to a heater duct or inside a chimney (Figure 5.13). Also, the exhaust duct must be of sufficient height to insure a proper pressure drop for draft requirements. At least one author contends that a two-story house provides the necessary height, leaving split-level homes with a problem.

Another major problem with the Clivus seems to be the presence of unpleasant odors. With the ducts arranged as we described above, natural ventilation should take all odors out through the exhaust duct. However, this seems to be a rather fragile process, subject to periodic difficulties. Toilet and garbage-disposal covers must be tightly closed when not in use to maintain the normal flow of air through the system. Although a down draft should theoretically exist, air entering from an open disposal cover will directly replace (into your home) the gases above the waste. And the channels that are specifically designed to facilitate aeration of the compost do not work if there is continual leakage of air through the disposal lids.

Pests should not pose much of a problem, since fine-meshed nets cover all air and gas intakes and outlets. However, these nets need constant inspection or your cellar could turn into a breeding ground for flies, worms, and other creatures. According to Lindstrom, a year at 90°F generally kills any undesirables in a Clivus; this seems to be an unproven statement, however, and warrants attention before you actually install one.

Cold locales seem to be at a disadvantage because the composting process is temperature-dependent and natural drafts often bring in air of subfreezing temperatures. Consequently, electric heaters often must be used to control the temperature and these accesories increase the cost. You must also watch them carefully: the heat cannot be inordinately high or the mediating bacteria will be unable to function. Additionally, the house itself must be well insulated from cold air drawn in from below.

Sewage facilities still seem to be a necessity since liquid wastes—bathwater, washing machine water, etc.—cannot be discharged into the Clivus. (Lindstrom has proposed a home "purifier," however, which sufficiently purifies soapy waters for use in backyard gardening, car washing, and so forth. Any excess possibly could be run into available storm sewers.)

The humus produced by the Clivus is a good soil conditioner of the following approximate composition:

Water	19.32%
Nitrogen	2.13%
Phosphorus	0.36%
Potash	1.04%
Organics	24.49%
Ash	52.26%

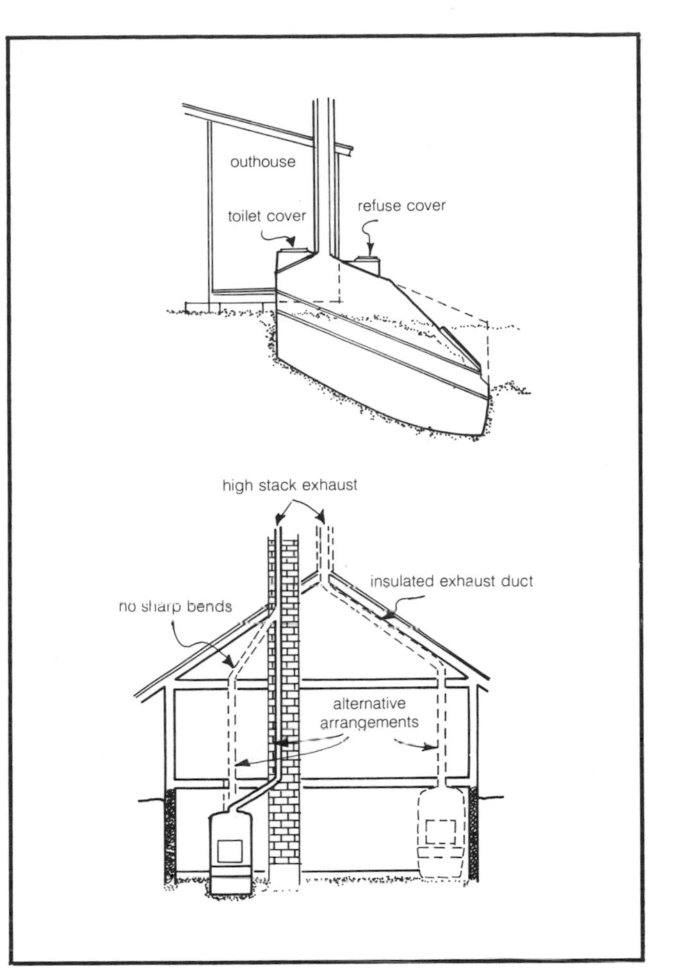

Figure 5.13 Alternative arrangements for placement of the Clivus Multrum.

By using the Clivus Multrum system, human wastes are kept out of waterways, thus reducing eutrophication, and a 45 percent saving in water consumption can be realized since no water is used to carry away the wastes. In addition, the expense of underground pipe networks and of periodic solid waste collection is reduced. And

Table 5.10 Water Use in Domestic Households[a]

Use	Percentage of Total
Bath	37
Toilet	41
Kitchen	6
Laundry	4
Cleaning	3
Drinking	5
Car wash	1
Lawn sprinkling	3

Notes: a. Adapted from "Household Wastewater Characterization," by K. Ligman, et al.

while the Clivus seems quite specific as to applicability, and the process itself appears quite fragile and dependent on many factors which can easily fail and produce drastic results from both aesthetic and health points of view, with proper design and operation these negative factors can be eliminated. Finally, while the capital cost of $2000 to $3000 (including purifier) is substantial, it is not prohibitive. For additional information, write: Clivus Multrum, Inc., Room 5500, 30 Rockefeller Plaza, New York, N.Y. 10020.

Outhouses

Pit Privy

The pit privy is by far the most widely used type of outhouse, due to its dependability and simplicity of design and construction. The system consists of a superstructure built over a pit into which the human wastes are deposited. The outhouse is built outside and downwind from the main dwelling, and preferably is placed distant and downhill from any water source, to prevent possible seepage and contamination.

A pit privy cannot be used if the water table is close enough to the pit to cause concern for possible contamination. Naturally, the soil type is of primary importance: too porous a soil allows seepage of the waste to great depths, possibly to groundwater level, while an impervious soil such as clay is totally unacceptable because the waste cannot undergo natural filtration. Especially dangerous is a pit over limestone or fissured rock, for the waste can seep into cracks and travel great distances unchecked. In a moderately pervious soil, the depth of penetration is generally about 3 feet; if the groundwater is at a sufficiently greater depth (10 feet), there is little worry of contamination.

Assuming wastes of approximately 2 pounds per person per day, the pit privy is constructed with a capacity of 1.5 cubic feet per person per year where liquids are allowed to leach away, and 19.5 cubic feet per person per year where watertight vaults are used. There are many construction designs, ranging from very elaborate types accentuating comfort, to the simplest, consisting of four walls, a roof, a seat, and a hole in the ground (see Figure 5.14). Of course, you also can choose to locate two or more latrines within the same building (over the same pit) and divide the building into individual compartments, thus permitting complete privacy and the separation of sexes. Brick or concrete floors are recommended since daily cleaning is essential. You also should make provisions to keep insects and animals out of the pit. Complete construction specifications are available in many of the references in the Bibliography.

The pit privy has a life of 5 to 15 years, depending upon its capacity (see Table 5.11). When the level of excreta rises to within approximately 20 inches of ground level, the waste should be covered with dirt and the structure moved over a new pit. After a period of about one

Table 5.11 Data for a Pit Privy[a]

Service Life (yrs)	Volume[b] (ft^3)	Depth (ft)
4 (minimum)	41	4.6
8	81	9.0
15 (maximum)	150	16.6

Notes: a. Adapted from *Excreta Disposal for Rural Areas and Small Communities* (Wagner and Lanoix). Data for a family of five.
b. Cross-sectional area of 9 ft^2.

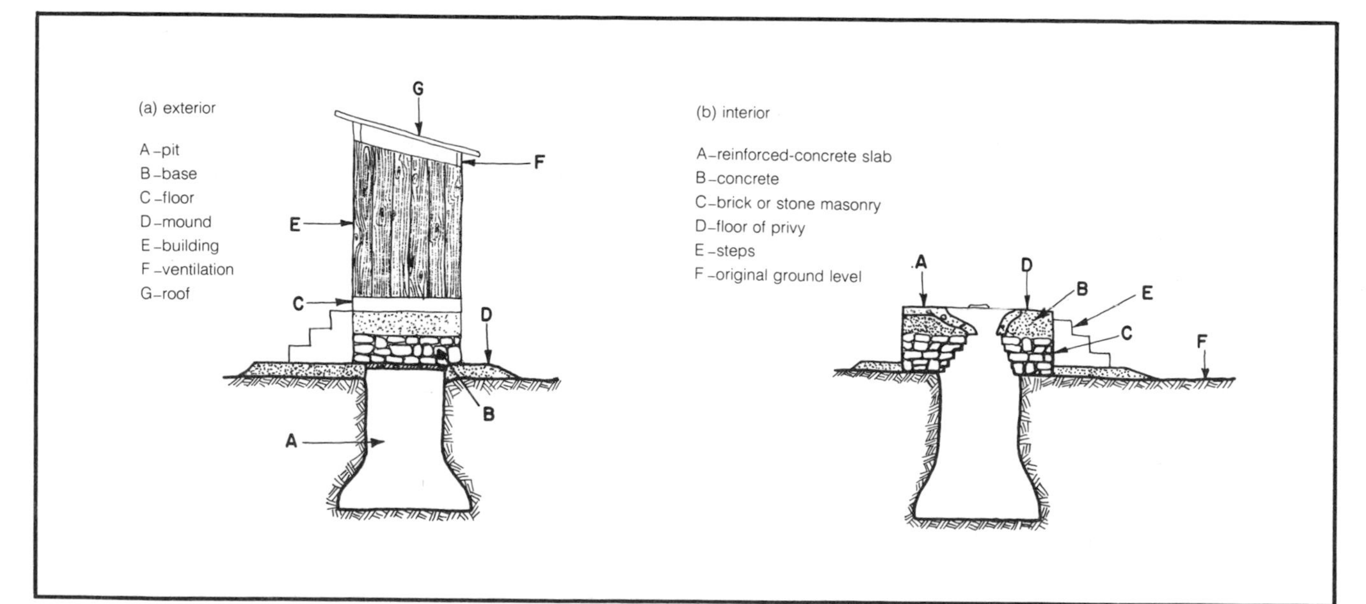

Figure 5.14 A rural pit privy.

year, the old waste pit should have completely decomposed anaerobically, leaving an end product that may be used as a soil conditioner in acreage which will not be used for food crops. Of course, after the humus has been removed, the pit may be used for the same purpose once again, thus requiring only two pits for a serviceable system.

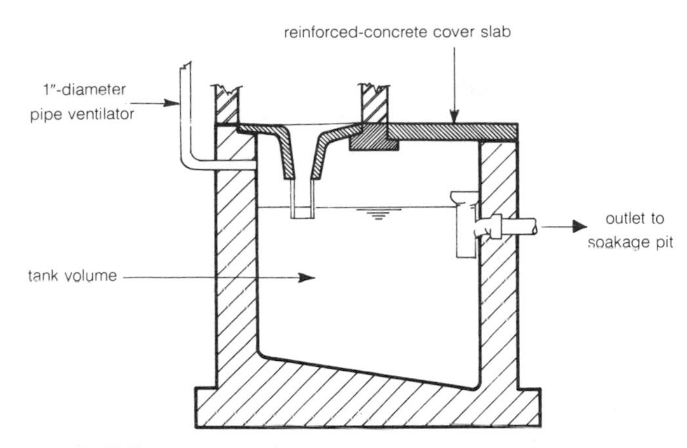

Figure 5.15 Cross section of an aqua privy.

Aqua Privy

Aqua privies, another widely used outhouse type, have the advantage of being permanent structures. A watertight tank replaces the earthen pit as the receptacle of the excreta. A slightly larger tank volume is needed compared to the pit privy, since liquids do not leach out. A drop pipe extends from the toilet seat into the liquid in the tank; the outlet is thus submerged, preventing any major quantity of gas from escaping into the interior of the outhouse. A vent is provided to allow the escape of these foul-smelling and toxic gases (see Figure 5.15). Care must be taken to keep the water level at a sufficient height; otherwise, in addition to foul odors, flies will reach and breed in the waste.

As with the pit privy, the sludge is considerably reduced in volume through the decomposition process, but, after approximately 6 to 8 years of use by an average family (four people), the tank will be approximately 40 percent full and should be emptied. The contents are removed through a manhole. This manhole should provide easy access not only to the sludge but also to the outlet tee and ventilation opening, both of which may need periodic cleaning; the manhole cover naturally should fit tightly to prevent the entrance of flies and mosquitoes.

Several problems, however, beset the aqua privy. Cleaning water is necessary on a daily basis and the unit is not workable in cold climates. A drainage trench also is necessary since approximately a gallon per day per person must be removed from the tank. Moreover, the drop pipe often becomes clogged and so presents a surface upon which flies can deposit their eggs.

While the cost of a pit privy can be made about as low as you desire through your choice of materials, such is not the case with the aqua privy. Materials must be purchased for the superstructure, for the leakproof tank, and for an outlet pipeline leading to the seepage pit. This cost is generally around several hundred dollars.

Septic Tanks

The septic tank is a most useful and dependable method for disposal of excreta and liquid wastes in a rural environment. The system is composed of a covered

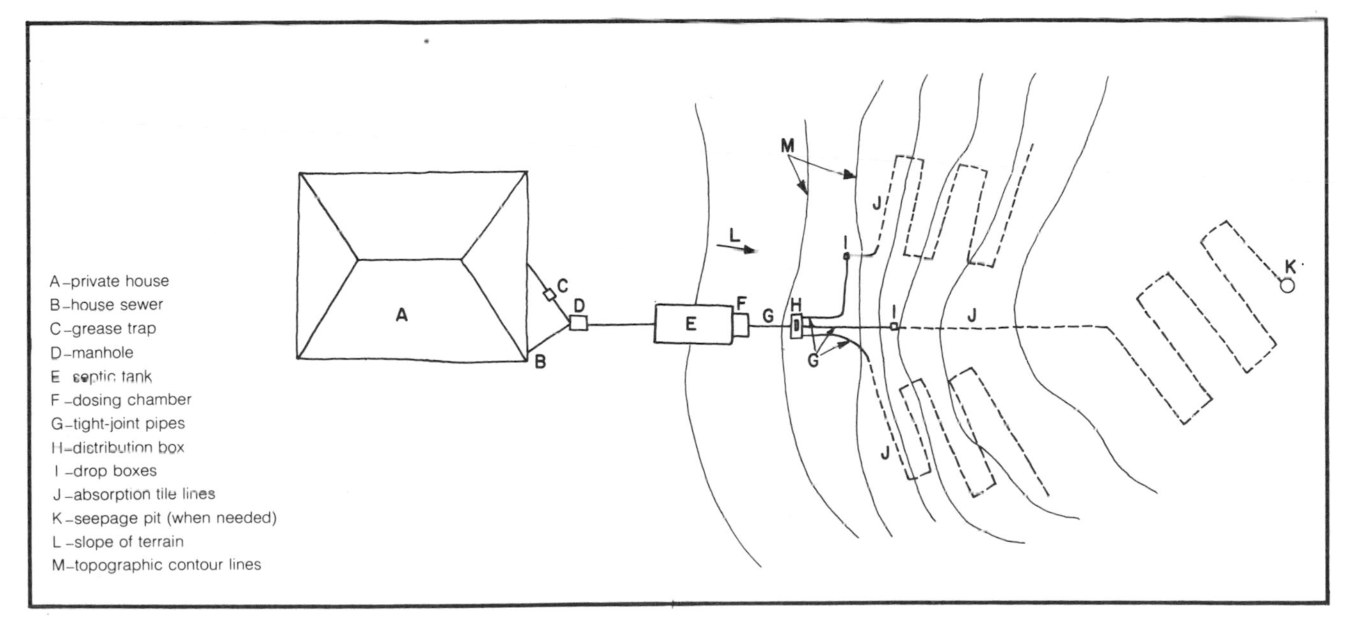

Figure 5.16 A septic-tank system with drainage field.

settling tank into which the waste flows and a drainage field for final disposal. After a sufficient detention time, the liquid waste flows out of the tank to a distribution box which apportions it to different areas of the drainage field, where it undergoes natural "secondary" treatment (see Figure 5.16).

Settling Tank

The detention time inside the tank generally runs from 3 to 5 days. The heavier solids settle to the bottom while the lighter wastes, including grease and fats, accumulate at the liquid surface and form a scum layer; the rest is carried away to the disposal area. The solids accumulating in the tank undergo anaerobic decomposition; this reduces the volume of the sludge and lengthens the period between cleanings of the tank to from 2 to 5 years. The effluent has a putrid odor, is slightly turbid, has a moderately high *BOD* (biochemical oxygen demand—the oxygen required by aerobic bacteria to oxidize waste material), and may contain pathogens, bacteria, cysts, or worm eggs. Obviously, it is a potential health hazard and you must give some thought to the design of the drainage field.

As the sludge decomposes, gases bubble up to the surface and carry with them organic particles which are vital to the putrification process. These particles eventually enlarge the scum layer. Once the scum becomes so thick and heavy that it sinks slightly, it impinges on the main sewage current; bounded below by the sludge, the flow area then becomes so narrow that adequate sedimentation becomes impossible. Consequently, your tank should be inspected and cleaned at regular intervals or the system will cease to operate properly.

The size and shape of the tank are very important, since these factors determine detention time, effective flow area, sludge space, dead space, and capacity. In addition, the tank volume must be large enough to keep turbulence and surge flows at a minimum, for these phenomena play havoc with the settling process. And finally, since the bubbling up of the sludge gas interferes with sedimentation, it is not uncommon to divide the tank into two compartments in series. In this arrange-

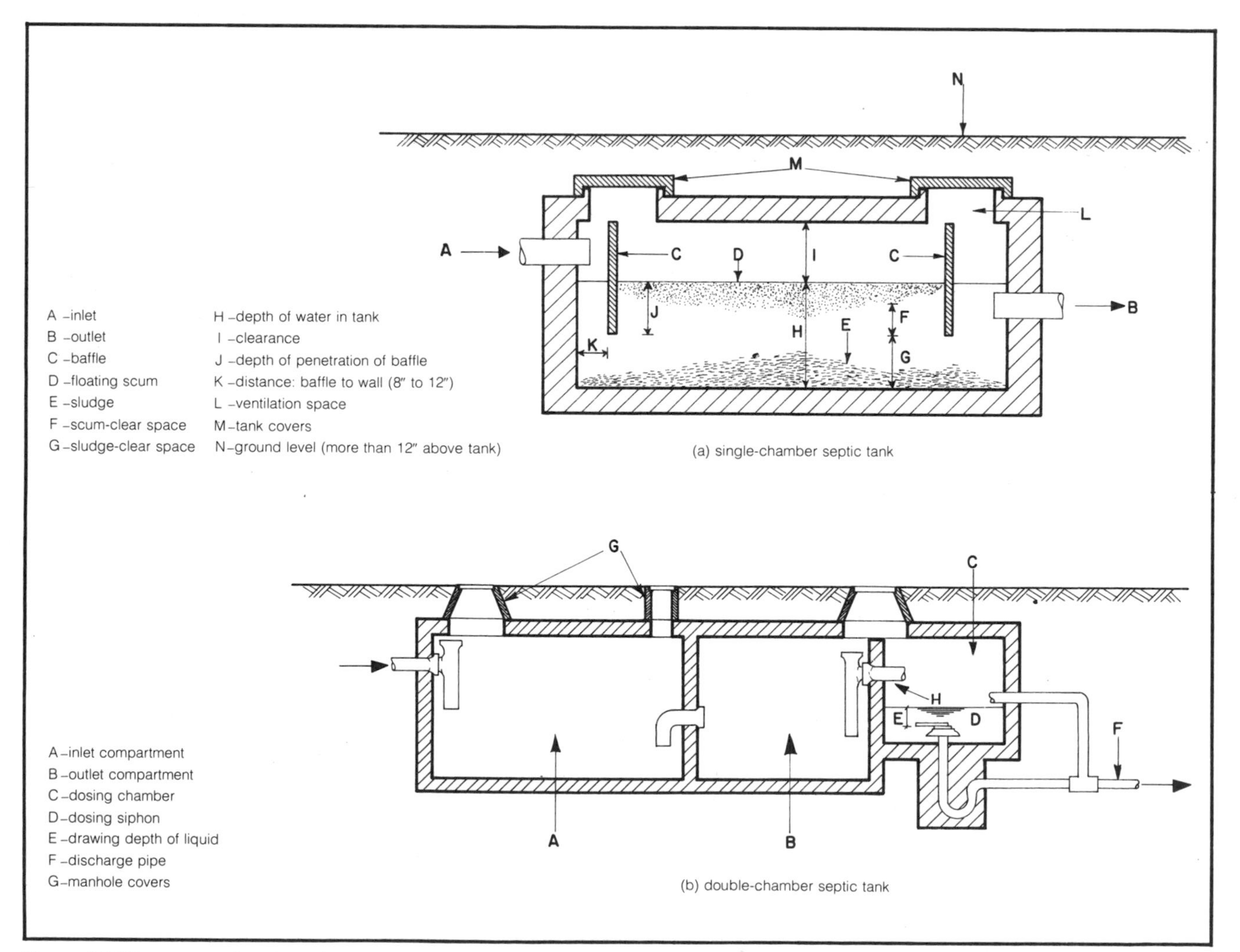

Figure 5.17 Two types of septic tanks.

ment, the bulk of the sludge decomposes in the first chamber, while the lighter suspended solids flow through to the second chamber where, with a minimum of bubbling interference, they settle out. Figure 5.17 shows a cross section of both a single- and double-chamber septic tank.

The design of septic tanks is relatively simple, and, since septic tanks have been in use for many years, there is considerable tabulated information concerning the volume required per person, drainage area, and other design parameters. Based on an assumption of 50 gallons per person per day, a tank capacity of 1500 gallons is needed for 16 people. Wagner and Lanoix have tabulated the dimensions of a tank to maximize efficiency—for a 1500-gallon tank, these dimensions are 4.5 feet wide, 10 feet long, and 5.5 feet deep. However, their estimate of daily wastewater production may be quite conservative; it is preferable to build a larger tank rather than risk the effects of surge charges, overloading, and turbulence.

There are several inlet and outlet techniques that may be used for septic tanks. The simple baffle arrangement shown in Figure 5.17 is most often used and plays an important role in the process: the depth of penetration of the baffles into the liquid helps control the volumes of clear space and sludge accumulation. Experience has shown that the inlet baffle should extend approximately 12 inches below water level while the outlet baffle should penetrate to about 40 percent of the liquid depth. Both baffles should extend no higher than an inch from the covers of the tank, to allow adequate tank ventilation. The inlet pipe should be 3 inches above water level, while an el pipe is used to allow flow between the compartments.

The tank should be buried approximately 18 inches below ground level and preferably located downhill from any water source or dwelling in case of leakage. Adequate inspection manholes must be provided. The covers of these manholes should be round rather than rectangular, to prevent you from accidentally dropping the cover into the tank.

Every 12 to 18 months, check that the distance from the bottom of the scum to the bottom of the baffle (scum-clear space) is greater than 3 inches and measure the depth of accumulation of sludge over the tank bottom; the total depth of the scum plus sludge should not exceed 20 inches. When cleaning is necessary, you can either bail out the sludge with a long-handled dipper bucket or else pump it out. This excess sludge then should be buried rather than used as fertilizer, since it will contain an undigested portion.

Drainage Field

Under noncleaning conditions, what becomes of the effluent after leaving the tank? It is vital that this effluent be evenly distributed throughout a drainage field. The device accomplishing this task is called the distribution box (Figure 5.18), and all its drainage lines should leave the box at the same level. The distribution box regulates the flow among the drainage tiles. If one area becomes oversaturated or clogged, the flow to that area can be either reduced or cut off entirely.

After leaving the distribution box, the effluent is divided among several tile lines laid at a slope of 2 to 4 inches per 100 feet. None of these lines should be longer than 100 feet or uneven distribution will result. Generally, a 4-inch tile laid with open jointing (0.5 to 0.25 inches) is sufficient to disperse an effluent from an average family within a reasonable area. If vehicles cross over the line, vitrified tile, which affords greater strength, should be used. The trenches in which the tiles are laid are 18 to 36 inches deep and approximately 24 inches wide at the bottom. As the pipe is being laid, tar paper or other suitable material must be placed over the joints to prevent the entrance of sand or other substances that might block the line. About 6 inches of coarse rock should be laid on the floor of the trench, followed by the pipe, and then 2 inches of drainage material (crushed rock or road sand); the trench is then filled to the surface with sand or loam.

Multiple tile lines must be used as the quantity of

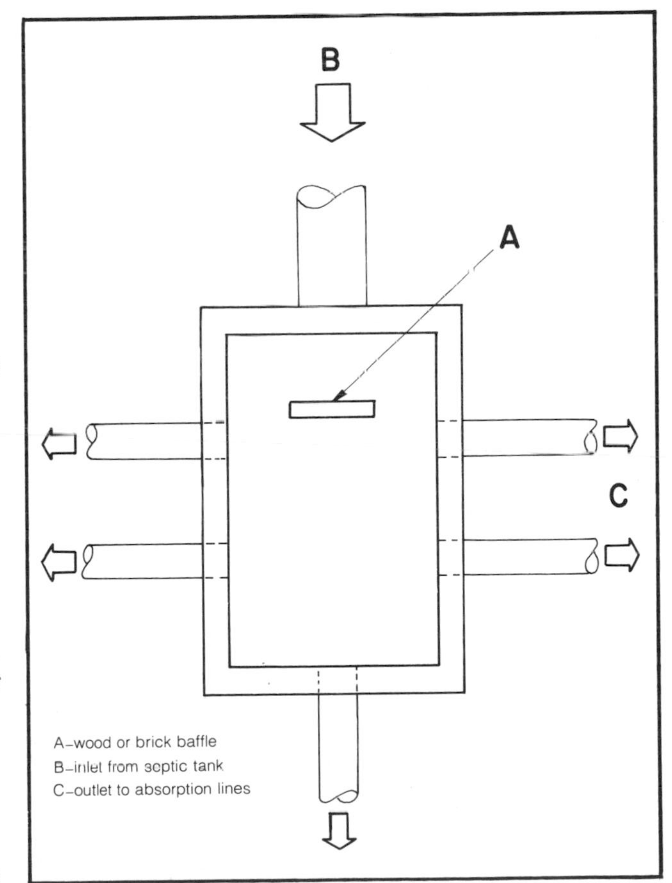

Figure 5.18 A septic-tank distribution box.

waste increases. The actual length and number of your tile lines must be determined by a test of the absorptive properties of the soil. See *Excreta Disposal for Rural Areas and Small Communities* (Wagner and Lanoix) for specific testing and evaluation details. A rule of thumb is to limit the length of tile lines to less than 100 feet, with a distance between lines of at least three times the trench bottom width, assuming a minimum of 6 feet.

If the water table is very low, drain tiles might not be necessary at all; here you can use, for example, a sand filter. If possible, tile pipes should be at least 10 feet from the water table; at any lesser distance, the soil may become saturated by capillary action, thus preventing air from entering. Since the decomposition requires oxygen, we must avoid this condition to operate properly. Assuming that your tile line is properly laid, overloading generally results from poor soil or a faulty tank.

The septic tank, when operated properly, is a very convenient, dependable, and practical manner of disposing of human excreta. Its main drawbacks are that it requires great quantities of flushing water, the proper type and size of drainage area, and considerable amounts of materials and labor to install, consequently increasing the capital cost (which, by the way, can run from $1000 to $3000). Also, if clogging does occur, getting the system back into operation can be a sizable and expensive headache.

Oxidation Ponds

The oxidation-pond process offers a very good low-cost, low-maintenance treatment method for domestic wastewater. Oxidation ponds are shallow basins used to treat wastewater by storage under conditions that favor the growth of algae. The process takes advantage of algae's ability to trap solar energy through photosynthesis and to accomplish this capture in a symbiotic relationship with bacteria in the pond which utilize organic waste as their energy source.

There has been considerable use of oxidation ponds throughout the world to treat raw wastewater, but most of these setups are fairly large. How can we utilize this technique on a small-scale basis? And what are some of the advantages and disadvantages of these ponds?

First, the good news: using an oxidation pond we can dispose of our wastewater; use the pond as an equalizing basin to absorb rapid fluctuations in the flow and strength of wastewater; produce algae for use as chicken feed; under appropriate conditions, provide ourselves with a duck and fish pond or a wildlife refuge; and accomplish all this at a low initial cost, when conditions are favorable and land is available.

But, in exchange for these advantages, we are stuck with a potential health hazard; with aesthetically unappealing conditions when maintenance is not proper; with possible contamination of groundwater and adjacent surface waters (pollution of surface waters can result from the accidental overflow or flooding of the pond); with the cost of maintenance and harvesting our algae; and, under certain conditions, with silting, overgrowth of algae and aquatic weeds, and the prolific breeding of mosquitoes and other flying insects.

Some of these problems are more easily dealt with than others. For example, the breeding of insects—particularly mosquitoes—can be prevented or controlled by raising top-feeding minnows in the pond (assuming there is sufficient oxygen available in the water). And we can avoid some of the dangers of using raw sewage by using only the effluent from a septic tank or methane digester; thus, we dispose of this treatment waste and yet capture the nutrient value still held in the waste effluent (see Figure 5.19). But under no circumstances should a small group consider using *raw sewage* as a *direct source* for the pond; the health hazards are too great.

Because of this risk, let's assume that we are utilizing the effluent from a digester as our main flow source for the oxidation pond. We also assume that the size of the community is about ten families with a total of maybe 40 people. But before moving on to any specific design details, we should spend a little more time exploring some of the process features to get a better conceptual idea of how this system works.

Bacteria and Algae

In shallow ponds, bacterial growth is supported by aerated water and the presence of organic waste. The

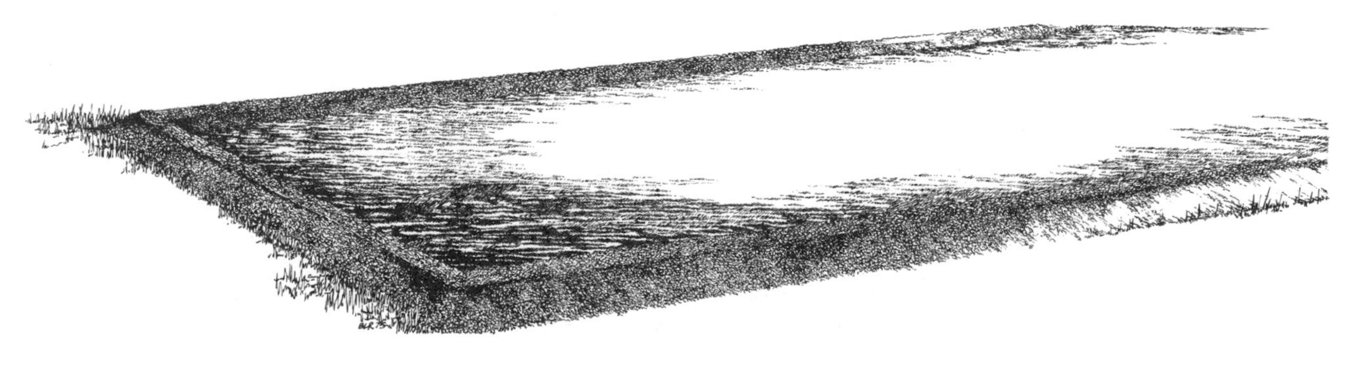

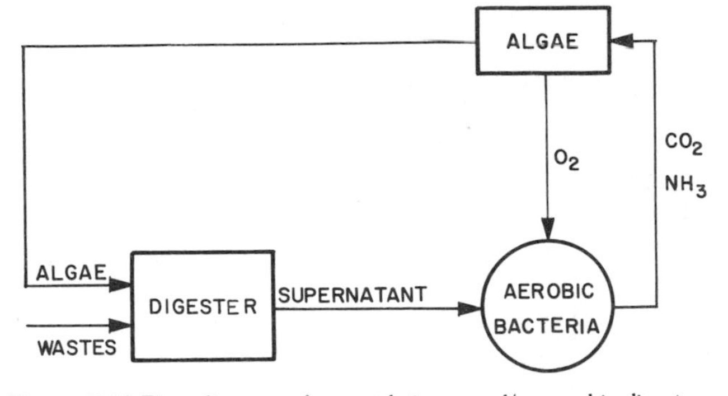

Figure 5.19 Flow diagram of an oxidation pond/anaerobic digester system.

bacteria aerobically oxidize the waste organics in the water, producing carbon dioxide and different mineral-nitrogen compounds. In the presence of light, algae will grow in the pond by using the bacterial by-products of carbon dioxide and mineral nutrients. The algae, in turn, release oxygen into the water. This process is known as photosynthesis and is shown in Figure 5.20.

Since photosynthesis can proceed only with sufficient solar radiation, it is obvious that the symbiosis cannot operate during the night. Because photosynthetic efficiency changes with the intensity of light, seasonal effects also are important.

During the day, aerobic decomposition of the waste occurs. At night, however, as the amount of carbon dioxide increases in the water, anaerobic oxidation takes place and the pH decreases (CO_2 in the water increases acidity) if the load of wastes in the pond is too heavy. When light returns, the algae consume the CO_2 and restore a favorable pH for aerobic action. By producing oxygen, these plants stop possible anaerobic oxidation.

Another source of oxygen is a daily cycle of gentle mixing and destratification by the actions of wind and temperature. The ratio of aeration by daily cycle to aeration by photosynthesis increases with the dimensions of the pond. Since we are considering a pond for a small community, it is assumed that the major source of oxygen is the oxygen produced by the algae. And, since we only consider aeration by photosynthesis in our design, there will be excess oxygen due to gas transfer from the air. This excess may allow the cultivation of fish in the pond (see Chapter 7 for further details).

Because algae play such a vital role, we must pay some attention to their requirements and the rewards they bestow on us. Nitrogen and phosphorus both stimulate algal growth and these two nutrients are important for favorable operation of oxidation ponds. A study of the nutritional requirements of algae in oxidation ponds by Oswald and Gotaas (1955; see Bibliography) determined that normal domestic sewage contains enough phosphorus to support an algae culture concentration of 400 ppm. Nor are magnesium and potassium

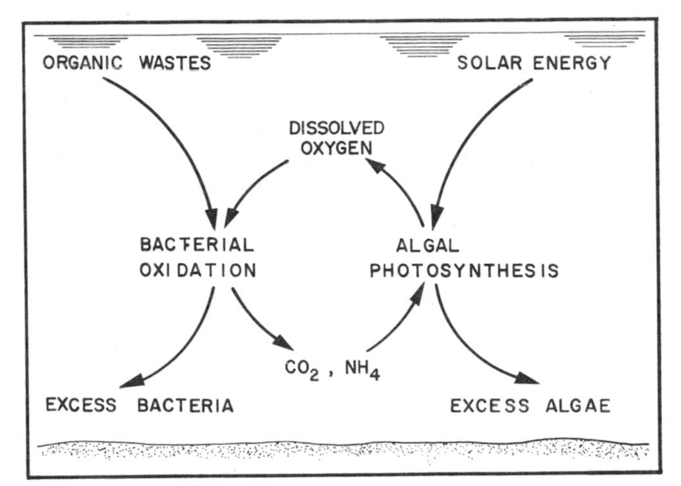

Figure 5.20 The symbiotic processes of bacteria and algae in oxidation ponds.

limiting elements, since normal domestic sewage contains sufficient magnesium and potassium to support a 500-ppm concentration of algae. These workers found that carbon is the usual limiting element. This condition is partially alleviated by the culture becoming basic due to photosynthesis, which in turn causes absorption of atmospheric carbon dioxide, a usable source of carbon. Nutrition seems to be the limiting factor up to 300-ppm concentrations of algae. Beyond that, the limiting factor is the amount of available light for photosynthesis.

The biochemical oxygen demand (*BOD*) is the amount of oxygen required to degrade or destroy organic material via bacterial action. Figure 5.21 (from Oswald and Gotaas) relates algal yield to the *BOD* of a cultural medium. An average *BOD* of 250 ppm can be assumed for our oxidation pond and this would produce concentrations of algae of 280 ppm, just below the limiting values imposed by nutritional or photosynthetic light demands (we will discuss the *BOD* more fully in a while).

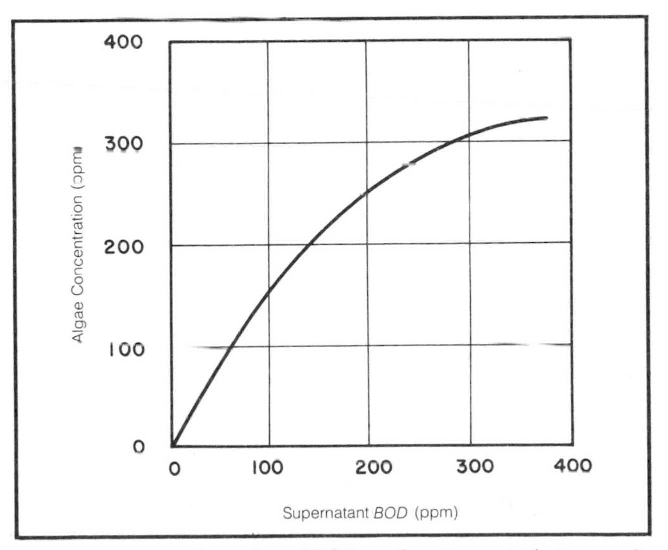

Figure 5.21 The relationship of *BOD* to algae in an oxidation pond.

The ability of algae to scavenge phosphorus, nitrogen, and *BOD* from effluents is highly useful to prevent contamination of water sources close to waste-treatment facilities. During high photosynthetic activity, both nitrogen and phosphorus are removed efficiently, providing an effluent water of a quality generally acceptable for most of the sources into which it may flow. But it is best and most safely used for irrigation purposes.

The quantity of water required to produce a pound of protein by using algae as feed can be less than a hundredth of that required by conventional agricultural methods. Wastewater-grown algae have been fed to a number of animals with no evidence of unsatisfactory results. For example, the value of this kind of food as a supplement to chicken feed is now approximately $250 per ton. The rate of algal yield may vary from 1 ton per acre per month in winter to 5 tons in summer. If we compare this yield to that of field crops, we find that it is twenty times the agricultural average.

The high protein content (more than 50 percent) is not the only important property of the algae. They may become an important source of vitamins, of raw products for organic synthesis, and also of such elements as germanium (which algae concentrate). Moreover, the fuel characteristics of dry algae are similar to those of medium-grade bituminous coal, although their heat content is somewhat less (ranging up to 10,000 Btu/pound). Algae also may be used as a carbon source for digesters producing methane by fermentation.

Wastewater and the *BOD*

Before we can begin our design calculations, we need to know some basic information about our wastewater characteristics and the quantity of flow. Earlier we mentioned the *BOD* (biochemical oxygen demand), because wastewater strength is generally measured in terms of the amount of oxygen required by aerobic bacteria to oxidize organic wastes biologically (to CO_2 and H_2O). The amount of oxygen required to completely oxidize the organic matter in a wastewater is called the *ultimate* or *maximum BOD.* For a family of four, the average waste load is about 0.695 pounds of *BOD* per day, and the wastewater volume produced is around 200 gallons; the concentration then is about 300 to 400 ppm of *BOD*. After the raw wastewater has been processed by a digester or septic tank, the effluent contains a concentration on the order of 75 to 150 ppm of *BOD*. (If animal wastes and other materials are being added to the digester, the concentration in the effluent may be higher, around 300 to 400 ppm.)

We can estimate the volume of wastewater flow at about 50 gallons per day per person. In the design we are considering, we have ten families with around forty people; this gives us an average daily flow of about 2000 gallons per day.

Climatic Considerations of Design

In temperate areas, winter temperatures can be low enough so that the rates of all biological reactions (photosynthesis, aerobic and anaerobic oxidation) fall severely, even if no ice cover occurs—little waste stabilization takes place beyond sedimentation. For an oxidation pond, consequently, concentration of wastes in winter remains higher than in summer. Because anaerobic oxidation (the primary source of odors, by the way) is also reduced, we design the pond for the winter period.

By way of contrast, in a tropical area during the summer (temperature over 73°F), stratification is intense and anaerobic conditions and fermentation are dominant in the lower two-thirds of the pond. If winter temperatures are high, stratification is absent and waste stabilization is high. In this case, the pond should be aerated mechanically during the summer.

As we have mentioned, a large oxidation pond receives the main part of its oxygen from the air; in a small pond, the oxygen comes from the biological process of photosynthesis. Concentrated wastes require a dense algal growth (which needs lots of light) and so the depth of the pond has to be shallow to allow a sufficient penetration of the light; dilute domestic wastewater may be processed at greater depth. In order to provide a detention period suitable for effective photosynthetic oxygen production during both winter and summer, certain compromises are necessary.

Perhaps a few broader remarks are now in order. We must design for the winter months because that is the period of slowest biological activity. In the summer, the efficiency of a pond designed for winter months is very low. Without proper variations in operating procedures, the result can be overproduction of algae, a part of which may die, decompose, and produce a pond effluent with a high supernatant *BOD*. On the other hand, if we design for midsummer months, our detention time will be very low (about a day or so) and, during the winter, the algae will be unable to grow fast enough to prevent being washed out of the pond. So, in order to provide a detention period suitable for effective photosynthetic oxygen production in both winter and summer and with some capacity to sustain changes in light, temperature, and shock loading, we reiterate: certain compromises are necessary.

In general, we can note that, for most conditions, detention times should not be less than a day for summer conditions nor more than 10 to 12 days for winter conditions. A pond having a detention period of about 3 days and a depth of 12 inches should, for example, satisfactorily produce adequate oxygen by photosynthesis more than 80 percent of the time (latitudes up to 40°N), so long as continuous ice cover does not occur.

Now we must get acquainted with the design equa-

Table 5.12 Solar Radiation: Probable Average Values of Insolation[a,b,e]

North Latitude		Month																								North Latitude	
Degree	Range	January		February		March		April		May		June		July		August		September		October		November		December		Range	Degree
		vis[c]	tot[d]	vis	tot	vis	tot	vis	tot	vis	tot	vis	tot	vis	tot	vis	tot	vis	tot	vis	tot	vis	tot	vis	tot		
0	max	255	685	266	700	271	708	266	690	249	645	236	626	238	630	252	666	269	690	265	694	256	683	253	667	max	0
	min	210	580	219	583	206	536	188	462	182	480	103	274	137	368	167	432	207	533	203	530	202	543	195	527	min	
2	max	250	670	263	693	271	706	267	697	253	655	241	642	244	646	255	673	269	693	262	688	251	666	249	646	max	2
	min	206	560	213	560	204	534	188	464	184	484	108	288	141	375	169	442	206	531	200	523	198	526	189	505	min	
4	max	244	650	259	688	270	704	268	701	258	665	247	656	250	657	258	678	269	695	260	680	246	650	244	628	max	4
	min	200	540	206	543	202	532	187	466	187	492	113	300	146	385	171	448	204	529	196	513	194	510	183	480	min	
6	max	238	630	254	675	268	702	270	705	262	675	252	668	255	669	261	683	269	697	256	670	240	634	238	610	max	6
	min	193	520	199	530	200	530	186	467	189	500	118	310	150	395	172	452	202	524	191	500	188	494	176	460	min	
8	max	230	610	249	665	267	700	270	709	266	685	258	678	260	680	263	688	267	695	252	660	234	616	231	590	max	8
	min	187	495	192	510	196	523	185	467	191	506	124	320	154	405	174	456	200	518	186	486	182	478	169	440	min	
10	max	223	595	244	655	264	694	271	711	270	694	262	688	265	690	266	693	266	693	248	650	228	600	225	570	max	10
	min	179	475	184	490	193	513	183	464	192	512	129	330	158	414	176	460	196	510	181	474	176	462	162	420	min	
12	max	216	572	239	645	262	690	271	710	273	702	267	700	269	700	267	697	264	691	244	640	221	585	217	550	max	12
	min	172	455	176	470	189	500	181	462	193	518	133	343	161	421	176	464	193	502	176	462	169	446	154	400	min	
14	max	208	555	233	630	258	680	271	709	276	710	272	710	273	708	269	700	262	688	240	627	214	567	209	536	max	14
	min	163	430	167	450	184	487	179	460	194	524	137	354	164	429	177	467	189	496	170	449	162	430	146	380	min	
16	max	200	530	226	610	255	670	272	707	279	718	276	720	277	715	270	703	259	684	234	615	206	554	200	520	max	16
	min	154	400	159	430	180	473	177	456	194	528	141	363	167	435	177	469	185	489	164	434	154	410	138	360	min	
18	max	192	515	220	590	250	664	272	705	232	723	280	728	280	723	272	705	256	680	229	605	198	538	192	500	max	18
	min	144	380	150	410	174	459	174	452	194	530	145	375	170	442	177	471	180	479	157	418	146	390	129	340	min	
20	max	183	500	213	575	246	652	271	703	284	730	284	738	282	729	272	706	252	674	224	596	190	520	182	480	max	20
	min	134	360	140	390	168	440	170	447	194	532	148	383	172	450	177	472	176	467	150	400	138	370	120	320	min	
22	max	174	480	206	560	241	644	270	701	286	734	286	747	285	736	273	707	248	668	218	582	183	500	172	460	max	22
	min	123	335	132	370	162	426	167	440	193	530	152	392	173	454	176	472	170	455	143	380	128	350	110	300	min	
24	max	166	460	200	545	236	625	268	697	288	738	290	753	287	742	273	708	244	659	212	568	175	480	161	440	max	24
	min	111	310	123	340	156	410	164	433	191	525	155	403	176	459	174	471	165	443	136	360	119	326	101	280	min	
26	max	156	440	192	530	230	615	266	690	288	741	292	760	288	749	273	706	240	652	205	552	166	460	149	420	max	26
	min	99	280	114	310	149	390	160	425	189	518	158	409	177	463	172	469	160	429	128	332	109	300	90	260	min	
28	max	146	420	184	510	224	603	264	683	289	743	294	764	288	755	272	704	236	635	199	537	157	440	138	400	max	28
	min	87	250	106	290	142	373	156	415	187	506	161	418	178	467	169	466	154	415	120	310	99	278	80	236	min	
30	max	136	400	176	490	213	587	261	675	290	744	296	768	289	759	271	702	231	625	192	524	148	420	126	380	max	30
	min	76	220	96	260	134	362	151	405	184	490	163	425	178	469	166	462	147	399	113	290	90	256	70	210	min	
32	max	126	380	169	470	212	570	258	663	290	744	296	772	289	761	269	700	226	615	185	510	138	400	114	360	max	32
	min	63	180	87	240	126	340	146	395	181	475	166	431	178	472	163	458	140	385	104	270	80	224	60	184	min	
34	max	114	360	160	450	204	553	254	657	290	743	297	775	289	763	267	696	221	602	178	490	128	380	101	338	max	34
	min	53	155	78	215	118	320	141	385	176	462	168	439	178	472	159	448	134	368	96	250	70	202	47	158	min	
36	max	103	335	150	430	196	538	250	650	288	741	298	776	289	765	264	690	215	590	170	470	118	360	88	314	max	36
	min	44	133	70	200	111	300	136	375	172	444	170	443	177	470	155	438	127	350	88	230	60	180	39	134	min	
38	max	90	310	140	415	159	520	246	640	287	738	298	778	288	766	262	684	210	576	162	450	106	336	77	290	max	38
	min	36	120	62	180	103	280	131	365	166	428	171	448	175	464	152	429	120	330	80	216	50	158	30	111	min	
40	max	80	280	130	390	131	500	241	630	286	732	298	778	288	765	258	680	203	532	152	430	95	313	66	270	max	40
	min	30	105	53	160	95	270	125	355	162	415	173	450	172	455	147	416	112	310	72	202	42	134	24	94	min	
42	max	68	255	119	370	172	485	236	618	283	728	298	777	287	761	254	670	196	547	144	410	84	289	56	244	max	42
	min	24	90	45	140	38	250	120	344	157	405	174	451	167	442	143	403	105	290	65	187	34	112	19	78	min	
44	max	55	228	106	340	155	470	230	607	280	722	298	777	285	755	250	660	189	530	132	390	72	263	47	218	max	44
	min	20	80	37	130	80	230	114	325	153	395	175	453	164	430	139	389	98	270	58	173	28	98	15	62	min	
46	max	45	200	94	315	156	450	224	598	278	716	298	776	284	749	245	650	181	512	122	370	61	238	39	194	max	46
	min	16	74	30	110	72	210	108	315	150	385	175	455	161	420	134	374	90	250	52	158	23	86	11	48	min	
48	max	35	180	82	290	149	430	218	582	274	710	297	776	282	740	241	640	174	496	111	350	50	210	32	170	max	48
	min	12	64	25	99	64	190	102	307	146	378	176	458	158	410	129	358	81	230	45	144	18	70	9	37	min	
50	max	28	164	70	265	141	410	210	568	271	703	297	776	280	733	236	625	166	480	100	329	40	183	26	144	max	50
	min	10	54	19	80	58	173	97	300	144	371	176	458	155	403	125	342	73	210	40	130	15	60	7	30	min	
52	max	22	140	60	240	134	390	202	555	267	695	296	776	278	725	232	615	158	460	87	307	32	160	21	121	max	52
	min	8	45	14	62	51	158	92	295	141	366	176	460	153	398	120	326	65	190	34	120	12	53	4	27	min	
54	max	16	120	50	215	126	370	194	542	263	687	296	776	276	720	224	602	150	440	76	285	25	140	16	99	max	54
	min	6	40	11	45	46	145	88	289	139	360	176	460	150	394	116	312	58	170	29	106	9	43	3	23	min	
56	max	12	102	43	200	120	350	188	528	258	680	295	775	273	714	218	587	141	420	64	261	20	120	12	78	max	56
	min	4	35	8	35	41	132	85	283	136	352	175	460	148	390	110	297	51	150	24	95	7	36	2	18	min	
58	max	9	80	37	170	113	330	182	516	254	670	294	774	270	710	212	575	134	402							max	58
	min	3	28	6	28	37	118	82	277	134	346	175	460	146	385	106	285	44	132							min	
60	max	7	64	32	150	107	310	176	500	249	660	294	773	268	708	205	556	126	386							max	60
	min	2	20	4	20	33	105	79	270	132	340	174	460	144	380	100	270	38	116							min	

Notes: a. From "Photosynthesis in Sewage Treatment" (1957) by W.J. Oswald and H.B. Gataas.
b. Direct and diffuse insolation on a horizontal surface at sea level (Langleys per day).
c. *Visible:* Radiation of wave lengths 4000 to 7000Å penetrating a smooth water surface.
d. Total: all wave lengths in solar spectrum.
e. For corrections due to elevation up to 10,000 feet and for cloudiness, see Equations 5.37, 5.38, and 5.42.

tions and the various parameters necessary for a successful design calculation.

Computation of Depth

We can obtain an approximation of the depth of an oxidation pond by using the following formula:

E. 5.36 $$d = \frac{\ln (I_i)}{C_c \alpha}$$

where d equals the depth (cm); I_i equals the incident light intensity (footcandles); ln indicates a mathematical operation—the natural logarithm of I_i; C_c equals the concentration of algal matter; and α equals a specific absorption coefficient.

The estimation of $ln\ (I_i)$ can be obtained by the following steps:

1. Find the maximum and minimum total solar radiation for the relevant latitude from Table 5.12.
2. Make necessary corrections for cloudiness and elevation (see Equations 5.37 and 5.38).
3. Multiply the resulting value by 10.
4. Multiply the result of the third step by the fraction of time the sun is visible for the approximate latitude and month as determined from Figure 5.22. This gives us a value for I_i.
5. Refer to Table 5.13 for logarithmic values, using I_i as the value of N. The listed value gives us $ln\ (I_i)$ for use in Equation 5.36.

The correction for cloudiness—that is, what percent of the time we actually have clear weather and thus available solar radiation—is given by

E. 5.37 $$\textit{Total Radiation} = min + (max - min) \times \left(\frac{\%\ time\ clear}{100}\right)$$

where minimum and maximum values also are given in Table 5.12 (use "total," not "visible"). This gives the corrected total for sea level. The correction for a particular elevation is given by

E. 5.38 $$\textit{Total Radiation} = (\textit{total at sea level}) \times [1 + (0.0185 \times elevation)]$$

where elevation is expressed in thousands of feet: that is, an elevation of 2500 feet would be 2.5 in the above equation. And summarizing from steps "3" and "4"

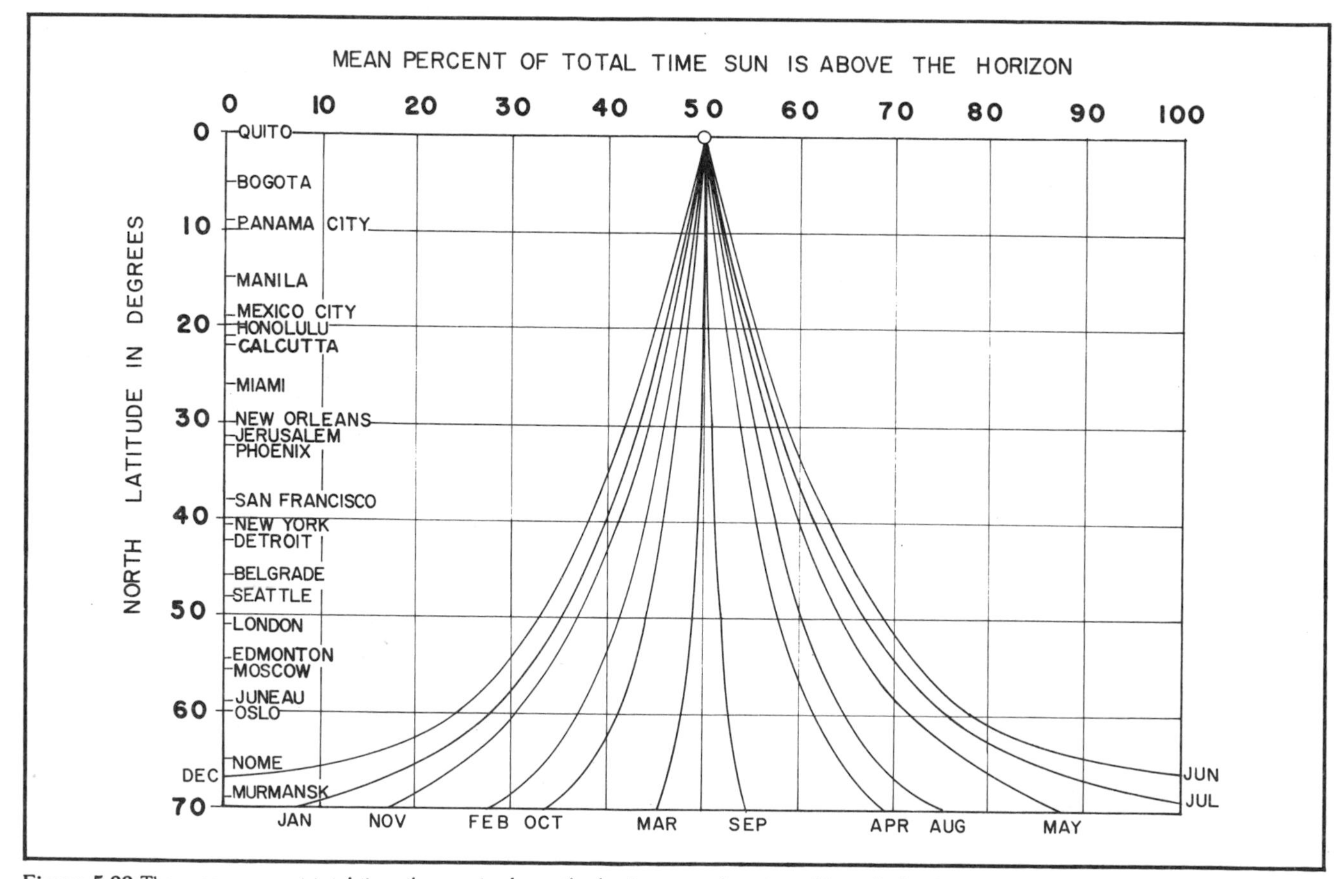

Figure 5.22 The mean percent total time the sun is above the horizon as a function of latitude for the northern hemisphere.

Table 5.13 Logarithmic Values (0-999)

N	0	1	2	3	4	5	6	7	8	9
0	− ∞	0.00000	0.69315	1.09861	.38629	.60944	.79176	.94591	*.07944	*.19722
1	2.30259	.39790	.48491	.56495	.63906	.70805	.77259	.83321	.89037	.94444
2	.99573	*.04452	*.09104	*.13549	*.17805	*.21888	*.25810	*.29584	*.33220	*.36730
3	3.40120	.43399	.46574	.49651	.52636	.55535	.58352	.61092	.63759	.66356
4	.68888	.71357	.73767	.76120	.78419	.80666	.82864	.85015	.87120	.89182
5	.91202	.93183	.95124	.97029	.98898	*.00733	*02535	*.04305	*.06044	*.07754
6	4.09434	.11087	.12713	.14313	.15888	.17439	.18965	.20469	.21951	.23411
7	.24850	.26268	.27667	.29046	.30407	.31749	.33073	34381	.35671	.36945
8	.38203	.39445	.40672	.41884	.43082	.44265	.45435	.46591	.47734	.48864
9	.49981	.51086	.52179	.53260	.54329	.55388	.56435	.57471	.58497	.59512
10	4.60517	.61512	.62497	.63473	.64439	.65396	.66344	.67283	.68213	.69135
11	.70048	.70953	.71850	.72739	.73620	.74493	.75359	.76217	.77068	.77912
12	.78749	.79579	.80402	.81218	.82028	.82831	.83628	.84419	.85203	.85981
13	.86753	.87520	.88280	.89035	.89784	.90527	.91265	.91998	.92725	.93447
14	.94164	.94876	.95583	.96284	.96981	.97673	.98361	.99043	.99721	*.00395
15	5.01064	.01728	.02388	.03044	.03695	.04343	.04986	.05625	.06260	.06890
16	.07517	.08140	.08760	.09375	.09987	.10595	.11199	.11799	.12396	.12990
17	.13580	.14166	.14749	.15329	.15906	.16479	.17048	.17615	.18178	.18739
18	.19296	.19850	.20401	.20949	.21494	.22036	.22575	.23111	.23644	.24175
19	.24702	.25227	.25750	.26269	.26786	.27300	.27811	.28320	.28827	.29330
20	5.29832	.30330	.30827	.31321	.31812	.32301	.32788	.33272	.33754	.34233
21	.34711	.35186	.35659	.36129	.36598	.37064	.37528	.37990	.38450	.38907
22	.39363	.39816	.40268	.40717	.41165	.41610	.42053	.42495	.42935	.43372
23	.43808	.44242	.44674	.45104	.45532	.45959	.46383	.46806	.47227	.47646
24	.48064	.48480	.48894	.49306	.49717	.50126	.50533	.50939	.51343	.51745
25	.52146	.52545	.52943	.53339	.53733	.54126	.54518	.54908	.55296	.55683
26	.56068	.56452	.56834	.57215	.57595	.57973	.58350	.58725	.59099	.59471
27	.59842	.60212	.60580	.60947	.61313	.61677	.62040	.62402	.62762	.63121
28	63479	.63835	.64191	.64545	.64897	.65249	.65599	.65948	.66296	.66643
29	.66988	.67332	.67675	.68017	.68358	68698	.69036	.69373	.69709	.70044
30	5.70378	.70711	.71043	.71373	.71703	.72031	.72359	.72685	.73010	.73334
31	.73657	.73979	.74300	.74620	.74939	.75257	.75574	.75890	.76205	.76519
32	.76832	.77144	.77455	.77765	.78074	.78383	.78690	.78996	.79301	.79606
33	.79909	.80212	.80513	.80814	.81114	.81413	.81711	.82008	.82305	.82600
34	.82895	.83188	.83481	.83773	.84064	.84354	.84644	.84932	.85220	.85507
35	.85793	.86079	.86363	.86647	.86930	.87212	.87493	.87774	.88053	.88332
36	.88610	.88888	.89164	.89440	.89715	.89990	.90263	.90536	.90808	.91080
37	.91350	.91620	.91889	.92158	.92426	.92693	.92959	.93225	.93489	.93754
38	.94017	.94280	.94542	.94803	.95064	.95324	.95584	.95842	.96101	.96358
39	.96615	.96871	.97126	.97381	.97635	.97889	.98141	.98394	.98645	.98896
40	5.99146	.99396	.99645	.99894	*.00141	*.00389	*.00635	*.00881	*.01127	*.01372
41	6.01616	.01859	.02102	.02345	.02587	.02828	.03069	.03309	.03548	.03787
42	.04025	.04263	.04501	.04737	.04973	.05209	.05444	.05678	.05912	.06146
43	.06379	.06611	.06843	.07074	.07304	.07535	.07764	.07993	.08222	.08450
44	.08677	.08904	.09131	.09357	.09582	.09807	.10032	.10256	.10479	.10702
45	.10925	.11147	.11368	.11589	.11810	.12030	.12249	.12468	.12687	.12905
46	.13123	.13340	.13556	.13773	.13988	.14204	.14419	.14633	.14847	.15060
47	.15273	.15486	.15698	.15910	.16121	.16331	.16542	.16752	.16961	.17170
48	.17379	.17587	.17794	.18002	.18208	.18415	.18621	.18826	.19032	.19236
49	.19441	.19644	.19848	.20051	.20254	.20456	.20658	.20859	.21060	.21261

Notes: a. Taken from *Standard Mathematical Tables*, 21st Edition, edited by S.M. Selby.

Table 5.13—*Concluded*

N	0	1	2	3	4	5	6	7	8	9
50	6.21461	.21661	.21860	.22059	.22258	.22456	.22654	.22851	.23048	.23245
51	.23441	.23637	.23832	.24028	.24222	.24417	.24611	.24804	.24998	.25190
52	.25383	.25575	.25767	.25958	.26149	.26340	.26530	.26720	.26910	.27099
53	.27288	.27476	.27664	.27852	.28040	.28227	.28413	.28600	.28786	.28972
54	.29157	.29342	.29527	.29711	.29895	.30079	.30262	.30445	.30628	.30810
55	.30992	.31173	.31355	.31536	.31716	.31897	.32077	.32257	.32436	.32615
56	.32794	.32972	.33150	.33328	.33505	.33683	.33859	.34036	.34212	.34388
57	.34564	.34739	.34914	.35089	.35263	.35437	.35611	.35784	.35957	.36130
58	.36303	.36475	.36647	.36819	.36990	.37161	.37332	.37502	.37673	.37843
59	.38012	.38182	.38351	.38519	.38688	.38856	.39024	.39192	.39359	.39526
60	6.39693	.39859	.40026	.40192	.40357	.40523	.40688	.40853	.41017	.41182
61	.41346	.41510	.41673	.41836	.41999	.42162	.42325	.42487	.42649	.42811
62	.42972	.43133	.43294	.43455	.43615	.43775	.43935	.44095	.44254	.44413
63	.44572	.44731	.44889	.45047	.45205	.45362	.45520	.45677	.45834	.45990
64	.46147	.46303	.46459	.46614	.46770	.46925	.47080	.47235	.47389	.47543
65	.47697	.47851	.48004	.48158	.48311	.48464	.48616	.48768	.48920	.49072
66	.49224	.49375	.49527	.49677	.49828	.49979	.50129	.50279	.50429	.50578
67	.50728	.50877	.51026	.51175	.51323	.51471	.51619	.51767	.51915	.52062
68	.52209	.52356	.52503	.52649	.52796	.52942	.53088	.53233	.53379	.53524
69	.53669	.53814	.53959	.54103	.54247	.54391	.54535	.54679	.54822	.54965
70	6.55108	.55251	.55393	.55536	.55678	.55820	.55962	.56103	.56244	.56386
71	.56526	.56667	.56808	.56948	.57088	.57228	.57368	.57508	.57647	.57786
72	.57925	.58064	.58203	.58341	.58479	.58617	.58755	.58893	.59030	.59167
73	.59304	.59441	.59578	.59715	.59851	.59987	.60123	.60259	.60394	.60530
74	.60665	.60800	.60935	.61070	.61204	.61338	.61473	.61607	.61740	.61874
75	.62007	.62141	.62274	.62407	.62539	.62672	.62804	.62936	.63068	.63200
76	.63332	.63463	.63595	.63726	.63857	.63988	.64118	.64249	.64379	.64509
77	.64639	.64769	.64898	.65028	.65157	.65286	.65415	.65544	.65673	.65801
78	.65929	.66058	.66185	.66313	.66441	.66568	.66696	.66823	.66950	.67077
79	.67203	.67330	.67456	.67582	.67708	.67834	.67960	.68085	.68211	.68336
80	6.68461	.68586	.68711	.68835	.68960	.69084	.69208	.69332	.69456	.69580
81	.69703	.69827	.69950	.70073	.70196	.70319	.70441	.70564	.70686	.70808
82	.70930	.71052	.71174	.71296	.71417	.71538	.71659	.71780	.71901	.72022
83	.72143	.72263	.72383	.72503	.72623	.72743	.72863	.72982	.73102	.73221
84	.73340	.73459	.73578	.73697	.73815	.73934	.74052	.74170	.74288	.74406
85	.74524	.74641	.74759	.74876	.74993	.75110	.75227	.75344	.75460	.75577
86	.75693	.75809	.75926	.76041	.76157	.76273	.76388	.76504	.76619	.76734
87	.76849	.76964	.77079	.77194	.77308	.77422	.77537	.77651	.77765	.77878
88	.77992	.78106	.78219	.78333	.78446	.78559	.78672	.78784	.78897	.79010
89	.79122	.79234	.79347	.79459	.79571	.79682	.79794	.79906	.80017	.80128
90	6.80239	.80351	.80461	.80572	.80683	.80793	.80904	.81014	.81124	.81235
91	.81344	.81454	.81564	.81674	.81783	.81892	.82002	.82111	.82220	.82329
92	.82437	.82546	.82655	.82763	.82871	.82979	.83087	.83195	.83303	.83411
93	.83518	.83626	.83733	.83841	.83948	.84055	.84162	.84268	.84375	.84482
94	.84588	.84694	.84801	.84907	.85013	.85118	.85224	.85330	.85435	.85541
95	.85646	.85751	.85857	.85961	.86066	.86171	.86276	.86380	.86485	.86589
96	.86693	.86797	.86901	.87005	.87109	.87213	.87316	.87420	.87523	.87626
97	.87730	.87833	.87936	.88038	.88141	.88244	.88346	.88449	.88551	.88653
98	.88755	.88857	.88959	.89061	.89163	.89264	.89366	.89467	.89568	.89669
99	.89770	.89871	.89972	.90073	.90174	.90274	.90375	.90475	.90575	.90675

Example: N = 92 ln(92) = 4.52179 N = 403 ln(403) = 5.99894

above, we can express I_i as

E. 5.39 $$I_i = 10 \times \left(\frac{\%\ time\ sun\ above\ horizon}{100}\right) \times Total\ Radiation$$

It also turns out that we have available figures for C_c and α: C_c is usually from 100 to 300 ppm; α is generally taken to be 0.0015. So, let's try an example to get a feel for all these equations and numbers.

Example: Designing for winter temperatures, we want to estimate the depth of an oxidation pond to be built near San Francisco (latitude 37°N) at an elevation of 1000 feet with clear weather 50 percent of the time in December.

Solution: From Table 5.12, we get maximum total radiation equals 290 Langleys per day and the minimum total radiation equals 111 Langleys per day. Using Equation 5.37 to correct for clear weather only 50 percent of the time, we find

$$Total\ Radiation = 111 + (290 - 111)\left(\frac{50}{100}\right)$$

$$= 200.5\ Langleys/day$$

The correction for elevation is made using Equation 5.38:

$$Total\ Radiation = (200.5) \times [1 + (0.0185 \times 1)]$$

$$= 204.21\ Langleys/day$$

Now using Equation 5.39, we must multiply our result by 10 and by the fraction of the time the sun is visible above the horizon (from Figure 5.22 we see that this fraction for December is nearly 40 percent):

$$I_i = 10 \times \frac{40\%}{100} \times 204.21$$

$$= 816.8\ footcandles$$

If you check for $ln(I_i)$ in Table 5.13, you find that the natural logarithm of 816.8 is about 6.705. Now we can calculate the depth of the pond, using 0.0015 for α and 200 ppm for C_c with Equation 5.36:

$$d = \frac{ln(I_i)}{C_c} = \frac{6.705}{200 \times 0.0015}$$

$$= 22.35\ cm = 8.80\ in$$

where 1 inch equals 2.54 centimeters.

Computation of Algal Concentration

The concentration of algal cells can be estimated from the following equation:

E. 5.40 $$C_c = \frac{L_t}{P}$$

where C_c equals the concentration of algae cells (ppm; in the depth calculation, we used only an average figure); L_t equals the *BOD* removal (ppm); and P equals the ratio of the weight of oxygen produced to the weight of algae produced. P has been found to be in the range 1.25 to 1.75, with a value of 1.64 recommended.

Computation of Detention Time and Area

The detention time is the average length of time the wastewater will stay in the oxidation pond and is given by the following formula:

E. 5.41 $$DT = \frac{(h \times L_t \times d)}{(F \times P \times T_c \times S)} \times 0.001$$

where DT equals the detention time (days); h equals the unit heat combustion of algae (6 cal/mg); L_t equals the *BOD* removal (ppm); d equals the depth (cm); F equals the efficiency of energy conversion (about 0.06); P equals 1.64; T_c equals a temperature coefficient from Table 5.14; and S equals the insolation (visible radiation) from Table 5.12.

The efficiency of light conversion (F) in outdoor oxidation ponds ranges between 1 and 10 percent, with most values in the narrow range of 3 to 7 percent. In general 6 percent is a good approximation.

Table 5.14 Temperature Coefficients for Chlorella

Mean Temperature (°F)	Photosynthetic Temperature Coefficient (T_c)
32	—
41	0.26
50	0.49
59	0.87
68	1.00
77	0.91
86	0.82
95	0.69
104	—

Notes: a. From "Photosynthesis in Sewage Treatment" (1957) by W.J. Oswald and H.B. Gataas. Data determined in pilot-plant studies.

The *visible* solar radiation (S) can be found in Table 5.12. For example, in Sacramento (38°N) in December, S equals 34 cal/day-cm². Corrections still need to be made for percent clearness and elevation. We still use Equation 5.37 for cloudiness corrections, but there is another formula for elevation corrections for visible radiation:

E. 5.42 $$S = (\text{total visible at sea level}) \times [1 + (0.00925 \times \text{elevation})]$$

Example: Consider a small group of 40 people using flush toilets and a methane digester where the digester effluent has a *BOD* of about 125 ppm. We want to design an oxidation pond for the winter conditions used in the previous example (the depth calculation in San Francisco, at 37°N latitude and an elevation of 1000 feet; clear weather 50 percent of the time in December). Let us also suppose that the mean temperature is 50°F.

Solution: We can compute C_c by knowing L_t and using the value for P of 1.64. In this case, let's suppose L_t equals 125 ppm. Then

$$C_c = \frac{L_t}{P} = \frac{125}{1.64}$$

$$= 76.2 \text{ ppm}$$

The depth of the oxidation pond can now be computed as before from Equation 5.36:

$$d = \frac{\ln(816.8)}{76.2 \times 0.0015}$$

$$= 58.71 \text{ cm} = 23.11 \text{ in}$$

Quite a change! The depth has gone from less than 9 inches to over 23 inches. Now we need to compute the detention time (DT) of the pond. Using Equations 5.37 and 5.42 with the value for *visible* radiation at 38°N latitude from Table 5.12, we correct for an elevation of 1000 feet for 50 percent cloudiness:

$$S = \min + (\max - \min)\left(\frac{\%\text{ time clear}}{100}\right)$$

$$= 30 + (77 - 30)\left(\frac{50}{100}\right)$$

$$= 53.5 \text{ Langleys/day}$$

Now correcting for altitude where elevation is in thousands of feet:

$$S = (\text{visible at sea level}) \times [1 + (0.00925 \times \text{elevation})]$$

$$= 53.5 \times [1 + (0.00925 \times 1)]$$

$$= 54.0 \text{ Langleys/day}$$

From Table 5.14, we find T_c equals 0.49 for 50°F mean temperature. Assuming we remove 80 percent of the *BOD* (a good average value) from our effluent, we can now compute the detention time with Equation 5.41:

$$DT = \frac{6 \times 100 \times 58.71}{0.06 \times 1.64 \times 0.49 \times 54.0} \times 0.001$$

$$= 13.5 \text{ days}$$

From Table 5.9, we estimate the average flow Q to be 2000 gallons per day for a community of 40.

We now know the detention time (DT), the depth (d), and the flow (Q); the surface area of the pond can be computed from the volume (V) which is given by the equation

E. 5.43 $$V = Q \times DT$$

$$= 2000 \times 13.5$$

$$= 27{,}000 \text{ gal} = 3609.6 \text{ ft}^3$$

The area (A) is then given by

E. 5.44 $$A = \frac{V}{d}$$

$$= \frac{3610}{23.1} \times 12$$

$$= 1875 \text{ ft}^2$$

or roughly a square 43 feet on a side—a fair-sized pond. Notice that the volume was given in cubic feet and the depth in inches; that is why we multiplied V/d by 12.

Construction and Maintenance

The ideal soil for pond construction is relatively impervious so that there will not be excessive seepage (concrete-lined ponds are actually best). Embankments around ponds should be constructed of compacted impervious material and have inside slopes of 2.5:1 as a maximum and 4:1 as a minimum, with outside slopes 2:1 as a minimum. The inside surface should be sodded as well as the outside; for small oxidation ponds, a plastic coat can be used. Top width should not be less than 8 feet and the freeboard should be 2 feet at a minimum, with more where considerable agitation is expected. Overflows may be constructed at the side. Care should be taken to prevent bank erosion at outlets or inlets if these also are in the embankments. The pond bottom should be level. As a final suggestion, you might enclose the pond to hinder access to animals and children. (For further information on construction details, see Chapters 3 and 7 and the literature listed in the Bibliography.)

Detention times should be kept near 3 or 4 days in the winter and 2 or 3 days in the summer. Ponds should be kept at a depth of 8 to 10 inches in the winter and 12 to 18 inches in the summer. Cultures should be mixed thoroughly once a day, but not continually. These figures assume climatic conditions of Richmond, California, but give a general idea of the ranges you will encounter. Average algal yield under these conditions is about a pound of dry algae per 500 gallons of supernatant from your digester.

Other maintenance consists of elimination of emergent vegetation, care of embankments, and control of possible insects.

Another aspect of the use of oxidation ponds is a reduction in the number of pathogenic organisms in waste. In the case of a single pond, it has been reported that 90 percent of the bacteria are killed during the first 6 days of detention time. But this is for *raw* sewage, not the already-treated effluent of a digester or septic tank. Even though we recommend detention times shorter than 6 days the prior use of the digester and/or septic tank has reduced the hazard significantly.

For those instances when the oxidation pond is fed raw sewage, the only way to obtain a better efficiency figure is to build ponds in series, which takes the wastewater of several thousand people to be viable. If, instead, we try to increase the detention time to treat the wastes longer, we run the risk of creating anaerobic conditions. Under such conditions, mortality of pathogenic fecal organisms appears to be very low. Rapid oxidation of organic matter accelerates the accumulation of sludge in a pond, especially in a small one. We can offset this sludging, which is a source of anaerobic conditions, by increasing the dilution factor of our supernatant input, controlling our algae crop, distributing the load through multiple inlets or through one in the middle of the pond, and orientating the pond to obtain the maximum benefit of wind mixing and aeration. Normally, the water at the output of the oxidation pond can be used for irrigation, or it can be placed in a stream or receiving water, if necessary; but we recommend you have a bacteriological test done on it first.

Harvesting and Processing Algae

Methods of harvesting and processing depend on the uses to which the algae will be applied. For example, a high-grade algae is required for use in a digester, and a higher grade yet is necessary when algae are used as a livestock feed.

Processing involves three steps: initial concentration or removal, dewatering, and final drying. The difficulty with harvesting algae lies in the small size and low specific gravity of the particles, characteristics which give them a slow settling rate.

Concentration can be accomplished most easily by precipitation and there are several means available. Cationic flocculants under the trade names Purifloc 601 and Purifloc 602 can be used to gather algae into a cohesive mass. At concentrations of 10 ppm (at pennies per ton of harvest), 100 percent removal of the algae is possible; at 3 ppm, 95 percent removal is possible. In terms of speed, with concentrations of 10 ppm, 90 percent removal is possible with a 4-hour settling time, while 98 percent removal is possible for a 24-hour settling time. The use of lime to raise pH above 11 causes a gelatinous coagulation of the algae. Use of 40 ppm of ferric sulphate and 120 ppm of lime gives the best harvesting results, but the slurry and supernatant then contain objectional amounts of iron. Alum also can be used at neutral pH for precipitation; concentrations of 90 ppm give 98 percent removal with settling complete in 2 to 3 hours.

Dewatering, the second step, can be accomplished by centrifugation, filtering, or by use of a sand bed (ranked in descending order of expense). Centrifugation involves far too substantial a cost for a small-scale operation. But industrial nylon filters give concentrations of 8 to 14 percent solids within 24 hours; the speed of filtering, however, rapidly decreases as an algae cake forms on the nylon. Also, a 2-inch slurry on paper filters on a sand bed dewatered and dried to 12 to 15 percent solids in 24 hours.

A sand bed can be used for both dewatering and drying through drainage evaporation. The amount of sand embedded in the dried product increases with sand particle size. Golueke and Oswald used sand which passed through a 50-mesh screen (opening 0.297 mm); with a slurry depth of 5 inches, they obtained 7 to 10 percent solids after 24 to 48 hours, and after 5 to 7 days the algae contained only 15 to 20 percent water. The dry algae chips then were collected by raking. These chips were sieved over a 0.16-cm mesh screen to remove most of the sand, 2 to 3 percent of the total dry weight. For a slurry of 1.6 percent solids, Golueke and Oswald estimated that a square foot of drying bed would be needed for each 7 square feet of pond area. Although sand dewatering and drying was the most economical of the processes they considered, they indicated that cost increased substantially for all processes as the scale of the operation decreased.

Final Comments

Hopefully, enough now has been said to allow you to select the most appropriate waste-handling technique or process for your needs—be it digester, septic tank, or Clivus Multrum. You will undoubtedly work through several designs as you evaluate the trade-offs between costs, benefits, convenience, and reliability; these systems are most suitable for a rural or suburban setting. But no matter which design (if any) you select, there are always a few things to keep in mind. Always design from a conservative position—overestimate your needs and underestimate your supply, be it your water supply or your gas or disposal needs. A second point is that when you move to the operation and maintenance of your systems, you must always be concerned with factors of health and safety. The potential for explosion with a methane digester is always present and contamination of a water supply is always possible—proper operation and maintenance procedures minimize, but do not remove, dangerous possibilities. With these few points in mind, you should be able to move toward a higher degree of self-sufficiency both safely and efficiently.

Bibliography

Methane Digesters

Barker, H.A. 1965. "Biological Formation of Methane." *Industrial and Engineering Chemistry* 48:1438–42.

Buswell, A.M. 1947. "Microbiology and Theory of Anaerobic Digestion." *Sewage Works Journal* 19:28–36.

———. 1954. "Fermentations in Waste Treatment." In *Industrial Fermentations*, eds. L.A. Underkofter and R.J. Hickey. New York: Chemical Publishing Co.

Buswell, A.M., and Mueller, H.F. 1952. "Mechanism of Methane Fermentation." *Industrial and Engineering Chemistry* 44:550–52.

DeTurk, E.E. 1935. "Adaptability of Sewage Sludge as a Fertilizer." *Sewage Works Journal* 7:597–610.

Eckenfelder, W.W., and O'Connor, D.J. 1961. *Biological Waste Treatment*. London: Pergamon Press.

Erickson, C.V. 1945. "Treatment and Disposal of Digestion Tank Supernatant Liquor." *Sewage Works Journal* 17:889–905.

Fischer, A.J. 1934. "Digester Overflow Liquor—Its Character and Effect on Plant Operation." *Sewage Works Journal* 6:956–65.

Fry, L.J. 1974. *Practical Building of Methane Power Plants for Rural Energy Independence*. Santa Barbara, Calif.: Standard Printing.

Gilbert, K. 1971. "How to Generate Power from Garbage." *Mother Earth News* 3.

Greely, S., and Valzy, C.R. 1936. "Operation of Sludge Gas Engines." *Sewage Works Journal* 8:57–63.

Kappe, S.E. 1958. "Digester Supernatant: Problems, Characteristics and Treatment." *Sewage and Industrial Wastes* 30:937–52.

Kelly, E.M. 1937. "Supernatant Liquor—A Separate Sludge Digestion Operating Problem." *Sewage Works Journal* 9:1038–42.

Klein, S.A. 1972. "Anaerobic Digestion of Solid Wastes." *Compost Science* 13:6–11.

McCabe, B.J., and Eckenfelder, W.W., Jr. 1958. *Biological Treatment of Sewage and Industrial Wastes*, vol 2. New York: Reinhold.

McCarty, P.L. 1964. "Anaerobic Waste Treatment Fundamentals." A series of monthly articles in *Public Works*, September–December 1964.

McNary, R.R., et al. 1951. "Methane Digestion Inhibition by Limonene." *Food Technology* 5.

Merrill, R., and Fry, L.J. 1973. *Methane Digesters for Fuel Gas and Fertilizer*. Santa Barbara, Calif.: New Alchemy Institute.

Merrill, R.; Missar, C.; Garge, T.; and Bukey, J., eds. 1974. *Energy Primer*. Menlo Park, Calif.: Portola Institute.

Metcalf and Eddy, Inc. 1972. *Wastewater Engineering—Collection, Treatment, Disposal*. New York: McGraw-Hill.

Morris, G. 1973. "Methane for Home Power Use." *Compost Science* 14:11.

Nishihara, S. 1935. "Digestion of Human Fecal Matter, with pH Adjustment by Air Control (Experiments with and without Garbage)." *Sewage Works Journal* 7:798–809.

O'Rouke, J.T. 1968. *Kinetics of Anaerobic Waste Treatment at Reduced Temperatures*. Ph.D. Dissertation, Stanford University.

Perry, J.H. 1963. *Chemical Engineers Handbook*. New York: McGraw-Hill.

Rudolfs, W., and Fontenelli, L.J. 1945. "Relation Between Loading and Supernatant Liquor in Digestion Tanks." *Sewage Works Journal* 17:538–49.

Sanders, F.A., and Bloodgood, D.E. 1965. "The Effect of Nitrogen-to-Carbon Ratio on Anaerobic Decomposition." *Journal Water Pollution Control Federation* 37:1741–52.

Sawyer, C.N., and McCarty, P.L. 1967. *Chemistry for Sanitary Engineers*. 2nd ed. New York: McGraw-Hill.

Singh, R.B. 1971. *Bio-gas Plant—Generating Methane from Organic Wastes*. Ajitmal, Etawah (U.P.) India: Gobar Gas Research Station.

Speece, R.E., and McCarty, P.L. 1964. "Nutrients Requirements and Biological Solids Accumulation in Anaerobic Digestion." *First International Conference on Water Pollution Research*. London: Pergamon Press.

Standard Methods. 1971. *Standard Methods for the Examination of Water and Wastewater*. 13th ed. Washington, D.C.: American Public Health Association.

Taiganides, E., and Hazen, T. 1966. "Properties of Farm Animal Excreta." *Transactions of American Society Agricultural Engineering* 9.

Other Waste-Handling Techniques

Babbit, H.E. 1937. *Engineering in Public Health*. New York: John Wiley and Sons.

Babbit, H.E., and Bauman, E. 1958. *Sewerage and Sewage Treatment*. New York: John Wiley and Sons.

Basbore, H.B. 1905. *The Sanitation of a Country House*. New York: John Wiley and Sons.

Caldwell, D.H. 1946. "Sewage Oxidation Ponds Performance, Operation and Design." *Sewage Works Journal* 18:433–58.

Canale, R.P., ed. 1971. *Biological Waste Treatment*. New York: Wiley-Interscience.

Clark, J.W.; Viessman, W.; and Hammer, M. 1971. *Water Supply and Pollution Control*. Scranton, Pa.: International Textbook Co.

Cox, C.R. 1946. *Laboratory Control of Water Purification*. New York: Case-Sheppard-Mann.

Ehler, V.M. 1927. *Municipal and Rural Sanitation*. New York: F.D. Davis Co.

Ehler, V.M., and Steele, E.W. 1950. *Municipal and Rural Sanitation*. 4th ed. New York: McGraw-Hill.

Fair, G.M., and Geyer, J.C. 1954. *Water Supply and Waste-Water Disposal*. New York: John Wiley and Sons.

Fair, G.M.; Geyer, J.C.; and Okun, D.A. 1968. *Water and Wastewater Engineering*. New York: John Wiley and Sons.

Gainey, P.L., and Lord, H. 1952. *Microbiology of Water and Sewage*. New York: Prentice-Hall.

Golueke, C.G. 1970. *Comprehensive Studies of Solid Waste Management; First and Second Annual Reports*. Washington, D.C.: Bureau of Solid Waste Management, Public Health Service, U.S. Department of Health, Education and Welfare.

Golueke, C.G., and Oswald, W.J. 1957. "Anaerobic Digestion of Algae." *Applied Microbiology* 5:47–55.

———. 1959. "Biological Conversion of Light Energy to the Chemical Energy of Methane." *Applied Microbiology* 7:219–27.

———. 1963. "Power from Solar Energy—Via Algae-Produced Methane." *Solar Energy* 7:86.

———. 1965. "Harvesting and Processing Sewage-Grown Planktonic Algae." *Journal Water Pollution Control Federation* 37:471–98.

Herring, S.A. 1909. *Domestic Sanitation and Plumbing*. New York: Gurney and Jackson.

Hills, L.D. 1972. "Sanitation for Conservation." *Ecologist*, November 1972.

Hopkins, E.S., and Elder, F.B. 1951. *The Practice of Sanitation*. 1st ed. Baltimore: Williams and Wilkins Co.

Hopkins, E.S., and Schulze, W.H. 1954. *The Practice of Sanitation*. 2nd ed. Baltimore: Williams and Wilkins Co.

Imhoff, C. 1956. *Sewage Treatment*. New York: John Wiley and Sons.

Jayangoudar, I.S.; Kothandaraman, V.; Thergaonkar, V.P.; and Shaik, S.G. 1970. "Rational Process Design Standards for Aerobic Oxidation Ponds in Ahmedabad." *Journal Water Pollution Control Federation* 42:1501–14.

Kibbey, C.H. 1965. *The Principles of Sanitation*. New York: McGraw-Hill.

Ligman, K.; Hutzler, N.; Boyle, W.C. 1974. "Household Wastewater Characterization." *Journal of the Environmental Engineering Division, ASCE* 100:201–13.

Marais, G.R. 1974. "Fecal Bacterial Kinetics in Stabilization Pond." *Journal of the Environmental Engineering Division, ASCE* 100:119–39.

Minimum Cost Housing Group. 1973. "Stop the Five Gallon Flush." Available from: School of Architecture, McGill University, P.O. Box 6070, Montreal 101, Quebec, Canada.

Mitchell, R. 1972. *Water Pollution Microbiology*. New York: Wiley-Interscience.

Ogden, H.R. 1911. *Rural Hygiene*. New York: MacMillan.

Oswald, W.J., and Gataas, H.B. 1955. "Photosynthesis in Sewage Treatment." *Proceedings American Society Civil Engineers*, Separate no. 686.

———. 1957. "Photosynthesis in Sewage Treatment." *Transactions American Society Civil Engineers* 122:73–97.

Rich, G. 1963. *Unit Processes of Sanitary Engineering*. New York: John Wiley and Sons.

Skinner, S.A., and Sykes, G., eds. 1972. *Microbial Aspects of Pollution*. Society for Applied Biology Symposium Series, no. 1. New York: Academic Press.

Steel, E.W. 1960. *Water Supply and Sewerage*. New York: McGraw-Hill.

Wagner, E.G., and Lanoix, J.N. 1958. *Excreta Disposal for Rural Areas and Small Communities*. Geneva, Switzerland: World Health Organization.

Wiley, J.S., and Kochtitzky, O.W. 1965. "Composting Developments in the U.S." International Research Group on Refuse Disposal, Bulletin no. 25. Washington, D.C.: U.S. Department of Health, Education and Welfare.

Wolman, A. 1967. *Water, Health and Society*. Bloomington: Indiana University Press.

WATER

Water and where to find it

Capture and storage techniques

Solar distillation and hydraulic rams

Figuring your water requirements

Making it fit to drink

Water Supply

In this section we discuss several alternative water supply systems, providing you with basic information required to evaluate your available water sources. No attempt has been made to be totally complete; rather, emphasis here is placed on very simple systems which can generate a small supply adequate for up to twenty or twenty-five people.

As you would expect, water-quality requirements vary considerably depending upon the expected use. If you are going to use the water as a domestic drinking supply, then the water must be free of water-borne diseases, toxic materials, or noxious mineral and organic matter. For the irrigation of crops, on the other hand, water-quality requirements are much lower and more easily met. Even here, however, there are cautions you must be aware of; for instance, high concentrations of sodium can be detrimental to some types of soils, causing them to become impermeable to water. The best approach in making a decision on the quality of a given water source for a particular use is to seek out the advice of a competent authority and then decide whether you can use the water source directly or whether the water must be treated. Later in this section we will discuss some simple treatment techniques.

Along with establishing the quality requirements for a water supply and assessing how far existent conditions meet these requirements, you must determine *how much* water is needed. Some typical requirements for different household and farm uses are given in Table 6.1. There is no foolproof method of determining the quantity of water needed per capita. Consumption fluctuates for different locations, climates, and conditions. For example, a person will use less water if he has to pump it by hand. But the estimates drawn from this table should be adequate to establish the basic size requirements for the water supply and distribution system.

Several additional pieces of information might be useful. First, if you have a garden and irrigate it using a three-quarter-inch hose with a quarter-inch nozzle, you need approximately 300 gallons per hour for this task alone. Sprinkling lawns will require another 600 gallons per day for every 1000 square feet. Another important consideration is fire protection. To fight a small fire you should always have available an adequate "first-aid" water supply giving a discharge of about 3 to 10 gal/min; 10 gal/min is sufficient for one quarter-inch nozzle and a storage tank of 600 gallons can supply water at that rate for an hour.

Once you have made a preliminary assessment of the amount of water needed, you must figure out whether there is enough water available to meet your needs. To answer this question may take more time and understanding than to evaluate the amount of water needed; but until you arrive at the answers to *both* questions, planning cannot proceed. It would be wise, too, to check into what state laws might apply in your area, at this time. Assuming you have arrived at a reasonable estimate of the quantity of water required

Table 6.1 Water Quantity Requirements [a]

Type of Use	Gallons Per Day
Domestic Use [b]	
Household with	
1 hand pump	10
1 pressure faucet	15
Hot and cold running water	50
Camps and schools	
Work camp, hot and cold running water	45
Camps, with flush toilets	25
Camps, without running water or flush toilets	5
Small dwellings and cottages	50
Single family dwelling	75
Farm Use (animal consumption)	
Horses or mules [c]	12
Sheep or goats [c]	2
Swine [c]	4
Brood sows, nursing [c]	6
Dairy cows, average [c]	20
Calves [c]	7
Chickens [d]	6
Turkeys [d]	20
Ducks [d]	22
Flushing floors for sanitation [e]	10

Notes: a. From *Planning for an Individual Water System* (AAVIM) and *Water Supply for Rural Areas and Small Communities* (Wagner and Lanoix).
b. Per person.
c. Per head.
d. Per hundred birds.
e. Per hundred square feet.

to meet your expected needs (from Table 6.1), we can proceed to an analysis of the various sources of water supply and the criteria for making the best source available.

SOURCES

The two basic criteria for the selection of a water source for your water supply are the quality of the water source and the relative location of the water source with respect to the area of intended use. Using these criteria, we can weigh water sources in general terms and rank them accordingly: 1) water which requires no treatment to meet bacteriological, physical, or chemical requirements, and can be delivered through a gravity system; 2) water which requires no treatment to meet bacteriological, physical, or chemical requirements, but must be pumped to consumers (well supplies fall into this category); 3) water which requires simple treatment before it meets requirements, but can be delivered by a gravity system; and 4) water which requires simple treatment, and must be pumped to the consumer (obviously the most expensive). These rankings are very general in nature and your actual costs in time, labor, and money must be examined carefully before a final decision is made.

Groundwater

Every time it rains, part of the rainfall percolates into the soil and is deposited in formations of pervious materials. These water-bearing formations are called *groundwater aquifers* or just *aquifers*. Aquifers may be confined between two impervious layers or may be unconfined, and the upper water surface of an unconfined aquifer is called the *water table*. The water table in most aquifers is not constant; it rises during rainy seasons and falls during dry seasons. Consequently, it is not unusual for a well to go dry during extended rainless periods. Another common cause of a low water table is excessive pumping of the well.

Groundwater is one of the main water sources in rural areas; in earlier times, an adequate supply was the limiting factor in many homestead regions. Groundwater has many advantages as a water supply. It is likely to be free of pathogenic bacteria and, in most cases, it can be used directly without further treatment. Groundwater aquifers can also be used as reservoirs to store excess water. Certain water districts in California use this technique to minimize evaporation losses of water normally stored behind dams.

Locating aquifers is often not an easy or inexpensive job. You can get a good indication of the existence of groundwater by examining the depth and productivity of other wells in the vicinity of your land. In addition, geological studies done by state and federal agencies (e.g., U.S. Geological Survey) can provide useful information. Proximity to rivers and lakes indicates a water table close to the ground surface. But in any case, you should seek the advice of an expert in this matter. Sinking a dry well can be a very expensive learning experience! Engineers have such special aids as seismic and resistivity surveys to help them locate aquifers.

Once a good groundwater source has been located, there are several ways of tapping the aquifer, including wells, infiltration galleries, and—if you are really fortunate—springs; your choice will depend upon the location and formation of the aquifer (Figure 6.1).

Wells

Drilled wells are the most common means of extracting water from aquifers. However, wells may also be

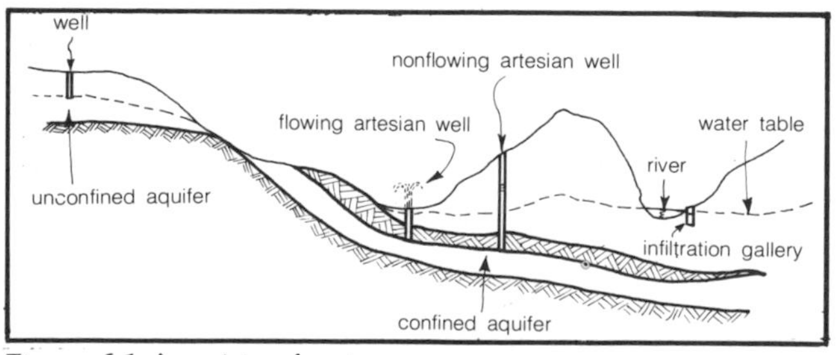

Figure 6.1 A variety of water sources.

constructed by such techniques as digging, jetting, and driving. Special equipment and experienced personnel are required for drilling, jetting, or driving a well.

Dug wells are relatively cheap, requiring simple equipment and little experience. They are seldom deeper than 50 feet, and a minimum diameter of 3 feet is required for construction purposes: all dug wells require a lining to protect the water source and to prevent the walls from collapsing, and usually this lining is built as the construction proceeds. Lining materials may be masonry, brickwork, steel, or timber. The construction of dug wells is a slow, tedious process, and it can be difficult to penetrate some hard strata or layers in the soil. An additional problem is the difficulty of digging far enough below the water table to allow a sufficiently high yield.

Driven wells are constructed by driving (literally pounding) a perforated pipe (2.5 inches or smaller in diameter) into an aquifer. These types of wells are typically constructed in unconsolidated materials (soft earth) and may be as deep as 60 feet.

Drilled wells, with which you are probably most familiar, are constructed using special machinery and techniques to drill holes from 2 to 60 inches in diameter and from a few feet to thousands of feet in depth. The cost, of course, is dependent upon the diameter, depth, and type of material to be penetrated. You will have to get cost estimates for specific sites and, in general, it is best to get at least two independent estimates—they can vary considerably from firm to firm.

Gravel-packed wells are very efficient and easily constructed once you have decided to drill for your water. These types of wells are constructed by drilling a hole larger than actually required for the pipe; a screened pipe casing then is placed into the drilled hole and gravel is packed between the casing and the well walls, as shown in Figure 6.2.

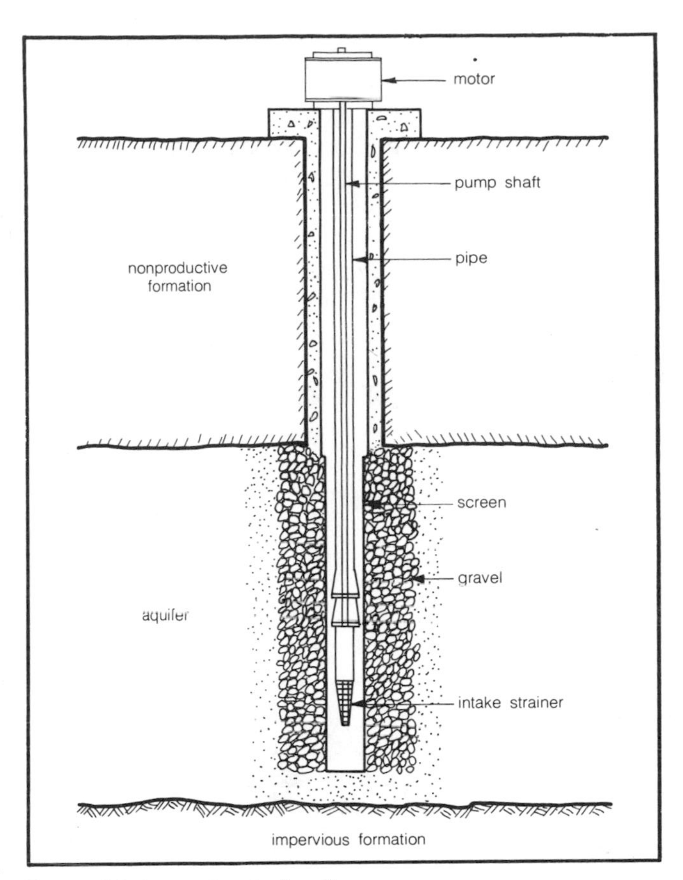

Figure 6.2 A gravel-packed well.

There are a number of precautions which must be observed in the siting, maintenance, and operation of a well, to insure that contamination does not occur. Many incidents of typhoid fever and other water-borne diseases have been caused by polluted wells. Keep your well at least 50 to 100 feet from any potential source of bacterial pollution; this means that animals should be kept clear of the well area. Lengthy discussions on details of this subject are given in *Water Supply for Rural Areas and Small Communities* (Wagner and Lanoix) and in *Planning for an Individual Water System* (AAVIM).

All outhouses, septic-tank lines, and sewer lines must be placed well away and downhill (if possible) from the well site. Although bacteria are usually removed from water percolating through soils, there is always the chance of short-circuiting and, hence, contamination of the water source. Wells can also become contaminated from surface water. Refuse, compost, and animal-feeding areas must be situated to avoid drainage toward the well. No chemicals, toxic materials, or petroleum products can be stored on the ground near the well site. You must be sure to seal the top of the well to protect it from possible surface contamination. Since a well can become contaminated during construction, it is common practice to disinfect it before use by adding chlorine so that a concentration of about 10 parts per million (ppm) is attained—at least four hours contact time is required to assure adequate disinfection. After this initial treatment, the first few gallons of pumped water will be undrinkable because of excess chlorine compounds and they should be discarded: use your own taste as a guide, and when the taste of chlorine has subsided to the point where it is aesthetically drinkable, your water should be all right.

Infiltration Galleries

Infiltration galleries were first used over a hundred years ago. They are horizontal wells which collect water over practically their entire length and are constructed almost always in close proximity to rivers, streams, or lakes. By digging a tunnel parallel to a river, but a safe distance inland from the shore, you use the sand and soil between the river and the tunnel as a kind of filtration medium for the water as it flows through the soil towards the gallery. If the gallery is located at a distance of 50 feet or more from the river or lake shore, you can collect good-quality water since this distance is sufficient to remove particulates, including bacteria. A typical cross section of an infiltration gallery is shown in Figure 6.3.

To construct a gallery, you must first dig a trench in which it will be located. The trench should be dug carefully and special care must be taken to support the side walls during work. Since the trench must be below the water table, you will probably require pumps during the construction phase. The gallery walls can be built of either masonry or concrete; or, you might wish to use a large perforated pipe instead of actually building a gallery. The length of the gallery or pipe depends upon the quantity of water to be collected and the rate at which infiltration occurs. Careful testing of the potential capacity of the gallery should be conducted before deciding to build. See the listing in the Bibliography under Wagner and Lanoix for details.

Springs

Natural flowing springs can be one of the best water sources if the flow is adequate. Springs are either *gravity-flow* or *artesian* in nature. In gravity springs, the water flows over an impermeable layer or stratum to the surface. The yield of a gravity spring will vary with any fluctuation in the height of the water table. It is not unusual for springs to run dry by the end of the summer, so be careful when selecting a spring as a primary source for water supply: you should know what the minimum yield is likely to be.

Springs, especially the gravity type, are subject to contamination unless adequate precautions are taken. Before using the water from a spring, you should have it tested for bacterial contamination and general chemical composition. Both public health agencies and independent companies can do these tests at little or no expense.

Safety measures similar to those for wells also apply to springs. A typical protection and collection structure is shown in Figure 6.4. Placing a ditch around the spring will divert surface runoff and thus protect it from possible contamination by surface waters; it is also advisable to exclude animals and buildings within about 300 feet.

Surface Water

Surface water originates primarily from the runoff portion of rainfall, although groundwater makes some contribution in certain locales. Those sources of surface water used to supply our needs can vary from a small stream, fed either by runoff or a spring, to a river, pond, or lake. You might also consider systems for the collection of rainfall.

The surface area draining into a stream is called the *watershed* for that particular stream. The geological and topographical features of the watershed, along with the type and density of vegetation and the human activities thereon, significantly affect the quantity and quality of the surface runoff. As water flows over the ground, it may pick up silt, organic matter, and bacteria from the topsoil.

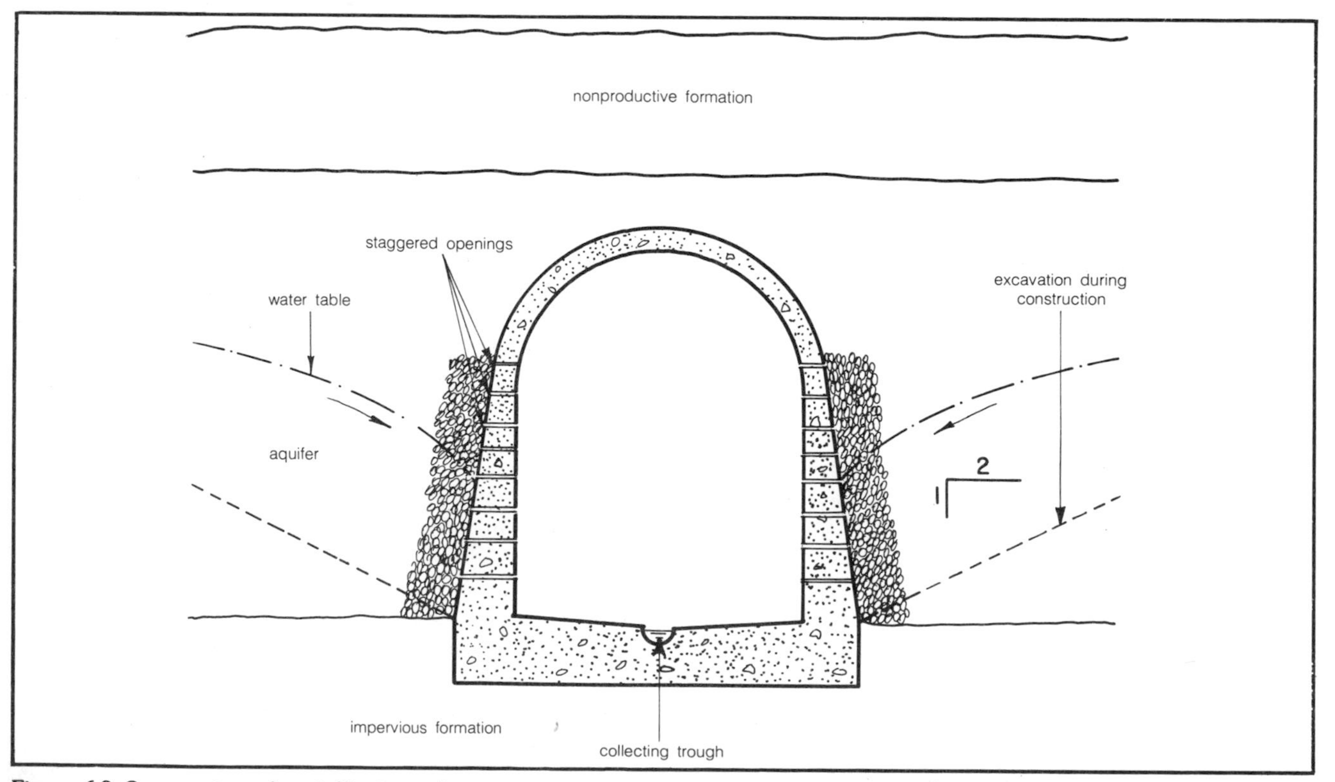

Figure 6.3 Cross section of an infiltration gallery.

In inhabitated watersheds, the water may contain industrial wastes and fecal material containing pathogenic organisms. Additional pollution results from pesticides used in the watershed. And the water will contain dissolved salts in amounts which depend upon the time of contact with and the type of soil, as well as the mineral content of any groundwater contributions. Rivers are often turbid and sometimes contain color from natural organic matter; in most cases, rivers should be considered polluted by both industrial and domestic wastes. Water from lakes is usually clearer (less turbid), but is not necessarily free of pollution. Obviously each case must be considered individually with appropriate tests.

If you decide to rely on a stream for your water needs, it is wise to take measurements at various times of the year so that you can evaluate the rate of flow during both the wet and dry seasons of the year. Techniques for these measurements are given in Chapter 3.

You can collect surface waters in artificial ponds and small reservoirs by building a dam. Dams and ponds are also discussed in some detail in the water-power section of Chapter 3. It is particularly important to prepare the reservoir site as it is there described; otherwise the water will acquire undesirable tastes and odors from decaying organic matter. As with streams, you should check for any sources of pollution or contamination in the watershed area. Actually, you will *always* want to be informed of any developments in the watershed, since it is possible that new pollution sources may appear after the construction of your pond.

Another way of providing yourself with small quantities of water is to collect rainfall from the roof of your home. This is an old technique still used in some rural areas in the United States and quite common in other countries. It is easy to estimate the average quantity to be expected if you have rainfall data for your area. This data is compiled by the U.S. National Weather Service in more than 13,000 stations throughout the United States, and is published monthly (see your local library). For example, if you live in an area where the rainfall is 30 inches per year on the average and your home has a roof with a total surface area of about 900 square feet (this is about right for a three-room house), then the quantity of water you can expect to collect over the year is

$$(900\ ft^2)\left(\frac{30\ in}{12\ in/ft}\right) = 2250\ ft^3 = 16{,}800\ gal$$

Assuming that you may have about 5 percent losses, the net available water (0.95 × 16,800) is about 16,000 gallons. This corresponds to a daily supply of 44 gallons—not bad, considering the ease of collection! This amount would cover a significant part of your domestic needs. The best practical guide we have seen on this subject is *Planning for an Individual Water System*, by the American Association for Vocational Instructional Materials (AAVIM).

Solar Distillation

Of the earth's total water mass, 97 percent is in the oceans of the world and about 2.5 percent is found in

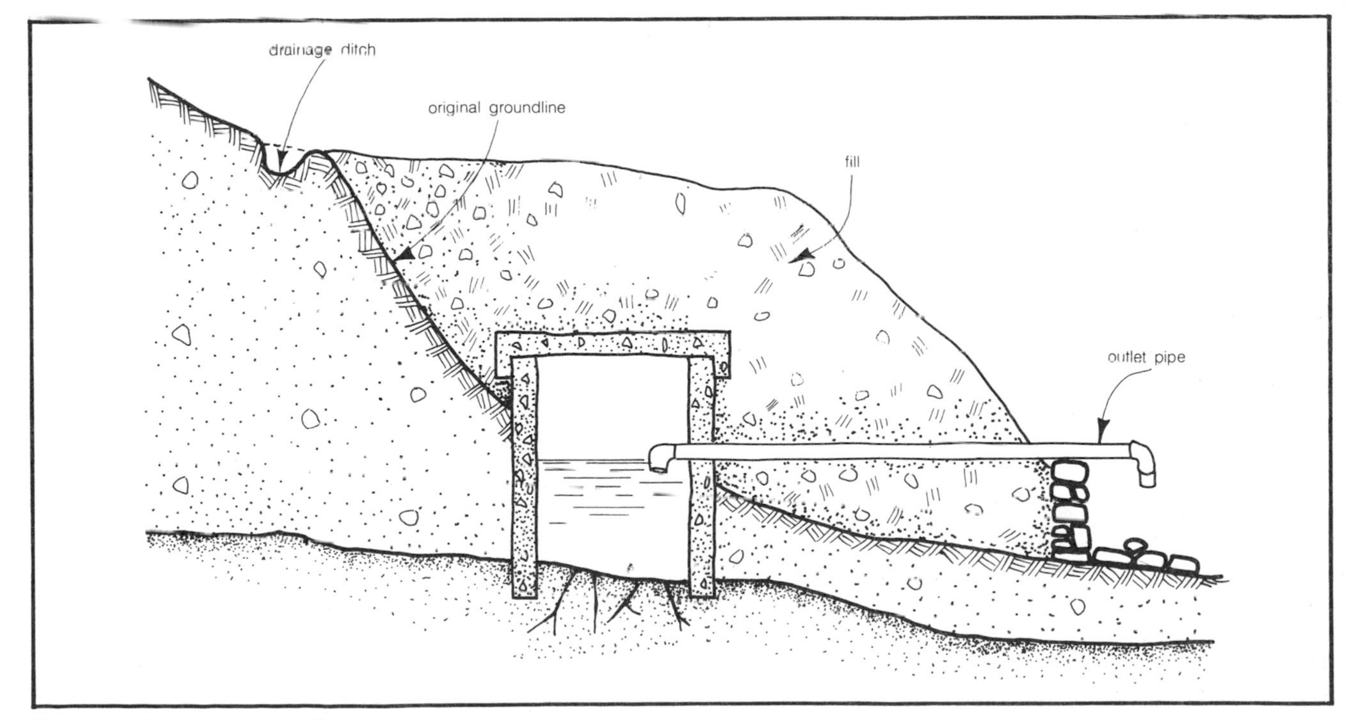

Figure 6.4 Protection and collection for a spring.

inland brackish waters. This leaves only about 0.5 percent as freshwater to be used and reused through the various methods of purification. Since solar energy is abundant in many places where freshwater is scarcest, the use of solar energy to obtain drinking water from saline sources is an attractive possibility. Mankind long ago started thinking about making the seas drinkable and Aristotle, in the fourth century B.C., described a method of evaporating ocean water to produce potable water. Much effort has been expended since then to develop desalination methods, but one characteristic of all proposed methods is the significant amount of energy required per unit of water produced. The possibility of using the endless and renewable energy of the sun to produce potable water is consequently intriguing. By the simple expedient of trapping solar energy in an enclosure containing brackish water, evaporating the water (which leaves the impurities behind), and then recondensing the water vapor in a collectable manner, we can have the sun do most of the work for us. In very simple terms, we have just described solar distillation; a detailed discussion of the availability of solar energy can be found in Chapter 4.

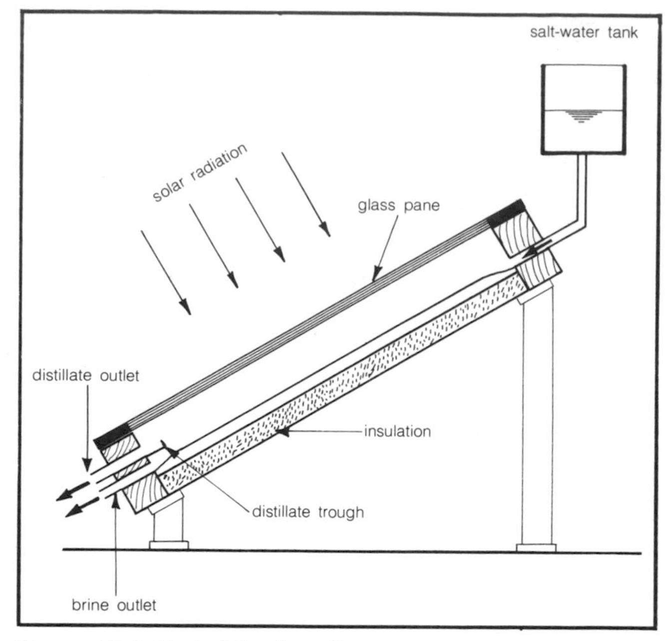

Figure 6.6 A tilted-wick solar still.

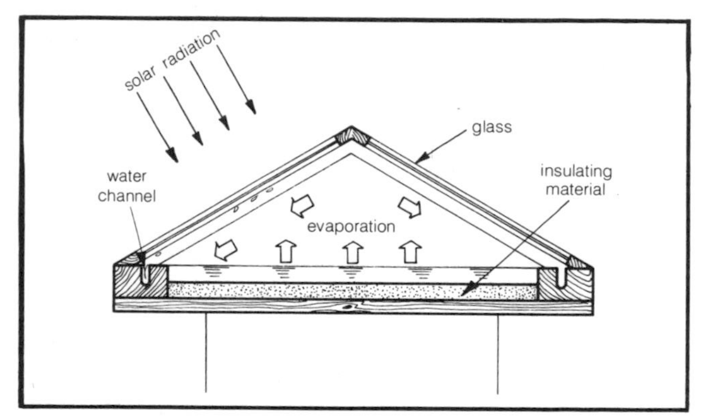

Figure 6.5 A solar still utilizing a basin.

All solar stills share the same basic concept. Figure 6.5 shows a basin solar still. The sun's rays pass through a glass cover and are absorbed by a blackened basin holding saline water. This water is heated by the energy reradiating from the basin. As the water vapor pressure increases from this heat, the liquid water evaporates and is condensed on the underside of the roof enclosure, from which it runs down into collection troughs.

Freshwater produced from a solar still is best removed by gravity flow to a storage tank of some sort, to keep the mechanical system as simple as possible. The depth of the water in the basin can vary from an inch to a foot, but since the bottom of the basin must be black in order to absorb solar radiation, you must flush out the brine from time to time to prevent the precipitation of light-reflecting salts. A reasonable rule of thumb is to replenish the brackish supply when half the volume has been evaporated.

There are a number of factors you must consider when thinking about solar distillation. The higher the latitude, the less the solar radiation that is intercepted by horizontal surfaces; unless you tilt your still toward the sun to intercept the radiation, it will lose efficiency. You may well ask, "How can we tilt a water surface?" Well, tilting a water surface can be done by providing a porous

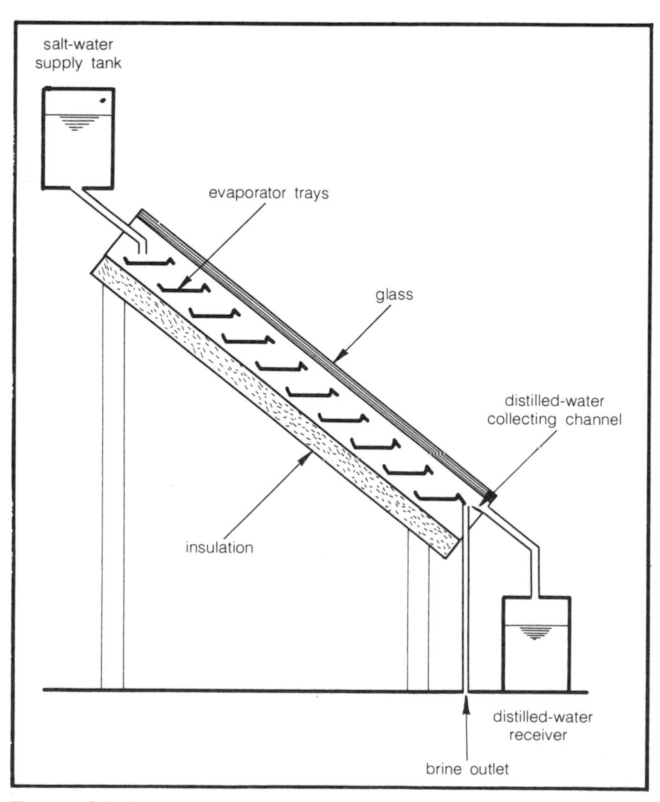

Figure 6.7 A multiple-tray tilted still.

wick or cloth (preferably black to absorb radiation) on an incline oriented toward the sun; Figure 6.6 shows an example. An inclined still with a wick or cloth allows evaporation to occur over the equivalent of a thin sheet of water as it flows down the incline. The inclined still depicted in Figure 6.7 consists of a series of shallow horizontal black trays. Brackish water overflows from tray to tray and is evaporated in the process. This last variety of stills is less efficient than the wick type and it is mostly used for brackish water rather than sea water.

Yet another type of solar still is shown in Figure 6.8. The parabolic mirror acts as a concentration reflector and provides a focusing effect for solar energy; the resulting high temperatures evaporate water rapidly. These stills must have a continuous supply of salt water to the central evaporating unit. Note that the water vapor goes into a condenser for cooling. Parabolic stills have an average productivity of about 0.5 to 0.6 gal/day-ft², but they are much more costly to build yourself (both in terms of time and money) and are presented here primarily as an indication of what is possible rather than what is likely.

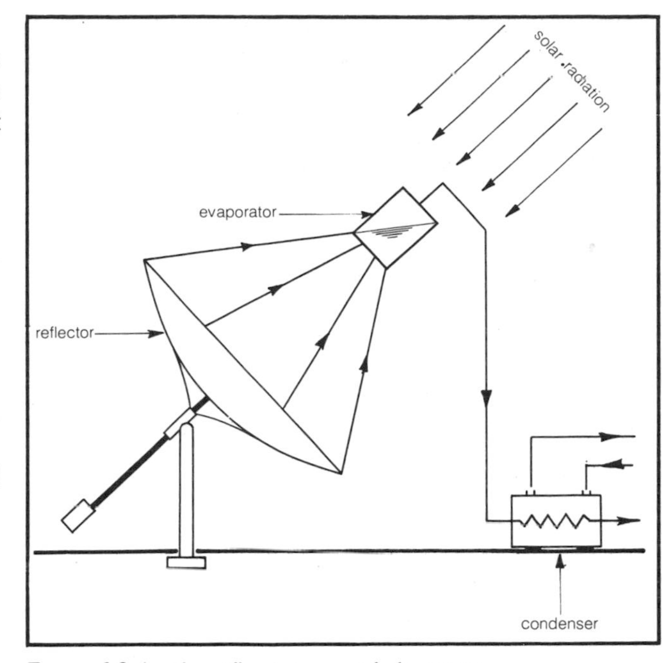

Figure 6.8 A solar still using a parabolic mirror.

Trombe and Foëx (see Bibliography) have come up with an interesting design of a solar house which incorporates solar distillation for both a water supply and air conditioning (Figure 6.9 shows a cross section). The roof is covered with a transparent plastic or glass, and immediately below are tanks filled with brackish water. Both illumination and temperature can be regulated by increasing or decreasing the surface of the tanks, through movable covers of either the tanks or the roof.

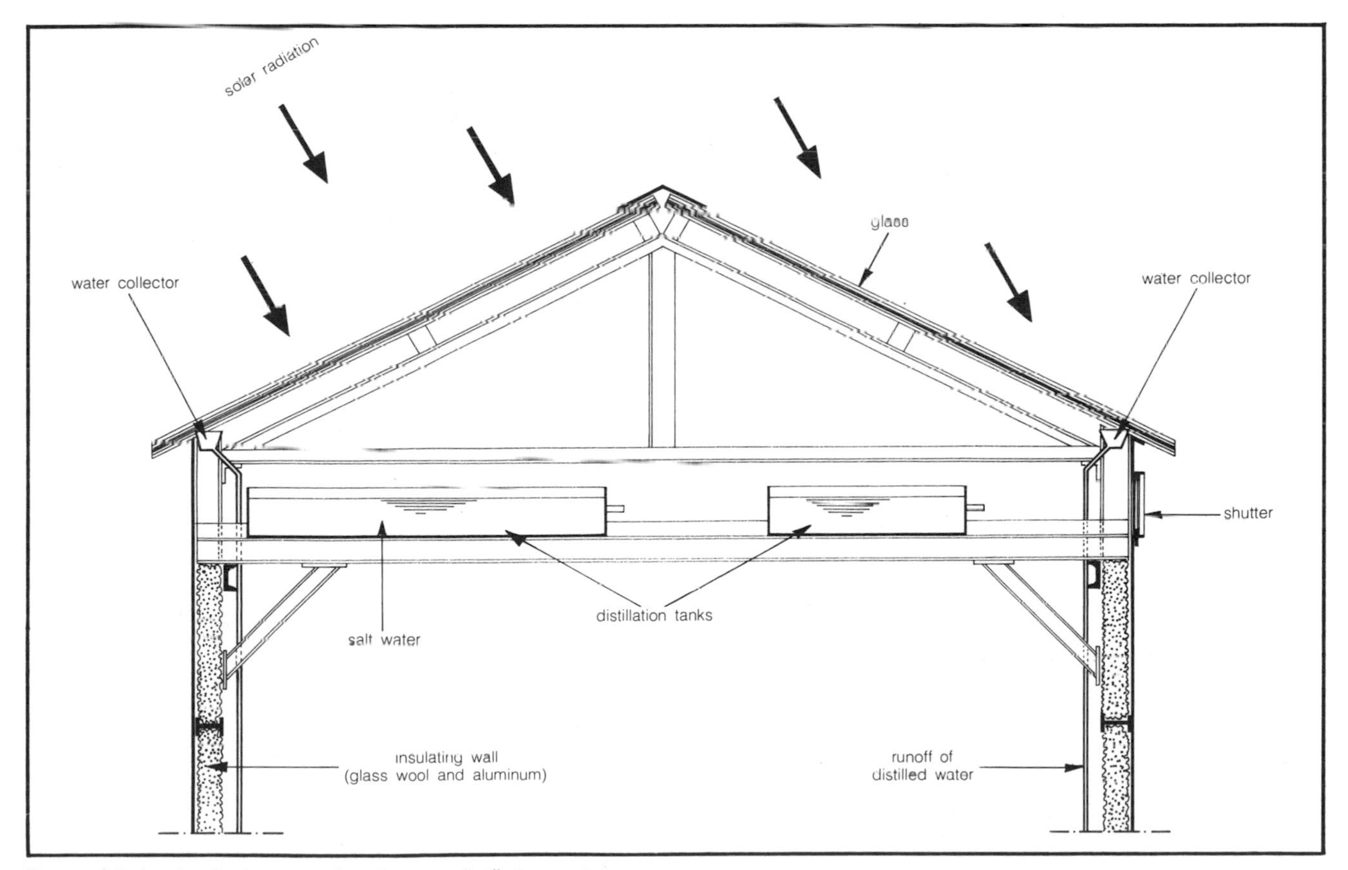

Figure 6.9 A solar hothouse with salt-water distillation system.

Table 6.2 Solar Still Construction Data [a]

Component and Material	Test Period (years)	Rating[b]	Life Expectancy (years)	Remarks
Side members				
Galv. iron	3	NR	2–10	Corrodes
Concrete	6	R	20+	
Asbestos	1	?	15+	
Basin liner				
Polyethylene	4	NR	1–4	Cracks
Butyl rubber	3	R	15+	
Distillate Trough				
Galv. iron	1	NR	1	Corrodes
Al. alloy 50/52	3	A	10+	
Al. alloy 1100	3	A	10+	
St. steel 316	3	A	10+	
St. steel 321	3	A	10+	
Cover				
Glass	13	R	20+	
Sealant				
Butyl molding	3	NR	10	
Butyl calk	3	NR	10	
Silicon	5	R	10+	
Insulation				
Polystyrene	3	A	10+	
Polyurethane	1	A	15+	

Notes: a. From *Design Philosophy and Operating Experience for Australian Solar Stills*, by P. Cooper and W. Read.
b. NR—not recommended; A—acceptable; R—recommended.

Efficiency and Production

You might think that efficiency considerations for a solar still would not be necessary because solar energy is "free." But when you stop to think about the fact that this form of energy is of low intensity and requires a lot of area to produce a usable quantity of water, it becomes clear that we had better pay some attention to the factors which can give us the highest possible efficiency for our system.

Efficiency is defined as the ratio of freshwater produced to the energy expended in producing it. Two of the important factors, our location and the salinity of our water, we are basically stuck with. Chapter 4 gives information for determining the available solar radiation of your specific locale. Areas with frequent dense cloudiness obviously will cause problems and the distillation process may become practically unfeasible.

The type of still that you choose is related to your particular needs and resources. Howe and Tleimat achieved 50 percent efficiency with a tilted-tray still yielding 0.12 gal/day-ft^2; they also claim that this unit is the most efficient design. However, the relatively complicated and expensive nature of this still has precluded common use. Current research is also investigating the use of multieffect stills. The multieffect still is composed of a number of stills in series that have a decreasing depth of water from top to bottom. This stepping produces a temperature gradient which increases the yield of freshwater. But again, while the efficiency increases, so does the cost. More information on this design is available from C. Gomella (see Bibliography).

There are several design features which can maximize efficiency for any particular type of still. Although the exact distance has not yet been pinned down, generally the closer your vaportight transparent cover is to the surface of the water, the greater the efficiency of the unit. And the higher the temperature of this glass or plastic cover, the better the efficiency. Glass temperature should be highest for units having the smallest ratio of glass area to evaporating area and also for those units with the best insulation under the tray bottom.

The greater the heat capacity of the still, the higher its efficiency. Choosing proper construction materials is thus critical. Table 6.2 lists a few commonly available materials with suitable remarks. There is, however, a sizable amount of in-progress research devoted to finding materials of greater thermal capacity, and you should check the literature frequently for the newest developments.

Reductions in efficiency also can arise from shading caused by opaque supports or neighboring stills. And if your construction is sloppy, leaks due to insufficient sealing can reduce your efficiency by half. In general, efficiencies of a solar still will range from an average of 35 percent to a maximum of 60 percent.

How large a still to construct is difficult to determine without the context of a specific plan, particularly when considering larger yields on the order of 1000 gallons per day. But in general, for a yield of 1 gal/day of freshwater, an area of 8 to 14 square feet is required; 10 square feet will provide a day's drinking water for one person. When we consider that the average per capita domestic use of freshwater in the United States is 50 gal/day, it becomes clear that a rooftop still on the order of 100 square feet, providing 10 gal/day, can hardly supply a household's per capita demand for water. On small islands in the Pacific, where freshwater is scarce and supply is dependent upon rainfall, the per capita demand is as low as 1.33 gal/day. Solar stills can play a significant part in supplying water to these isolated areas. On the other hand, where a nonpotable water supply is available for most nondrinking purposes, you can use a solar still to good advantage as a supplementary water supply. Another possible consideration is the production of high-quality water (high purity) for special uses.

The productivity of solar stills depends on the geographic location as well as the still type. On the average, glasshouse-type stills can have average annual yields of from 35 to 50 gallons per square foot. Figure 6.10 gives the daily productivity of two solar stills in Florida as a function of solar radiation. From these curves, you can estimate productivity (gal/ft^2-day); knowing your water demand you can then estimate the size of the still required to meet your needs.

Construction and Economy

The construction of a basin solar still is relatively simple. Do-it-yourself kits are available for individual stills from Solar Sunstill (Setauket, New York 11733) and Sunwater Company (1112 Pioneer Way, El Cajon, CA 92020). Materials for construction have been improved over the last ten years. For the roofs, horticultural glass, fiberglass, and new hard plastics are available. Glass is highly durable and, in the process of distillation, the distillate will form a continuous film on the inside of the glass and thus will maintain a continuous flow into the collection trough; but glass is also heavy and relatively difficult to seal. Plastics, on the other hand, are less durable and tend to tear in high winds. The transmittance of plastic is less than that of glass, mainly because plastic is a hydrophobic material. The distillate forms beads on the roof which tend to drip back into the undistilled water, reducing efficiency by about 10 percent. However, chemicals such as Sun Clear (available from Solar Sunstill) can be used to produce a wet film on the plastic rather than droplets, allowing for increased light transmittance. It appears that present designs should use glass, but that future designs will take advantage of the low cost and weight of plastic covers.

Polystyrene or foamed glass insulation is optional in the design of your still. Insulation increases material costs by about 16 percent, but the advantage is a decrease in heat loss to the ground (and hence, greater efficiency). For *all* stills, you need a black cover for the bottom of your basin or tray, to absorb the sun's radiant heat. You can paint the bottom or else line it with black earth, pebbles, sand, or cloth. Supports for your still should be kept to a minimum; metal along the ridges or concrete around the base can form shadows in the still and decrease the absorption of light. This is especially true of smaller stills.

Horace McCracken has researched various available still pans. Of the five types—wood and wood-fiber products with coatings or impregnates, thermosetting plastics, thermoplastic materials, asbestos-cement with impregnates and coatings, and metals with coatings—he concluded that porcelain-enameled steel is the optimum choice. It imposes little taste on the water, it is least affected by the water's corrosive properties (durability of 10 years or more), and its weight is a minor obstacle. In Table 6.2, you should note that durability is emphasized since life expectancy is a major factor in incurred costs.

The economics of solar distillation are for the most part based on capital investment. While taxes, land costs, financing, and labor vary from case to case, nevertheless existing setups cost on the order of \$3 to \$6 per 1000 gallons. Is it worthwhile for you to invest in a solar still? Consider the following simple example: If a solar still of

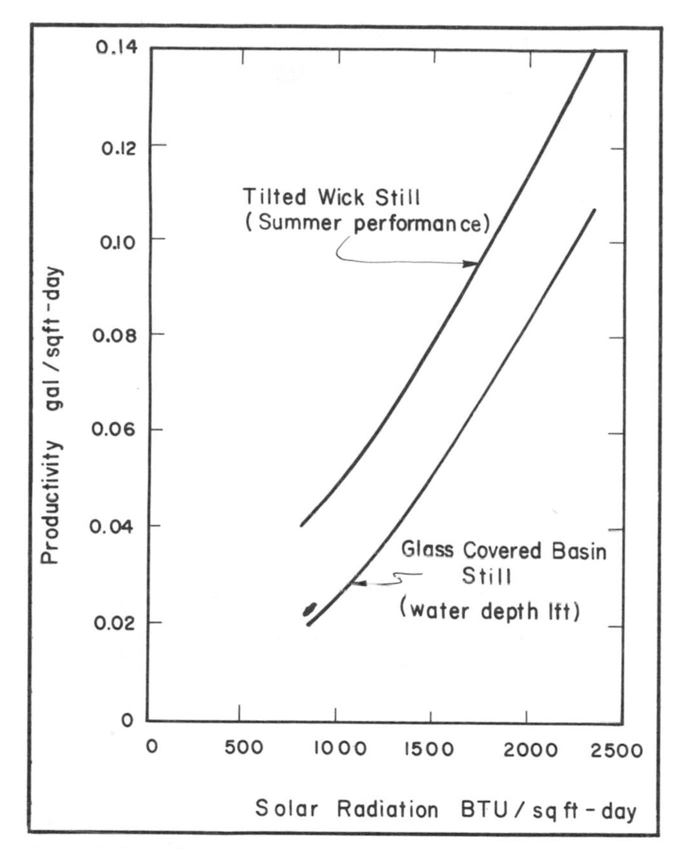

Figure 6.10 Productivity of solar stills as a function of solar radiation.

10 square feet produces 1 gal/day for 333 days a year, it will produce 1000 gallons in 3 years. Let's assume the still we are considering has a life expectancy of 15 years and we estimate it will cost us \$25 to build. It is clear that we are paying \$25 for 5000 gallons, or \$5/1000 gallons, plus labor and operating expenses to keep our still functioning properly. The usefulness of computing cost per 1000 gallons is for comparison with other sources of water and between various still designs.

Sometimes the costs for a still design are given in dollars per square foot. This unit cost for our example above (\$25/10 ft^2) would be \$2.50/ft^2.

The factors that you will have to balance are your water requirements, how much water is produced per square foot of still (in your climatic conditions) to meet these needs, how much it will cost to build a still of this required size, and how long the whole setup can be expected to keep producing at the levels you require. Longer life, higher efficiencies, and mass production of still units are the necessary general factors for cost reduction.

In the Bibliography are listings which provide cost data for particular designs. William Edmondson, for example, offers a still design with a fiberglass cover and a useful life of 20 years, giving an estimated crude unit-construction cost of 42 cents a square foot (these and all prices immediately following are for 1975). Strobel gives a deep-basin still design of 3000 square feet with a useful life expectancy of 30 years, for an estimated cost of $4.75 per square foot and a unit production cost of $10.72 per 1000 gallons. And Daniels gives cost comparisons for three solar stills that were exposed to the same conditions and yielded equal amounts of freshwater:

Still Type	*Life Expectancy (years)*	*Unit Cost ($/ft²)*
Glass-covered basin	50	$3.94
Tilted tray	10	$2.56
Plastic-covered	3	$0.87

Finally you should probably be aware of several general economic considerations which favor the solar distillation process: 1) Unit construction is not appreciably affected by the size of the still; 2) Power considerations are almost nonexistent, with the possible exception of pumps; 3) Solar stills can be constructed on-site with semi-skilled labor and you can handle the operation and maintenance without any technical training; 4) You can readily find the materials necessary for construction and they are generally very durable; and 5) Still designs are modular and easily handled by anyone.

TRANSPORT AND STORAGE

Sizing a Pump

If you have a well or your house is located above a lake or river, you will need some type of pumping device to move the water to the point of use. In order to select the proper pump for your water supply system, you must first determine what pumping capacity is required to meet your needs. The pump capacity depends upon both your rate of water consumption and the size of the storage tank you have in your system. If you have no substantial storage capacity in your system, your pump must be able to deliver water at a rate commensurate with your greatest need—usually called the *peak demand*.

For many years there was no satisfactory method of determining the pump capacity needed to supply adequate water for a particular set of conditions. At best, there was considerable guessing. The guessing now has been reduced to a minimum due to studies conducted by the U.S. Department of Agriculture regarding water use for the home, for appliances, for watering livestock and poultry, for irrigation, and for general cleaning purposes outside the home. It is now possible to determine the pump capacity needed with considerable accuracy.

There are two kinds of water usage—intermittent (5 minutes or less) and sustained (more than 10 minutes). You can determine the peak demand for your situation by adding up the various rates of water demand for all the uses you expect to encounter simultaneously. Table 6.3 gives the peak demand allowances for various uses as well as the average individual-fixture flow rate. In arriving at a peak demand for your pump, the procedures can be divided into three related groups of uses, each requiring a slightly different set of steps: 1) pump capacity needed for household uses; 2) pump capacity needed for irrigation,

Table 6.3 Peak Demand Allowance and Individual-Fixture Flow Rate for Various Uses[a]

Water Uses	Peak Demand Allowance (gal/min)	Individual-Fixture Flow Rate (gal/min)
1. *Household*		
Lavatory	0.5	2.0
Dishwasher	0.5	2.0
Toilet	0.75	3.0
Sink (no garbage disposal)	1.0	4.0
Shower only	1.0	4.0
Laundry sink	1.5	6.0
Bathtub or tub-shower combo	2.0	8.0
Clotheswasher	2.0	8.0
2. *Irrigation and Cleaning*		
Swimming pool	2.5	5.0
Lawn irrigation (per sprinkler)	2.5	5.0
Garden irrigation (per sprinkler)	2.5	5.0
Automobile washing	2.5	5.0
Equipment washing (tractor)	2.5	5.0
Cleaning milling equipment and storage areas	4.0	8.0
Flushing driveways and walkways	5.0	10.0
Hose cleaning barn floors, ramps, etc.	5.0	10.0
3. *Livestock Drinking Demands (All open-lot housing)*		
Horse, mule or steer (10 per watering space)	0.75 per space	1.5
Dairy cows (8 per watering space)	0.75 per space	1.5
Hogs (25 per watering space)	0.25 per space	0.5
Sheep (40 per watering space	0.25 per space	0.5
Chickens (100 per waterer)	0.12 per waterer	0.25
Turkeys (100 per waterer)	0.4 per waterer	0.8
4. *Garden, Fire Extinction and Other*		
Garden Hose—⅝ inch	3.5	
Garden Hose—¾ inch ¼-inch nozzle)	5.0	
Fire Hose—1½ inch (½-inch nozzle)	40.0	
Continuous flow drinking fountain	1.25	

Notes: a. Adapted from *Water Supply for Rural Areas and Small Communities* (Wagner and Lanoix) and *Planning for an Individual Water System* (AAVIM).

cleaning, and miscellaneous uses; and 3) pump capacity needed for watering livestock and poultry. As you design your system, you will need to select the group (or groups) of uses that fits your needs and follow the procedures given below for each category.

First, determine the pump capacity needed for your household uses by listing all of your home uses and the peak demand allowance for each from Table 6.3. Let's say you end up with the following list:

Home Use	*Peak Demand Allowance (gal/min)*
Tub and shower	2.0
Toilet	0.75
Kitchen sink	1.0
Clotheswasher	2.0
Total:	5.75

Make sure that all demands are accounted for, including duplicates (for example, *all* sinks). The total demand allowance for the uses listed above is 5.75 gallons per minute. If you are designing a system for a cluster of houses, use the same procedure for each house; but, after you have determined the demand allowance for each house, divide the total by 2. (Each additional house will tend to spread out the peak demand over time; this division accounts for the time spread.)

Now list all of the lawn, garden, and miscellaneous uses you have or expect to have and the water demand for each use. Let's suppose you have a small rural home with several acres available for gardening. Again using Table 6.3 for your estimated peak demand allowance you have:

Irrigation Uses	*Peak Demand Allowance (gal/min)*
Garden irrigation (3 sprinklers)	7.5
Lawn irrigation (1 sprinkler)	2.5
Hose cleaning barn floors, ramps, etc.	5.0
Tractor and equipment washing	2.5
Total:	17.5

For this category of usage, your total is 17.5 gal/min. But now you must determine which of the listed uses are *competing* with each other. Your judgment is very important here in determining which of the demands overlap each other. For example, you must consider whether both lawn and garden sprinkler systems will be operated at the same time, and whether, in addition, you will be running your hose cleaning units simultaneously. Once you have estimated the likely overlap, you can ignore the other noncompeting demands in the rest of the analysis. Let us assume that three sprinklers are likely to be on at the same time as the barn cleaning operation. We then modify our list to arrive at the following:

Irrigation Uses (competing)	*Peak Demand Allowance (gal/min)*
3 sprinklers	7.5
Barn cleaning use	5.0
Total:	12.5

Now let us assume that you have some livestock on your small acreage and you must determine the required pump capacity for this use as well. List all of the watering units you have in use or expect to have in use and the corresponding demand for each (again from Table 6.3). Let us assume you have a small operation to start with (you can expand it if you like!):

Livestock	*Peak Demand Allowance (gal/min)*
2 horses (open lot)	0.75
1 dairy cow (open lot)	0.75
30 laying hens	0.12
Total:	1.62

Notice that the table is set up per watering space or per waterer. But even though we have only one cow or two horses, we still use the figure for one entire watering space.

The next step is to determine, for each category, which use requires the *greatest fixture flow*. This step is somewhat tricky. We go back, say, to household uses and look at the second column of Table 6.2. There we see that *both* the bath/shower combination *and* the clotheswasher have individual flow rates of 8 gal/min. In the replacement step to follow, we will only replace *one* of them (it doesn't matter which). For purposes of irrigation and cleaning, the largest (competing) fixture flow is 10 gal/min for cleaning the barn. And for watering your livestock the largest fixture flow is 1.5 gal/min for the horses and cow. We then modify our list again, by putting in these values for our greatest fixture flow in place of our demand allowances:

Use	Peak Demand Allowance (gal/min)
Home	
Tub and shower	8.0
Toilet	0.75
Kitchen sink	1.0
Clotheswasher	2.0
Irrigation (competing)	
3 sprinklers	7.5
Barn cleaning use	10.0
Livestock	
2 horses (open lot)	1.5
1 dairy cow (open lot)	0.75
30 laying hens	0.12
Total:	31.62

So the total peak demand on this particular pump will be 31.62 gallons per minute. If you follow these steps carefully, you should be able to determine the pump capacity required for your own special case.

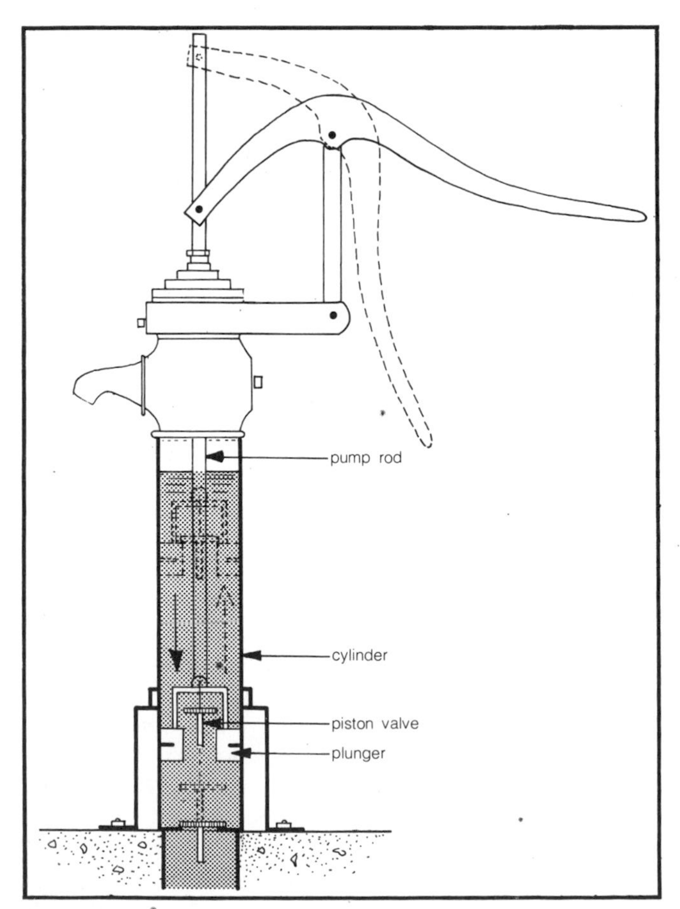

Figure 6.11 A hand-operated pump.

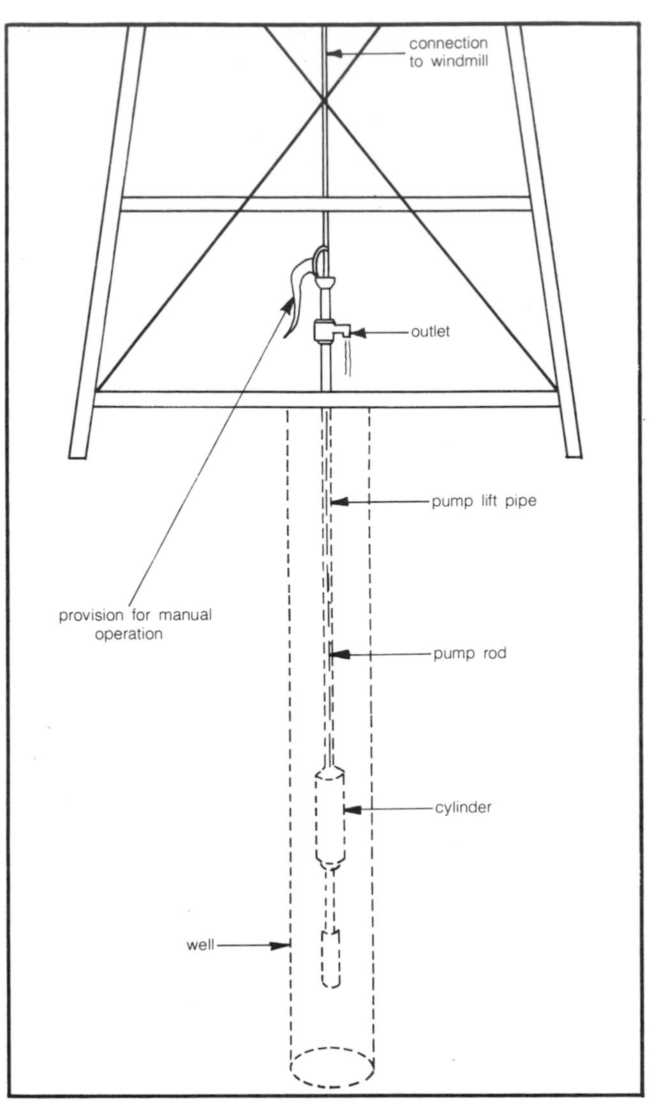

Figure 6.12 A water-pumping windmill with hand-operated pump unit.

Since most household uses do not last longer than 5 minutes, a small storage tank can cover most of your needs during the short peak demand periods and thus prevent any overload in the future. A storage tank also helps to deliver water during any small fire emergency. In fact, if you consider *real* fire protection in your pump sizing calculations, it will be the dominant requirement (see Table 6.3).

There are many kinds of pumps available and selecting one may be troublesome if you are not familiar with the advantages and disadvantages of the various types. This is the time to seek the advice of someone with lengthy practical experience in dealing with pumps (a repairman or mechanic, for instance).

Pumps can be classified according to the mode of power used to move the water: hand-powered, motor-powered, wind-powered (windmills), and water-powered (hydraulic ram). Let's consider a few of the design and

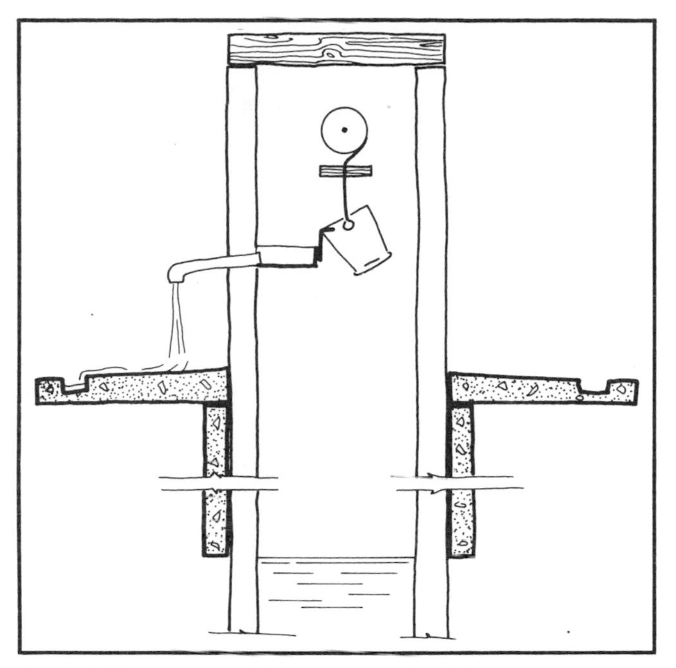

Figure 6.13 A rope and bucket system for lifting water.

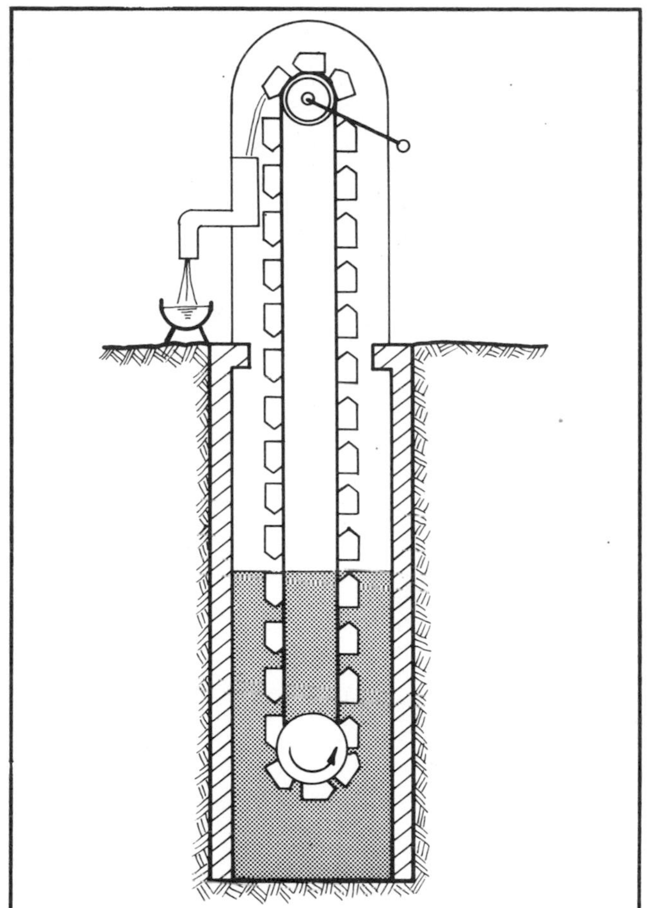

Figure 6.14 A continuous belt-bucket system.

efficiency aspects of these various types.

Hand-powered Pumps

The most popular hand-powered pump is the piston type shown in Figure 6.11. The two basic operational steps are also shown. During the downstroke, the valve in the plunger opens and the cylinder above the plunger is filled with water. At the same time, a vacuum is formed below the plunger, drawing the water into the suction pipe to open the check valve and fill the cylinder from below. This process is repeated each stroke. If the distance between the pump and water surface in your well is less than 15 or 20 feet, the cylinder can be used above ground level as the figure shows; otherwise the cylinder must be placed down the well shaft (Figure 6.12). Although plunger-type pumps have low efficiencies (from about 25 to 60 percent), they operate in a positive manner, have low maintenance requirements, and can deliver up to 10 to 15 gallons per minute.

There are many other kinds of hand-powered systems for moving water. Some of these designs are thousands of years old and are still used extensively in those places where energy has always been expensive for the common man. In many cases, animal power (horse, mule, or ox) is used to activate the system. The most simple mechanical device is probably the rope and bucket system (Figure 6.13). The design shown uses a hook to catch and tilt the bucket, discharging the water into a trough. Another simple system is the continuous belt-bucket design, shown in Figure 6.14, where a series of metal buckets are attached to a chain. The buckets are emptied as they reach the top of the chain. Many variations on this theme are evident in rural areas world-wide.

Motor-powered Pumps

There are a large number of different designs and sizes available in motor-powered pump units. We will discuss briefly three basic types: reciprocating pumps, centrifugal pumps, and jet pumps. More detailed information is available in specialty books for those intrigued by such things (see, for example, the listing for Hicks and Edwards in the Bibliography). Performance data are available from pump manufacturers or retailers and should be consulted for specific information.

The operation of reciprocating pumps is based on the same principle as that of the plunger-type hand-powered pumps. Simple single-action reciprocal pumps discharge water only on alternate piston strokes, while improved double-action pumps discharge water on each stroke, thus providing a more uniform flow. The cylinder can either sit on the ground or be seated in the well itself, depending on the depth of the water surface in your well. The efficiency of reciprocating pumps is low (25 to 60 percent) and they are usually used for relatively low flow rates of from 10 to 30 gallons per minute. Despite their

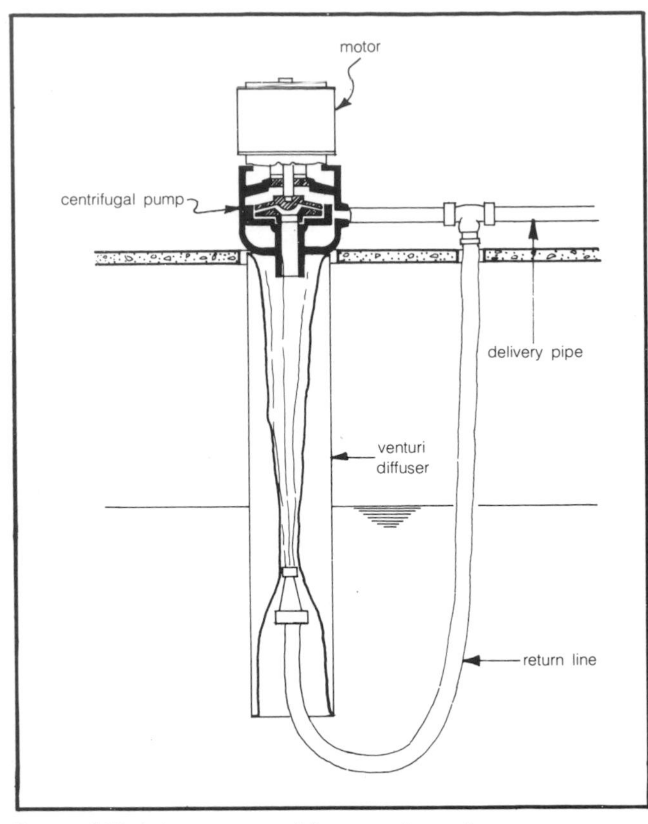

Figure 6.15 A jet pump providing pumping action.

poor efficiency, these pumps are simple, and thus easy to operate and maintain.

Centrifugal pumps operate quite differently. They have a rapidly rotating part (the impeller) which transforms kinetic energy into pressure. There are many different impeller designs. Centrifugal pumps which move water to great heights above the pump have two or more impellers, one above the other, and are termed multistage pumps. Operation of centrifugal pumps requires that they start with the casing full of water; that is, you have to prime the pump to make it work. Priming the pump removes any air trapped in the pump casing. The efficiency of centrifugal pumps varies from 50 to 95 percent and they may be used for a great range of lifts, pushing water up to 1500 feet above the level of the pump.

In jet pumps, the water is discharged from a nozzle at high velocity and is forced through a conical apparatus called a diffuser (Figure 6.15). The water velocity is increased significantly as it passes through the diffuser and this results in a drop in pressure in the pipe. The drop in pressure creates a suction, thus drawing water up into the pipe. The efficiency of these pumps is relatively low, varying from 40 to 60 percent. Jet pumps can be combined with centrifugal pumps (as shown in the figure) to increase the lift or height to which water may be driven. Used by themselves, they can move water to only relatively low heights above the pump.

Wind-powered Pumps

Chapter 3 discussed how we can use the wind to generate electricity. However, long before man thought to use wind to produce electrical energy, he was using the wind to pump water out of the ground mechanically. A simple arrangement for a windmill pump is shown in Figure 6.12. Here the energy of the wind is transformed into mechanical movement to drive a rod and piston assembly up and down in a well shaft. If you consider putting in a windmill, you might also consider providing a handle on the pump (at the base) so that water can be drawn by hand when there is no wind. The ratio of piston strokes per revolution of the windmill can be varied—from 1 stroke per revolution for high wind velocities to 1 stroke per 4 revolutions for lower wind speeds. The water-pumping capacity of a windmill varies with its design and size; typical capacities for traditional multiblade fans are given in Table 6.4.

Table 6.4 Water-Pumping Capacities for Multiblade Windmills [a]

	Pump Discharge for 15-mph Wind Velocity (gal/hr)	
Distance Water Is Lifted (ft)	**6-ft Fan Diameter**	**12-ft Fan Diameter**
25	350	24,000
50	200	1425
100	150	600
150	—	525
300	—	200

Notes: a. Adapted from *Water Supply for Rural Areas and Small Communities* by Wagner and Lanoix.

Hydraulic Rams

If you have a fairly large stream or have access to a river, you can take advantage of one of the truly fine engineering holdovers from the nineteenth century—the hydraulic ram (Figure 6.16). Here, power is derived from the inherent energy of flowing water itself. The force of the water is captured in a chamber where air is compressed; when the compressed air expands, it pushes a small amount of the water to a higher elevation than that from which it originally came. The water which provided the energy is then released to flow on its way downstream.

A typical hydraulic ram installation consists of two pipes (supply and delivery), an air chamber, and two valves (waste and delivery). You set the hydraulic ram into motion by merely opening the waste valve and allowing the water to run through the ram unit. As the flow is accelerated, the force exerted on the waste valve causes it to close, causing a so-called "water hammer" or pressure build-up. This pressure build-up forces open the delivery valve and water flows into the air chamber, compressing the air inside. Because water is flowing into the air chamber, the pressure on the waste valve is lessened; it then opens, and the delivery valve consequently closes, trapping behind it a small quantity of water and compressed air. As the compressed air expands, it shoves this water into the delivery pipe and, as this process is repeated endlessly, you find—presto!—the water is at a higher elevation, ready to use.

Rife Hydraulic Engine Manufacturing Company (P.O. Box 367, Millburn, N.J. 07041) is the only American-based supplier of hydraulic-ram pumps. They can provide you with pumps taking inlet and outlet pipe sizes from 1.25 to 8 inches. Costs run between \$300 up to \$3000 (FOB factory).

A few comments concerning the use of the hydraulic ram are in order. First of all, there is a spring on the waste valve of most hydraulic rams which regulates the pressure required to close the waste valve. You can regulate the quantity of water and the height you raise it by varying the tension on this spring. You will have to make a few experimental runs at first to decide what the best setting is for your particular needs. You can compute the quantity of water pumped by a hydraulic ram using the following equation:

E. 6.1
$$\frac{Q_p}{Q_s} = \frac{H_s}{H_p} e$$

where Q_p is the flow pumped to a new location (gal/min); Q_s is the flow supplied to the hydraulic ram (gal/min); H_p is the pumping head, i.e., the vertical distance between the ram and the water surface in the tank where the water is pumped (ft); H_s is the power head, i.e., the vertical distance between the free water surface of the supply water and the ram (ft); and e is the efficiency of the ram, usually about 50 percent.

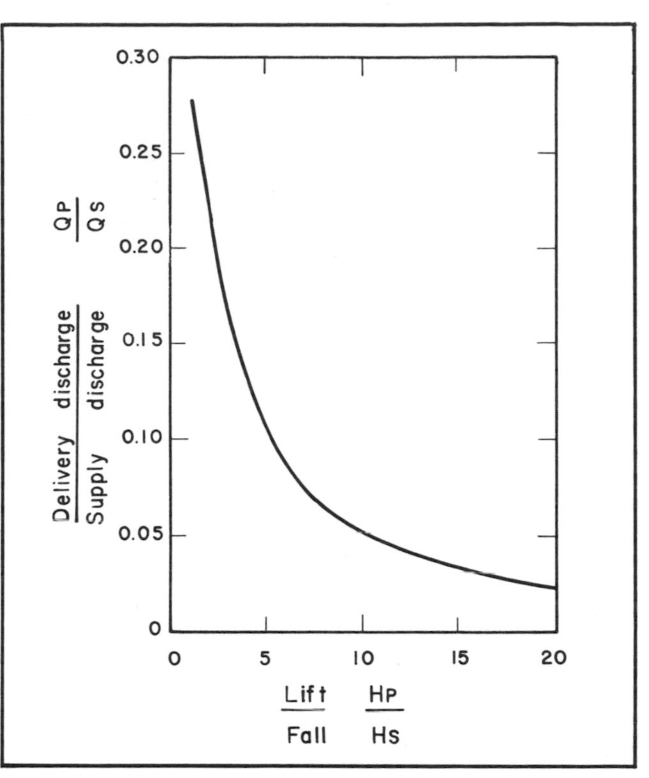

Figure 6.17 Efficiency curve for a hydraulic ram.

Figure 6.17 gives directly the ratio Q_p/Q_s as a function of the ratio H_p/H_s for typical hydraulic rams. If the supply and delivery pipes are long, you must take into consideration the hydraulic losses in these pipes (described in the water-power section of Chapter 3). H_s should be reduced by the equivalent losses in the supply pipe and H_p should be increased by the equivalent losses in the delivery pipe. Here we essentially are accounting

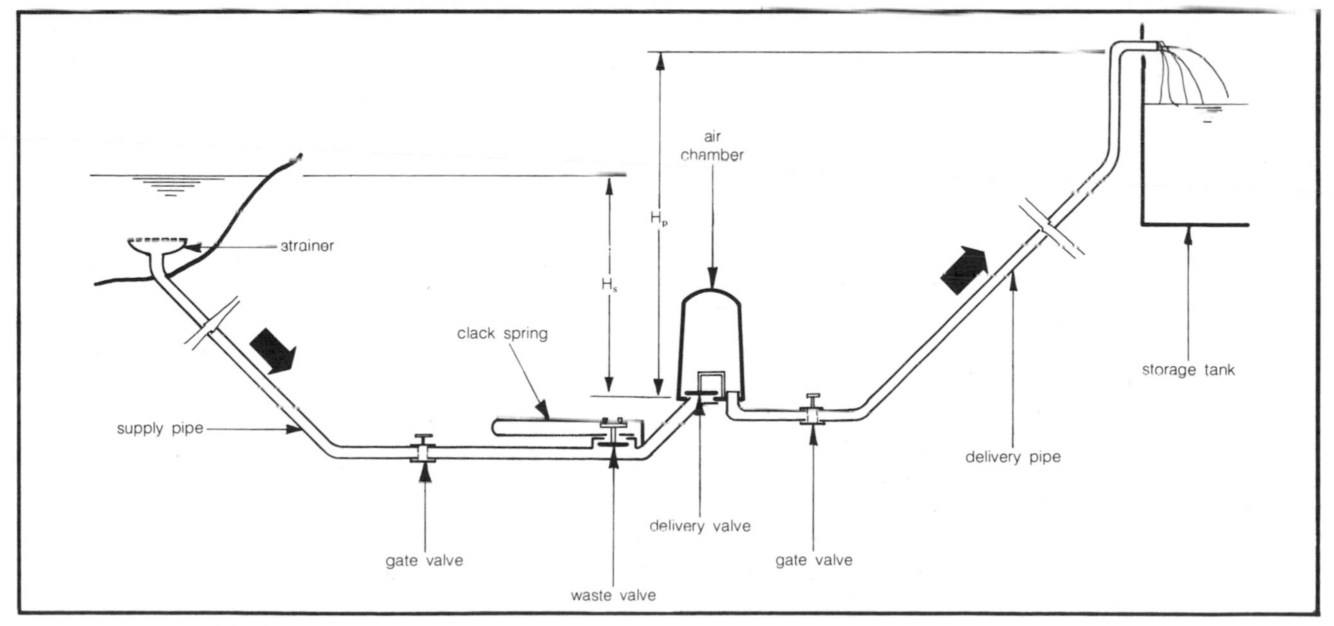

Figure 6.16 A typical hydraulic-ram installation.

for the energy lost due to friction in the pipes.

Example: You have a stream flowing at 10 gal/min on your property and you want to pump water to a tank 8 feet above the level of the stream. If you put the hydraulic ram 2 feet below the surface of the stream, how many gallons per minute can your hydraulic ram deliver to the tank?

Solution: The power head, H_s, is 2 feet and the pumping head, H_p, is 10 feet (8 + 2). From Figure 6.17 you find that for

$$\frac{H_p}{H_s} = \frac{10}{2} = 5$$

you have

$$\frac{Q_p}{Q_s} = 0.1$$

and

$$Q_p = 0.1 \times Q_s = 0.1 \times 10 = 1 \text{ gal/min}$$

Thus we see that we must use 10 gal/min to pump 1 gal/min to a height of 8 feet; the other 9 gallons per minute are returned to the stream. If your ram is allowed to pump continuously, you will pump daily (1 × 60 × 24) about 1440 gallons per day. Not bad considering that the water provides its own energy for pumping!

As you can see from this example, the hydraulic ram requires a lot of water to operate. The higher the ratio of the pumping head to the power head, the more water the ram uses for energy purposes. However, if large flows are available, the hydraulic ram may be a very good choice for your pumping needs: it is durable and inexpensive, requires little maintenance, and can be operated continuously. You should be aware of the fact that the ram is fairly noisy; but with proper siting, this should not be a major problem.

Storage Systems

The amount of water you must have in storage depends upon both a realistic assessment of your needs and a good estimate of the period of time during which your source can be expected to be dry or unavailable. The average value for domestic daily water consumption is about 50 gallons per person, 49 of which are devoted to washing and waste disposal; we drink and cook with the sole remaining gallon. Thus, for example, if you expect your water source to be dry for a maximum of one month at a time, you will need to have a storage volume capable of meeting your normal demands for 30 days. Assuming you have a family of four and use, on the average, 50 gallons per day per person, your required storage volume will be

$$30 \text{ days} \times 50 \text{ gallons/day-person} \times 4 \text{ persons} = 6000 \text{ gallons}$$

The area required to store this quantity of water is

$$6000 \text{ gallons} \times 0.1337 \text{ ft}^3\text{/gallon} = 802 \text{ ft}^3$$

So you will need a tank capable of storing at least 800 cubic feet of water—a tank 10 feet square and 8 feet deep would give the minimum capacity required.

A storage tank is one way of storing water; a reservoir or pond is another (both Chapters 3 and 7 discuss certain aspects of dam construction and maintenance). Of course, there are problems associated with open-surface water-bodies. For instance, the growth of algae may cause deterioration of water quality over the summer months; and the inadvertent introduction of materials into the water-body by animals, birds or wind also can reduce its purity. Because of these problems it is almost always necessary to treat surface water to insure that the water is hygienically acceptable.

In order to avoid pollution, it is advisable to store your drinking water in a covered tank of some sort. A water storage system must be kept reliably watertight to prevent leakage or contamination. This means that seepage, warping, and corrosion must be guarded against and that pipelines to and from the storage container must be sound. To avoid odors, tastes, or toxic materials in the water, the appropriate surfaces need to be constructed or coated with inert materials such as wood or some plastics and paints. In addition, there must be provision for easy access to the storage container above the maximum water height, to permit visual checks of the water level and convenient maintenance when necessary.

Water storage tanks come in a variety of sizes, shapes, and materials. The three main construction materials are wood, steel, and concrete. Wooden tanks are virtually maintenance-free and leave no taste or odor in the water. However, they must be kept from drying out for more than two months at a time; the wood may otherwise warp and allow leakage after refill. Steel tanks are also available. These containers tend to corrode or leave a metallic taste in the water if not properly lined; but minimum maintenance and proper lining can eliminate these problems. New steel tanks can be fabricated to meet specific criteria, while used ones are also available. The third type available, concrete tanks, are usually constructed in place. They are quite inert and require little upkeep. Some small concrete tanks are available commercially.

Before you select any container, you should consult several sources to get comparative prices and construction methods. Your best source of tanks is generally a local supplier; sometimes good bargains can be made by checking with local salvage companies. Although prices change very rapidly these days, it is still of some use for you to have a rough idea of the price range for different sizes and types of tanks—Table 6.5 gives prices for redwood tanks of various sizes. These prices do not include installation costs, which can vary considerably from site to site. If a special foundation is required for the tank, the cost will be significantly more. This is especially true for installations on slopes or areas subject to slides or slippage. Although pressure tanks are available, they are not recommended because of the added cost and maintenance; however, if your location is not readily suited to a gravity-type tank, you may not have any choice.

Gravity tanks are located at a higher elevation than the point where the water is to be used. The added height gives the water the necessary pressure for your supply system to work conveniently. The pressure (P) in pounds per square inch (psi) at a given point in your system is given by the formula

E. 6.2
$$P = \frac{H}{2.3}$$

where H is the elevation difference or head (in feet) between the water surface in the tank and the point where you want to know the pressure. If you have a long pipe from the tank to the point at which you wish to know the pressure (or, more likely, where you wish to use the water) you will need to reduce H in the above equation;

Table 6.5 Redwood Storage Tanks[a]

Gallons	Standard 2″ Redwood Tanks With Chime Joists Dia.	Height	Weight (lbs)	Price ($)	Flat Covers Weight (lbs)	Price ($)
200	4′0″	3′	260	207	65	57
300	4′0″	4′	330	247	65	57
400	4′8″	4′	410	282	75	69
500	5′2″	4′	470	333	85	74
600	5′0″	5′	525	356	85	74
900	6′0″	5′	660	489	100	92
1000	6′0″	6′	760	557	100	92
1500	7′0″	6′	930	690	175	109
2000	8′0″	6′	1095	747	225	155
2500	8′4″	7′	1400	897	260	161
3000	9′0″	7′	1600	943	300	181
4000	10′0″	8′	1965	1140	365	209
5000	11′0″	8′	2235	1253	435	238
5000	10′0″	9′	2100	1253	365	209
6000	12′0″	8′	2460	1448	510	247
8000	12′0″	10′	2850	1645	510	247
10,000	13′8″	10′	3550	1978	625	340
12,000	13′8″	12′	4050	2202	625	340

Gallons	Standard 3″ Redwood Tanks With Chime Joists Dia.	Height	Weight (lbs)	Price ($)	Flat Covers Weight (lbs)	Price ($)
10,000	14′0″	10′	5500	2865	670	340
12,000	14′0″	12′	6350	3289	670	340
15,000	15′6″	12′	7100	3634	800	360
20,000	18′0″	12′	8700	4416	1060	529
20,000	16′6″	14′	8600	4416	940	483
25,000	18′0″	14′	9700	5048	1060	529
30,000	20′0″	14′	10200	5695	1450	575
40,000	23′0″	14′	13650	6688	1910	685
50,000	24′0″	16′	15900	8120	2140	960
60,000	26′0″	16′	17900	9178	2370	1093
70,000	28′0″	16′	20000	10,290	2850	1205
75,000	29′0″	16′	21100	10,706	3070	1325
80,000	30′0″	16′	22300	11,270	3200	1403
90,000	30′0″	18′	24400	12,368	3200	1403
100,000	32′0″	18′	28700	13,588	3660	1541
100,000	30′0″	20′	26400	13,588	3200	1403

Notes: a. All dimensions are outside measurements: exact stave length is 1″ shorter than the nominal length in feet. Capacities are approximate: weights are estimated. 3″ × 4″ chime joists are used with tanks 2000 gallons and smaller; 4″ × 6″ chime joists are used with larger tanks. Approximate prices as of January 1, 1974. T.E. Brown, Inc., 14361 Washington Ave., San Leandro, California.

this reduction accounts for friction and methods to estimate this loss of head are described in the water-power section of Chapter 3. You will need to maintain a minimum pressure of 15 to 20 psi if you have any automatic water-using machines in your home (such as a dishwasher). The minimum elevation difference for such a pressure is about 35 to 45 feet, which you can verify from Equation 6.2. As we mentioned a moment ago, you may have to use a pressure tank if your location does not allow a natural elevation difference of 40 feet or so; in this case, you will need to check with other references for details (see, for instance, the very fine discussion in *Planning for an Individual Water System (AAVIM)*.

QUALITY AND CONTROL

First of all, let's recognize that water quality varies by degrees, not by absolutes. That is to say, we never have pure water in the true meaning of the word *pure*. What we essentially have are acceptable levels of impurities; when these levels are exceeded, it can often lead to health problems.

We should start this discussion by considering first those characteristics most easily recognized. Many rivers and lakes contain particulate material that makes their waters turbid. The turbidity may be due to silt and clay particles, but this condition also can be caused by organic matter. Organic matter is often responsible for taste and odor problems, as well.

Natural waters contain many dissolved salts, most of which have no deleterious effects. A common characteristic of most groundwaters and some surface waters is *hardness*. Hardness is caused by an excess of dissolved calcium and magnesium, metal ions which tend to interfere with the use of soaps by combining with the soap molecules. An additional problem sometimes associated with hard water is scaling in hot-water heaters, thus reducing their heat-transfer efficiency. When you use hard water, you find that you cannot easily form a lather with soap and your skin tends to roughen from washing. However, hard water is not harmful to your health.

Iron and manganese in a water source can cause staining of clothes and give water a characteristic metallic taste. Even when iron is present in low concentrations, red stains in clothes and on porcelain plumbing may appear. Manganese typically causes a brownish-black stain. Preparing your coffee and tea with water containing substantial manganese may result in a bitter tasting brew! Obviously you are going to want to remove these materials if you find them in your water supply.

It is worth mentioning a few other substances which are commonly found in natural waters to varying degrees. Sulfate is always found in water to some extent. However, a high sulfate concentration can have laxative effects if your body is not accustomed to it. There is very little you can do easily to correct this condition, so if your body objects too strenuously, you may have to choose another water source. Another common constituent of natural waters is carbon dioxide. Excess carbon dioxide imparts a degree of acidity to the water and may increase the rate of corrosion of metal pipes and fixtures in your supply system.

There are many potentially toxic or hazardous substances which can be found in a natural water—such things as arsenic, barium, cadmium, lead, and zinc. Water supply sources close to, or coming from, industrial areas or mining regions require careful examination by experts to insure that the water is healthful before it is developed as a drinking source. (If not, again look elsewhere; these conditions are untreatable by our methods.) This warning includes, of course, the most obvious danger—bacteria or microorganisms capable of transmitting diseases.

Several water-borne diseases can arise from untreated water and you should be aware of them. Troublesome bacterial or protozoan diseases include typhoid fever and both bacillary and amoebic dysentery. Of the viral infections that can contaminate water supplies, infectious hepatitis and poliomyelitis are the most likely. Salmonella-caused diarrhea can also result from drinking polluted waters.

Water is dangerous only when contaminated by some external source, and almost invariably pathogens are spread through the feces of infected hosts. This fact is important to remember in any community which also is involved with alternative methods of waste disposal. Since pathogenic bacteria do not normally multiply in water, they present a hazard only through recent or continuous contamination. Have your water tested *regularly* and *professionally* to assure that it meets safe standards.

Water testing is based on the assumption that any water which is contaminated by feces contains intestinal pathogens and is therefore unpotable. It would be too costly and time-consuming to test for all possible pathogens. In the United States, *Escherichia coli* is used as an indicator of fecal contamination because it only occurs in feces and is never free-living in nature. If water tests do show the presence of *E. coli*, it is advisable to find another water supply or to obtain professional help in treating it.

Other microorganisms can pollute water supplies without causing contamination. When algae are present in large quantities, they may cause turbidity and change the water color. Algae cause no known human diseases, but, as we have mentioned, they can contribute unpleasant tastes and odors.

Occasionally water contains small worms, blood-red or greyish in color. The red worms are the larvae of Chironomus flies; the others are larvae of related midge flies. Remedial measures involve draining and cleaning your basins to eliminate existing larvae; you must then use insect-tight covers on the basins to prevent reinfestation.

Disinfection

We have spent some time discussing what the water-quality problems might be; now let's turn to a few techniques that can help us, if necessary, to solve them.

Disinfection of water does not mean *sterilization;* we mean by disinfection the adequate destruction of water-borne pathogenic microorganisms including bacteria, viruses, and protozoa. (Sterilization also destroys certain bacterial spores; disinfection does not.) There are many ways to disinfect water but only a few of them have any real practical application for the development of your own water supply system.

We should start with the easiest technique for small quantities of water—simply boil it. Heating water to a temperature of 140°F for fifteen minutes is sufficient for disinfecting; boiling for a longer period tends to sterilize it unnecessarily and wastes fuel. Because of the energy required and the small quantities of water that can be handled easily, boiling can only be considered as an emergency technique.

The most reliable type of disinfectant for small-scale applications is chlorine or a chlorine compound. Chlorine solutions must be kept in brown or green bottles and stored in the dark to prevent photodecomposition of the active agent. Storage of chlorine solutions for long periods is not recommended since they can lose over half their activity in one year. Laundry bleaches should not be used as a chlorine source since most commercial products contain other additives which may cause taste and odor problems. And solutions containing more than 10 percent chlorine by weight are unstable and should not be used under any circumstances.

Now we can begin to make a few specific suggestions. But first we need to discuss some of the types of materials available and how to use them.

If you need to disinfect a small quantity of water for some reason—say, 20 gallons—and you don't have a regular disinfection system operating, you can use chlorinated lime. A stock solution of chlorinated lime can be mixed at home by adding a teaspoon of chlorinated lime (40 percent chlorine by weight) to a quart of water. After it is well mixed, one teaspoon of this stock solution for every two gallons of water usually provides adequate disinfection after half an hour. So, for 20 gallons, you would use 10 teaspoons of stock.

Calcium hypochlorite is one of the best and most common chlorine compounds around. It is available commercially in powder and tablet form containing 30 to 75 percent active chlorine by weight; usually the percentage of chlorine is found on the container. Common trade names are B-K Powder, Percholoron, and Pittchlor. If you purchase this material you will need to store the cans in a cool, dark area—away from any direct sunlight. The calcium hypochlorite powder can be used to make up stock solutions for disinfection in the same manner (same proportions) as with chlorinated lime compounds.

Now, how do we go about disinfecting a large volume of water on a batch basis? First, we need to know that the required concentration of chlorine in solution for adequate disinfection is about 1 to 5 parts per million (ppm). Typically you will mix up a stock solution of about 5 percent chlorine (about the same strength as found in common laundry bleach). The volume of water that can be disinfected by a specific volume of stock chlorine solution is given by the following equation:

E. 6.3
$$V_w = \frac{10{,}000\ PV_{Cl}}{C}$$

where V_w is the volume of water disinfected (gallons); V_{Cl} is the volume of stock chlorine solution (gallons); P is the percent, by weight, of chlorine in solution; and C is the concentration of chlorine in the final mixture (usually about 1 to 5 ppm).

Example: What quantity of water can you disinfect with one gallon of 5 percent chlorine solution if you want a chlorine concentration in the disinfected water to be 2 ppm?

Solution: Using Equation 6.3 with P equal to 5 percent, C at 2 ppm and V_{Cl} at 1 gallon, we have

$$V_w = \frac{10{,}000 \times 5 \times 1}{2} = 25{,}000\ gal$$

Now that is a whole lot of water! It is obvious that a little stock chlorine solution can go a long way in disinfection processes.

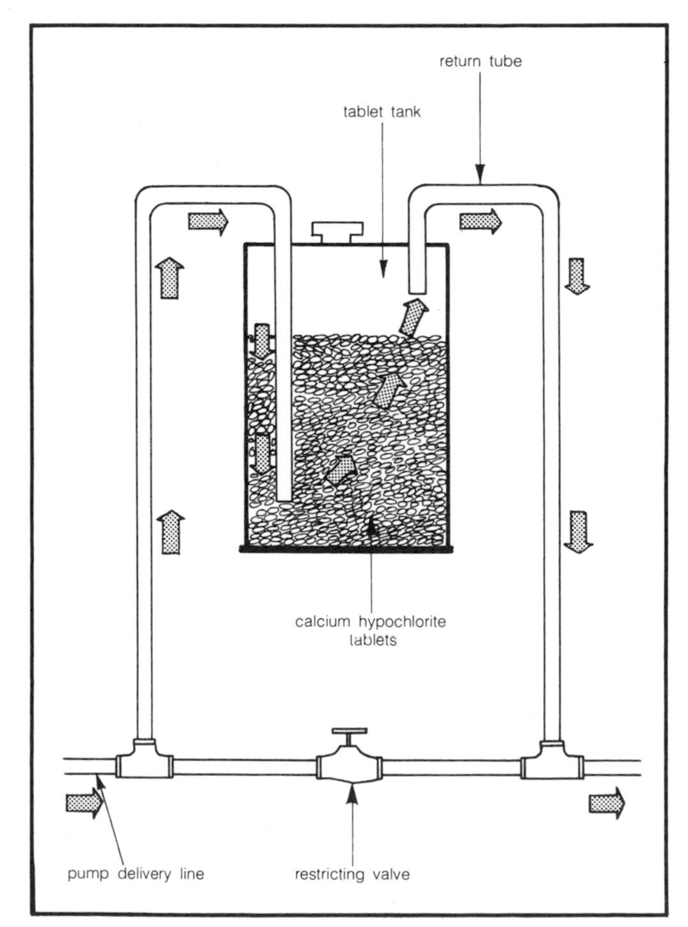

Figure 6.18 A chlorination unit using calcium hypochlorite tablets.

One application of batch disinfection might be in your storage tank, especially when you first start up your system. The process also cannot hurt if applied after any situation which you think might possibly have contaminated your source, such as after a big storm.

What do you do if you want the disinfection process to be automatic? Small chlorination units have been designed for use by individuals or small living groups. The main types are the so-called pump, injector, and tablet types. The pump type is designed to deliver a predetermined amount of chlorine solution at each stroke of a chlorination pump. A chlorinator of the injector type consists of a nozzle at some point along the pipe between pump and storage tank. As the water flows through the nozzle at high velocity, a suction develops in a line connected to the chlorine solution container and chlorine solution is injected into the main pipe in proportion to the flow in the main. The tablet form of chlorinator consists of a tank of calcium hypochlorite tablets through which water is circulated. Figure 6.18 shows the bypass line running through the calcium hypochlorite tablets. The best working information we have seen on this type of chlorinator can be found in *Planning for an Individual Water System* (AAVIM).

Filtration of Turbid Water

Filtration is probably the oldest and most easily understood process of water treatment available. It is used to remove suspended particles from water and also to remove some bacteria, although it cannot be counted upon to do this effectively. The most commonly used filters in household water supplies are sand and ceramic varieties.

You can easily build a sand filter at home; a design by the U.S. Department of Agriculture for home use is shown in Figure 6.19. The filter is made from two 10-inch-diameter sections of vitrified tile pipe set in concrete. The upper section contains the filter material (sand) and the lower section is used to collect the filtered water. It is important to note that the sand varies in grain size: it is finer in the upper part of the filter and becomes coarser as you get deeper into the filter bed. You can use either fine and coarse sand (your local sand and gravel man can help you here), or fine sand and crushed charcoal or coal. If you use activated charcoal in your filter, you will have the additional benefit of taste and odor removal.

The filter shown in Figure 6.19 has an overflow pipe on the left side and a glass tube on the right side to show pressure losses in the filter due to clogging. When the filter is clean, the level in the tube is the same as the level in the tile pipe. As the sand becomes clogged, resistance builds up to flow through the filter and the level in the glass tube becomes lower than the water level in the filter.

To clear a clogged filter, you have to backwash it by flushing clean water back through the sand; this flush water is drained off above the filter material. You may have to repeat this washing process several times until you get the same water levels in the glass tube and the filter. If you use charcoal in your filter, you will have to replace it occasionally (about once a year); if you are filtering organically laden surface water, more frequent replacement is necessary because bacteria may collect in the charcoal and reproduce there. The capacity of sand filters depends mainly on the surface area, height, and grain size of the sand. For example, the 30-inch-high sand filter shown in Figure 6.19 has a capacity of about 4 gallons per hour.

The pressure filter or rapid sand filter (Figure 6.20) operates on a slightly different basis: it is a closed tank containing sand of various grain sizes and the water comes in from the pump under pressure. The capacity of the rapid sand filter is high, about 2 to 3 gal/min per square foot of surface area, but its effectiveness in removing very small particles is limited and it is not recommended for waters with large quantities of fine matter.

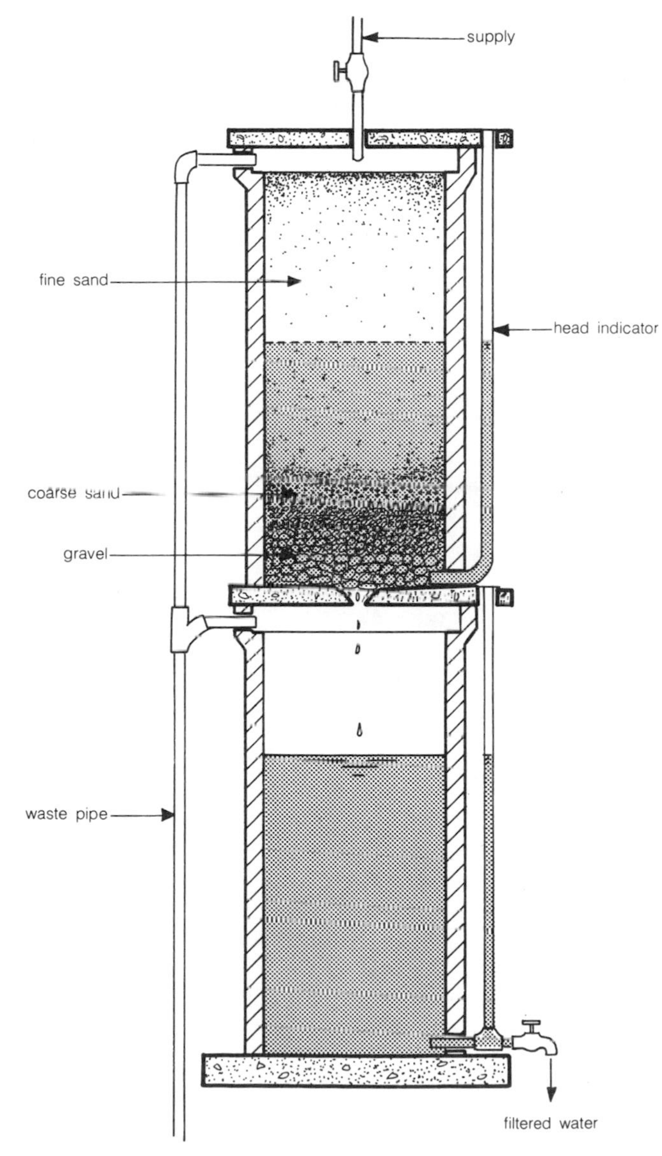

Figure 6.19 A slow sand filter made from clay pipe.

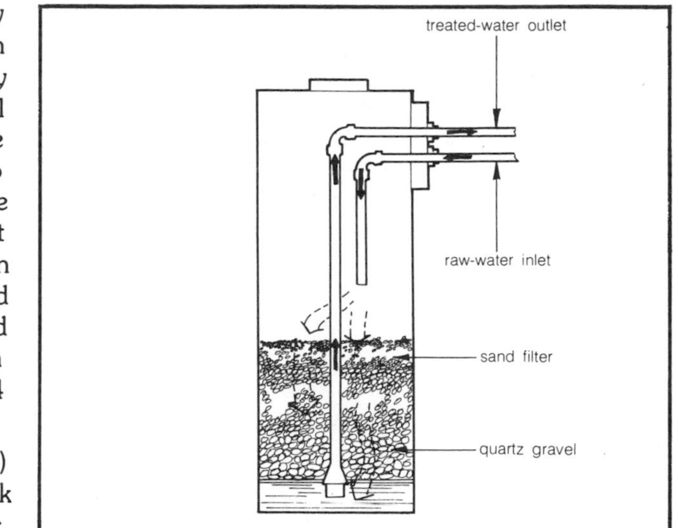

Figure 6.20 A pressurized rapid sand filter.

Ceramic filters constitute a class of filters well suited to small-scale operations. Figure 6.21 shows a typical ceramic filter. Water is filtered as it flows through a ceramic medium (candle) of very small pore size. The filter candle is contained in a tank which may or may not be under pressure. These filters are very efficient in removing fine particles, but we do not recommend their use with very turbid water since clogging then occurs very

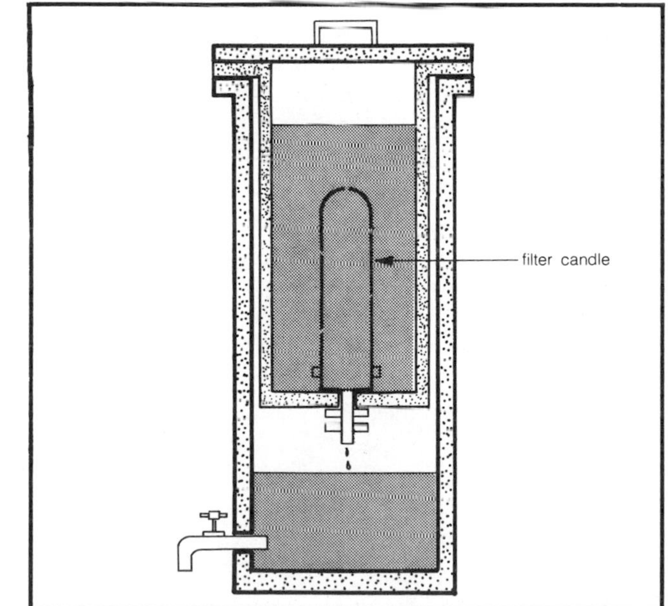

Figure 6.21 A Berkefeld ceramic filter with ceramic candle in place.

rapidly. Ceramic filter candles are made of various clays (including porcelain) and sometimes can be coated with diatomaceous earth. The Berkefeld filter is shown in Figure 6.21. The ceramic candles come in coarse, normal, and fine pore sizes. Careful maintenance of ceramic filters is important to avoid short-circuiting of water through cracks. With the exception of the porcelain variety, water processed through any type of filter should be disinfected before use.

Ion-Exchange Process

Problems with water quality associated with hard water—dissolved ions such as calcium and magnesium—are rather common, especially in groundwater sources. The presence of dissolved iron and manganese is also not unusual. There are numerous ways of dealing with these unwanted materials but, unfortunately, most of these processes are not suitable for small-scale home application. One gratifying exception is the ion-exchange process.

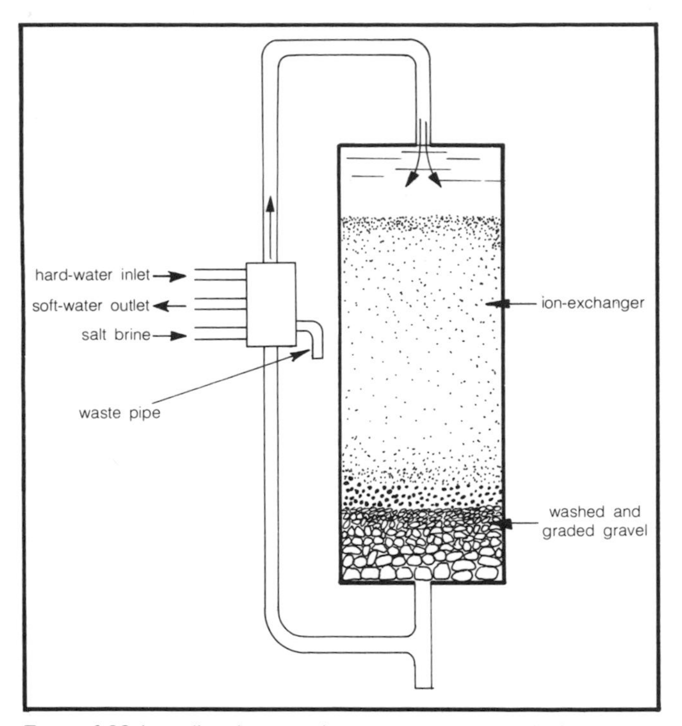

Figure 6.22 A small-scale ion-exchange unit appropriate for home use.

Figure 6.22 shows a typical ion-exchange unit. These units operate on the principle of exchanging one kind of ion (usually sodium) provided by the exchanger for another, unwanted kind of ion in solution in the water (calcium, for example); troublesome ions are thus replaced with a type more generally acceptable, and the water is "softened."

When all the sodium ions in the ion-exchanger have been replaced by calcium, the exchanger must be recharged—a process which requires a very concentrated solution of sodium chloride, common table salt. The ion-exchange units are generally placed directly before that point in the water line where softened or treated water is to be used—before the kitchen or bathroom, for example—to avoid unnecessarily treating water used for gardening and other such purposes. Small units available from local suppliers are capable of treating moderately hard water (60 to 80 ppm of calcium; 20 to 40 ppm of magnesium) on an economical basis, *if* you use them to treat only the water you need for washing and cooking. An example may be helpful here: for a family of four using 1000 gallons of water per month (for cooking and washing), this water containing 40 ppm of calcium, then one cubic foot of ion-exchange resin would need to be recharged only once a month, two cubic feet of resin would last two months, and so forth.

We are not presenting design calculations here because they can get very cumbersome. However, once testing has determined the particulate composition of your water in parts per million, a consultation with a local supplier should enable you to size an ion-exchange unit fairly easily. For those of you interested in doing it all yourself, we suggest you look into the publications put out by the American Water Works Association and also into *Planning for an Individual Water System* (AAVIM).

If you have a small piece of land in the country and a source for a water supply, you may have found some useful information and suggestions here. We hope that you will consider carefully your needs and your resources before deciding to alter your landscape. And whether you use groundwater or a lake or a stream for your water source, remember that it is always cheaper and safer to protect a good, clean source of water than it is to treat the water to make it acceptable. With a little foresight and care, you should be able to enjoy a healthy, adequate supply of fine-tasting water.

Bibliography

American Association for Vocational Instructional Materials. 1973. *Planning for an Individual Water System*. Athens, Georgia: AAVIM.

American Water Works Association. 1941. *Manual of Water Quality and Treatment*. New York: AWWA.

______. 1969. *Water Treatment Plant Design*. New York: AWWA.

______. 1971. *Water Quality and Treatment*. 3rd ed. New York: AWWA.

Anderson, K.E. 1966. *Water Well Handbook*. Rolla, Missouri: Missouri Water Well Drillers Association.

Baum, V.A. 1961. "Solar Distillers." From *Proceedings of the U.N. Conference on New Sources of Energy*, Rome, vol. 6. New York: United Nations.

Bloemer, J.W. 1965. "A Practical Basin-Type Solar Still." *Solar Energy* 9:197.

Brinkworth, B.J. 1972. *Solar Energy for Man*. New York: John Wiley and Sons.

Cooper, P.I. 1969. "Digital Simulation of Transient Solar Still Processes." *Solar Energy* 12:313.

Cooper, P.I., and Read, W.R.W. 1974. "Design Philosophy and Operating Experience for Australian Solar Stills." *Solar Energy* 16:1–8.

Daniels, F. 1964. *Direct Use of the Sun's Energy*. New Haven, Conn.: Yale University Press.

Duffie, A., and Beckman, W. 1974. *Solar Energy and Thermal Processes*. New York: John Wiley and Sons.

Edmondson, W.B. 1974. "Vertical Floating Solar Stills." *Solar Energy Digest*. 3:12.

______. 1974. "Solar Water Stills." *Solar Energy Digest* 3:25.

Gibson, U.P., and Singer, R.D. 1971. *Water Well Manual*. Berkeley, Calif.: Premier Press.

Gomella, C. 1961. "Use of Solar Energy for the Production of Fresh Water." From *Proceedings of the U.N. Conference on New Sources of Energy,* Rome, vol. 6. New York: United Nations.

Grune, W.N. 1961. "Forced-Convection, Multiple-Effect Solar Still for Desalting Sea and Brackish Waters." From *Proceedings of the U.N. Conference on New Sources of Energy,* Rome, vol. 6. New York: United Nations.

Harr, M.E. 1962. *Groundwater and Seepage*. New York: McGraw-Hill.

Hazen, A. 1916. *Clean Water and How to Get It*. New York: John Wiley and Sons.

Hem, J.D. 1970. *Study and Interpretation of the Chemical Characteristics of Natural Water.* Washington, D.C.: U.S. Government Printing Office.

Hicks, T.G. and Edwards, T. 1971. *Pump Application Engineering*. New York: McGraw-Hill.

Holdren, W.S. 1970. *Water Treatment and Examination*. Baltimore: Williams and Wilkins.

Howe, E.D. 1961. "Solar Distillation Research at the University of California." From *Proceedings of the U.N. Conference on New Sources of Energy,* Rome, vol. 6. New York: United Nations.

______. 1964. "A Combined Solar Still and Rainfall Collector." *Solar Energy* 8:23.

Howe, E.D. and Tleimat, B.W. 1974. "Twenty Years of Work on Solar Distillation at the University of California." *Solar Energy* 16:97.

Howood, M.P. 1932. *The Sanitation of Water Supply*. New York: C.C. Thomas.

Hummel, R.L. 1961. "A Large Scale, Low Cost, Solar Heat Collector and Its Application to Sea Water Conversion." From *Proceedings of the U.N. Conference on New Sources of Energy,* Rome, vol. 6. New York: United Nations.

Lawand, T.A. 1965. *Simple Solar Still for the Production of Distilled Water*. Test Report no. 17, Brace Research Institute. Quebec: McGill University Press.

Löf, G.O.G. 1961. "Application of Theoretical Principles in Improving the Performance of Basin-type Solar Distillers." From *Proceedings of the U.N. Conference on New Sources of Energy,* Rome, vol. 6. New York: United Nations.

______. 1966. "Solar Distillation." From *Principles of Desalination*, ed. K.S. Spiegler. New York: Academic Press.

McCracken, H. 1964. "Solar Still Pans." *Solar Energy* 9:201.

New York State Department of Health. 1966. *Rural Water Supply*. Albany, New York: State Department of Health.

Selcuk, M.K. 1964. "A Multiple-Effect Tilted Solar Distillation Unit." *Solar Energy* 8:23.

Skeat, W.O. 1969. *Manual of British Water Engineering Practise*. London: W. Heffer and Sons, Ltd.

Strobel, J.J. 1961. "Developments in Solar Distillation." From *Proceedings of the U.N. Conference on New Sources of Energy*, Rome, vol. 6. New York: United Nations.

Trombe, F., and Foëx, M. 1964. "Utilization of Solar Energy for Simultaneous Distillation of Brackish Water and Air-Conditioning of Hot-Houses in Arid Regions." *Proceedings of the U.N. Conference on New Sources of Energy,* Rome, vol. 6. New York: United Nations.

United Nations. 1970. *Solar Distillation as a Means of Meeting Small Water Demands*. Sales No. E 70.11.B.1 from Sales Section, United Nations, N.Y. 10017.

Wagner, E., and Lanoix, J. 1959. *Water Supply for Rural Areas and Small Communities*. Geneva, Switzerland: World Health Organization.

AGRICULTURE/AQUACULTURE

Food chains

Farming, by land and sea

On the transcendence
and transformation of garbage

Gardens for both tenements
and the great outdoors

The purchase and care of livestock

Fishpond construction and
the bounties therefrom

Agriculture and Aquaculture

INTRODUCTION

"Everything is connected to everything else" is the first "law" of ecology coined by the ecologist Barry Commoner. This chapter is no different. Like electricity and solar heating, agriculture and aquaculture are also processes of energy conversion, transforming light and chemicals into food for our consumption. Those converted forms which we don't utilize (both before and after consumption) are of direct concern in the waste and water chapters. And, of course, like architecture, the practical aspects of the process of food raising cannot be separated from the aesthetic quality it adds to our homes and our lives.

We start by looking briefly at two ecological principles which are important for anyone wanting to work with the natural environment. The chapter then surveys basic information you'll need to farm both the land (agriculture) and also the lakes, streams, or oceans (aquaculture). In each case, your individual situation, whether it be urban, suburban, or rural, will determine what information is directly applicable.

For example, vegetables literally can be grown *anywhere*, even on the window ledge of an apartment. And if you're fortunate enough to have a sunny spot of ground, the tiniest plot can produce impressive yields with the Biodynamic/French Intensive method of crop production. Even urban animals can be raised in other than sylvan surroundings. Rabbits need as much room as it would take to store two bicycles. Gil Masters raises chickens in downtown Palo Alto and the roof of an apartment is a perfect location for a beehive—out of the way, but accessible.

Of course, if you own land, whether it be a large backyard or several acres, the possibilities multiply: both agriculture and aquaculture become available. Simple practices—blocking off a small inlet along the coast (which may require a permit from various agencies such as the Department of Fish and Game) or constructing several rock pools in a mountain stream—can yield increases in your fish harvest. If enough land is available, a low-maintenance fish pond for recreation and food-producing purposes can be stocked free of charge by the state or federal government; with proper management, this source can yield 200 pounds or more per acre every year. We discuss the farm pond possibilities in some detail because of its low-maintenance, low-cost attributes. We also provide some general information on fish species and culturing for both freshwater and marine habitats. We encourage you to be imaginative and end our chapter with an outline of two experimental aquaculture systems which are presently under development. They are significant in their attempt to integrate several concepts—food growth with wastewater handling, and the combined use of recycled water, solar heating, and windmill power.

Because the literature is vast, we attempt here to provide you with a collection of basic information, highlighting methods which are energy-conservative, low in waste, and adaptable to individual needs and locations.

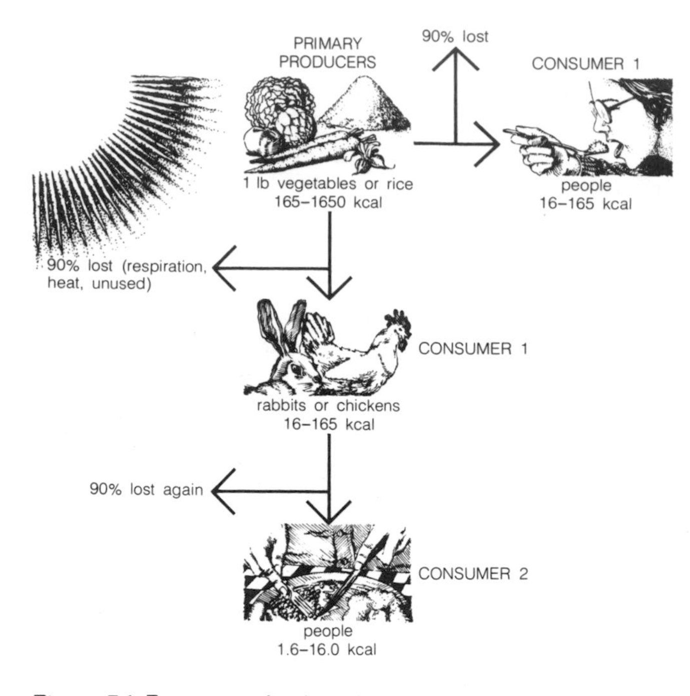

Figure 7.1 Energy transfer through a food chain.

Although these individual needs can range from total self-sufficiency to supplemental food raising, we hope this chapter will be useful no matter what your situation.

Let us begin by looking at two fundamental concepts of ecology which are important to both agricultural and aquacultural practices. The first is the concept of the food chain, which refers to a specific and sequential transfer of the energy available in food from one organism to another. The first members of the chain, plants or algae ("primary producers"), are consumed by a group of animals ("consumer level 1"), which in turn is consumed by a second group ("consumer level 2"), and so on. The primary producers are the most important members of this sequence. Only they have the unique capacity to transform solar energy or incident light into food energy or plant tissue through the fixation of inorganic CO_2. The basic chemical reaction is

$$CO_2 + H_2O \xrightarrow{\text{solar energy}} \underset{\text{plant tissue}}{(CH_2O)} + O_2$$

All other organisms essentially depend upon this transformation for their energy. Of the solar energy available to primary producers, only about 1 percent is captured in plant tissue. Of this 1 percent, only about 0.1 to 0.5 percent is available to the first level of consumers; the remainder has been used by the plant for its own respiration. With each subsequent transfer in the food chain, a major portion (about 90 percent) of the remaining available energy is lost, mostly as heat and respiration. Consequently, the longer the food chain, the less energy available from a given number of primary producers. For example, Figure 7.1 illustrates that one pound of vegetables or rice can provide from 165 to 1650 kilocalories of energy. If the pound were fed to rabbits or chickens, only about 16 to 165 kilocalories would be conserved and hence would be available for our own consumption. If we ourselves had consumed the vegetables or rice, we would have had available the original 165 to 1650 kilocalories.

A second concept to keep in mind is that, generally, a diverse system tends to be more stable than a simple system. For example, if only one species of fish inhabits a fish pond, the invasion of a single parasite can rapidly and completely eliminate the fish population. On the other hand, if polyculture is practiced, and several compatible species are grown together, the pathogenic organism, which tends to be specific for each species, can only attack a fraction of the total population. In addition, since the number of individuals of the victimized species is fewer, the disease would tend to spread more slowly. Large areas of land planted with a single genetic strain of but one crop also tend to be less stable when faced with a stress such as disease. Consequently, variety is not only the spice of life; it's also a healthy component.

We might add that a diverse ecosystem is also more productive at energy conversion. If, say, you cut down a rain forest (with all its complex interrelationships) and planted a single crop of some sort, the total amount of solar energy converted to biomass and available to various food chains would drop drastically. Of course, since we don't eat rain forests, much of the original forest biomass was not available for human consumption. Production efficiencies are dealt with at greater length when we discuss aquatic monoculture and polyculture techniques. But as a general principle to hold in mind, diverse ecosystems better utilize and fix available solar energy than monocultural ecosystems.

AGRICULTURE

The objectives of the agriculture section are many. Of course, the most important aims are to produce high-quality, nutritious, and delicious food for the household, and to recycle its solid and liquid wastes. But there are others—the beautification of the home, sun and wind protection—which also must be considered in the agricultural realm. These objectives must be balanced against three important factors: the labor cost, in terms of both quantity and quality; the economic expense; and perhaps most importantly, the ecological costs.

We have attempted to present the information necessary to allow the reader to plan for himself what will work best in his particular situation. The focus and the sizing considerations were determined for the use of a small living group; intensive agriculture of large acreage was not considered, and practices requiring high-maintenance, specialized techniques, and energy-consuming machinery were also de-emphasized.

We consider production of food from both plants and animals. Because of the vast amount of literature on the subject, no effort was made to describe the step-by-step procedures of traditional cultivation. However, since we consider them important, such techniques as composting and the Biodynamic/French Intensive method are described in more detail. The material is presented in a simple (hopefully coherent) manner, and should be both an adequate introduction to "small-time" agriculture and also a reference guide for the more experienced. It goes without saying that you should experiment with other plants, different species, and various techniques to find what best suits your own backyard.

Crops

Much of our information was drawn from *First-Time Farmer's Guide* by B. Kaysing and from texts by J.I. Rodale (see Bibliography). We recommend you consult them if you want more data.

Vegetables

Several major considerations go into the planning of a vegetable garden. Climate, soil, and a host of various environmental factors must be taken into account in the choice of plants. Nutritional value should not be ignored, in terms of calories, protein, and vitamin and mineral content. And, of course, personal preference may be the deciding factor in many cases. Hundreds of books, each hundreds of pages long, have been written on the topic of vegetable gardening. In order to present a great deal of information concisely, we have used a number of tables, while specific plants are discussed with an eye towards pointers. Although they appear later, Tables 7.6 and 7.7 on "companion planting" are also relevant here.

Alfalfa Sprouts or Bean Sprouts: Great for salads and sandwiches. Soak alfalfa seeds, soybeans, or mung beans overnight at room temperature. Rinse in a wide-mouthed jar with cheesecloth fastened over the mouth, then drain by inverting jar. Wash at least three times a day. It's ready to eat in three to four days—quick and nutritious.

Lima Beans: Beans are easy to grow, have both a high yield and a high caloric content, and are an excellent source of protein. Lima beans may be grown up a pole or as a bush, with comparable yields. To grow pole beans, any convenient pole, fence, or even cornstalk may be used. Care should be taken to avoid shading other plants.

Snap Beans: Also called string beans, they come in either green or yellow varieties, with either flat or round pods. Culturing of bush beans is the same as pole beans, but without any poles. Keep in mind that all beans grow well in nitrogen-poor soil since they have nitrogen-fixing bacteria growing in nodules on their roots.

Beets: A useful plant since both the red root and the green tops can be eaten. Beets are good indicators of soil acidity; a good beet crop is evidence that your soil is not too acid. They grow quickly, so that crops can be successively replanted several times during the growing season. They can be easily transplanted, are relatively free of disease and pests, and can be grown into the winter. Consequently, beets are a highly desirable crop.

Broccoli: A large plant, usually two feet high. Besides the flowering heads, stems and leaves are also edible. After cutting the central flower stalk, new flower heads will develop. But watch out for cabbage worm.

Cabbage: Easy to grow, store, and cook. It is nutritious, a hardy plant which plays an important role in many gardens. Cabbage likes a lot of sun, but does not like tomatoes nearby. Unfortunately, it is a prime target for insects.

Carrots: A sure thing for your garden. They are easy to grow, resistant to insects and disease, and very nutritious. Carrots can be grown year-round in warm climates. Plant some every month.

Celery: A bit choosy, with a preference for cool surroundings and rich soils. They have a fairly shallow root system and need to be kept moist.

Chard: Another fast and easy-to-grow plant with a long growing season. Chard can be harvested continuously one leaf at a time. They are susceptible to the same worms that eat cabbage, so don't put these two kinds of plants next to each other.

Corn: In addition to its obvious aesthetic appeal, corn is not hard to grow and has a high caloric content. Stalks may be used for cattle fodder or as poles for pole beans. Its only drawback is a low yield per foot. Corn should be planted in adjacent rows to insure pollination and may be planted in succession for a continuous harvest. More to come about corn when we discuss grain.

Cucumber: A rapid-growing, warm-season vegetable which needs lots of water. The vine can grow up onto a fence to save space and the cucumber may appear overnight, so be prepared for a pleasant surprise.

Eggplant: One of the most attractive plants and vegetables to look at, from its lavender flowers to its deep-magenta fruit. It likes a long, hot growing season and a rich soil. Just a few plants provide more than enough eggplant for a family.

Lettuce: Easy to grow and full of vitamins. Plant in

succession for a continuous crop, or grow the loose-leaf variety and continuously pick off the outer leaves. Who can eat a salad without lettuce?

Melons: A vegetable not to forget. The melon needs sun, heat, and a lot of space in which to grow. A good place might be around cornstalks. Don't be in a hurry—they need three or four months.

Onions: A hardy and readily grown plant. The simplest and quickest method is from bulbs. Onions have few pests and don't mind cold weather. An onion has stopped growing when its top falls over, but it need not be dug up right away. For green onions, pick them before they form bulbs.

Peas: A gourmet treat that rapidly deteriorates as it is stored. Peas should be planted early in the spring, and at ten-day intervals through the summer for a continuous crop. Trouble with pests is largely avoided by picking the peas when young and tender. Don't overlook the Snow Peas and Sugar Peas, which are eaten with their tender pods and thus create less wastage.

Peppers: Very nutritious and relatively free of diseases and insect pests. Plants can be started either in a hotbed, in a bed of rich soil covered with glass, or in pots indoors. They are transplanted in from eight to ten weeks, once the frost danger is past.

Radishes: An infallible crop that requires little care, grows rapidly, and can be planted just about anywhere. Their shallow roots need nutrients and moisture near the surface.

Spinach: A vigorous and attractive plant that contains a lot of vitamins and iron. It grows quickly (forty days from seed to maturity), prefers cool temperatures, and can be a winter crop in temperate regions. It is sown early and, like other leafy crops, can be seeded in succession to provide virtually a year-round crop. Harvest the whole plant, or individual leaves one at a time.

Soybeans: The most complete protein source of all vegetables. They are easily grown like other beans and have few insect enemies, but watch out for rabbits. They can be picked green, to shell and eat like peas; or pick them already dried on the stalk for storage, milling, sprouting, and eating. The entire plant is an excellent animal feed and nitrogen source for composting.

Squash: Produces phenomenal yields of nutritious food. The traditional Hubbard squash is a sure bet for most gardens. Squash need elbow-room and, like melons, can be grown interspersed with corn. It is worthwhile to know that squash can be eaten at any stage of development (though the summer squashes tend to be best eaten young), and all varieties keep very well.

Tomatoes: Because of the many ways in which tomatoes are used, they will probably figure prominently in any garden. Insect pests pose a real problem, but one that can be overcome. If chickens are available, they make ideal hunters for the tomato horn worm. Tomatoes can be started as seedlings in either glassed-in seedbeds or indoor pots. They are transplanted when two inches tall into separate pots and should be moved into the ground when all danger of frost is past. They can be either staked or unstaked. In the latter case, it is advisable to use a lot of mulch (an insulating layer of wood chips, sawdust, or vegetable matter) to keep the fruit clean and dry. Don't overwater; flowers drop, and so does your yield.

This is not a comprehensive list; it deals with a few of the more common, reliable vegetables. Table 7.1 summarizes the approximate planting periods for a variety of vegetables and may help you plan your garden for the year (data applicable to all climates similar to Palo Alto, California). Anyone doing gardening is advised to buy and keep for reference any of the fine books on organic gardening (see Bibliography), and to ascertain planting periods for your specific climatic situation.

Berries

Berries are an important element in your garden, as anyone who has ever eaten blueberry pie or peanut butter and jelly sandwiches can attest. And if you're a thirsty winebibber. . . .

Strawberries: Strawberries grow best in a slightly acidic environment (pH 6), with rich organic soil. It is a shallow-rooted plant; good drainage is important, as well as lots of moisture when the fruits are growing. Strawberries *need* sun; they should not be near trees or other plants with extensive roots (such as grapes). If different types of species are intermixed—early-season, mid-season, and late-season varieties, for example—a long harvest is possible. Plants will cost about $10 per one hundred (this and all other prices quoted herein are estimates; prices vary according to where and when purchased, the quality and age of the plant, and the quantity). Strawberries should be planted on a cloudy spring day to protect them from drying. The roots should always be kept moist; spread them well when you plant them in the soil. Pruning is suggested for the first year to insure good rooting. Strawberries will send out runners to propagate, and a few score plants will soon turn into a few hundred!

These berries are a good choice for an urban garden of limited size. Not much space is needed if a tiered pyramid of soil is fashioned for the strawberry patch. They can even be grown on a sunny patio in a strawberry pot or a barrel with holes uniformly spaced along the sides; the individual plants are each placed in a hole. Good drainage is very important, and the pot or barrel should be rotated so that all plants get sufficient sunlight.

Blueberries: Blueberry bushes need a lot of sun, a rich, loose, acidic (pH 5) soil, and fair amounts of water. They grow slowly and bear the most fruit after eight or ten years.

When buying blueberry bushes, be certain of two things: buy three-year-old plants, and buy at least three different types to insure cross-pollination. Three-year-old plants (twelve to eighteen inches) will cost about $2.50 each. Spring planting is best; water immediately after planting, then mulch with sawdust if the soil is too alkaline. It is more important in the first year for the plant to develop good roots than to blossom: all the blossoms should be pruned off the first year, and any suckers (shoots growing from the roots or stem) should also be cut. Mulching each year keeps the soil acid, and pruning each year increases yield. The bush will bear two years after transplanting. Yield will be about 10 quarts per bush, although the first harvest will probably be smaller.

Table 7.1 Vegetable Planting Periods[a]

Season	Vegetable	Planting Period
Sp	Asparagus	1/1–3/1
Sp/Su	Beans, Bush	3/10–8/1
Sp/Su	Beans, Lima	3/15–7/15
Sp/Su	Beans, Pole	3/10–7/15
A	Beets	All year
Sp/F	Broccoli	2/1–3/1 & 8/1–11/15
Sp/F	Brussels Sprouts	2/1–3/1 & 8/1–11/15
Sp/F	Cabbage, Early	1/15–2/25 & 8/1–11/15
Sp/F	Cabbage, Late	1/15–2/25 & 8/1–11/15
Sp	Cantaloupes	3/15–5/15
A	Carrots	All year
Sp/F	Cauliflower	1/20–2/20 & 8/1–11/15
Sp/F	Celery	2/1–3/1 & 8/1–11/15
A	Chard	All year
Sp	Chives	1/1–2/15[d]
Sp/F	Collards	1/15–3/15 & 8/1–11/15
Sp/Su	Corn	3/1–7/1[e]
Sp/Su	Cucumbers	3/1–7/1
Sp	Eggplant	3/10–4/15
Sp/F	Garlic	2/1–3/1 & 8/1–11/15
Sp	Horseradish	3/1–4/1
Sp/F	Kale	2/1–2/20 & 8/1–11/15
Sp/F	Kohlrabi	2/1–2/20 & 8/1–11/15
Sp/F	Leeks	1/15–2/15 & 8/1–11/15
Sp/F	Lettuce, Head	1/15–2/15 & 8/1–11/15
A	Lettuce, Leaf	All year[b]
Sp/F	Mustard	2/1–3/1 & 8/1–11/15
Sp	Okra	3/10–6/1
Sp	Onions	1/1–2/15 & 8/1–11/15
Sp/F	Parsnips	1/15–2/15 & 8/1–11/15
Sp/F	Peas, Bush	1/15–3/1 & 8/1–11/15
Sp/F	Peas, Pole	1/15–3/1 & 8/1–11/15
Sp	Peppers	3/15–5/1
Sp	Potatoes	1/15–3/1
Sp	Potatoes, Sweet	3/20–6/1
Sp	Pumpkins	3/15–5/15
A	Radishes	All year[c]
Sp	Rhubarb	2/1–3/7
A	Rutabagas	All year
A	Salsify	All year
Sp	Scallions	1/1–2/15[d]
Sp	Shallots	1/1–2/15[d]
Sp	Soybeans	3/20–6/30
Sp/F	Spinach	1/1–3/1 & 8/1–11/15
Sp/Su	Squash, Summer	3/15–7/1
Sp	Squash, Winter	3/15–5/15
Sp	Tomatoes	3/10–6/1[f]
A	Turnips	All year
Sp	Watermelons	3/15–5/15

Notes: a. From Common Ground Organic Supply, Palo Alto, Calif. Data relevant for climates similar to Palo Alto.
b. Some protection may be required in winter and summer.
c. Some protection may be required in summer.
d. March 1 for sets.
e. August 1 for early varieties.
f. July 1 for early varieties.

Blackberries and Raspberries: These tough and hardy bramble bushes grow happily almost anywhere, but more fruit is harvested when they are planted in deep, loamy soils (having porous consistency and good drainage) at pH 6. They prefer some shade and protection from winds, and also need extra water when the berries are maturing.

Buy one-year-old stock from a local nursery and plant in the early spring. The costs run about $28 for twenty-five small blackberry plants and $13 for an equal number of small raspberry plants. Plants should be placed from three to ten feet apart since they have extensive root systems. Mulching should be performed after planting, using ten inches of straw or rotted leaves, or three inches of sawdust.

Pruning for raspberries and blackberries is highly recommended. Left alone, they can take over your entire garden. All shoots should be cut back to ten inches in length after planting; each year, after harvesting, the fruit-bearing canes should be pruned entirely away, since they will no longer bear fruit. It is best to prune back to about half a dozen of the sturdiest year-old canes after the winter. Blackberries and raspberries self-propagate by sending up root suckers. To slow down the propagation, simply treat a sucker as you would any weed. One last thing: there is virtually no pest damage if you have good soil drainage.

Grapes: Grapes grow best in rich, well-drained, neutral soil (pH 6 to 8). They need plenty of sun and air, but can withstand dry weather very well because they are a deep-rooting vine, delving down from six to eight feet. They are long-lived (fifty to sixty years) and require from three to five years for full productivity.

Buy one- or two-year-old plants. The cost for five plants will range from $6 to $12, with the seedless varieties being more expensive. Plants should be spaced from four to ten feet apart, depending on the pruning and training procedures. They do well in sunny, urban gardens when trained for trellises or patios, but remember: it will take several years. It's also nice to know that they are relatively pest-free.

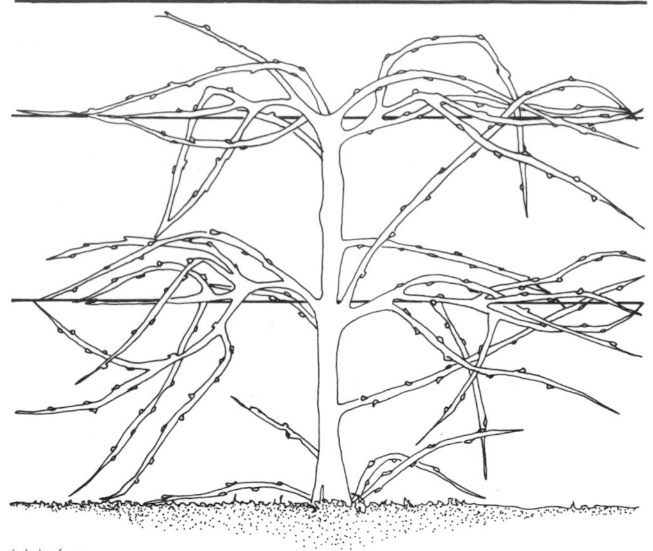

(a) before

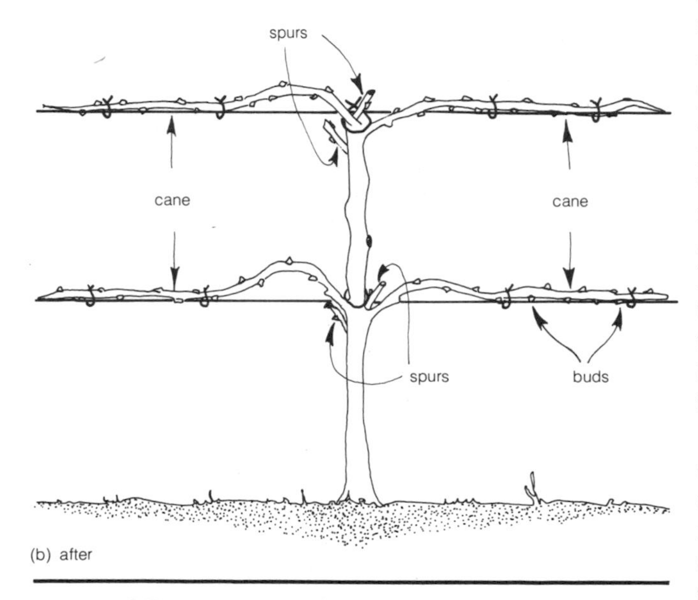

(b) after

Figure 7.2 The four-arm Kniffen method of pruning.

Careful pruning is not as critical for a decorative grape arbor as it is for a bountiful grape harvest. Pruning controls the old branches since fruits are carried only by new growth. Among the many methods, one example is described here and illustrated in Figure 7.2—the four-arm Kniffen method. Just before or after planting, trim away all the shoots until only one remains. This will form the trunk of the vine. Then, late the next winter, after the plant has grown out some, choose four canes, two growing in each direction, and prune away all other canes. These four canes should then be attached to a wire strung between the stakes. After the second winter, prune all but twelve buds away from the four canes; but also leave one spur, with one bud, next to each cane. A spur is a smaller branch, cut back almost to the trunk. It is these spurs which will form the secondary canes or laterals. Each successive year, leave a few more laterals on the plant; by the fifth year, the plant will be at its most desirable size.

Trees

When people think of gardening, they usually think only of plants which must be planted every year. There is no reason why trees can't be included in a backyard garden if the area is sufficiently large; dwarf varieties are perfect for these backyard situations. With enough land, of course, you can plan a nice orchard. Fruit- and nut-bearing trees also should not be overlooked when you consider landscaping or wind and sun protection; beyond beauty and shelter, they provide a regular source of food. The important nutritional value of these crops cannot be overlooked. Table 7.2 provides a summary of the nutritional aspects of a variety of fruits and nuts.

Fruit Trees: There are two very common methods for setting up an orchard: straight rows, or rows which curve along the contours of a hill. No matter which method you choose, the following recommendations should be followed. Fruit trees should be planted on higher land to prevent frost from killing the fruit and to insure adequate water drainage. They need well-drained soil (important for air circulation to the roots), a medium loam rich in organic material (3 to 5 percent), and a pH of about 6. Deep soil is also a must; depth can be determined by taking a boring with a soil tube. Plant on a northern slope to delay the blooming season and a southern slope to hasten it. Planting distances for the various trees are listed in Table 7.3.

All fruit trees should be watered only once a week; overwatering delays maturity of young trees and damages the fruits. Composting and mulching are essential for good growth. Make a ring of compost around the tree (but out away from the trunk to prevent rodent destruction) and then cover this ring with a foot of hay, leaves, or orchard-grass mulch. Pruning is also essential to increase the crop, improve fruit quality, and repair damages. Figure 7.3 summarizes the four basic pruning steps, along with some other techniques important for tree maintenance.

Purchase your stock from a local nursery; be sure to obtain types which do best in your area. It is important to remember that for the best-quality fruit (and for most trees to develop fruit at all), more than one variety must be planted in the same vicinity to insure cross-pollination, necessary for most fruit trees. Nursery stock usually are two- and three-year-old trees. The average cost will be about $3 to $5 per tree.

Propagation of most trees can be accomplished by layering; pears, apples, cherries, peaches, and plums are exceptions. Layering consists of bending and burying part of the length of a branch from three to six inches in

Table 7.2 Nutritional Value of Fresh Fruits and Nuts[a]

Fruit or Nuts	Quantity	Calories (Energy Value)	Protein (gm)	Fat (gm)	Calcium (mg)	Iron (mg)	Vitamin A (Internat. Units)	Vitamin B_1 Thiamin (mg)	Vitamin B_2 Riboflavin (mg)	Niacin (mg)	Vitamin C Ascorbic Acid (mg)
Almonds	1 cup (shelled)	850	26.0	77.0	332	6.7	—	0.34	1.31	5.0	—
Apples	1 medium	76	0.4	0.5	8	0.4	120	0.05	0.04	0.2	6
Apricots	3	54	1.1	0.1	17	0.5	2990	0.03	0.05	0.9	7
Avocados	1/2 (peeled)	279	1.9	30.1	11	0.7	330	0.07	0.15	1.3	18
Bananas	1 large	119	1.6	0.3	11	0.8	570	0.06	0.06	1.0	13
Blackberries	1 cup	82	1.7	1.4	46	1.3	280	0.05	0.06	0.5	30
Blueberries	1 cup	85	0.8	0.8	22	1.1	400	0.04	0.03	0.4	23
Cashews	1 cup (shelled)	1312	41.0	110.0	104	11.2	—	1.40	0.40	4.8	—
Cherries	1 cup	65	1.2	0.5	19	0.4	660	0.05	0.06	0.4	9
Currants	1 cup	60	1.3	0.2	40	1.0	130	0.04	0	0	40
Dates (fresh or dried)	1 cup (pitted)	505	3.9	1.1	128	3.7	100	0.16	0.17	3.9	0
Figs	4 large	79	1.4	0.4	54	0.6	80	0.06	0.05	0.5	2
Grapefruit	1/2 medium	75	0.9	0.4	41	0.4	20	0.07	0.04	0.4	76
Grapes (American)	1 cup (raw)	84	1.7	1.7	20	0.7	90	0.07	0.05	0.3	5
Guavas	1 medium	49	0.7	0.4	21	0.5	180	0.05	0.03	0.8	212
Lemons	1 medium	20	0.6	0.4	25	0.4	0	0.03	tr.[b]	0.1	31
Limes	1 medium	18	0.4	0.1	21	0.3	0	0.02	tr.[b]	0.1	14
Loganberries	1 cup	90	1.4	0.9	50	1.7	280	0.04	0.10	0.1	34
Mangoes	1 medium	87	0.9	0.3	12	0.3	8380	0.08	0.07	1.2	55
Oranges	1 medium	70	1.4	0.3	51	0.6	290	0.12	0.04	0.4	77
Papayas	1 cup	71	1.1	0.2	36	0.5	3190	0.06	0.07	0.5	102
Peaches	1 medium	46	0.5	0.1	8	0.6	880	0.02	0.05	0.9	8
Peanuts	1 cup (shelled)	1272	61.0	100.0	168	4.0	—	0.72	0.32	37.0	—
Pears	1 (peeled)	95	1.1	0.6	20	0.5	30	0.03	0.06	0.2	6
Pecans	1 cup (shelled halves)	696	9.4	73.0	74	2.4	50	0.72	0.11	0.9	2
Persimmons	1 medium (seedless)	95	1.0	0.5	7	0.4	3270	0.06	0.05	tr.[b]	13
Pineapple	1 cup (diced)	74	0.6	0.3	22	0.4	180	0.12	0.04	0.3	33
Plums	1 medium	29	0.4	0.1	10	0.3	200	0.04	0.02	0.3	3
Prunes (dried)	4 medium (unsulfured)	73	0.6	0.2	15	1.1	510	0.03	0.04	0.5	1
Raisins (dried)	1 cup (unsulfured)	429	3.7	0.8	125	5.3	80	0.24	0.13	0.5	tr.[b]
Raspberries (black)	1 cup	74	1.5	2.1	54	1.2	0	0.03	0.09	0.4	32
Raspberries (red)	1 cup	70	1.5	0.5	49	1.1	160	0.03	0.08	0.4	29
Strawberries	1 cup	54	1.2	0.7	42	1.2	90	0.04	0.10	0.4	89
Tangerines	1 medium	35	0.6	0.2	27	0.3	340	0.06	0.02	0.2	25
Walnuts (English)	100 gm.	650	15.0	64.0	99	3.1	30	0.33	0.13	0.9	3

Notes: a. From *The Encyclopedia of Organic Gardening* and *How to Grow Vegetables and Fruits by the Organic Method*, by J.I. Rodale. b. Trace.

the ground. The top of the branch becomes the new plant. Roots will form where it is buried; the connection with the parent tree then can be cut. This technique produces a new seedling quickly. Pear, apple, cherry, peach, and plum trees can be propagated by growing seedlings and then budding or grafting them onto a root stock (a length of root taken from an adult plant). Detailed information on these propagation techniques can be found in almost every reference on orchards. Harvesting, of course, is the most fun; optimum ripeness is determined by your taste. The approximate yields and bearing ages of various fruit trees are listed in Table 7.3.

Dwarf Trees: Many species of fruit trees also can be obtained as dwarf trees which are from five to eight feet tall; there are even smaller "double dwarf" trees which are from four to six feet tall. They produce, however, the same size fruit as a standard tree. The advantages of dwarf trees are many: they save space if land is at a premium (in urban gardens); they fruit sooner (see Table 7.3); and yield is sometimes better than that from a regular tree, because dwarf trees are easier to care for

Table 7.3 Fruit Tree Data[a]

	Min. Distance Between Plants (feet)	Approx. Yield per Plant (bushels)	Bearing Age (years)
Apple, standard	35	8.0	6–10
Apple, double dwarf	20	2.0	4–6
Apple, dwarf	12	0.5	2–3
Pear, standard	25	3.0	5–8
Pear, dwarf	12	0.5	3–4
Peach	20	4.0	3–4
Plum	20	2.0	4–5
Quince	15	1.0	5–6
Cherry, sour	20	2.0	4–5
Cherry, sweet	25	2.5	5–7

Notes: a. From J.I. Rodale, *How to Grow Vegetables and Fruits by the Organic Method.* Prepared by Virginia Polytechnic Institute.

and easier to harvest, with less consequent waste and fallen fruit. However, dwarf trees have less extensive root systems and so are more susceptible to wind damage. They also have shorter bearing lifespans (twenty to thirty years) than standard trees (forty to fifty years), and they cost more initially.

The trees from a local nursery will run about $8 apiece. Soil and climatic requirements are the same as for normal trees, but planting is a bit different. The dwarf tree must be planted with the graft-union above ground—the graft-union is a small knob on the trunk where the bark is a different color. These trees also need cross-pollination: more than two species *must* be planted. After planting, trim off any suckers or branches within a foot of the ground at the trunk. Maintenance, mulching, and harvesting are the same as for the regular trees, and the approximate yields are in Table 7.3.

Nut Trees: The best trees for edible nuts are Chinese chestnut, hickory, pecan, walnut, and filbert. Certain types grow better in specific climates, as do certain species of each type, so check in advance with your local nursery or Agricultural Extension Service.

Some basic advice will suffice for all the many types and species. A northern slope with deep soil and good drainage is best. Trees should be planted from twenty to thirty feet apart; this distance is adequate for their deep, extensive root system, yet close enough for cross-pollination to occur. Again, to insure cross-pollination, at least two species of any single type must be planted. Planting requires a large hole for the extensive taproot (the main root which grows straight down). Pack soil around the tree; it then should be watered and a mulch of straw or rotted leaves spread around the trunk. Maintenance and pruning are the same as for fruit trees, and propagation can be achieved by layering.

One final note: these trees grow very slowly and do not always bear. Even when they bear, the yield is small unless conditions are almost perfect, so don't plant these trees with exaggerated hopes. Approximate costs are as follows (apiece): chestnut—$4.50; walnut—$8.00; pecan—$5.50; filbert—$5.50; and hickory—$6.00.

Grain

If your farm is to be self-sufficient, it will need at least a few acres of grain. The grain is used not only for human consumption, but also for fodder, winter bedding for your animals, covercropping, mulching, and feed for your methane digester. Obviously, it is not a good choice for the urban garden of limited size.

Except for corn, the procedures for planting and harvesting the different grains are very similar. Grain is planted by two basic methods: either get a grain drill, drill holes, and drop the seed in; or simply broadcast the seeds uniformly over the ground and rake to cover them over. Harvesting for any grain (except corn) is a back-breaking, time-consuming labor. Stiff muscles may cause you to wonder just how self-sufficient you really want to be. Grain-bearing grass is cut with either a normal or cradle scythe. A cradle scythe is one with an arrangement of fingerlike rods which catch the grass and enable you to lay it evenly on the ground; alternatively, the grass can be propped up and tied into bundles. For threshing, the grass is laid out on a hard surface, and a flail is used to beat grain loose from the straw. To separate the grain from the straw, you must winnow: the threshed grass is thrown up in the air when the wind is blowing, and the straw is blown away as the heavier grain drops straight down. The grain can then be stored until milling. It should be well protected from insects.

Buckwheat: Buckwheat can be used as an ingredient in breads or for livestock feed. It grows best in a cool, moist climate; but it also grows on soil which does not support other grains very well because of buckwheat's ability to utilize nutrients from the soil more effectively. The two major varieties are the Japanese and the silverhull, which are often sown together; but check with your local Agricultural Extension to determine the proper types for your area. Be sure the seeds you purchase are certified disease-free. The best planting time is ten weeks before the first fall frost. If you have a grain drill, or can borrow one, drilling is preferred. Use about 4 pecks (1 peck = 0.25 bushel = 8 quarts) per acre. After harvesting, allow your grain to field-dry for ten days. It can then be stored for winter fodder or else be threshed in preparation for milling. After threshing, the straw can be used for mulch or bedding. Yield is between 20 and 30 bushels per acre.

Oats: Oats can be used for bedding, for fodder, and, of course, for cereals and breads to eat yourself. Make certain you buy the right type for your area. You will need about 5 to 10 pecks per acre, depending on regional conditions. Spring seeding is preferred, and should take place about a week before the last spring frost. Adequate water is very important. Harvest the oats when the kernels are soft enough to be dented slightly by

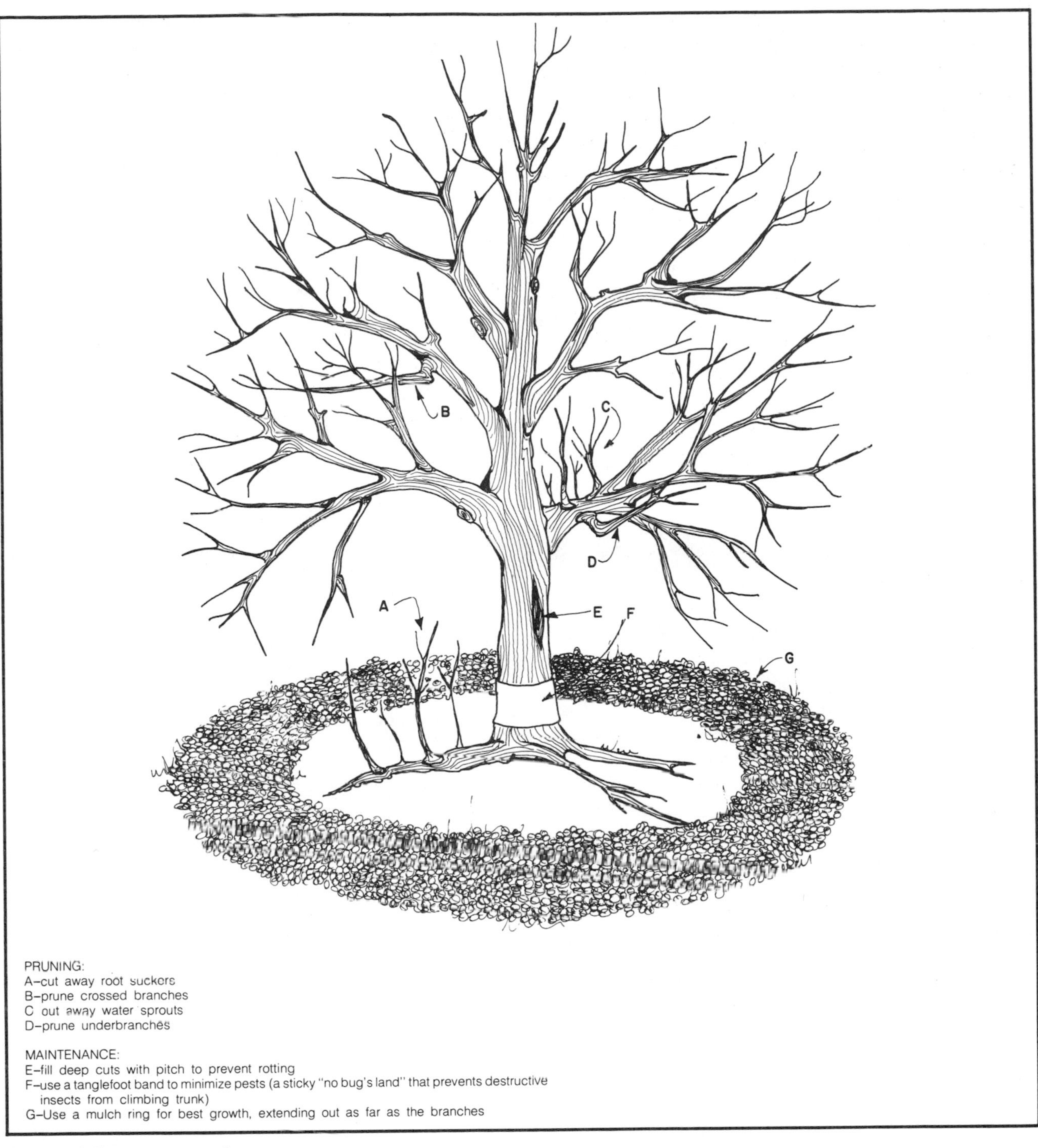

Figure 7.3 Tree pruning and maintenance.

your fingernail. Cut it with a cradle scythe, let it dry for a day, and then bind it into twelve-inch-diameter bundles; stand these bundles upright to field-cure for two weeks. Yield will be about 25 to 40 bushels per acre.

Rye: Because rye will grow in almost any type of soil, it is useful as a cover crop or as green manure (a plant which is plowed back into the soil so that it will decompose into fertilizer). Rye also can be used for making bread. Plant rye in either fall or spring, at about the time of the first or last frost, respectively. Use 2 bushels of seed per acre for green manuring, and 1 bushel per acre for grain. Harvest as you would wheat; yield will be about 15 bushels per acre.

Wheat: Wheat does best where there is a cool, moist growing season followed by a warm, dry period for curing. Spring wheat is usually planted in the northern

United States, and winter wheat in the South. There are many types of wheat—hard red, soft red, winter, hard red spring, and durum, among others. Buy the one that best suits your growing area. Since wheat is self-fertilizing, you will be able to use your own seed the next year. Use about 6 pecks per acre when sowing. Plant winter wheat about the time of the first frost, and spring wheat about the time of the last killing frost. Harvest, as with rye and oats, when you can dent the grain kernel with your fingernail. Yield will be about 20 bushels per acre. To harvest, cut the wheat using a cradle scythe, bind it into bundles ten inches in diameter, and let it field-dry for ten days. For flour, thresh and winnow the grain, and then grind it as needed to make bread. The straw can be used either for mulch or for bedding your livestock.

Corn: Plant dent corn or flint corn for stock feed, ground meal, and silage (green fodder preserved in a silo until it ferments). Plant sweet corn for yourself. All these types grow best in deep, rich, neutral-pH soil with lots of loam and adequate water.

Buy yellow corn seeds as opposed to white varieties, and open-pollinated types as opposed to hybrids. With the nonhybrid varieties, you will be able to use your own homegrown seeds for the following year's planting. In addition, the open-pollinated varieties tend to be sweeter and higher in protein. Use three different varieties in one planting to insure a long harvest, or else plant rows a week apart for from four to six weeks.

You should plant your corn a week after the last spring frost. The soil temperature required for germination is 60°F. Seeds should be deposited about 1.5 inches into the soil, in at least several rows to insure pollination. Rows should be spaced from thirty to forty inches apart. Mulch with hay when the sprouts are eight inches tall; or, if you prefer, cultivate in between the cornstalks such low-growing plants as melons or beans. Corn should be fed with compost or organic fertilizer when the lower leaves begin to lose their color. Harvest as soon as it is ripe; ripeness is indicated by brown silk and the appearance of a milky liquid when the kernel is dented with your fingernail. Yield is about thirty ears for a fifty-foot row.

Livestock

As we pointed out earlier, all animals ultimately derive their energy from plants. In theory, energy conservation could be best effected by eating only plants. But we consider small farm livestock as an integral part of any agricultural system both because of their useful conversion to palatable food of plant materials (grass, garbage, chicken feed) most of us find unappetizing and also because they are a good source of manure, indispensable to a good organic garden.

For the most part, unfortunately, raising animals is not as viable as gardening on small urban "homesteads." But this is not to say there are no urban possibilities; these will be pointed out along the way. There are also a number of pamphlets put out by the state and federal government which are useful, concise, and free (see Bibliography).

Chickens

Raising chickens for eggs and meat, as well as for their highly valuable manure, is relatively inexpensive and easy. With your own homemade feed, you can say goodbye to the hormone-injected monsters weighed out in today's markets. To have an efficient, steadily producing flock rather than half-wild birds, it is important to have a good breed, good food and shelter, regular cleaning, and accurate laying records.

A dozen good laying hens will produce an average of six eggs per day. An easily managed project would be to start off with fifty one-day-old chicks and raise them, gradually selecting a dozen good pullets (young hens) for egg production and using the rest as fryers. The next year you start a new flock of egg layers by buying and raising a new brood of day-old chicks. Year-old hens lose about 15 to 30 percent of their egg-laying ability in their second year; they should be replaced gradually by newly selected pullets. This procedure presents a problem in housing since both groups need a coop; but if the chick raising is performed in the spring, extensive shelter will not be needed.

The shelter should take into consideration climate, feeding, egg collection, and maintenance. Based on a figure of 4 square feet per laying hen, a dozen laying hens will need a 6-by-8-foot house, with a small yard for exercise.

A chicken roost eighteen inches high, with slats to provide a frame and covered with one-inch chicken mesh, can be used within the coop (see Figure 7.4). The use of this roost can cut cleaning to three or four times a year, unless odors and flies become objectionable. Only three nests are required for twelve hens. To ease disturbance during egg collection, a rear hinged door can be installed for each nest so that you can reach into the coop quickly and quietly.

Breeds and Brooding: The breeds that you select should be the best ones available for year-round egg production and also for meat. The Rhode Island Red is best in this respect, with the Barred Plymouth Rock as a close second. Cross-breeding between these two breeds also provides excellent results.

To prevent the development of fatal pullorum disease in chicks, the breed selected should be rated "USDA pullorum-clean." To obtain the best egg layers, the chicks should be rated "U.S. Record of Performance Chicks" or "U.S. Certified Chicks." The latter two are National Poultry Improvement Plan ratings.

High-quality chicks run about 50 cents apiece, so that a flock of fifty will cost about $25. It is best to buy a flock of both sexes as the cockerels (young males) can be used for meat at the end of three months, when they weigh around three pounds.

The most self-sufficient plan for raising chickens is to acquire both an incubator for hatching and a brooder for raising the chicks. (You will need, of course, a rooster as well.) Provided that adequate energy is available, an incubator can be used for the twenty-day hatching period. The hens also can be used to hatch the eggs; however, this maternal care takes up egg-laying time. In most cases, it may be better to purchase day-old chicks each April when starting a new flock rather than hatch your own. Brooders (heated shelters—see Figure 7.5) should be used for about six weeks with increased space provided according to the size of the chicks. The brooder is set inside an enclosure with accessible water jars and feeders. A 60- to 100-watt bulb, a gallon jug of hot water, or hot-water pipes can provide adequate heat to the brooder. The area should be free from drafts. The temperature is right when the chicks stretch out uniformly under the hover (a heat-producing cover). The amount of heat should be reduced as rapidly as possible since lower temperatures help the chicks to feather faster. Some heat, however, is generally required for the first six weeks. Litter—peat moss, sand, wood shavings, and so on—should be placed underneath the brooder to absorb waste. It is advisable that the chicks be brooded away from older hens to prevent possible disease transmission. A basement or side room in your house will provide adequate warmth to maintain the 90°F temperature under the hover.

Feeding: For the first three days, the chicks can be fed a chick grain of finely cut corn, wheat, and oats, with chick-size insoluble grit or coarse sand also available in a feeder alongside. They then should be switched to starter mash—a mixture of corn, oats, and warm water, supplemented with grain meal and vitamins. About six weeks later, scratch—a mixture of corn, wheat, and oats without water—should be added in increasing proportions until, at four months, the chicks are receiving equal servings of scratch and mash. Calcium is an important nutrient for laying hens and is supplied by additions of ground limestone or oyster shells to the mash. The mash and scratch should be kept before the chickens in self-feeding hoppers nearby to gallon-size waterers. Save on purchased feed by using such kitchen wastes as leftover bread, vegetable peelings, leftover meat, fish, cheese, and culled potatoes. These scraps are boiled in a minimum of water and mixed with a sufficient amount of mash to make a crumbly mixture that is an excellent feed. Surplus greens—lawn clippings, kale, cabbage—can also be used as a feed supplement. For scratch feed, an acre of field corn produces 75 percent of the total feed requirements for twenty-five chickens. A small corn sheller and feed grinder or a small feed mill can be purchased for $10. The only feed items that you will have to buy are ground oats, bran, soybean oil meal, corn gluten meal, limestone, and salt; these items are far cheaper to purchase than to grow on a small farm. Over the year, you will need about forty-five or fifty pounds each of both mash and grain for every fowl. Generally, a good hen will

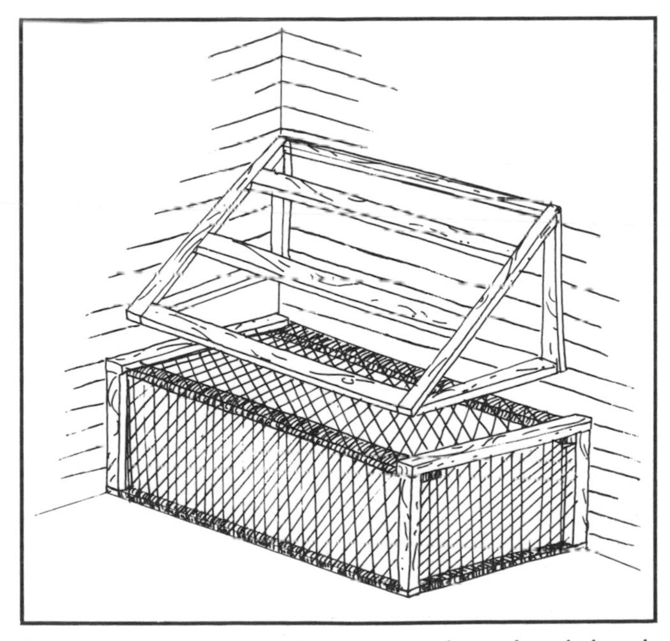

Figure 7.4 The chicken roost should be located away from drafts and against a wall. It can be constructed of supporting slats and chicken wire; if desired, a pit can be dug under the roosting area.

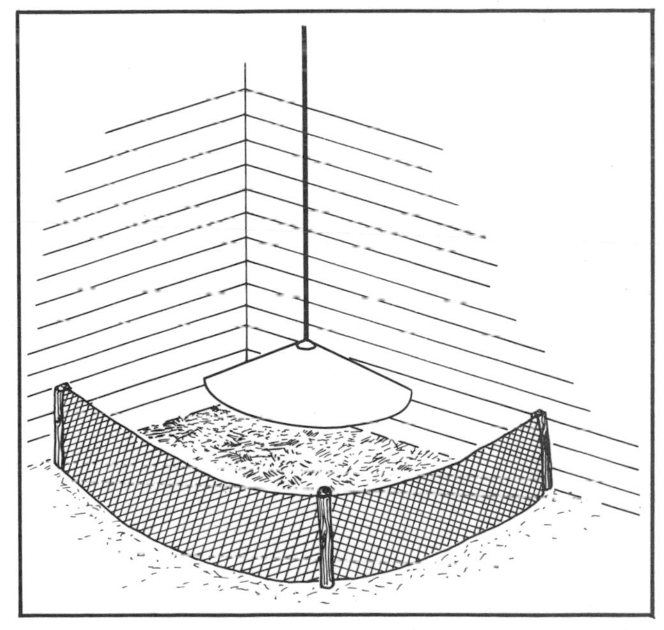

Figure 7.5 A brooder should be located in an area well protected from drafts. The area needs to be fenced to keep the little chicks in and the larger chickens out. Provide food, water, and heat from a hover.

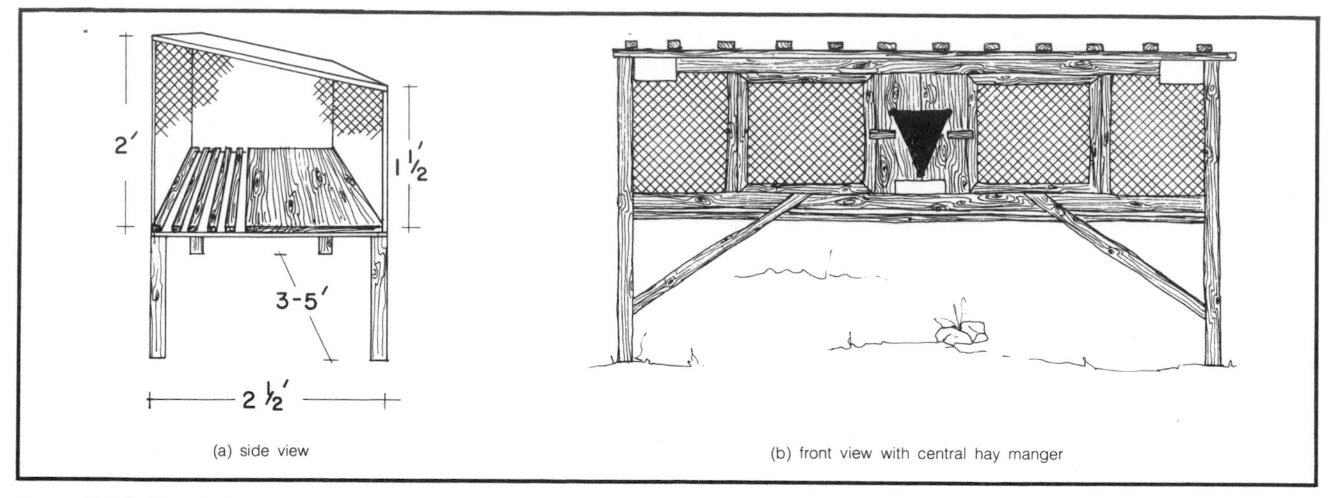

Figure 7.6 Rabbit shelters.

produce between 200 and 240 eggs a year. Keep in mind that some hens end up as nonlayers; these should be weeded out and used as fryers.

It is advisable not to feed the fowl during the twelve hours before slaughter. A small hand guillotine is the most humane and cleanest method; a stretch and twist of the neck is also fast.

Because of their relatively small size and their adaptability to enclosures, chickens can be raised in a large backyard. The amount of space available will limit the number of chickens which can be accommodated. Remember, also, that a dozen or so chickens can get noisy and may not win friends among your neighbors. It's probably a good idea to check for community restrictions; you may not be "zoned" for poultry. For further information, see listings for J. Morley and G. Klein in the Bibliography.

Rabbits

Rabbits are among the easiest animals to raise and care for; they are also quite suitable for both rural and urban regions. Irate neighbors are certainly less probable than with chickens. The medium and large breeds are best for food requirements. Some breeds to consider are the French Angora, which grows to over eight pounds and can be sheared for wool every ten weeks; California rabbits, which grow to eight or ten pounds; the American Chinchilla, which grows to ten pounds; and the New Zealand breed, which grows to a comparable size. As with goats or poultry, you should check on the reliability of your dealer when you go into rabbit raising; that is, he should stand behind his product and guarantee the health and productivity of his animals. One buck and from two to four does are sufficient for a starting herd. The gestation period is very short, thirty days, and a good doe can produce four litters (five to eight young each litter) every year. Consequently, you have either a steady supply of food or a rapidly increasing rabbit herd! The young are marketable at four pounds, attained in about two months.

Shelter: Adequate shelter from the rain and occasional frost should be provided, as well as adequate shade and ventilation during the summer months. The dimensions for a medium-sized hutch (three to six rabbits) are generally 2 feet high, 2.5 feet wide, and 3 to 5 feet long (see Figure 7.6). Two hutches with a central storage compartment form a portable unit that can be taken under a larger shelter or put outside for better ventilation. The floors should be slatted or wire meshed on one side for easier cleaning and for the collection of manure. A slight slope is best for good drainage. Nest boxes should be provided for seclusion when a doe gives birth; strong cardboard boxes with a small nest of hay will do. These should be placed in the hutch a few days before the doe gives birth.

Feeding: In order to minimize your labor, the holders should have the capacity for several feedings. The use of a central hay manger is very efficient. Grain hoppers made from five-gallon cans are also excellent, especially for pregnant does, since you want to minimize disturbances during gestation. Fresh water should be in continuous supply, particularly in hot weather, when a doe and her litter can drink up to a gallon a day.

The best hays to feed rabbits are such broad-leafed, green, short-stemmed hays as alfalfa, clover, lespedeza, and peanut; all of these are high in protein and need no supplement. Stemmier grasses—timothy, prairie, Johnson, Sudan, or carpet grasses—are fine, but you will need to add additional protein. High-protein diets may be necessary if your herd is doing poorly. For example, cracked grains can be supplemented with a soybean meal. Salt, too, should be provided; you can use blocks or else add it to the meals in a 0.5 or 1 percent concentration. Vegetable table scraps are excellent feed supplements and what the rabbits can't finish, the compost heap can.

Breeding: The medium-sized breeds can be mated between the ages of five and six months and the does will continue to reproduce for up to thirty months thereafter. Each litter requires a four-week gestation period and two months of nursing; as we mentioned, a healthy doe can produce four litters a year.

Slaughter: Rabbits grow rapidly and will be ready for slaughter after eight weeks. Killing is best effected by a stretch and twist of the neck or a sharp blow to the base of the skull. To skin the carcass, it should be hooked through the right rear leg. The tail, head, front feet, and free rear leg (up to the hock joint) should be severed. A slit is cut across the rear and the skin is pulled off like a glove with as much fat left on the carcass as possible. To clean the meat, the belly should be opened and the entrails and gall bladder removed. The carcass is then washed off and kept cool. The skin must be stretched while it is still warm. The use of barrel stretchers and patient work in a ventilated area will produce good results.

For further information on rabbits in general, see the listings under B. Angier, B. Kaysing, and *Raising Livestock on Small Farms* (U.S. Department of Agriculture) in the Bibliography.

Goats

Goats are a good source of milk and cheese on a small farm. The better dairy goats provide 2 to 4 quarts per day for ten months of the year, eat less than a cow, and are easier to care for. The two dry months are taken up by pregnancy.

When selecting a particular breed, check with several dealers so that you see a variety of animals. There are five main milking breeds in the United States: American La Mancha, Nubian, Saanen, French Alpine, and Loggenburg. You should purchase two does, both to insure a continuous flow of milk and to give your goat company. A good doe costs from $35 to $75 and two provide enough milk for eight or ten people.

Housing: Special housing is not required, although any simple shelter should take into consideration dryness, cleanliness, and adequate warmth and shade. You must have space to handle hay, grain, water, and manure, and about 15 square feet of moving room per animal. For most of the year, however, goats can be put out to pasture.

Breeding: Unless you care to raise a larger herd, it is advisable to have the does bred at the dealer rather than raise a buck. Bucks can develop objectionable odors and they give little return for their feed. Also, a well-recorded buck insures excellent milkers in its offspring, and dealers keep such records. The gestation period is about 150 days and each doe generally produces two kids. They breed between September and February and give no milk for two months in early pregnancy.

Feeding: Goats are perhaps the best of all livestock for feeding on unimproved pasture. They should not be let out, however, on spring pasture until there is at least three or four inches of growth. The best range includes alfalfa, brome grass, clover, Sudan grass, or millet, the latter two being adequate for summer feed. An excellent spring or fall pasture is rye, wheat, or barley.

Goats need loose salt mixed in with their feed since they do not eat sufficient amounts of salt if it is in block form. Also, if the pasture is predominantly grass, they will need calcium and phosphorus supplements.

A milking doe will need about 450 pounds of grain and 500 pounds of hay each year, as well as vegetable scraps and any available root crops such as carrots, beets, and parsnips. When she is on a good pasture that is balanced in grasses and legumes, she should be fed one pound of grain for every four pounds of milk that she gives. To be on the safe side, equal quantities of salt and di-calcium phosphate should be easily available in separate, weatherproof containers. Yearlings should be given enough hay and grain to keep them content, but not so much that they become fat and you can no longer feel their ribs.

For the Best Milk: To obtain the best-tasting milk, the following guides should be followed:

1. Don't feed strong-flavored feeds (onions, silage, cabbage) within four hours of milking.
2. The udders and flanks should be clipped to prevent hair from getting in the milk.
3. Wash udders with warm water and iodophor or a solution of quaternary ammonium prior to eack milking.
4. Use a milking stand with a stanchion (a bar that holds the animal's head in place) and stainless-steel milking pails.
5. The milk should be strained when poured into the milk cans.
6. To retard bacterial growth, the cans should be immersed in coolers that are at a temperature of 33 to 35°F.
7. Pasteurization can be done easily by rapid heating to from 150 to 180°F over the stove and then cooling.

For further information, see the listings in the Bibliography under B. Colby and *Raising Livestock for Small Farms* (U.S. Department of Agriculture).

Cows

The advantages of raising a cow need hardly be recounted—all the milk, cream, yoghurt, and butter you can eat. In addition, cow manure is one of the finest additions your garden or compost heap can ever see. You will have the best results if pasture and hay are plentiful, shelter and daily care can be given, and breeding is available.

Dairy cows vary in the amount of milk they can produce. A cow with moderate production can be purchased for $300 or $400, kept five years, and then sold for $100; this would amount to about $50 or $60 of "depreciation" each year. This animal will produce from 3000 to 5000 quarts per year, roughly enough milk for from eight to sixteen people. An upper estimate of total yearly costs, including the cost for feed and bedding, would be $460—roughly 15 cents per quart! Actual costs would probably be less.

Housing: The cow needs only a simple shelter with a small amount of moving space. A three-sided stable with sunny southern exposure is probably adequate in a moderate climate. The cow can be either stanchioned or left untied in a box stall about ten feet square. Bedding of straw, cornstalks, or similar material is required—about 800 to 1600 pounds a year.

Feeding: A milk cow can eat up to 25 pounds of hay per day, which adds up to 3 or 4 tons per year. She also needs one or two tons of grain supplements each year. These costs can run about $30 to $60 a ton for hay, and around $40 to $80 a ton for grain; naturally, you can save money if you grow feed grain and hay yourself. Two acres of pastureland supplying about six months of food can cut the cost in half. A rough estimate of yearly food and bedding costs for a cow is from $100 to $350.

Care: A cow must be tended every day and generally is milked both morning and evening; the guidelines for goat's milk also applies here. A small electric home pasteurizer costs about $50 and can be used for either goat's or cow's milk. For further information, see listings for B. Angier and B. Kaysing in the Bibliography.

Bees

Beekeeping (apiculture) is a venture that requires only a small initial investment and a minimal amount of labor, and is adaptable to both urban and rural settings. From very modest efforts, you can expect a yield of approximately 75 to 100 pounds of honey a year per hive. In addition to being a source of food, bees also pollinate flowering crops and play an important part in increasing crop yields. And, they sting! To produce a honey crop, you need to make about twenty visits to the hives a year; in contrast to raising animals, no feed need be provided.

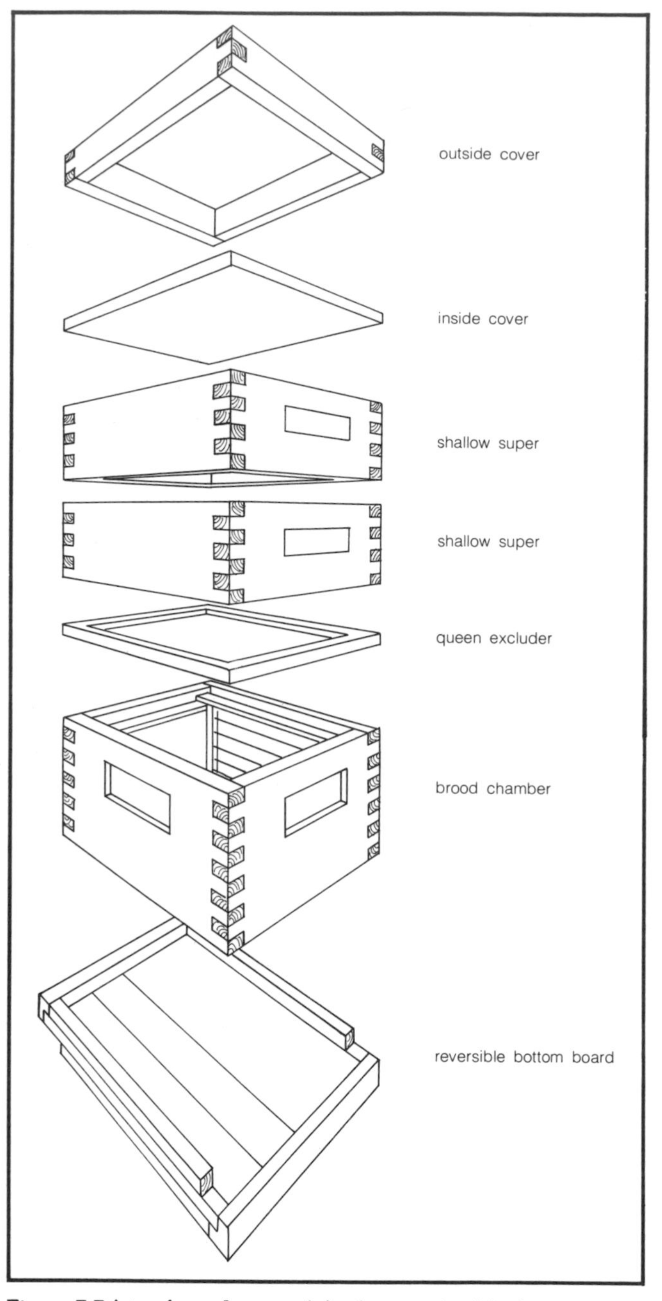

Figure 7.7 A ten-frame Langstroth beehive: each of the honey supers and the brood chamber can hold ten wax frames hung vertically.

There are five basic pieces of "equipment" to which you will need access: a hive, bees, protective clothing, a smoker, and an extractor. For a start, you can buy the bees for from $10 to $25 from a supplier (who can certify that they are disease-free), invest in one hive, and then rent the extractor.

Housing: Bees and a large secondhand hive can be purchased for around $70. The standard hive produced in the United States is called a ten-frame Langstroth. As illustrated in Figure 7.7, it includes a cover, two honey supers, a queen excluder, a brood chamber, and a bottom stand. Ten wax frames, embossed with a pattern of hexagonal cells to provide a base for the bees to build regular cells, are placed perpendicularly into each of the honey supers and the brood chamber. These are reusable and can last up to forty years. The queen excluder is a sievelike screen that allows the smaller workers to pass through, but confines the queen to the brood chamber. This isolation insures that no eggs will be laid in the honey supers.

When handling the hive, you should wear light-colored, smooth clothing because this attire seems to calm angry bees. Wear hightop boots with the cuffs of your pants tucked in securely. A fine mesh or screen should be used to protect your face and neck. But bear in mind, no matter how good your protective clothing is or how careful you are, you still can expect to get stung occasionally.

Care: The best time to start beekeeping is in the spring. If the bees are properly managed, the major honey flow will occur in late June or early July. A second honey flow will occur in the fall. The bees should have access both to water and to fifteen or twenty pounds of reserve honey at any time during the year. Since a single frame of honeycomb usually contains from five to seven pounds of honey, this means three frames should be left in the hive as surplus. But to over-winter a hive in cold climates, your bees will need approximately sixty pounds of honey or supplementary sugar water to last until spring.

Bees are subject to diseases that can reduce honey yields. These diseases generally appear in the hive when the bees are under stress. Consequently, removing the stress usuually eliminates the disease. If you suspect a problem, it is a good idea to consult the local apiary inspector.

To extract honey, first use the smoker to blow smoke into the hive; this pacifies the bees. Then the frames are removed and placed in the extractor. The extractor spins around and centrifugal force throws the honey against the walls. The honey then can be collected from the bottom of the extractor.

After gaining experience and discovering your limitations, you might expand until you have up to four hives. Unless you are going into business, four hives produce about as much honey as most small groups can consume. For further information, see listings in the Bibliography for R.A. Morse and A.I. Root.

Compost and Fertilizer

The concept of fertilization is fundamental to the process of recycling nutrients through the biosphere. As a plant grows, it extracts nutrients from the soil, the most important ones being nitrogen, phosphorus, and potassium. Without restoration of its nutrients, the soil would eventually be depleted and be unable to support plant growth.

There are four arguments against the use of chemical fertilizers to resupply the soil. First, the nutrients found in chemical fertilizers have been "displaced" from some other "home" in the natural world; the redistribution of these nutrients (away from their home) somewhere disturbs the balance of natural, complex cycling processes. Secondly, chemical fertilizers do not improve soil texture or soil's ability to hold moisture. Nutrients are leached away by rainfall; large runoffs of nitrogen fertilizer can cause algal blooms in rivers and streams, which lead to lower dissolved-oxygen levels in the water, which can, in turn, kill fish. Thirdly, the manufacture of chemical fertilizer consumes great quantities of valuable energy. And finally, the use of chemical fertilizers implies that natural sources of fertilizer are not being exploited and, most probably, are being disposed of in some polluting manner. A typical municipal sewer system is an example of real waste and pollution through the disposal of "wastes" that, properly handled, could have been utilized as fertilizer. Human, animal, and vegetable wastes can be recycled through a field of food crops repeatedly without seriously upsetting the natural mineral balance of the soil.

Composting

Composting is one of the most commonly practiced methods of recycling nutrients through crop lands on a small scale and is useful for both urban and rural gardens. Although there are many various techniques, the basic composting process is the mixture of organic wastes and soil into a pile where the temperature can be maintained at around 140 to 160°F. This high temperature enables bacteria and fungi from the soil to decompose the organic wastes into chemical nutrients that can be used by plants.

Almost any organic matter can be used as composting material (see Table 7.4). Exceptions include such hard calcareous organic matter as bone or seashell. The primary sources of material for the compost pile are animal wastes (manure), vegetable or crop wastes, kitchen wastes, and soil. All green material should be partially withered.

Most materials will be dry or, at most, damp. Each component must be sprinkled with water for the process to occur. It is important, however, not to make the pile too wet, as excessive moisture will hinder the process. If the materials are quite wet, they should be spread out to dry somewhat before adding to the compost heap.

Besides these basic ingredients, additives can improve the pH and the nutrient content of the final product. Ground limestone, commercial lime, or ashes can be mixed in to increase the pH (lower the acidity) of the compost; satisfactory compost has a pH between 7 and 8. There are also nitrogen-rich additives—blood meal or sewage sludge—that can increase its nitrogen

Table 7.4 Composition of Various Materials for Compost[a]

Material	Nitrogen (%)	Phosphoric Acid (%)	Potash (%)
Alfalfa hay	2.45	0.50	2.10
Apple (fruit)	0.05	0.02	0.10
Apple (leaves)	1.00	0.15	0.35
Apple pomace	0.20	0.02	0.15
Apple skins (ash)	—	3.08	11.74
Banana skins (ash)	—	3.25	41.76
Banana stalk (ash)	—	2.34	49.40
Barley (grain)	1.75	0.75	0.50
Bat guano	1.0–12.0	2.5–16.0	—
Beet roots	0.25	0.10	0.50
Bone (ground, burned)	—	34.70	—
Brewer's grains (wet)	0.90	0.50	0.05
Brigham tea (ash)	—	—	5.94
Cantaloupe rinds (ash)	—	9.77	12.21
Castor-bean pomace	5.0–6.0	2.0–2.5	1.0–1.25
Cattail reed and stems of water lily	2.02	0.81	3.43
Cattail seed	0.98	0.39	1.71
Coal ash (anthracite)	—	0.1–0.15	0.1–0.15
Coal ash (bituminous)	—	0.4–0.5	0.4–0.5
Cocoa-shell dust	1.04	1.49	2.71
Coffee grounds	2.08	0.32	0.28
Coffee grounds (dried)	1.99	0.36	0.67
Common crab	1.95	3.60	0.20
Corncobs (ground, charred)	—	—	2.01
Corncob (ash)	—	—	50.00
Corn (grain)	1.65	0.65	0.40
Corn (green forage)	0.30	0.13	0.33
Cottonseed	3.15	1.25	1.15
Cottonseed-hull (ash)	—	7.0–10.0	15.0–30.0
Cowpeas (green forage)	0.45	0.12	0.45
Cowpeas (seed)	3.10	1.00	1.20
Crab grass (green)	0.66	0.19	0.71
Cucumber skins (ash)	—	11.28	27.20
Dog manure	1.97	9.95	0.30
Dried jellyfish	4.60	—	—
Dried mussel mud	0.72	0.35	—
Duck manure (fresh)	1.12	1.44	0.49
Eggs	2.25	0.40	0.15
Eggshells (burned)	—	0.43	0.29
Eggshells	1.19	0.38	0.14
Feathers	15.30	—	—
Field bean (seed)	4.00	1.20	1.30
Field bean (shells)	1.70	0.30	0.35
Fire-pit ashes from smokehouses	—	—	4.96
Fish scrap (red snapper and grouper)	7.76	13.00	0.38
Fish scrap (fresh)	2.0–7.5	1.5–6.0	—
Freshwater mud	1.37	0.26	0.22
Garbage rubbish (New York City)	3.4–3.7	0.1–1.47	2.25–4.25
Greasewood ashes	—	—	12.61
Garden beans (beans and pods)	0.25	0.08	0.30
Gluten feed	4.0–5.0	—	—
Greensand	—	1.6–2.0	5.00
Grapes, fruit	0.15	0.07	0.30
Grapefruit skins (ash)	—	3.58	30.60
Hair	12.0–16.0	—	—
Harbor mud	0.99	0.77	0.05
Hoof meal and horn dust	10.0–15.0	1.5–2.0	—
Incinerator ash	0.24	5.15	2.33
Kentucky bluegrass (green)	0.66	0.19	0.71
Kentucky bluegrass (hay)	1.20	0.40	1.55
King crab (dried and ground)	10.00	0.26	0.06
King crab (fresh)	2.0–2.5	—	—
Leather (acidulated)	7.0–8.0	—	—
Leather (ground)	10.0–12.0	—	—
Leather, scrap (ash)	—	2.16	0.35
Lemon culls, California	0.15	0.06	0.26
Lemon skins (ash)	—	6.30	31.00
Lobster refuse	4.50	3.50	—
Lobster shells	4.60	3.52	—
Milk	0.50	0.30	0.18
Mussels	0.90	0.12	0.13
Molasses residue in manufacturing of alcohol	0.70	—	5.32
Oak leaves	0.80	0.35	0.15
Oats, grain	2.00	0.80	0.60
Olive pomace	1.15	0.78	1.26
Olive refuse	1.22	0.18	0.32
Orange culls	0.20	0.13	0.21
Orange skins (ash)	—	2.90	27.00
Pea pods (ash)	—	1.79	9.00
Peanuts, seeds or kernels	3.60	0.70	0.45
Peanut shells	0.80	0.15	0.50
Peanut shells (ash)	—	1.23	6.45
Pigeon manure (fresh)	4.19	2.24	1.41
Pigweed, rough	0.60	0.16	—
Pine needles	0.46	0.12	0.03
Potatoes, tubers	0.35	0.15	0.50
Potatoes, leaves and stalks	0.60	0.15	0.45
Potato skins, raw (ash)	—	5.18	27.50
Prune refuse	0.18	0.07	0.31
Pumpkin, flesh	0.16	0.07	0.26
Pumpkin seeds	0.87	0.50	0.45
Rabbit-brush ashes	—	—	13.04
Ragweed, great	0.76	0.26	—
Red clover, hay	2.10	0.50	2.00
Redtop hay	1.20	0.35	1.00
Residuum from raw sugar	1.14	8.33	—
Rockweed	1.90	0.25	3.68
Roses, flower	0.30	0.10	0.40
Rhubarb, stems	0.10	0.04	0.35
Rock and mussel deposits from sea	0.22	0.09	1.78
Salt-marsh hay	1.10	0.25	0.75
Salt mud	0.40	—	—
Sardine scrap	7.97	7.11	—
Seaweed (Atlantic City, N.J.)	1.68	0.75	4.93
Sewage sludge from filter beds	0.74	0.33	0.24
Shoddy and felt	4.0–12.0	—	—
Shrimp heads (dried)	7.82	4.20	—
Shrimp waste	2.87	9.95	—
Siftings from oyster-shell mound	0.36	10.38	0.09
Silkworm cocoons	9.42	1.82	1.08
Soot from chimney flues	0.5–11.0	1.05	0.35
Spanish moss	0.60	0.10	0.55
Starfish	1.80	0.20	0.25
String bean strings and stems (ash)	—	4.99	18.03

Table 7.4—*Continued*

Material	Nitrogen (%)	Phosphoric Acid (%)	Potash (%)
Sunflower seeds	2.25	1.25	0.79
Sweet potato skins, boiled (ash)	—	3.29	13.89
Sweet potatoes	0.25	0.10	0.50
Tanbark ash	—	0.24	0.38
Tanbark ash (spent)	—	1.5–2.0	1.5–2.5
Tea grounds	4.15	0.62	0.40
Tea-leaf ash	—	1.60	0.44
Timothy hay	1.25	0.55	1.00
Tobacco leaves	4.00	0.50	6.00
Tobacco stalks	3.70	0.65	4.50
Tobacco stems	2.50	0.90	7.00
Tomatoes, fruit	0.20	0.07	0.35
Tomatoes, leaves	0.35	0.10	0.40
Tomatoes, stalks	0.35	0.10	0.50
Waste from hares and rabbits	7.00	1.7–3.1	0.60
Waste from felt-hat factory	13.80	—	0.96
Waste product from paint manufacture	0.028	39.50	—
Waste silt	8.0–11.0	—	—
Wheat, bran	2.65	2.90	1.60
Wheat, grain	2.00	0.85	0.50
Wheat, straw	0.50	0.15	0.60
White clover (green)	0.50	0.20	0.30
White sage (ashes)	—	—	13.77
Wood ashes (leached)	—	1.0–1.5	1.0–3.0
Wood ashes (unleached)	—	1.0–2.0	4.0–10.0
Wool waste	5.0–6.0	2.0–4.0	1.0–3.0

Notes: a. From *How to Grow Vegetables and Fruits by the Organic Method*, by J.I. Rodale.

content, while bone meal can be used to increase the compost's phosphorus content. Seaweed and sawdust provide trace minerals. Many of these additives can be obtained for free from certain industries; or, you can gather them easily from nature. In a small living group, you can use your vegetable wastes, manure from your livestock, and your household wastes in the form of garbage and excrement. (Human excrement probably is most safely handled with initial treatment by a Clivus Multrum or methane digester; see Chapter 5.) These available ingredients, plus additives and soil, are the components of a recycling system that can be supplied almost entirely by the living processes of a single small community.

A sheltered site with a windbreak for the north and west is best—particularly for smaller piles—since strong winds cool the pile and stop fermentation. For these smaller piles, protection should be given on three sides by means of walls or hedges where possible, but the compost must never be banked up against a wall. The site also should have good drainage on all sides so that rain does not create a giant vegetable mud pie!

Compost piles are built either into or above the ground. Above-ground compost piles are enclosed by some supporting structure (of brick or wood) or else formed in natural mounds. The minimum size of a pile should be three feet high by three feet square. This sizing is necessary to insure proper insulation so the pile can maintain the high temperatures required by the decomposition process. The pile should not be heaped too high; otherwise, it will become compacted due to its own weight. When a pile is built into the ground, it is necessary to cover the pit with either a plastic tarp or a wire screen to keep out vermin. If plastic is used, the decomposition process is called *anaerobic* (without oxygen). It is important to maintain the proper moisture balance, about that of a squeezed-out sponge; in an above-ground pile, this is accomplished by covering it with a protective material. A layer of straw works well and also acts to insulate the pile.

Since the compost from an above-ground pile is formed by *aerobic* (with free oxygen available) decomposition, it is important that the pile be well ventilated. Aeration can be provided by either regular turning of the material undergoing decomposition, a forced aeration system, or a combination of both. The pile breathes from above and below, and also at the sides. It must not, of course, be trampled on or pressed down. To increase air access from the bottom, a thin layer of twigs or branch material can be used to lift the heap a little. Or, you can thrust a few metal pipes open to the air into the core of the pile. If abundant oxygen is not present, putrefaction may occur. A properly constructed compost pile should never putrefy, and it should never breed maggots or flies, or emit an unpleasant smell. If these do occur, they are infallible signs of faulty conditions. The cause is usually excessive wetness or some form of imperfect aeration. The cure is to re-admit air by turning. Above-ground compost is best for a permanent household because it has several advantages: it is easier to turn, easier to move when you want to spread it on the fields, and, when decomposing, it obtains higher temperatures than pit setups. This higher temperature speeds decomposition, kills pathogens from waste material, and aids in the elimination of pests. The minimum condition for pathogen destruction is eighteen to twenty-one days at temperatures consistently above 130°F. Such temperatures are reached in normal operation, and temperatures of up to 170°F are reached in larger piles.

To service your compost pile you will need plastic bags to store the composting material until it is time to construct a pile; a shovel and pitchfork to load the compost and turn the pile; and a wheelbarrow to transport the compost to its ultimate destination. A shredder is a useful (but not essential) item. Shredding the composting material speeds the decomposition process by exposing more surface area to the activities of decomposing bacteria. There are both power and manual shredders, but, as a matter of fact, a manual rotary mower serves the purpose well.

There are numerous methods of composting that

advocate different mixtures of composting materials and types of additives to supplement the compost. For example, the Biodynamic method calls for 70 percent green matter, 20 percent manure, and 10 percent soil. Four of the most widely used techniques include the Indore, the Rodale, the aforementioned Biodynamic, and the Santa Cruz French Intensive (we will discuss these last two in more detail soon). All these methods require from two to three months and yield approximately 1000 pounds per cubic yard, enough to cover 2000 square feet of garden. The differences in these methods seem to involve basically "cultish" preferences; none has a proven advantage in terms of efficiency of recycling or increased crop yield. The basic concept of composting is that anything organic can be composted and, although some composted materials—shredded paper, for example—have few concrete benefits, they do no harm and are disposed of in a satisfactory manner. For all methods, the decomposition process can be accelerated by increasing the aeration of the pile, by adding nitrogenous materials, and by shredding the materials to increase the surface area exposed to decomposition bacteria. The important thing to remember is that nitrogen, usually in the form of manure, is essential.

Two somewhat different methods of composting are *sheet composting* and *green manuring*. Sheet composting involves the spread of garbage, vegetation, and manure on crop land (usually in late summer) and then the tillage of these wastes into the soil; they decompose right in the crop land. For green manuring, you till a crop (usually a legume) into the soil and let the crop decompose. Both of these methods require either power machinery or intense labor. They are also more applicable when acres of crops are being grown rather than square feet.

The best time to start a compost pile is at the end of a growing season, when plenty of vegetable wastes are available. Also, if the pile is started at the end of a growing season, it can be ready to apply before the next year's growing season begins; it is easiest to apply compost to a bare field during a slack period, when you are not otherwise busy. Even so, labor on a compost pile averages only five man-hours a week, including construction, gathering material, tending, and spreading the final product.

Although timing varies, the composting process should last about three months. The most intense activity occurs during the first three weeks, at the end of which time the pile should be turned (with a pitchfork) so that all parts are thoroughly mixed. This turning assists and accelerates the process. About five weeks later, the pile should be turned again. That's all that needs to be done! If everything has gone well, you needn't thrust in any metal pipes or poke any holes (for aeration), although it wouldn't hurt. Four weeks later, or about three months after the process began, the compost is ready to use and can be applied to the soil. Compost is ready to apply when it is dark and crumbly; not all of its component materials need to be broken down completely.

Orchard crops and perennials (plants which live for more than one year, like asparagus and artichokes) can be fertilized at any time of the year. With annual crops (plants which must be planted every year—most garden crops are annuals), it is best to apply compost before they are planted; this way you don't disturb any seeds while you make nutrients available for the important seedling stage. If you have enough, apply compost twice a year to crop lands. Spread from one to three inches over the soil (approximately 50 pounds per 100 square feet). The compost then should be mixed with the top four inches of topsoil, so chemicals and nutrients don't evaporate. This treatment will improve soil aeration and moisture-holding ability, as well as make nutrients available to the crops. It thus can be used to upgrade marginally productive land for future gardens. For further general information, see the listings under B. Kaysing, J.I. Rodale, and R. Merrill, et al. in the Bibliography.

Clivus Multrum

A device known as the Clivus Multrum presents another opportunity to dispose of organic household wastes in a productive way. Details regarding its usage, design, and construction can be found in Chapter 5. The Clivus requires a fairly large initial investment ($1500 purchased; less if constructed), but so do septic tanks and pipes to plug into a municipal disposal system. A Clivus will last indefinitely; there are no moving parts to wear out or complex pipe systems to get choked or broken. Moreover, it recycles nutrients efficiently without pollution, does not use water (an important point), and produces an odor-free compost. Before installing one, however, it is best to check for any problems with local health and building codes.

Digester Effluent

Another source of recycled fertilizer comes from the anaerobic decomposition which occurs in a methane digester. The sludge that accumulates in a methane digester is similar to municipal sewage sludge, which has been widely used as a fertilizer. Physical and chemical characteristics of both the liquid effluent and the solid sludge are described in detail in Chapter 5. Like the Clivus, the digester requires a fairly large initial investment. Potential health problems have created state regulations controlling its use. It should be used as a source of fertilizer only for crops where the actual food crop does not come in direct contact with the sludge or effluent. Application to orchards or to crops which are to be plowed under for green manure would be acceptable.

Biodynamic/French Intensive Method

To increase the self-sufficiency of a living group in an ecologically sound manner, your methods of fertilizing and nutrient regeneration need to take into account the type of community in which you are living. On a farm or in a rural area, it is possible to use all three—composting, the Clivus, and the methane digester—simultaneously. Garden and animal waste can go into compost and/or the methane digester, while household waste can be decomposed in the Clivus. This integrated program would provide a steady supply of fertilizers for your crops.

On the other hand, in an urban or suburban area, the Clivus and methane digester may not be feasible. Composting, however, can be easily used and provides excellent garden nutrients.

Fertilizer collected from compost piles, a Clivus, or a digester is fundamental to the Biodynamic/French Intensive method (B/FI) of crop production. This highly productive, nontraditional method, developed over the last decade by combining two slightly older practices, is here given a more detailed discussion because of its adaptability to any type of community. Since the B/FI method is based on the principle of intensive cultivation of small land areas, it is particularly useful in the urban backyard. One outstanding and important aspect is that its yields (determined for annual vegetable crops) are reported to be 1.5 to 7 times greater than by traditional methods. It requires a little more dedication and intensive labor, but the results appear to be well worth it.

The Biodynamic/French Intensive method is based on the theory that natural biological control is better for agriculture than man-made chemical control: modern agriculture is thought to extract nutrients from the soil (which inorganic fertilizers do not replace correctly), while organic fertilizers are thought to keep the nutrients balanced. The B/FI method also incorporates the practice of growing plants on raised beds of loosened and fertilized soil. A natural weeding process is encouraged by growing plants very close together, so that the foliage of your crop shades the ground and forms a "living mulch." Another worthy aspect of the method is that water conservation is promoted; estimates indicate that about half as much watering is needed, compared to conventional gardening, since the closely grown plants tend to discourage evaporation from the soil. Insect pests are controlled by companion planting and by encouraging predaceous insects.

Preparations: Preparing the raised bed is the most exhausting part of this farming method. The theory is that a mound warms quickly and also drains and aerates the soil best. Beds are from three to five feet wide and about twenty feet long; they should run north and south. One such bed can provide the vegetable needs for four people (one pound per day each) indefinitely.

The first step is to prepare the site. Hard and dry soils should be well watered (about one or two hours), then allowed to drain for a few days. Loosen the soil on the plot (down to twelve inches deep) and add from one to three inches of compost or manure (remember that manure must age a few months before use to get rid of some of its ammonia). The suggested B/FI proportions for compost are equal portions of green vegetation, kitchen scraps, and soil; the nitrogen is provided by the green vegetation. However, other ways of composting should do fine.

Now the digging starts. As illustrated in Figure 7.8, ideally a trench twelve inches square (cross section) should be dug across the head of the bed. If the soil is too hard, then dig as deep as possible; put the soil you've dug out into a wheelbarrow. Now, along the entire bottom of the trench, drive your shovel or fork into the ground and wiggle it back and forth. Don't take any soil out, just loosen it. Ideally your shovel should loosen the soil to a depth of another foot. This procedure is called "double digging."

Next dig another trench directly behind the first and push the soil from the second trench forward into the first trench. Double dig! Dig a third trench behind the second with double digging and continue this back-breaking labor until you reach the foot of the bed. Dump the soil from the wheelbarrow into your last trench. Then spread from two to four pounds each of both bone meal and wood ash and from two to four cubic feet of aged manure over the entire bed, mix the top few inches with a fork, level and shape, and (voila!) you have a raised bed. Put a flat board over the bed if you must walk on it. This process needs to be repeated after every crop, but it should get easier each time: new bed preparations take from six to twelve hours and repeat bed preparations take only from four to six hours. Preparation steps are summarized in Table 7.5.

Planting: Before sowing your seed, you should study *companion planting* possibilities. Some plants get along well with other plants: each grows better in the presence of the other. Some plants repel insect pests, lure pests away from a more vulnerable crop, or attract insect predators. Some plants provide nutrients for other plants. Still other plants are grown together because they make an efficient use of space (one example is growing shallow-rooted plants together with deep-rooted plants). Tables 7.6 and 7.7 summarize the information on who likes whom and the effect of herbs.

Now it's time for planting. Requirements vary for different plants, but the key is to place them close enough, so that their full-grown leaves just touch. Plant in staggered rows so that each plant has six nearest neighbors which are the same distance away (see Figure 7.9). Putting a chicken-wire "template" over the mound

Table 7.5 Biodynamic/French Intensive Bed Preparation

1. Initial preparation: soak 1–2 hours for dry, hard soils; drain and dry 1–2 days; loosen soil to 12″ and add texturizers—manure and/or seed.

2. Double digging: spread 1–3″ compost; dig and remove soil from trench 12″ wide by 12″ deep, extending across end of bed; double dig by loosening soil 12″ deeper at trench bottom; dig second trench by filling first trench; double dig; continue to end; add soil from first trench to the very last trench.

3. Finishing touches: level and shape raised bed; add 2–4 lbs bone meal, 2–4 lbs wood ash; 2–4 ft^3 aged manure.

4. Plant and sow.

5. Repeat from Step 2 for continuing bed.

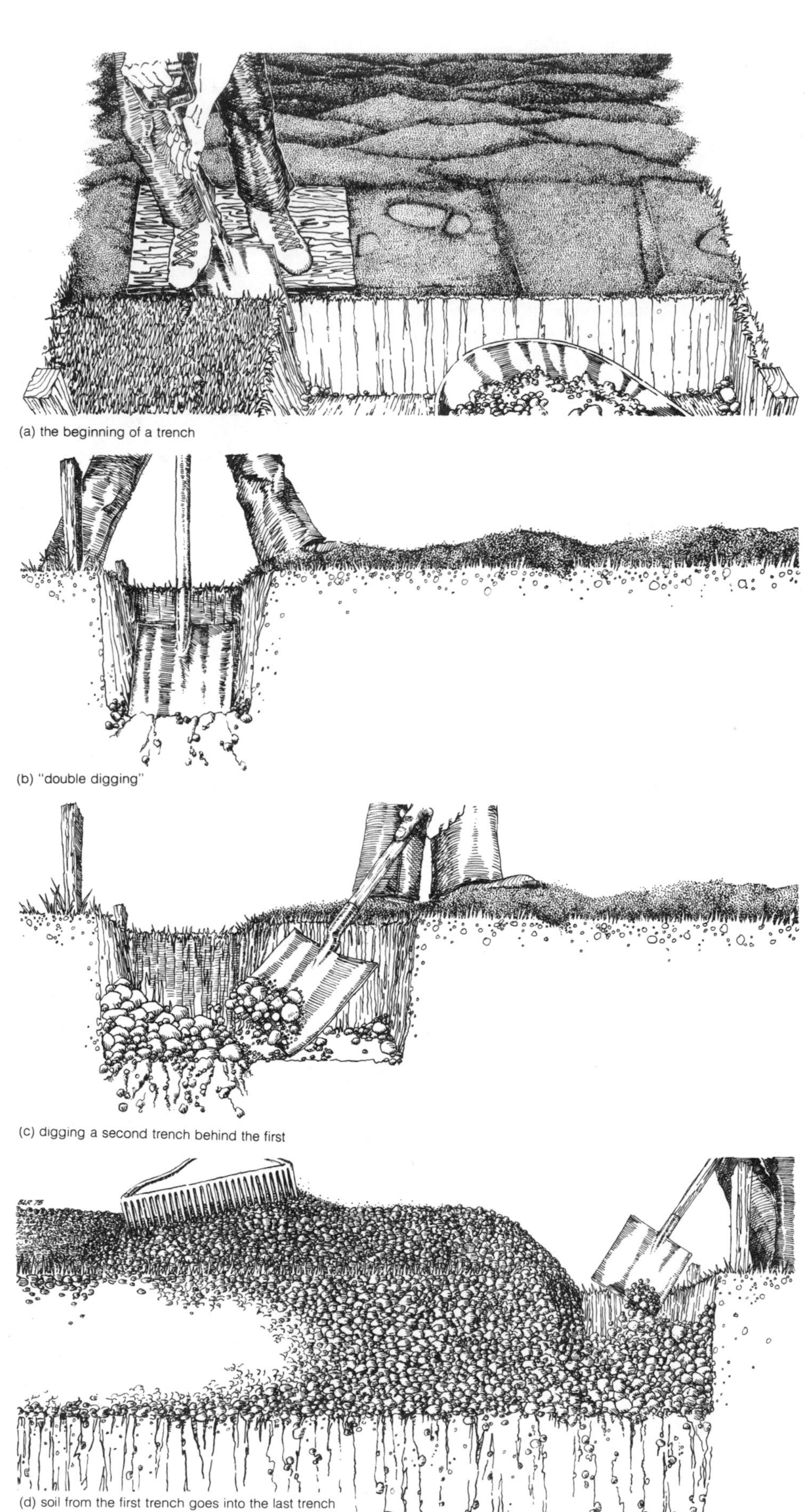

(a) the beginning of a trench

(b) "double digging"

(c) digging a second trench behind the first

(d) soil from the first trench goes into the last trench

Figure 7.8 Bed preparation for Biodynamic/French Intensive gardening (see Table 7.5 for complete summary).

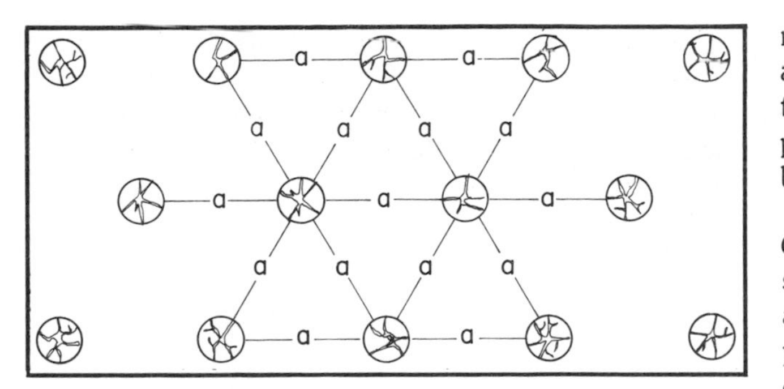

Figure 7.9 Plant spacing for Biodynamic/French Intensive gardening. Staggered rows provide each plant with an even distance from each neighboring plant.

and planting in the meshes might help your spacing technique. After some experience, you can throw the seeds on the mound like a real farmer. If your plants do grow too closely, they can always be thinned; but you can't do anything if they're too far apart.

Seedlings can be transplanted, but don't plant them near seeds; the seedlings will gobble up all the nutrients and your seeds won't germinate. Also, be careful with their roots. It is recommended that planting be done by phases of the moon; this precaution might actually help, but its logic is today still somewhat controversial.

Care: Water daily with a sprinkler a few hours before sunset; if you've watered enough, the soil stays shiny for around five seconds before the water sinks in. Because of plant placement and shading, weeds shouldn't be much of a problem after the plants are grown; but some weeds will always grow even where it doesn't seem light can penetrate.

As we mentioned, companion planting discourages pests. In general, growing a variety of plants on one bed encourages insect predators and inhibits insect pests (an example of how diversity can promote health). Insect pests are best discouraged by keeping plants healthy, but you can always pick them off plants by hand, put up sticky, impassable insect barriers, spray garlic juice on

Table 7.6 Companion Planting[a]

Vegetables	Likes	Dislikes
Asparagus	Tomato, parsley, basil	
Bean	Potato, carrot, cucumber, cauliflower, cabbage, summer savory, most other vegetables and herbs	Onion, garlic, gladiolus
Bush Bean	Potato, cucumber, corn, strawberry, celery, summer savory	Onion
Pole Bean	Corn, summer savory	Onion, beet, kohlrabi, sunflower
Beet	Onion, kohlrabi	Pole bean
Cabbage Family (cabbage, cauliflower, kale, kohlrabi, broccoli)	Aromatic plants, potato, celery, dill, camomile, sage, peppermint, rosemary, beet, onion	Strawberry, tomato, pole bean
Carrot	Pea, leaf lettuce, chive, onion, leek, rosemary, sage, tomato	Dill
Celery	Leek, tomato, bush bean, cauliflower, cabbage	
Chive	Carrot	Pea, bean
Corn	Potato, pea, bean, cucumber, pumpkin, squash	
Cucumber	Bean, corn, pea, radish, sunflower	Potato, aromatic herbs
Eggplant	Bean	
Leek	Onion, celery, carrot	
Lettuce	Carrot, radish (lettuce, carrots, and radishes make a strong team grown together), strawberry, cucumber	
Onion (and garlic)	Beet, strawberry, tomato, lettuce, summer savory, camomile (sparsely)	Pea, bean
Parsley	Tomato, asparagus	
Pea	Carrot, turnip, radish, cucumber, corn, bean, most vegetables and herbs	Onion, garlic, gladiolus, potato
Potato	Bean, corn, cabbage, horseradish (should be planted at corners of patch), marigold, eggplant (as a lure for Colorado potato beetle)	Pumpkin, squash, cucumber, sunflower, tomato, raspberry
Pumpkin	Corn	Potato
Radish	Pea, nasturtium, lettuce, cucumber	
Soybean	Grows with anything, helps everything	
Spinach	Strawberry	
Squash	Nasturtium, corn	
Strawberry	Bush bean, spinach, borage, lettuce (as a border)	Cabbage
Sunflower	Cucumber	Potato
Tomato	Chive, onion, parsley, asparagus, marigold, nasturtium, carrot	Kohlrabi, potato, fennel, cabbage
Turnip	Pea	

Notes: a. From *Organic Gardening and Farming*, February 1972, p. 54.

infested portions, or set up deterrent fences of fresh ash.

The B/FI method of crop production should be easier to do each time. For further information, see *How to Grow More Vegetables (than you ever thought possible on less land than you can imagine)*, put out for $4 by Ecology Action (2225 El Camino, Palo Alto, CA 94306). This and the progress report put out by J. Jeavons are listed in the Bibliography.

Hydroponics

Now that you've been thoroughly convinced of the benefits of organic farming, and now that you've been told how it's best to let natural biological controls solve all your problems, let's look at another highly productive method of farming which is *not* natural and, in its pure form, uses nothing but chemicals. Hydroponics may be taken loosely to include all forms of agriculture that do not rely on soil as a planting medium. Sand, gravel, cinders, and other such materials can be included as hydroponic media.

The main advantage of hydroponic agriculture is increased production with low maintenance costs. Production is high because growth can be controlled easily through careful regulation of the nutrients applied to the plants. Disadvantages include the high initial cost of plumbing, the use of processed chemical nutrients, and the great volume of water needed. Water can be conserved if a recycling system is designed, but this feature also adds to the initial plumbing costs.

There are as many ways to apply the nutrients as there are types of media. Nutrient solutions may be prepared and applied by hand or by pump. The solution can be sprayed onto the plants, dripped onto the plants, or supplied to the plant bed by irrigation or subirrigation. The beds may be located either outdoors or in greenhouses. Obviously, building a greenhouse would be a significant expense that would have to be balanced against its higher yield.

Table 7.7 Herbal Companions and Effects[a]

Herb	Companions and Effects
Basil	Companion to tomatoes; dislikes rue intensely; improves growth and flavor; repels flies and mosquitoes.
Borage	Companion to tomatoes, squash, and strawberries; deters tomato worm; improves growth and flavor.
Caraway	Plant here and there; loosens soil.
Catnip	Plant in borders; deters flea beetle.
Camomile	Companion to cabbages and onions; improves growth and flavor.
Chervil	Companion to radishes; improves growth and flavor.
Chive	Companion to carrots; improves growth and flavor.
Dead Nettle	Companion to potatoes; deters potato bug; improves growth and flavor.
Dill	Companion to cabbage; dislikes carrots; improves growth and health of cabbage.
Fennel	Plant away from gardens; disliked by most plants.
Flax	Companion to carrots, potatoes; deters potato bug; improves growth and flavor.
Garlic	Plant near roses and raspberries; deters Japanese beetle; improves growth and health.
Henbit	General insect repellent.
Horseradish	Plant at corners of potato patch to deter potato bug.
Hyssop	Deters cabbage moth; companion to cabbage and grapes; keep away from radishes.
Lamb's Quarters	This edible weed should be allowed to grow in moderate amounts in the garden, especially in corn.
Lemon Balm	Sprinkle throughout garden.
Marigold	The workhorse of the pest deterrents; plant throughout garden to discourage Mexican bean beetles, nematodes, and other insects.
Marjoram	Here and there in garden; improves flavors.
Mint	Companion to cabbage and tomatoes; improves health and flavor; deters white cabbage moth.
Mole Plant	Deters moles and mice if planted here and there.
Nasturtium	Companion to radishes, cabbage, and cucurbits[b]; plant under fruit trees; deters aphids, squash bugs, striped pumpkin beetles, improves growth and flavor.
Petunia	Protects beans.
Peppermint	Planted among cabbages, it repels the white cabbage butterfly.
Pigweed	One of the best weeds for pumping nutrients from the subsoil; especially beneficial to potatoes, onions, and corn; keep these weeds thinned.
Pot Marigold	Companion to tomatoes, but also plant elsewhere; deters asparagus beetle, tomato worm, and general garden pests.
Rosemary	Companion to cabbage, bean, carrots, and sage; deters cabbage moth, bean beetles, and carrot fly.
Rue	Keep far away from Sweet Basil; plant near roses and raspberries; deters Japanese beetle.
Sage	Plant with rosemary, cabbage, and carrots; keep away from cucumbers; deters cabbage moth, carrot fly.
Southernwood	Plant here and there in garden; companion to cabbage, improves growth and flavor; deters cabbage moth.
Sowthistle	This weed in moderate amounts can help tomatoes, onions, and corn.
Summer Savory	Plant with beans and onions; improves growth and flavor; deters bean beetles.
Tansy	Plant under fruit trees; companion to roses and raspberries; deters flying insects, Japanese beetles, striped cucumber beetles, squash bugs, ants.
Tarragon	Good throughout garden.
Thyme	Here and there in garden; it deters cabbage worm.
Valerian	Good anywhere in garden.
Wormwood	As a border, it keeps animals from the garden.
Yarrow	Plant along borders, paths, near aromatic herbs; enhances essential oil production.

Notes: a. From *Organic Gardening and Farming*, February 1972, pp. 52–53. Also includes a few weeds and flowers. b. Plants in the gourd family.

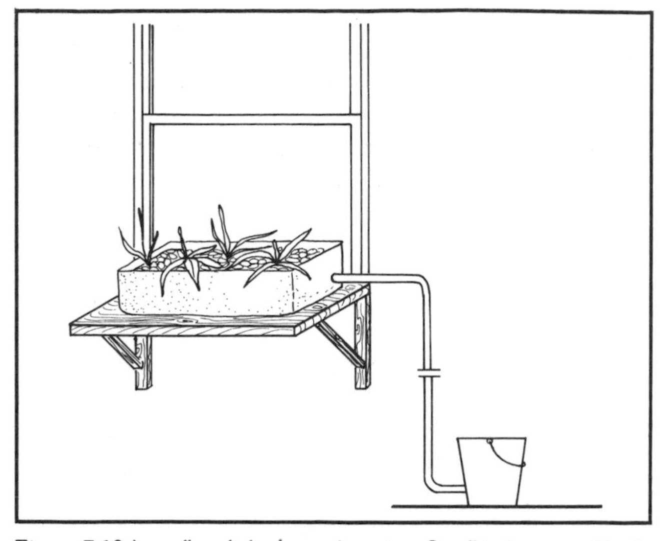

Figure 7.10 A small-scale hydroponics setup: Small indoor gravel beds can support healthy growing vegetables. The bucket containing the feed solution is lifted to water the vegetables, then lowered to allow drainage. This is done twice a day.

In general, the more complicated and mechanized the hydroponic system is, the more expensive the installation costs will be. The simplest setup is a sand or gravel system where the nutrients are applied dry and then watered in by a hand-held hose. If gravity is used, no pump will be needed to drain the nutrients. (Feeding should last about fifteen minutes; after this time, the nutrient must be removed somehow, by some form of drainage.) Moderately coarse silicate sand from the beach is ideal as a medium. The sand can be placed directly into a shallow pit lined with a wooden border. Such a setup can be used to grow virtually all plants that are grown in soil, with root crops doing the best. Your major expense is then only the chemical nutrients; industrial grade salts are recommended since they also supply needed trace elements. A typical mixture is:

Sodium nitrate	13.50 oz.
Potassium sulphate	4.00 oz.
Superphosphate 16% P_2O_5	10.00 oz.
Magnesium sulphate	4.00 oz.
Ferrous sulphate	0.25 oz.
Water	100 gal.

Adequate stock solution can be made up once a week; the amount of time involved depends upon your dexterity. The solution should be applied at least once a day. A significant time-saving feature of hydroponics is that almost no time has to be spent weeding. A hydroponics system that requires no inorganic chemical additives and is used in conjunction with a fish-farming scheme is mentioned at the end of our discussion of aquaculture.

Small-scale hydroponics can be practiced in urban communities and even in apartments to grow beans, tomatoes, and other small vegetables. The easiest procedure is to place three or four inches of clean gravel into several plastic dishpans (see Figure 7.10). A small-diameter plastic hose is fixed into a side hole along the bottom edge of each pan. The other end of the hose connects up with a pail of nutrient solution. Feeding and watering is done by raising and lowering the pail twice a day. When you raise the pail above the level of your pans, the solution percolates into the gravel; after fifteen minutes, when the pail is lowered below the dishpans, the solution percolates back into your pail and drainage is complete. For further information on hydroponic designs, see the listings under R. Bridwell, T. Saunby, and C.E. Ticquet in the Bibliography.

AQUACULTURE

A good percentage of the food for your household can come from aquaculture, the art of applying agricultural principles to control the raising of aquatic organisms; it includes the cultivation of fish, shellfish, and algae. This art has been practiced worldwide for thousands of years, most extensively in the Far East. Aquaculture *does* exist in the United States—primarily in the form of catfish farms and trout hatcheries—but terrestrial agriculture still accounts for most of our food production. We don't mean to insult agriculture, which is a perfectly respectable way to grow food, but aquaculture is generally *more productive* than agriculture and *more efficient* than raising livestock. (Figures for chicken production are comparable to some aquaculture figures, but aquaculture is more efficient at converting feed into biomass.) Any person who is trying to get the maximum food yield out of his land would be wise to consider using part or all of it for an aquaculture operation.

Aquaculture's superior productivity is due to three characteristics of its medium. First, water absorbs and retains the sun's energy better than land; a pond thus will have more energy available for organisms to use than a comparable land area. Second, water can diffuse material throughout itself, an ability dry land lacks. This ability to mix materials within itself results in a much more even distribution of nutrients than that of soil; on land, one spot might be deficient in a vital nutrient that is at toxic concentration in a spot but a few feet away. The third (and main) advantageous characteristic of water is that it occupies a three-dimensional space: different organisms can live at different depths under the same area of surface. Land is virtually two-dimensional, with life only on, and a short distance above and below, the surface. Thus, an area of water is more productive than an equal area of land because there is more habitable volume. The ancient Chinese found that using all three dimensions of water made a very productive system. Realizing that different species of fish live at different levels in the water column and eat different food ("occupy different niches" in biological terms; see Figure 7.11), they deduced that it might be possible to culture many different species of fish in the same pond; the fish would not only not interfere with each other, but would even complement each other. For example, an algae-feeding species near the surface and a bottom feeder, both with high reproductive rates, can harmoniously coexist; their young are controlled by a predator species, and the waste products of all three species could in turn fertilize the water, thus supporting the growth of algae for the herbivorous species. This culturing of more than one species is called *polyculture*. Obviously, different types of fish must be chosen with care, as two random species do not necessarily complement each other.

In spite of the greater efficiency of polyculture, often economic and social factors confine only one type of fish to a pond. This practice (called *monoculture*) is less efficient with both food and space, and consequently less productive than polyculture.

Already we've mentioned twice that aquaculture is "more efficient" than agriculture, but we haven't really explained what we mean. Efficiency, in this case, is the number of pounds of food an animal must be fed for it to grow one pound heavier. It takes more than one pound of food to grow one pound of animal because some food is burned up as the energy the animal uses to live. Aquatic animals need less of this energy than land animals because water supports their bodies. In addition, most aquatic animals are cold-blooded, and less energy is expended by cold-blooded animals than warm-blooded animals. Since our domestic land animals are warm-blooded, it follows that aquatic animals burn less food into energy than land animals, and can convert more of this food into meat. Thus, in principle, aquaculture is more efficient than culturing land animals.

Another theoretical concept sheds light on this question of efficiency: the food chain. In a generalized ecological system, production occurs when tiny one-celled plants, phytoplankton, use energy from the sun and nutrients from the water to grow and produce more biomass. They are eaten by microscopic animals, zooplankton, which are in turn eaten by bigger and bigger animals until the final consumer is a large fish or land predator. You will remember that organisms at each level use up some (indeed, *most*) of the energy they absorb to carry on their own life functions; thus less energy is available to other consumers higher up the chain. Consequently, the most efficient theoretical way of utilizing the original solar energy input into a system is to cultivate the lower-order consumers: herbivores or zooplankton feeders. But there is a limit to how low you can go before dining becomes unpalatable, or inefficient, or both. Most human societies prefer fish to chironomid larvae and steak to grass; even if you were so inclined, it would take hours to catch a mouthful of copepods. In such circumstances, it's better to let a higher-order consumer convert food to an edible state. A good discussion on energy needs and fish yields for various types of aquaculture practices can be found in the *Energy Primer* (see Bibliography, under Merrill, et al.).

The amount of biomass produced is limited by the concentrations of nutrients and vital molecules present in the system, some of the most important being nitrogen, phosphates, carbon dioxide, and oxygen. In a given situation, nitrogen and phosphate concentrations may be too low to permit further production of biomass. When organic waste is added to this limited-nutrient system, bacteria decompose it, releasing nitrogen compounds, phosphates, and other nutrients. Thus, adding *garbage* will add nutrients to a pond, which may stimulate plant growth, which stimulates zooplankton blooms, which in turn increase the food supply and therefore the growth of the organism which the culturist is attempting to raise! (For further information, see "Oxidation Ponds" in Chapter 5.) However, caution must be exercised when applying any such "fertilizer"; the decomposition that produces these nutrients also uses up vital oxygen and makes the water more turbid. It is possible that the end result may be the death of a fish crop, rather than its enhanced growth. Inorganic fertilizers like superphos-

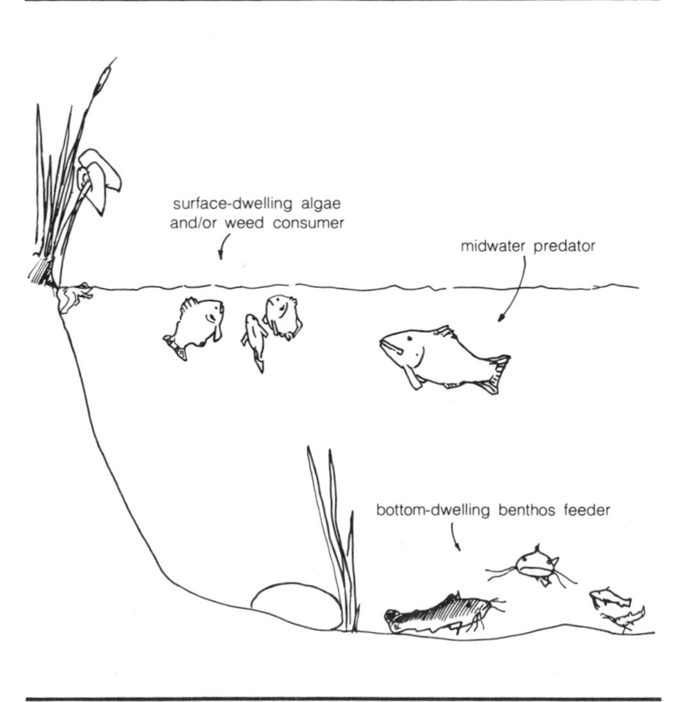

Figure 7.11 Polyculture in a fish pond: efficient utilization of the three-dimensional pond environment can be promoted by culturing compatible, noncompetitive species which have differing feeding niches and habitat niches.

phates can also stimulate growth and, since bacterial decomposition isn't needed, avert any danger of an induced oxygen shortage. But inorganic fertilizers cost money, while garbage is cheap and plentiful.

Before going any further, we must differentiate intensive from nonintensive aquaculture. *Intensive aquaculture*, as the name implies, involves feeding the fish, controlling their breeding, and much more; in general, you spend a good deal of your time and labor shaping the environment and life patterns of fish. *Nonintensive culture* is a minimal-labor operation; the fish are not controlled except for an occasional cleanup of debris in the pond, fertilization of the pond, and the harvest itself of the fish. As you might imagine, nonintensive culture is generally less productive, but it is also less expensive and time-consuming.

Production figures which follow in our discussion are given in kilograms per hectare per year (kg/ha-yr). A kilogram (kg) is 2.2 pounds, and a hectare (ha) is 10,000 square meters or 2.47 acres. Thus, a kg/ha-yr is a little less than a pound/acre-yr.

Please realize that all we can do is summarize some of the information applicable to the question of aquaculture. For a more detailed look at specific problems, you are encouraged to consult other references and sources, many of which are listed in the Bibliography.

Some Fundamental Considerations

One cannot, so to speak, just plunge into an aquaculture operation. There are several important questions which must be faced beforehand. Hopefully, those preceding pages where we described the superiority of aquaculture haven't made you excessively optimistic. Success still requires a great deal of work and the first job is to forget anything you've heard about how easy aquaculture is. A few success stories have led to books being rushed into print, describing the ease with which fish can be raised and their riches reaped. These stories should be filed in the same place as the early-1929 stories of how anyone and everyone could get rich in the stock market. If aquaculture was such a sure bet, everyone would be doing it. Although it has great potential, aquaculture requires as much care and maintenance as gardening, and, like gardening, not every site is equally suitable. The weeds still must be pulled, the water checked, and the fish fed. And as an added bonus, fish, too, are menaced by a great variety of predators and diseases just itching to get at them.

The central questions you must answer concern the physical features of your site. If a body of water can't exist on a site, obviously aquaculture is impossible. The soil and subsoil must be tested for water-retaining qualities and fertility. Existing topographical features should allow complete pond drainage, and the legalities of draining into existing streams must be investigated. Other important factors include the accessibility of the pond and nearby insecticide concentrations. Of *critical* importance are adequate primary and emergency sources of water, the temperature and quality of this water, the organisms present in it, the likelihood of floods, and the need for water permits.

The next level of questioning is related to the specific organism you want to culture and where to get it. Obviously, the question of species must be resolved first. Then you must decide whether to initiate the stock with mature breeders or with a younger stage (fry or fingerlings, in the case of fish). Remember that, in many cases, it's better to buy fish that are already hatched than to try to hatch them yourself, even if it means you can't be self-sufficient. Breeding and caring for eggs and fry is an art—an extremely difficult art for some species— and it can require additional expensive equipment. If you want to breed the fish yourself, there must be a supplier from whom you can purchase breeders. Then the age, size, and number of breeders of each sex all need to be determined. Furthermore, the culturist should know at what temperature spawning occurs and how long the breeders must be in breeding surroundings before they spawn, so that you don't harvest them too early and wipe out your stock. And with fingerlings, the culturist must know what size and number to stock and how to stock them without causing thermal shock. The aquaculturist must also do all the necessary preparatory work on the pond water before stocking.

And yet there is more: you must determine and stock the proper food for the fish. The culturist should determine the fertility of the pond, the type and amount of fertilizer that will induce the desired plankton population, the feasibility of artificial feed, and what is needed for the winter.

Answers to these questions are beyond the scope of this chapter, but we refer you to some sources of information. For general reference information, the best English-language encyclopedia is *Aquaculture: the Farming and Husbandry of Freshwater and Marine Organisms* by John Bardach, John Ryther, and William McLarney. It contains a great deal of information on practices throughout the world and information on fish not commonly cultured in the United States. A very good book on trout culture and fish diseases is *Culture and Diseases of Game Fish* by H.S. Davis. These and other books on fish culture are listed at the end of this chapter, in the Bibliography.

For more specific questions, help can be obtained from federal and state agencies—fish and game commissions, the Fish and Wildlife Service of the Department of the Interior, and conservation agencies. Other sources might be nearby universities and colleges, especially those involved in Sea Grant programs. Still other sources are those people actively involved in existent commercial ventures: fish farmers, fish-farming associations, and distributors or researchers for fish-food companies. Helpful publications are put out by federal and state governments, universities, and the United Nations Food and Agriculture Organization (FAO). Information is also supplied by such periodicals as *The American Fish Farmer, Farm Pond Harvest,* and *The Progressive Fish-Culturist,* as well as in research reports in scientific and industrial journals. However, one note of caution is necessary. If you have a problem, understand that advice and solutions can vary from one source to another. As a case in point, a fish breeder once had a disease problem he couldn't solve. He asked the state fish and game agency for advice and sent them some diseased fish. Evidently they didn't examine the fish too carefully because, each week for several successive weeks, they told the breeder to try different solutions, all of which were expensive and none of which worked; the breeder lost most of his fish. He later learned that his problem was due to flukes, a parasite easily controlled by one application of the right chemical. The moral of the story is that aquaculture is an inexact science; don't expect anyone to know all the answers.

Living Quarters for Fish

Don't forget! Fish live in water. This means that successful aquaculture requires an adequate water supply and a container to hold the water. Obvious though this may seem, it is sometimes forgotten, and the results are both costly and embarrassing.

An adequate water supply is one that furnishes water during the driest part of the season, which is usually when it's needed the most. For freshwater aquaculture, this water can come from either wells, surface runoff, streams, or springs. Well water has the advantage of being a fairly reliable, relatively pollution-free source, and is also free of undesirable aquatic organisms; however, it needs to be pumped and is low in oxygen. Surface water may not have to be pumped, but it usually contains undesirable fish which are extremely adept at crawling through filters and infesting your pond. Springs are clean water sources that don't need to be pumped and are preferred when available.

Whatever the source, this water must be contained in some manner, in either ponds, troughs, raceways, circular ponds, or silos. Fish also can be isolated in larger water-bodies by means of floating nets or cages. The type of container you select depends upon local geology, your aesthetic taste, the type of fish you're raising, the amount you want to produce, and your pocketbook. If you want to raise fish in a sylvan setting reminiscent of Walden Pond, concrete circular ponds will never do; you should opt for a farm pond (and lower production). Or, if you live downtown and your backyard only has room for garbage cans and a clothesline, then a farm pond is not practical; you should look into fiberglass silos.

Ponds: Ponds are used mostly for warm-water fish, but they can also be used for trout. For a dirt-bottom pond, the soil must retain water. The Soil Conservation Service of the Department of Agriculture can check the moisture-retaining properties of your soil. If the earth doesn't retain water, a sealant such as bentontite can be used, or a vinyl sheet can be laid between water and ground; but both these methods add considerably to your cost. As we mentioned earlier, be sure that the pond is constructed so that it can be drained completely. Details of construction and illustrations are presented in our discussion of farm pond programs. If there is already a pond present which you wish to use, care must be taken to insure that no undesirable fish are in it. This "care" usually means draining your pond and starting from scratch.

Raceways and Circular Ponds: Pond water, because it is stationary, can be enriched with fertilizer; the pond is thus applicable for warm-water fish which usually live in "productive" water (turbid with phytoplankton, for example). However, salmonid fish (trout and salmon) prefer clear, nonproductive water. Because running water carries away the growth-inhibiting waste products of these fish, the best production of salmonids occurs in either raceways or circular ponds. Expenses can mount up for these types, since they usually need to be constructed and maintained with pumps and filters.

Raceways are long troughs which are slanted or stepped downhill. Water enters through the uphill end and leaves from the downhill end (see Figure 7.12). Since water is continuously flowing through the raceway, great quantities are needed to operate it. American raceways are usually built with concrete, and this material is not cheap. Earth raceways are less expensive and provide natural food for the fish. The soil must retain water, however, and the probability of disease is increased. If you happen to have a clear, cold-water stream on your property, you have a natural raceway and may need only to dam it up a little or build several small pools.

In a circular pond, water is shot out of a pipe on the rim of the pond and the water flows around the pond in a spiral, finally exiting by a drain in the center (see Figure 7.12). This approach uses much less water than a raceway, but it still consumes a considerable amount. Because there must be a permanent slope down to the drain, concrete must be used for the bottom and sides, a sacrifice of both natural beauty and many dollars. Construction of a pond 2.5 feet deep and 20 feet in diameter would cost between $1500 and $2000.

If you don't have enough room, water, or money for raceways or circular ponds, high production also has

Figure 7.12 Most frequently found in hatcheries, raceways and circular ponds both require large quantities of water and also pumping equipment.

been attained in barrels. In *Aquaculture* (Bardach, et al.), it is reported that Pennsylvania Fish Commission biologists at Bellefonte, Pennsylvania, once raised 2720 kilograms of trout in a fiberglass silo 5 meters high and 2.3 meters in diameter. It was merely an experimental approach, but you might look into it.

Cages and Nets: Fish are grown inside cages and nets when they are cultured in the ocean, in powerplant effluents, or in large streams. This kind of isolation makes harvesting easy, yet provides a constant supply of circulating water to the fish. The fish, however, must be hatched and raised through early youth in another type of container or else young fry must be collected from some other source. In addition, any use of structures on public waterways first must be cleared with legal authorities.

Freshwater Aquaculture

Now that we've surveyed the promise, problems, and structures used in aquaculture, we finally can get down to discussing the fish. Much of the data which follows has been drawn from *Aquaculture* by J. Bardach, et al. Bear in mind that great climatic differences occur within the expanse of the United States; some forms of aquaculture which are applicable to one area are wholly inappropriate in another. Most aquaculture activity in this country is carried out in the Southeast, where the warm climate makes possible a long growing season and rapid growth during this season. In the northern part of the United States, the temperature during the summer can be every bit as warm as in the South; but the warm season is shorter, resulting in less annual growth for fish, and the winters are extremely cold, often resulting in *no* growth. Tropical breeds like *Tilapia*, which are very successful in warm climates, cannot survive in the North without special care. And such eurythermic breeds (tolerant of wide temperature ranges) as channel catfish and bass *can* be cultivated in both the North and South, but grow more slowly in the colder climate.

Do not, however, be misled: aquaculture is still (and equally) feasible in the northern United States. It has been carried on for centuries with great success in the extreme climates of Japan, the north of China, and Eastern Europe. Species native to these regions can survive the local extremes of temperature; years of experience have taught the people which species grow and taste the best, and are the most amenable to aquacultural techniques.

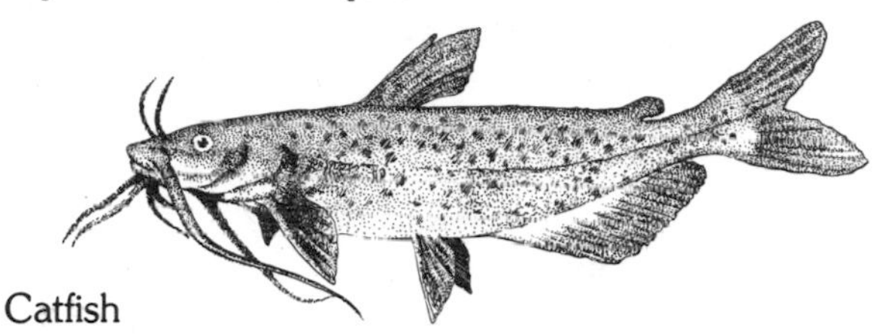

Catfish

There is a great deal of data on commercial catfish farming and so it is a good first species to take under consideration. Catfish do well under intensive culture because they adapt well to crowded conditions and have an excellent feed-coversion ratio (usually about 2 pounds of feed for every pound of fish, but often as low as 1.3:1). Large fish farms often spawn catfish in aquariums or tanks. The small fish are placed in ponds at concentrations of anywhere between 20,000 and 100,000 per acre. After the first season, when the fish are between three and eight inches long, they are restocked at reduced concentration levels—usually between 1500 and 2000 fish per acre. They are ready for market in another seven or eight months, weighing about a pound each. Hence, an average yield is about 1700 pounds per acre.

Optimum water temperature for catfish is about 85°F. Below 60°F and above 90°F, feeding and growth drop off sharply. The growing period, however, lasts as long as the temperature is within this range.

The dissolved-oxygen content of the water is of vital importance. Greater quantities of fish per acre make heavier demands on available food and oxygen, so management becomes more difficult. A large part of the oxygen in a pond is produced by algae; it is sometimes necessary to use fertilizer to encourage algal growth if you start a new pond. Later, the problem becomes one of controlling excess algae, since at night the algae become oxygen consumers and compete with the fish for the available oxygen. Overfeeding, too, can lead to serious problems in this area, since uneaten food can stimulate algal growth and also consumes oxygen as it decays. Oxygen content is highest late in the afternoon and lowest just before dawn. It is not unusual for the entire population of a pond to die in one night due to oxygen

Table 7.8 Catfish Diet Guidelines[a]	
Diet Element	Percent of Total Feed
Crude protein, more than	30
Digestible protein, more than	25
Animal protein, more than	14
Fish meal protein, more than	5
Crude fat, more than	6
Crude fiber, less than	20

Notes: a. Taken from F. Meyer, *Second Report to the Fish Farmer.*

depletion. In the summer, oxygen levels should be measured every day early in the morning and should be kept well above 3 parts per million (consult fish and game agencies for measurement techniques). A serious shortage can be alleviated temporarily by partially draining the pond from the bottom, where there is less oxygen, and then adding more water. The best way to add this water is by spraying it through the air where it picks up a great deal of oxygen. In the long run, however, the problem must be solved by careful pond maintenance and control.

Catfish are fed about 3 to 5 percent of their body weight each day. During a seven-month growing period, the cost of feed is between $100 and $150 per acre for a pond with 2000 fish per acre. During the winter, only about one-third as much food is consumed per day; the cost is proportionally less.

The amount your fish will grow is directly related to proper diet. The ingredients, of course, can be varied. Catfish have been raised using soybean as the only source of protein, although its utilization is very poor at lower temperatures. Table 7.8 presents a set of diet guidelines, and Table 7.9 illustrates a suggested feed formula. Balance in the feed content is stressed.

If you want to avoid commercial fish food, you can raise soybeans as a protein source; but processing to make them acceptable to the fish must be tried out experimentally. Worms and insect larvae also can be raised in a compost pile. Alternate sources of protein are minnows or other herbivorous fish raised in the same pond. But with these diets, don't expect the same growth rates as from commerial feed.

Catfish have been raised in densities of up to 600 per cubic yard in cages and raceways. Cages anchored in lakes and rivers rely on natural wind and water currents to carry out wastes and to carry in an adequate oxygen supply, while raceways require continuous circulation of fresh water to provide the same two maintenance tasks. However, because extremely well balanced diets and highly skilled management are necessary, these high-density culture methods are still largely in the experimental stage.

Production is directly related to management capabilities. The more a catfish is fed, the faster it will grow. On a commercial fish farm, a catfish is brought to marketable size (about a pound) in two growing seasons. Since fish are stocked at a rate of between 1500 and 2000 per acre, an average commercial yield is between 1500 and 2000 pounds. The total cost for rearing a one-pound catfish on a commercial basis is about 35 cents, of which 10 or 15 cents is spent for feed.

The problems of intensive culturing are many, some of which cannot be handled adequately in a small living unit or backyard. If, for example, you have only a limited supply of electricity, it might be impossible to harvest and preserve the fish all at once. This limitation would mean feeding a large population of nongrowing fish through the winter. There is also the matter of reproduction. Most small-scale catfish farmers find it more economical to buy fingerlings directly from hatcheries than to attempt the difficult task of raising their own. Large numbers of small fish do not compete well in the same pond with significant numbers of larger fish. And diseases can wipe out the farmer overnight. If you're still interested in catfish, it is best to start with a small pond and then expand as you learn more about it. This, in fact, is a good rule for any aquaculture operation you try. (For more information, see the listing under F. Meyer in the Bibliography; polyculture of catfish is discussed in the next section.)

Tilapia

Tilapia is a species of small tropical fish which is omnivorous, feeding happily on algae, plants, or insects. The New Alchemy Institute conducted an experiment where several individuals in different parts of the country attempted to raise *Tilapia* in small ponds; details are described in the *Backyard Fish Farm Workbook for 1973* (see listing under W.O. McLarney in the Bibliography), and a brief summary is presented here.

Tilapia require a water temperature above 60°F to survive; for breeding and maximum growth, the best temperature is in the mid-eighties. Consequently, most of

Table 7.9 A Suggested Feed Formula[a]	
Ingredients	Percent
Fish meal (menhaden)	12.0
Soybean meal	20.0
Blood meal	10.0
Distillers solubles	8.0
Rice bran	35.0
Rice by-products	10.0
Alfalfa meal	4.5
Vitamin premix	0.5
	100.0

Notes: a. Taken from F. Meyer, *Second Report to the Fish Farmer.*

the ponds used a plastic dome to retain heat and extend the growing season. The plans for making the dome are available for $5 from *Popular Science Magazine* (355 Lexington Avenue, New York 10017). The ponds described in the experiment were approximately twenty feet in diameter. (Commercial swimming pools can be used if it is not possible to dig a pond.) The earthen ponds were lined with heavy plastic sheets to prevent water leakage. Ordinary tap water can be used but, prior to the addition of fish, it must stand for a day or two to permit dissipation of the chlorine residue.

The ponds were fertilized with manure to induce algal growth. The manure was not added directly to the pond, since this direct application would have made the water too turbid. Instead, it was placed in a burlap bag which then was placed into the pond. Another method of fertilizing involved the preparation of enriched "manure tea"; this stew was separated from the pond proper by a partition which only allowed the passage of liquids.

A pump and filter for recirculating the water is desirable because fish produce growth-inhibiting *metabolites* as waste products. These metabolites stunt the size of other fish of the same species. Commerical filters for small ponds are available, but they are expensive and use a lot of electricity. The New Alchemy report describes a filter in which the water was pumped first into a settling tank, next over broken oyster or clam shells (which removed the metabolites), and finally back into the pond.

The chief source of food for the experimental *Tilapia* was algae. The experimenters also fed them greens, soybeans, earthworms, insects attracted by ultraviolet light, and mosquito larvae. No attempt, it seemed, was made to be particularly quantitative about the amounts.

None of the participants in the New Alchemy project spent over $200. As an example of their results, one pond stocked 119 fish in May and then added 168 later in the season. Total weight of the fish stocked was 1319 grams or 2.85 pounds. In November, 316 fish were harvested, weighing a total of 6140 grams or 13.5 pounds, a total weight increase of about 360 percent. Only twenty-five of the fish harvested, however, were large enough to eat. There are plans for a repeat experiment and better results are expected.

It would be feasible for a small group with limited space to attempt to raise *Tilapia,* but, as of this writing, it is illegal to import *Tilapia* into California except by special permit for exhibition purposes. You should check with your own state fish and game agencies about local regulations. They might possibly make an exception for a research project.

One promising polyculture procedure is the combination of catfish with *Tilapia*. One discouraging result of the New Alchemy *Tilapia* experiment was the large number of small fish in their harvest. The catfish could use these small fish as a food source and, at the same time, control their population. The *Tilapia*, in turn, could control pond fertility by consuming excess algae and waste matter. In an experiment at Auburn University, 500 *Tilapia* and 1800 catfish were stocked per acre. The result was 400 pounds more catfish per acre than from a pond stocked with catfish alone. On many fish farms, catfish and minnows are raised together. One drawback, however, is that catfish do not become active predators until they weigh a pound or more.

Carp

Carp (*Cyprinus carpio* and other species) are the most heavily cultivated fish in the world. They have a bad reputation in the United States, but it's not entirely the carps' fault. They were transplanted here during the last century by someone who wanted to culture carp in a country that was then up to its gills in wild fish and had no demand for cultured carp. Ultimately, the unwanted carp were released into the wild, where in a few short years they reverted to their wild state and erased the tasty characteristics that had been established by centuries of careful breeding. However, their descendants still have many traits which make them desirable for culture; it also might be possible to get some of the untainted foreign species if reverted "natives" displease you.

Carp are tough fish, highly tolerant of temperature and water conditions; they spawn easily in captivity and can be cultured over most of the United States. Breeding carp spawn in the spring, when water temperature warms up to about 60°F. In the Far East, fiber mats are anchored just below the surface of the water and carp lay their eggs against the bottom side of these mats. In intensive-culture procedures, the eggs then are removed from the pond to be hatched and raised past the fry stage in separate containers. They feed on zooplankton, algae, and detritus, and do especially well in cages in sewage effluents.

In some parts of the world, subsistence (nonintensive) culture of carp is carried out; breeders and eggs are left alone in the pond and the new fry survive on naturally produced food. Here, as elsewhere, nonintensive yields are less than those of intensive culturing. Polyculture with a predator fish (trout, pike, perch, bass) to control the number of young carp can increase the growth of the survivors. In China, polyculture has produced incredible yields in small ponds. Different species of carp have been bred so that one species feeds on macroalgae, a second feeds on phytoplankton, another on zooplankton, and yet another on bottom detritus. In this highly efficient system, each species complements the others and none are competing. Yields can reach 7500 kg/ha-yr.

On the other hand, subsistence farming in Haiti produced 550 kg/ha-yr. Fingerlings were stocked which took from seven to nine months to reach maturity. With intensive culture in sewage effluent in Java, up to 1,000,000 kg/ha-yr have been produced, a very impressive production figure.

Raising carp, however, does present a few problems. For example, because the temperature fluctuates in the temperate zone, it's possible that carp can be induced to spawn too early and a late cold spell will kill the fry. Competing species usually make survival difficult for young carp. Also, carp eggs are particularly susceptible to fungus outbreaks, and the grown fish easily fall victim to diseases. As with *Tilapia*, metabolites produced by carp limit their own growth (these wastes can be eliminated by similar filtering systems, as well). In spite of any problems, carp culture, particularly a Chinese polyculture arrangement, is well worth considering.

Pike

The pike is a carnivorous, large game fish, highly prized by sport fishermen. Pike are cultivated commercially in the United States, to be released into the wild and then hooked by the fortunate angler.

Intensive culture is accomplished by catching breeders in the spring when the water reaches 50°F, stripping out and then mixing together the eggs and sperm. Fertilized eggs are hatched in jars, and these fry must be fed intensively with zooplankton and small fish.

In nonintensive culture, a number of fingerlings can be planted in a pond in spring and left to their own devices until you fish the survivors out in the fall or the next year (only 2 to 5 percent of the fry actually last until fall).

Common problems of pike culturing are the difficulty of producing fry, their high mortality, and (because they are further along in the food chain) their low production. Pike have a low conversion efficiency and produce only about 10 kg/ha-yr. In a polyculture arrangement, pike are suggested for use in controlling an out-of-hand carp population; they may or may not prove effective in this role.

Perch and Walleye

Perch and walleye are carnivorous fish of the same family. The former is commonly cultivated in Europe and the latter in the United States. Techniques are similar and both can be cultured in any region of the United States.

In intensive culture, the farmer builds a nest and induces spawning in the breeders by pituitary hormone injection. Eggs are hatched in jars and, after the fry consume their yolk sacs, they are fed zooplankton. While still in the fry stage, they can be stocked in a pond with carp.

Fertilizing the ponds of perch and walleye with garbage may present some problems. For example, a Minnesota walleye pond was fertilized with barnyard manure in the spring and a desirable cladoceran (a crustacean) bloom resulted. However, summer fertilization resulted only in algal scum growth and no cladocerans; the fish ran out of food. Other growers found that sheep manure and brewer's yeast applied together kept the cladocerans blooming through the summer. Yeast is expensive, however, and its role is not exactly clear. There also have been reports of perch grown in sewage effluent.

In nonintensive culture, a few males and females can be stocked together and left unattended; young fingerlings should appear by fall. Some growers remove the fish from the pond over the winter; it takes these growers about 150 kilograms of fish to feed 100 kilograms of perch. Another solution to the over-wintering problem is to dig the pond deep enough for them to survive by themselves (twelve to fifteen feet in at least one-quarter of the pond).

Production of perch and walleye is higher than that of pike. They are, however, more difficult to handle and the questions of maintaining a carnivorous food supply and over-wintering must be dealt with. These fish also are suggested for polyculture with carp, even though their effectiveness as predators is not clear.

The Air-breathing Labyrinth Fish

Labyrinth fish (family Anabantidae) are air-breathing fish which can crawl over land when their pond dries up. They have been cultivated successfully (both intensively and nonintensively) in Southeast Asia, usually in a polyculture system with carp. Labyrinth fish are omnivorous, feeding on zooplankton, insects, and plants. They also flourish in water which has a low oxygen concentration, and thus avert a major headache for growers who use small stagnant ponds. Production depends on the species, but usually averages a few hundred kg/ha-yr.

The difficulty with applying labyrinth fish culture to our local situation is that the fish are not native to the United States, which raises certain problems. An ecological objection is that introducing a foreign fish which could escape might upset the balance of our local ecosystems. There's also a likelihood that its introduction is illegal. Finally, breeders would be very difficult to find. Even if these obstacles could be overcome, the utilization of labyrinth fish is still in experimental stages. Culture of the gourami (a labyrinth variety) was attempted in France unsuccessfully, and thus doubt remains as to whether this tropical fish can survive in more temperate zones.

Eels

Eels are catadromous fishes; that is, they spawn at sea but the young mature inland in freshwater. They are cultivated primarily in Japan and Taiwan, where they are considered a delicacy.

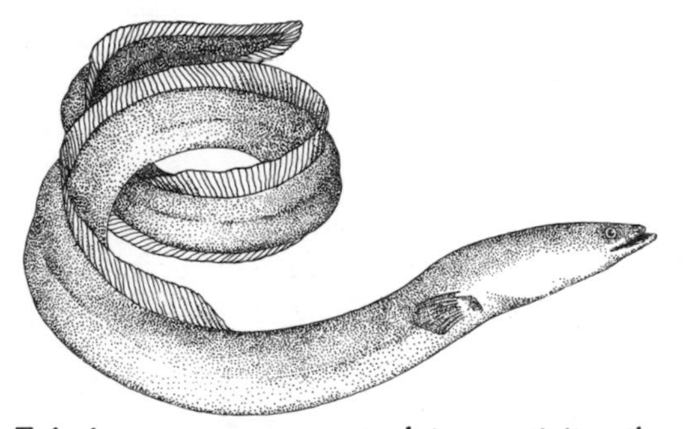

Eels have never spawned in captivity; the stock, therefore, must be replenished each year. Elvers (newly hatched eels) are caught with nets when they migrate to the mouth of a river and these captives are stocked in small ponds at a concentration of 600,000 per hectare for three months, after which they are thinned and then transferred to bigger ponds.

Eels are sloppy and inefficient feeders; they swim into cages filled with worms, minced fish, crushed mollusks, and animal guts, and thrash about in this food until a sufficient amount manages to get into their mouths. The food which they thrash out of the cage need not be wasted. Different species of carp can be stocked which feed on this "excess" and yet another species can be planted to feed on the plankton and algae which bloom after the excess food decomposes. With natural food, the protein conversion efficiency for eels is 6:1, which is very low; artificial food can raise this ratio to 2:1.

The best production is obtained in running water or in aerated and recirculated ponds. Optimum temperature is from 68 to 82°F. When polycultured with carp, total production averages 8000 kg/ha-yr. In well-developed running-water systems, production *averages* 26,000 kg/ha-yr (Japan). The European eel proves more difficult to culture because it is susceptible to diseases to which the Japanese species is resistant.

Drawbacks to the culture of eels include their inefficient protein conversion rate and the extremely high cost and quantity of their food. And, since they don't spawn in captivity, you must replenish your stock on a yearly basis. To make matters worse, elvers in America are only common on the Atlantic coast, and methods of capture have not been well developed.

Trout

These fish are among the most popular game fish in the world. A large proportion of the fish culture in the United States is devoted to trout, which are usually raised for the purpose of stocking fishing grounds.

Intensive culture is a laborious and difficult process. Trout are unusual in that they grow best in cold, sterile water and are more efficient in monoculture than polyculture. Trout won't spawn in captivity without help; they must be artificially fertilized by stripping eggs and sperm. Eggs are hatched in trays, and the fry are raised in shallow raceways and then moved to larger raceways as they grow; circular ponds and silos can also be used. In any case, great quantities of clean water must be available. Water temperatures are normally maintained between 50 and 65°F, and oxygen concentrations must be no less than 5 ppm (parts per million). Trout feed on invertebrates or artificial food. One-year-old trout can be polycultured as predators with carp, although they prefer cleaner water. Intensive culture can produce up to 60,000 kg/ha-yr under the best conditions.

It is also possible to raise trout nonintensively, in cold-water ponds. In general, such ponds exist only in the northern tier of states, in Canada, and on the Pacific Coast. To supply your family, a pond should be between one-half to one acre in area, with a degree of both inflow and outflow; at least a quarter of the pond should be from twelve to fifteen feet deep if it freezes in winter. Water temperature during summer *must* be between 50 and 70°F and culture should not be attempted where the water is warmer. The top six inches or so can exceed this limit of 70°F, but if the warm zone delves much deeper, there may not be enough oxygen to sustain the fish.

Trout rarely spawn in these ponds. Consequently, fingerlings (two to four inches) must be stocked at least every other year. Depending on the fertility of the pond, the stocking rate should be 300 to 600 fingerlings per acre; the cost would be only about $20 per acre. Restocking should take place in the fall or late summer, after most large fish have been harvested, to avoid any cannibalism. The pond should be able to produce enough food to sustain the trout; in cold or infertile ponds, supplementary feeding or fertilizing might be required. The danger of fertilization is that the decomposing organic matter uses up oxygen, a very critical factor for trout.

Fishing can begin as soon as the fish reach six or eight inches in length and must be done regularly or the pond will never reach its full carrying capacity. Production averages 50 kg/ha-yr without supplementary feeding.

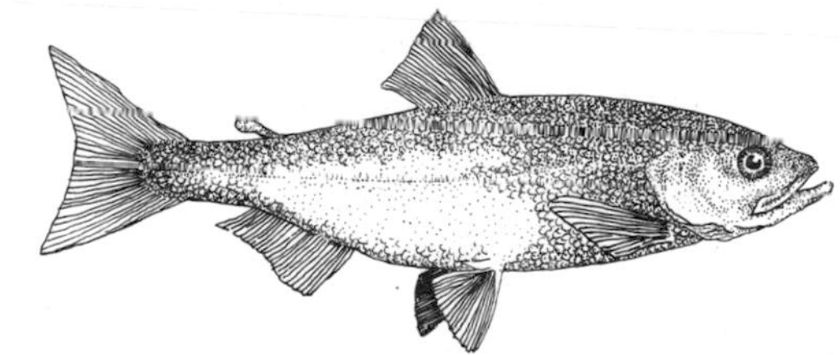

Salmon

Salmon belong to the same family as trout and, like them, are anadromous (they spawn in freshwater and mature at sea). Although trout *can* grow in freshwater, salmon *must* mature in brine.

Salmon are cultured intensively, if at all. Like trout, spawning is achieved by stripping; unlike trout, the parents always die after spawning. The fry are raised in raceways until they reach the smolt stage (about two years), after which they must be put out to sea. Fingerlings which have been grown experimentally in cages floating in the ocean are fed on pellet food for six to nine months and then harvested. Growth is rapid and this method shows some promise, but obviously it can be used only in selected areas.

Sturgeon

Sturgeon are huge anadromous fish which are highly prized in Russia for food and caviar (their eggs). They were once plentiful in the Great Lakes but were needlessly and unjustly destroyed for preying on "more desirable" species; today, they are cultured for the most part by fish and game agencies for the purpose of replenishing depleted sturgeon fisheries.

Sturgeon are difficult to spawn; they, too, require a pituitary hormone injection and are stripped by incision. The fry are kept in troughs, fed on cladocerans for ten days, and then put into ponds. Here they are fed *Daphnia* and oligochaetes for several weeks, and finally are released into brackish water. Up to 30 percent of the eggs survive through this stage.

A freshwater sturgeon is called a sterlet. These, of course, are not released into brackish water and can be cultured with carp. The laborious spawning process limits their use for food production. They also don't grow very rapidly; some sturgeon in your pond might make great conversation pieces, but don't count on them for food.

Crayfish

Crayfish are freshwater decapod crustaceans which are cultured mostly in Louisiana and France, but they have also been grown as far north as Wisconsin and Washington. Necessary for their culture is a flat-bottomed, shallow pond with some plant cover along the edges. This is flooded in May or June and two weeks later, adult crayfish are added. The crayfish pair up and dig burrows into the sides of the pond, where they mate. Two weeks after adding the crayfish, the pond is drained at a rate calculated to take about thirty days. During this period, the crayfish live in their burrows and forage for their young. The pond is reflooded in September. The young mature in six months, and harvesting can begin in November.

It is possible to integrate crayfish culture with a rice crop in a two-year cycle. Rice is planted in the spring and the field is flooded in May, stocked with crayfish, and then drained from July to August. The rice is harvested from December to June. The field then can be drained again and turned into pasture until the following spring. Crayfish production is from 400 to 700 kg/ha-yr in rice fields and less under other circumstances. American crayfish, fortunately, have no major disease problems. However, there may be a difficulty with deoxygeneration in the burrows.

Clams

Cultivation of marine clams in the Far East takes place by collecting larvae at sea and planting them in clam beds, where they set in and grow. There is little available material on freshwater clam culture, although it is known that some clam species require a fish host at one stage in their larval development. At Auburn University, a few clams of the species *Lampsilis claibornensis* were stocked in the polyculture with bluegill, redear sunfish, and bass, and then ignored until the annual harvest. The average yield for clams over six years was 1010 kg/ha-yr (318 kilograms without the shells); the fish contributed 464 kg/ha-yr. In a control pond without clams, only 317 kilograms of fish were harvested; clams may promote additional growth in a fish culture by their water-filtering capability. If freshwater clams are native to your area, there doesn't seem to be any harm in adding a few adult clams to a fish pond, and there may be a great deal to gain.

Sunfish and Largemouth Bass

Sunfish, primarily the bluegill (genus *Lepomis*), are excellent food and game fish which can tolerate a wide temperature range. While in some fish cultures adults are reluctant to breed, the major problem with sunfish is that they breed too easily. These mad propagators eat invertebrates and some algae, and their high reproductive rate, if unchecked, causes competition and depletion of food sources; the culturist will end up with many small fish, no one of which can make up a meal.

Reproductive excesses of sunfish can be attacked in various ways. It is possible to breed hybrids which, when crossed, will produce sterile offspring or offspring of only one sex. Another technique is to raise fish in floating cages so that the eggs fall through the mesh; the parents are unable to care for the young, which then die. These methods seem to be both wasteful of life and nonproductive; they also don't encourage a self-sustaining fish population. A better way to control reproduction is to polyculture sunfish with a predator such as the largemouth bass, which eats enough young sunfish to control their numbers and which itself can be harvested for food.

The largemouth bass is an excellent game fish of large size, rated as one of the most intelligent fish in the country. Culturists cannot force it to spawn at different times of the year, even with hormone injections; but when left alone in a pond during the spring, a pair of bass is only too anxious to be accommodating. They are solely piscivorous (fish-eating) and are used to control populations of catfish, buffalofish, bluegill, redear sunfish, and carp. Largemouth bass are, in fact, ideal for nonintensive polyculture with a prolific breeder like bluegill.

The Farm Pond Program

The nonintensive polyculture systems mentioned above are often used in ponds developed under the farm pond program, a scheme initiated by the government to conserve water and wildlife, which reached its peak popularity in the 1950s, but which is still widely used. Under this program, federal or state agencies supply and stock noncommerical ponds one time with bass and bluegill fingerlings, *free of charge*. Sunfish or catfish are legal substitutes. Maintenance and labor is minimal; after a few months, you merely face the challenge of catching your own pan-sized fish!

These ponds are considered "recreational" and, while they can't produce enough for commercial purposes, they can contribute significantly to the food supply for a family living unit. For the purposes of a small self-sufficient living group in temperate climates, a bass-bluegill farm pond culture is considered the best combination. Production figures of 250 to 450 kilograms per hectare have appeared, but it has not been made clear whether or not this is a sustained yield.

Maintenance is minimal once stocking is completed, but there are several problems. A pond of this nature has to be fished regularly or it will become overstocked. Poor pond design has often resulted in too rapid a flow rate or floods which wash out the fish. Weeds, while not too difficult to control, may require some attention.

Constructing the Farm Pond

A farm pond is not a backyard undertaking; an area larger than a quarter of an acre is necessary to produce sufficient fish biomass. Details of pond construction outlined here are also applicable to any other type of aquaculture that uses ponds.

The best type of site is a small valley or depression with steep sides and gradually sloping floors. The steep sides help to assure fairly deep water at the edge of the pond, which is useful in weed control. One mistake often made is selecting a site through which *too much* water flows. This surplus flow carries silt into and fish out of the pond. Depth should be from two to eight feet. Wells or springs are excellent water sources.

The pond can be either excavated or else constructed by building levees or dams around a natural depression. Construction is preferable since drainage can be accomplished without pumping. In general, the dam should be designed with a 2:1 slope on both sides. The width of the base should be about four times the height (plus the width of the top). The top of the dam can be about seven to twelve feet wide. Soil for this purpose should contain a large percentage of clay. In addition, to prevent water seepage, a clay core wall should be built beneath the dam. This is done by digging a trench four to ten feet wide down to "watertight" soil and then refilling it with a clay soil (see Figures 7.13 and 7.14). Of course, soil for the dam itself can be taken from the pond bed. If your dam is more than twelve feet high, you had better consult a civil engineer. In addition, don't forget to include a drain and maybe even a spillway. After construction, the dam tends to settle several inches. You also might check in Chapter 3 for further discussion of dams in terms of electricity. For visual attractiveness and for erosion control, plant your vegetation as soon as possible. Don't limit yourself to grass; consider all the possibilities—fruit or nut trees, vegetables, berries. . . .

The cost of a dam varies. Soil can be moved with a dragpan pulled by a team or tractor, or be pushed with bulldozers, or be hauled in trucks. If you own your own equipment, you can construct a pond fairly cheaply. Otherwise, a bulldozer operator will cost approximately $20 an hour to hire.

Weed Control

Weeds should be kept to a minimum; one suggested figure is 25 percent or less of the surface area. Since young bluegill can escape from bass by hiding in the weeds, excess vegetation tends to slow down the growth of bass and contributes to an overpopulation of competing small bluegill, few of which can grow to edible size. Shallow-water weeds such as cattails and marsh grass

Figure 7.13 A side view of a pond.

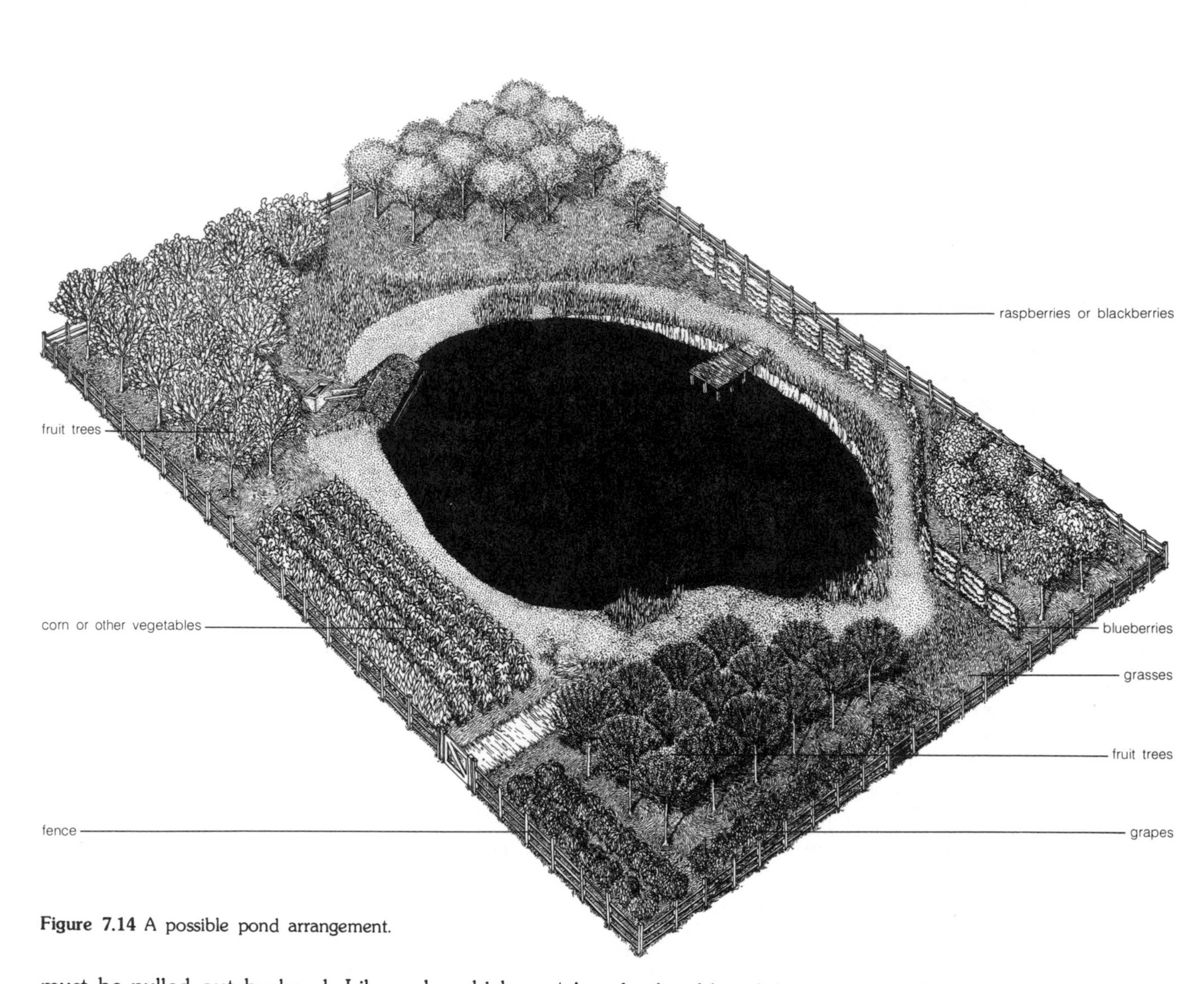

Figure 7.14 A possible pond arrangement.

must be pulled out by hand. Lily pods, which root in deep water but have leaves that rest on the surface, are best controlled by cutting the leaves; some plants also have leaves beneath the surface of the water, which, too, need to be cut. Floating plants such as duckweed must be raked out. Filamentous algae can be minimized by successfully controlling the other weeds. And any harvest of unwanted water weeds is an excellent addition to your compost pile.

Stocking and Fishing

In California, the Department of Fish and Game will provide free first-time stocking for any noncommercial pond (check your own state agency for local regulations). The fish should spawn soon after they are stocked. A sand and gravel bottom is best for spawning. If your pond has a mud bottom, good spawning grounds can be created by adding a few bushels of gravel at ten-foot intervals around the margin of the pond.

Stocking must be done correctly the first time around or the pond will never reach its maximum carrying capacity. Largemouth bass should be stocked at a rate of 50 to 100 fingerlings per acre, depending on the pond's fertility; bluegill fingerlings should be stocked at a rate of at least ten times that of the bass. Pond fertility for ponds which are not to be fertilized is based on the soil type. For example, forest soils support the least fertile ponds, light-colored soils are intermediate, and black-colored soils support the most fertile ponds. Table 7.10 summarizes the guidelines for stocking.

Stocking is done in the spring and, as we have mentioned, can be done free of charge by a federal or state agency. It is best not to fish for the first year, and only bluegill should be taken the second year. Bass do not reproduce until they are two years old; they should not be fished until June of the third year—that is, after their spawning season. Only bass over twelve inches long should be taken because overfishing small bass will result in a high concentration of small bluegill. Table 7.11 provides a guide for harvesting. It is important that fishing be done regularly; in spite of the greater attraction that larger bass have for fishermen, four pounds of bluegill should be taken for every pound of bass, or the population will fluctuate (see Figure 7.15).

The stability of the pond population can be measured either by seining and counting or by recording the progress of your fish catch. A desirable population has catches of both bluegill (at least six inches) and bass (one to two pounds). Any other yield indicates that one of the species is overcrowded. If you start to catch crappies, bullheads, carp, buffalofish, suckers, or green sunfish, then potential problems may develop since these fish compete with bass and sunfish for food and may also prey on their young. Carp are excellent fish for culture, but they don't contribute to a bass-bluegill system. They can be used, however, as substitutes for bluegill to form a bass-carp polyculture system.

Pond Fertility

To increase the carrying capacity of your pond,

Table 7.10 Fish Stocking for Unfertilized Ponds[a]

Fish Species	Black-colored Soils	Light-colored Soils	Forest Soils
Largemouth Bass	100	75	50
Bluegill	1000	750	500
Bluegill	700	500	350
Redear	300	250	150
Channel Catfish	100	75	50

Notes: a. Numbers indicate populations of fingerlings to be stocked per acre; from A.C. Lopinot, *Pond Fish and Fishing in Illinois*.

Table 7.11 Recommended Maximum Angling Harvests[a]

	Species of Fish					
	Largemouth Bass			Bluegill and/or Redear Sunfish		
Carrying Capacity of Pond (pounds per acre)	25	50	100	75	200	400
1st Year harvest						
Total number	None	None	None	None	None	None
or						
Total pounds	None	None	None	None	None	None
2nd Year harvest						
Total number	None	None	None	120[b]	320[b]	640[b]
or						
Total pounds	None	None	None	30	80	160
3rd Year harvest						
Total number	10	20	40	120[b]	320[b]	640[b]
or						
Total pounds	10	20	40	30	80	160
Each succeeding Year harvest						
Total number	10[c]	20[c]	40[c]	120[b]	320[b]	640[b]
or						
Total pounds	10	20	40	30	80	160

Notes: a. Based on pond size of 1 acre; from A.C. Lopinot, *Pond Fish and Fishing in Illinois*.
b. 6 inches and larger.
c. After quota is reached, all bass over 18 inches can be harvested.

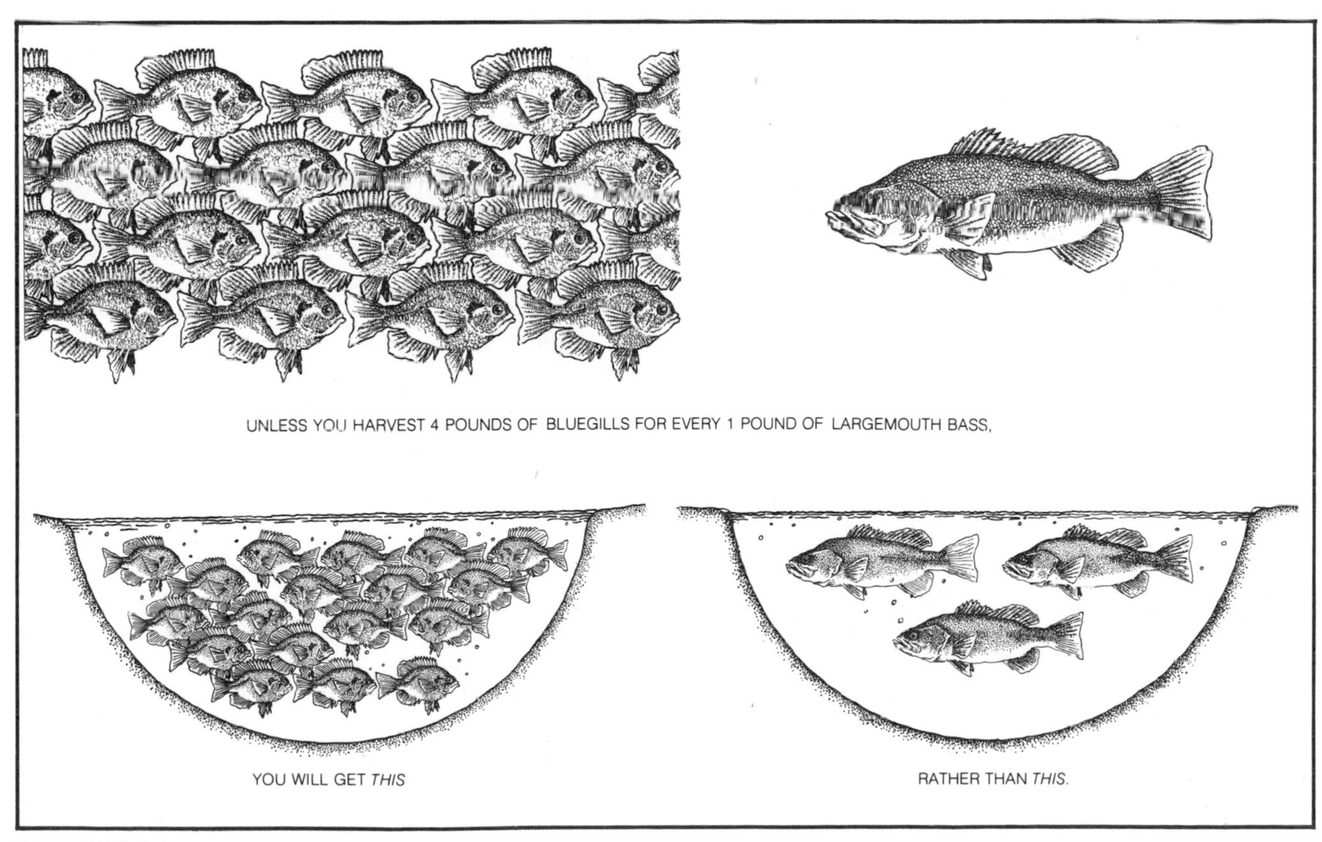

Figure 7.15 A harvesting pointer.

fertilizers for algal growth can be used. Since algae form the base of this aquatic food chain, in theory, more algae will yield more and larger fish. Fertilizing is a difficult technique to control, however, whether it be in the form of organic wastes or inorganic compunds. To repeat a few points made earlier, decomposition of organic waste (including organic fertilizer) uses up oxygen and this oxygen depletion may kill fish. Or fertilizing can overstimulate algae and aquatic plants; their subsequent decomposition also causes oxygen depletion and kills fish. Sometimes only a green algal scum (rather than a zooplankton bloom) results. An important point to remember is that carp are much more tolerant of both low oxygen concentration and high turbidity than bass or bluegill; consequently, fertilizer can be used with carp more successfully than with the latter two species.

Using a pump or some similar device to add oxygen to fertilized water would eliminate the problem of oxygen depletion, or so it would seem. Then an unlimited amount of fertilizer could be added safely and fish production would be correspondingly greater; but, in fact, such success does not follow automatically. As fish become overcrowded, they display a greater susceptibility to disease.

One safe and (potentially) highly productive way of fertilizing your pond is to add in a few native adult clams. If they flourish, they will be a significant source of food for the fish. Furthermore, their filter-feeding capabilities may improve your water quality. If the clams don't live, very little will have been lost. Further information on the farm pond program can be found in two publications listed in the Bibliography, by the California Department of Fish and Game and the U.S. Department of Agriculture, from which much of the above material was drawn.

Mariculture

The prospect of farming the oceans for food has been widely discussed for many years. Mariculture is best carried out in protected salt water, in a salt-water pond, slough, or protected bay. Tidal flow would be the most likely source of circulation, although pumps could be used in some cases. If circulation is adequate, a pond of only one-quarter of an acre is fine for most mariculture operations, some of which are so productive that we wonder why there isn't more research in this field. In the United States, there are legal restrictions on the use of public waters; this limitation may explain the lagging research, to some degree. By and large, mariculture practices require a combined sense of art, science, and experimentation, and the harvests should be considered useful food supplements rather than staples. Explore the possiblities if you live near the ocean.

Invertebrates

In the United States, more is known about culturing invertebrates than vertebrates. A difficulty with all mariculture, and particularly invertebrate culture, is that the animals undergo a sequence of different larval stages before maturity; each stage requires special (often unknown) environmental conditions. Nonetheless, we can discuss a few of the more prominent crops.

Crustaceans: Crustaceous animals grow by molting (shedding their exoskelton), a process which wastes a lot of energy; thus their food conversion efficiency is low. In most countries, shrimp culture consists of opening a sluice gate at high tide and letting the water (which hopefully contains shrimp) flow into a pond. The gate is then closed and any shrimp present are trapped. Captured shrimp grow in the pond until they are harvested. Yields of 300 to 800 kg/ha-yr are reported, but these yields can fluctuate dramatically. Only the kuruma shrimp of Japan is cultured intensively, and this culture involves very close attention to the water quality and the feeding at each stage of the shrimp's life cycle.

Malaysian prawns are often mentioned as a promising food source for culturing since they are durable and live in everything from brackish concentrations to freshwater. Unlike many invertebrates, they can be raised in captivity from egg to adulthood. For food, larval prawns require brine shrimp and fish eggs; older prawns eat small pieces of meat. Control of the temperature and salinity of the water is necessary for all life stages. As we will discuss later, Malaysian prawns can be polycultured with numerous types of herbivorous fish.

Bivalve Mollusks: Bivalves are filter feeders and require no additional feeding. They also are adept at concentrating pollutants in their biomass at toxic levels, so know the quality of your water. The two most popularly cultured bivalves are oysters and mussels. Site selection is vital in bivalve culture; they like areas with high algal productivity, strong tidal currents, and few violent waves.

Free-swimming oyster larvae metamorphose and settle on a solid surface. The young, settled oysters ("spat") then grow to maturity. The trick in oyster farming is to catch the spat, which requires that you be in the right place at the right time; your local government biologist can advise you on improving your chances. Spat are caught by putting out the shells of other bivalves as tempting solid surfaces on which the spat hopefully will settle. These shells are either piled on the bottom or suspended from rafts. Maintenance consists of checking for predators and cleaning off silt. Estimated yields are around 50,000 kg/ha-yr for rafts and 500 kg/ha-yr for bottom culture.

Mussel culture is similar to that of oysters, but the larvae are caught using rope rather than shells. This rope is spiraled around a stake driven into the bottom. Mussels sometimes get so heavy and crowded along it that they fall off; thinning is necessary periodically. Environmental requirements are the same as for oysters. Production in the Galician bays of Spain, where conditions are evidently perfect, reaches the astonishing figure of 600,000 kg/ha-yr. Unless conditions in your neighborhood are really perfect, don't expect yields like that; many other places, in fact, get very low yields.

Fish

Several different species of marine fish have been cultured commercially as a food source. Among these are yellowtail, Pacific mackerel, sardines, anchovies, milkfish, and mullet. The two major species are mullet and yellowtail, and a description of the practices by which they are currently raised adequately illustrates the status of the art. Neither species is native to America (and can't be cultured here), but the techniques used in their culture can be applied to native fish if you're enterprising.

Mullet: Mullet are brackish-water fish which also can survive in freshwater. Spawning can be induced only by injections of carp pituitary, and even if spawning is achieved, the fry seldom last a week. Culturing is still experimental and utterly dependent on getting advanced fry from fishermen. In Israel, these fry are planted in fertilized ponds with carp and *Tilapia*. The best yield from these ponds was 1155 kg/ha-yr, 44 percent of which were mullet. Culture in other countries is less complicated and consists of blocking off bays, letting the fish grow, and then harvesting.

Yellowtail: Yellowtail are pelagic fish of the Western Pacific which also cannot spawn in captivity. Larvae are picked up at sea and sold to fry specialists, who sell the fry in turn to culturists. In Japan, yellowtail are kept in floating nets and are fed reject shrimp and fish. Reported production is incredible, reaching 500,000 kg/ha-yr. The major obstacle to yellowtail culture in the United States is that they aren't available here; consequently, fry cannot be obtained. There are other likely pelagic species off American waters, but it's still an unresolved question whether or not these other crops can be grown in captivity.

Algae

Culturing algae (freshwater or marine) in an enclosure—for example, a small blocked bay or a backyard tank—can add another layer to the recycling operations of a small community. Algae can grow on the soluble nutrients in wastewater from your anaerobic digester. Not only can algae be used to remove certain pollutants from your water, but, when harvested, they can be used as either compost, fertilizer, animal feed, or fuel to cycle back into the digester. Such a scheme has been proposed by Golueke and Oswald in Berkeley (see Bibliography). Using algae as a tertiary treatment for secondary effluent of domestic wastewaters has also been proposed by J.H. Ryther and his coworkers. Details of the characteristics of the effluent from the digester, its use in algae production, factors affecting algal yields, and the biochemical characteristics of algae can be found in "Oxidation Ponds" (Chapter 5).

Experimental Aquaculture Systems

Several fairly complex experiments on recycling waste in an aquaculture system are now in progress. The goals are twofold: the elimination of potential pollutants from the environment and the generation of useful products. They are described here more as an illustration of possible future developments than as projects for you to tackle.

Tertiary Biological Treatment

A research group headed by John Ryther at Woods Hole Oceanographic Institution, Woods Hole, Massachusetts, is investigating the effectiveness of growing marine organisms on secondary effluent from domestic wastes. Planktonic algae have been found to remove all available nitrogen from mixtures of wastewater to seawater in the range of 1:5 to 1:20. The algae are then fed to shellfish in a second tank; under optimal operating conditions, 85 percent of the algae are removed by the oysters. The water from the oyster tank then flows into a third system containing macroalgae such as *Chondrus* and *Ulva*, which remove the nitrogenous wastes generated by the oysters. Ryther's group is also experimenting with macroalgae as food for abalone. The actual usefulness of such a system is going to depend upon further research to determine whether hazardous levels of viruses, microorganisms, heavy metals, and trace contaminants accumulate in the food products. Figure 7.16 schematically summarizes the proposed system.

The Ark

The New Alchemy Institute is experimenting with a greenhouse-enclosed, recirculating aquaculture system applicable for use by a small living group. Water flows by gravity through three ponds in sequence. The uppermost pond filters the water through crushed shells on which grows a slime layer of microorganisms. Here wastes are filtered by the shell and detoxified by the microorganisms. Ammonia is oxidized to nitrites and nitrates which are used by algae growing in a separate area of the first pond. The algae are fed to *Daphnia*, a crustacean growing in the second pond. Both Daphnia and algae are fed into the third pond in which *Tilapia* and catfish are grown in polyculture. The water is then recirculated to the first pond by means of a wind-driven pump; it also passes through a solar heater which restores optimum tempera-

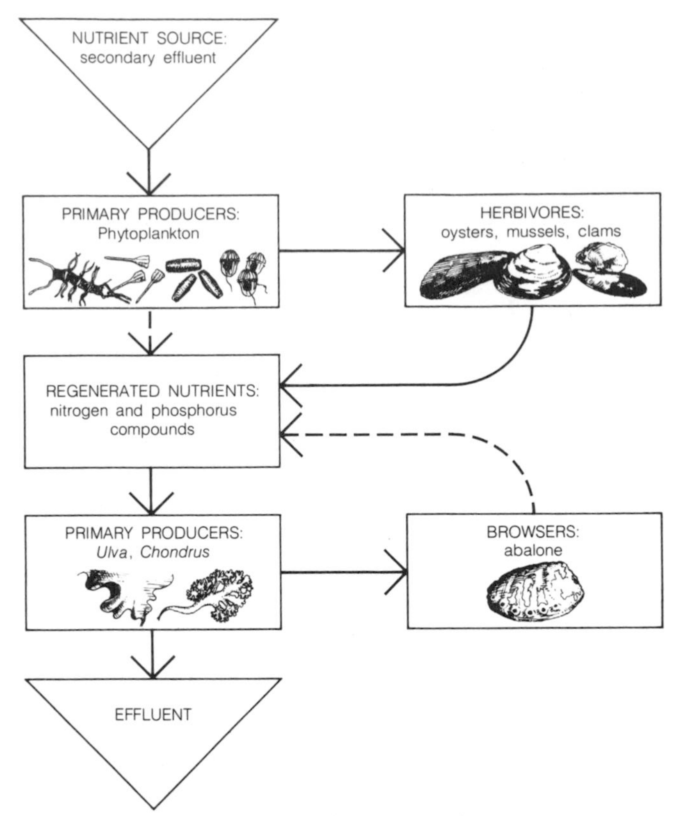

Figure 7.16 A tertiary biological treatment scheme.

tures. Costs have been estimated to be about $2300 for materials. Several modifications of the basic three-pond ecosystem are possible. One proposal is to grow vegetables in gravel hydroponic tanks fed by water from the fish pond. Another suggestion is a horizontal division of the fish tank by screens, to permit the layered culture of the great Malaysian prawn in the top section and carp in the bottom. Further information can be found in the *Energy Primer* (Merrill, et al.) and in an article by N. Wade, both listed in the Bibliography.

And so this chapter, like the whole book, is a beginning. Food, with all of its processes, possibilities, delightful mysteries, and ramifications, is a vast subject. You now have some basic information which you can expand through further reading and through the process of "growing your own." There are new avenues to be explored, avenues which will not unbalance any more of this earth than has already been thrown awry, which will not lead to consumption of any more resources than have already been depleted, which will not waste any more "waste." And what is considered "experimental" today may be common practice in a few years. So grow, harvest, and enjoy the fruits of your labor; "double dig" it!

Bibliography

Agriculture

Angier, B. 1972. *One Acre and Security*. New York: Vintage Books.

Bell, R.G. 1973. "The Role of Compost and Composting in Modern Agriculture." *Compost Science*, November–December 1973.

Bridwell, R. 1974. *Hydroponic Gardening*. Santa Barbara, Calif.: Woodbridge Press.

Cienciala, M. 1966. "Factors in the Maturing Process of Some Swiss Refuse Composts." International Research Group on Refuse Disposal, Bulletin no. 28. Washington, D.C.: U.S. Dept. of HEW.

Colby, B. 1972. *Dairy Goats: Breeding/Feeding/Management*. Amherst, Mass.: American Dairy Goat Association.

Commoner, B. 1971. *The Closing Circle*. New York: Alfred A. Knopf.

Edinger, P., ed. 1971. *Sunset Guide to Organic Gardening*. Menlo Park, Calif.: Lane Books.

Golueke, C.G. 1973. "Latest Methods in Composting and Recycling." *Compost Science*, July–August 1973.

Heck, A.F. 1931. "The Availability of the Nitrogen in Farm Manure under Field Conditions." *Soil Science* 31:467–81.

Heck, A.F. 1931. "Conservation and Availability of the Nitrogen in Farm Manure." *Soil Science* 31:335–63.

Howard, L.E. 1947. *Earth's Green Carpet*. Emmaus, Pa.: Rodale Press.

Jeavons, J. 1973. *The Lifegiving Biodynamic/French Intensive Method of Agriculture*. Palo Alto, Calif.: Ecology Action of the Midpeninsula.

———. 1974. *How to Grow More Vegetables*. Palo Alto, Calif.: Ecology Action of the Midpeninsula.

Jeris, J.S., and Regan, R.W. 1973. "Controlling Environmental Parameters for Optimum Composting." *Compost Science*, January–February 1973.

Jull, M. 1961. *Producing Eggs and Chickens with the Minimum of Purchased Feed*. Bulletin no. 16. Charlotte, Vermont: Garden Way Publishing.

Kaysing, B. 1971. *First-Time Farmer's Guide*. San Francisco: Straight Arrow Books.

Klein, G. 1947. *Starting Right with Poultry*. Charlotte, Vermont: Garden Way Publishing.

Merrill, R.; Misser, G.C.; Gage, T.; and Bukey, J. 1975. *Energy Primer*. Menlo Park, Calif.: Portola Institute.

Morse, R.A. 1972. *The Complete Guide to Beekeeping*. New York: Dutton Press.

Obrist, W. 1967. "Experiments with Window Composting of Cominuted Domestic Refuse." International Research Group on Refuse Disposal, Bulletin no. 29. Washington, D.C.: U.S. Dept. of HEW.

Odum, E.P. 1971. *Fundamentals of Ecology*. 2nd ed. Philadelphia: W.B. Saunders Company.

Philbrick, J., and Philbrick, H. 1971. *Gardening for Health and Nutrition*. New York: Rudolph Steiner Publications.

Poincelot, R.R. 1972. *The Biochemistry and Methodology of Composting*. New Haven, Conn.: Connecticut Agricultural Experiment Station.

Richardson, H.L., and Wang, Y. 1942. "Nitrogen Conservation of Night Soil in Central China." *Soil Science* 54:381–89.

Ried, G. 1948. *Practical Sanitation*. New York: C. Griffin.

Rodale, J.I. 1961. *How to Grow Vegetables and Fruits by the Organic Method*. Emmaus, Pa.: Rodale Press.

———. 1966. *The Complete Book of Composting*. Emmaus, Pa.: Rodale Press.

———. 1973. *The Encyclopedia of Organic Gardening*. Emmaus, Pa.: Rodale Press.

Rolle, G., and Orsanic, B. 1964. "New Method of Determining Decompostable and Resistant Organic Matter in Refuse and Refuse Compost." International Research Group on Refuse Disposal, Bulletin no. 21. Washington, D.C.: U.S. Dept. of HEW.

Root, A.I. 1973. *Starting Right with Bees: A Beginners Handbook in Beekeeping*. Medina, Ohio: A.I. Root Company.

Root, A.I., and Root, E.R. 1966. *The ABC and XYZ of Beekeeping*. Medina, Ohio: A.I. Root Company.

Rubins, E.J., and Bear, F.E. 1942. "Carbon-Nitrogen Ratios in Organic Fertilizer Materials in Relation to the Availability of Their Nitrogen." *Soil Science* 54:411–23.

Saunby, T. 1953. *Soilless Culture*. New York: Transatlantic Arts.

Sunset Magazine. 1972. "Getting Started with the Biodynamic-French Intensive Gardening." *Sunset*, September 1972, p. 168.

Ticquet, C.E. 1956. *Successful Gardening Without Soil*. New York: Chemical Publishing.

U.S. Department of Agriculture. 1966. *Raising Livestock on Small Farms*. Farmers Bulletin no. 2224. Washington, D.C.: Soil Conservation Service, U.S. Department of Agriculture.

———. 1971. *Beekeeping for Beginners*. Home and Garden Bulletin no. 158. Washington, D.C.: Soil Conservation Service, U.S. Department of Agriculture.

Aquaculture

Bardach, J.E.; Ryther, J.H.; and McLarney, W.O. 1972. *Aquaculture: The Farming and Husbandry of Freshwater and Marine Organisms*. New York: Wiley-Interscience.

Borrell, A., and Scheffer, P. 1966. *Trout in Farm and Ranch Farms*. Farmers Bulletin no. 2154. Washington, D.C.: Soil Conservation Service, U.S. Department of Agriculture.

California Department of Fish and Game. 1965. *The Fish Pond—How to Stock and Manage it*. Sacramento, California: California Department of Fish and Game.

Davis, H.S. 1970. *Culture and Disease of Game Fish*. Berkeley, Calif.: University of California Press.

Dobie, J., et al. 1956. *Raising Bait Fishes*. Circular 35. Washington, D.C.: Fish and Wildlife Service, U.S. Department of Interior.

Edminister, F.C. 1947. *Fish Ponds for the Farm*. New York: Charles Scribner.

Golueke, C., and Oswald, W. 1965. "Harvesting and Processing Sewage Grown Plankton Algae." *Journal of Water Pollution Control Federation* 37:471–98.

Jones, W. 1972. "How to Get Started." *American Fish Farmer*, December 1972.

Lopinot, A.C. 1972. *Pond Fish and Fishing in Illinois*. Fishing Bulletin no. 5. Springfield, Ill.: Illinois Department of Conservation.

McLarney, W.O. 1971. "Aquaculture on the Organic Farm and Homestead." *Organic Gardening and Farming*, August 1971, pp. 71–77.

———, ed. 1973. *The Backyard Fish Farm Workbook for 1973*. Woods Hole, Mass.: New Alchemy Institute.

Maloy, C., and Willoughby, H. 1967. *Marketable Channel Catfish in Ponds*. Resources Publication vol. 31, January 1967. Washington, D.C.: U.S. Department of Interior.

Meyer, F. 1973. *Second Report to the Fish Farmers*. Washington, D.C.: Fish and Wildlife Service, U.S. Department of Interior.

Ryther, J.H.; Dunstan, W.M.; Tonore, K.R.; and Huguenin, J.E. 1972. "Controlled Eutrophication—Increasing Food Production from the Sea by Recycling Human Wastes." *Bioscience* 22:144–52.

Ryther, J.H.; Tenore, K.R.; Dunstan, W.M.; Goldman, J.C.; Prince, J.S.; Vreeland, V.; Kerfoot, W.B.; Corwin, N.; Huguenin, J.E.; and Vaughn, J.M. 1972. "The Use of Flowering Biological Systems in Agriculture, Sewage Treatment, Pollution Assay and Food Chain Studies." Progress Report. NSF-RANN GI 32140. Woods Hole, Mass.: Woods Hole Oceanographic Institution.

U.S. Department of Agriculture. 1948. *Farm Fish Ponds*. Farmers Bulletin no. 1983. Washington, D.C.: Soil Conservation Service, U.S. Department of Agriculture.

Wade, N. 1975. "New Alchemy Institute: Search for an Alternative Agriculture." *Science* 187:727-29.

TABLE OF CONVERSION FACTORS

Because different conventions historically have been used to measure various quantities, the following tables have been compiled to sort out the different units. This first table identifies the units typically used for describing a particular quantity. For example, speed might be measured in "miles/hour".

Most quantities can also be described in terms of the following three basic dimensions:

length	L
mass	M
time	T

For example, speed is given in terms of length divided by time, which can be written as "L/T". This description, called "dimensional analysis", is useful in determining whether an equation is correct. The product of the dimensions on each side of the equal sign must match. For example:

$$\text{Distance} = \text{Speed} \times \text{Time}$$
$$L = L/T \times T$$

The dimension on the left side of the equal sign is length, L. On the right side of the equal sign, the product of L/T times T is L, which matches the left side of the equation.

The second table is a Conversion Table, showing how to convert from one set of units to another. It might be necessary to take the reciprocal of the conversion factor or to make more than one conversion to get the desired results.

Measured Quantities and Their Common Units

Length(L)	Area(L^2)	Volume(L^3)
mile(mi.)	sq. mile(mi^2)	gallon(gal.)
yard(yd.)	sq. yard(yd^2)	quart(qt.)
foot(ft.)	sq. foot(ft^2)	pint(pt.)
inch(in.)	sq. inch(in^2)	ounce(oz.)
fathom(fath.)	acre	cu. foot(ft^3)
kilometer(km.)	sq. kilometer(km^2)	cu. yard(yd^3)
meter(m.)	sq. meter(m^2)	cu. inch(in^3)
centimeter(cm.)	sq. centimeter(cm^2)	liter(l)
micron(μ)		cu. centimeter(cm^3)
angstrom(Å)		acre-foot
		cord(cd)
		cord-foot
		barrel(bbl.)

Mass(M)	Speed(L/T)	Flow Rate(L^3/T)
pound(lb.)	feet/minute (ft./min.)	cu. feet/min.
ton(short)	feet/sec.	cu. meter/min.
ton(long)	mile/hour	liters/sec.
ton(metric)	mile/min.	gallons/min.
gram(g.)	kilometer/hr.	gallons/sec.
kilogram(kg.)	kilometer/min.	
	kilometer/sec.	

Pressure($M/L/T^2$)	Energy(ML^2/T^2)	Power(ML^2/T^3)
atmosphere(atm.)	British thermal unit(Btu.)	Btu./min.
pounds/sq. inch(psi)	calories(cal.)	Btu./hour
inches of mercury	foot-pound	watt
cm. of mercury	joule	joule/sec.
feet of water	kilowatt-hour (kw-hr.)	cal./min.
	horsepower-hour (hp.-hr.)	horsepower(hp.)

Time(T)	Energy Density(M/T^2)	Power Density(M/T^3)
year(yr)	calories/sq. cm.	cal./sq. cm./min.
month	Btu./sq. foot	Btu./sq. foot/hr
day	langley	langley/min.
hour(hr.)	watthr./sq. foot	watt/sq. cm.
minute(min.)		
second(sec.)		

Table of Conversion Factors

MULTIPLY	BY	TO OBTAIN:
Acres	43560	Sq. feet
"	0.004047	Sq. kilometers
"	4047	Sq. meters
"	0.0015625	Sq. miles
"	4840	Sq. yards
Acre-feet	43560	Cu. feet
"	1233.5	Cu. meters
"	1613.3	Cu. yards
Angstroms(Å)	1×10^{-8}	Centimeters
"	3.937×10^{-9}	Inches
"	0.0001	Microns
Atmospheres(atm.)	76	Cm. of Hg(0°C)
"	1033.3	Cm. of H_2O(4°C)
"	33.8995	Ft. of H_2O(39.2°F)
"	29.92	In. of Hg(32°F)
"	14.696	Pounds/sq. inch(psi)
Barrels(petroleum, U.S.)(bbl.)	5.6146	Cu. feet
"	35	Gallons(Imperial)
"	42	Gallons(U.S.)
"	158.98	Liters
British Thermal Unit(Btu)	251.99	Calories, gm
"	777.649	Foot-pounds
"	0.00039275	Horsepower-hours
"	1054.35	Joules
"	0.000292875	Kilowatt-hours
"	1054.35	Watt-seconds
Btu/hr.	4.2	Calories/min.
"	777.65	Foot-pounds/hr.
"	0.0003927	Horsepower
"	0.000292875	Kilowatts
"	0.292875	Watts(or joule/sec.)
Btu/lb.	7.25×10^{-4}	Cal/gram
Btu/sq. ft.	0.271246	Calories/sq. cm. (or langleys)
"	0.292875	Watt-hour/sq. foot
Btu/sq. ft./hour	3.15×10^{-7}	Kilowatts/sq. meter
"	4.51×10^{-3}	Cal./sq. cm./min(or langleys/min)
"	3.15×10^{-8}	Watts/sq. cm.
Calories(cal.)	0.003968	Btu.
"	3.08596	Foot-pounds

"	1.55857 × 10^{-6}	Horsepower-hours
"	4.184	Joules (or watt-sec s)
"	1.1622 × 10^{-6}	Kilowatt-hours
Calories, food unit (Cal.)	1000	Calories
Calories/min.	0.003968	Btu/min.
"	0.06973	Watts
Calories/sq. cm.	3.68669	Btu/sq. ft.
"	1.0797	Watt-hr/sq. foot
Cal./sq. cm./min.	796320.	Btu/sq. foot/hr.
"	251.04	Watts/sq. cm.
Candle power (spherical)	12.566	Lumens
Centimeters(cm.)	0.032808	Feet
"	0.3937	Inches
"	0.01	Meters
"	10.000	Microns
Cm. of Hg(0°C)	0.0131579	Atmospheres
"	0.44605	Ft. of H_2O(4°C)
"	0.19337	Pounds/sq. inch
Cm. of H_2O(4°C)	0.0009678	Atmospheres
"	0.01422	Pounds/sq. inch
Cm./sec.	0.032808	Feet/sec.
"	0.022369	Miles/hr.
Cords	8	Cord-feet
"	128(or 4×4×8)	Cu. feet
Cu. centimeters	3.5314667	Cu. feet
"	0.06102	Cu. inches
"	1 × 10^{-6}	Cu. meters
"	0.001	Liters
"	0.0338	Ounces(U.S. fluid)
Cu. feet(ft.^{3})	0.02831685	Cu. meters
"	7.4805	Gallons(U.S., liq.)
"	28.31685	Liters
"	29.922	Quarts(U.S., liq.)
Cu. ft. of H_2O (60°F)	62.366	Pounds of H_2O
Cu. feet/min.	471.947	Cu. cm./sec.
Cu. inches(in.^{3})	16.387	Cu. cm.
"	0.0005787	Cu. feet
"	0.004329	Gallons(U.S., liq.)
"	0.5541	Ounces(U.S., fluid)
Cu. meters	1 × 10^6	Cu. centimeters
"	35.314667	Cu. feet
"	264.172	Gallons(U.S., liq.)
"	1000	Liters
Cu. yard	27	Cu. feet
"	0.76455	Cu. meters
"	201.97	Gallons(U.S., liq.)
Cubits	18	Inches
Fathoms	6	Feet
"	1.8288	Meters
Feet(ft.)	30.48	Centimeters
"	12	Inches
"	0.00018939	Miles(statute)
Feet of H_2O(4°C)	0.029499	Atmospheres
"	2.2419	Cm. of Hg(0°C)
"	0.433515	Pounds/sq. inch
Feet/min.	0.508	Centimeters/second
"	0.018288	Kilometers/hr.
"	0.0113636	Miles/hr.
Foot-candles	1	Lumens/sq. foot
Foot-pounds	0.001285	Btu.
"	0.324048	Calories
"	5.0505 × 10^{-7}	Horsepower-hours
"	3.76616 × 10^{-7}	Kilowatt-hours
Furlong	220	Yards
Gallons(U.S., dry)	1.163647	Gallons(U.S., liq.)
Gallons(U.S., liq.)	3785.4	Cu. centimeters
"	0.13368	Cu. feet
"	231	Cu. inches
"	0.0037854	Cu. meters
"	3.7854	Liters
"	8	Pints(U.S., liq.)
"	4	Quarts(U.S., liq.)
Gallons/min.	2.228 × 10^{-3}	Cu. feet/sec.
"	0.06308	Liters/sec.
Grams	0.035274	Ounces(avdp.)
"	0.002205	Pounds(avdp.)
Grams-cm.	9.3011 × 10^{-8}	Btu.
Grams/meter^2	3.98	Short ton/acre
"	8.92	lbs./acre
Horsepower	42.4356	Btu./min.
"	550	Foot-pounds/sec.
"	745.7	Watts
Horsepower-hrs.	2546.14	Btu.
"	641616	Calories
"	1.98 × 10^6	Foot-pounds
"	0.7457	Kilowatt-hours
Inches	2.54	Centimeters
"	0.83333	Feet
In. of Hg(32°F)	0.03342	Atmospheres
"	1.133	Feet of H_2O
"	0.4912	Pounds/sq. inch
In. of Water(4°C)	0.002458	Atmospheres
"	0.07355	In. of Mercury(32°F)
"	0.03613	Pounds/sq. inch
Joules	0.0009485	Btu.
"	0.73756	Foot-pounds
"	0.0002778	Watt-hours
"	1	Watt-sec.
Kilo calories/gram	1378.54	Btu/lb
Kilograms	2.2046	Pounds(avdp.)
Kilograms/hectare	.893	lbs/acre
Kilograms/hectare	.0004465	Short ton/acre
Kilometers	1000	Meters
"	0.62137	Miles(statute)
Kilometer/hr.	54.68	Feet/min.
Kilowatts	3414.43	Btu /hr.
"	737.56	Foot-pounds/sec.
"	1.34102	Horsepower
Kilowatt-hours	3414.43	Btu.
"	1.34102	Horsepower-hours
Knots	51.44	Centimeter/sec.
"	1	Mile(nautical)/hr.
"	1.15078	Miles(Statute)/hr.
Langleys	1	Calories/sq. cm.
Liters	1000	Cu. centimeters
"	0.0353	Cu. feet
"	0.2642	Gallons(U.S., liq.)
"	1.0567	Quarts(U.S., liq.)
Lbs./acre	.0005	Short ton/acre
Liters/min.	0.0353	Cu. feet/min.
"	0.2642	Gallons(U.S., liq.)/min.
Lumens	0.079577	Candle power(spherical)
Lumens(at 5550Å)	0.0014706	Watts
Meters	3.2808	Feet
"	39.37	Inches
"	1.0936	Yards
Meters/sec.	2.24	Miles/hr.
Micron	10000	Angstroms
"	0.0001	Centimeters
Miles(statute)	5280	Feet

"	1.6093	Kilometers
"	1760	Yards
Miles/hour	44.704	Centimeter/sec.
"	88	Feet/min.
"	1.6093	Kilometer/hr.
"	0.447	Meters/second
Milliliter	1	Cu. centimeter
Millimeter	0.1	Centimeter
Ounces(avdp.)	0.0625	Pounds(avdp.)
Ounces(U.S., liq.)	29.57	Cu. centimeters
"	1.8047	Cu. inches
"	0.0625(or 1/16)	Pint(U.S., liq.)
Pints(U.S., liq.)	473.18	Cu. centimeters
"	28.875	Cu. inches
"	0.5	Quarts(U.S., liq.)
Pounds(avdp.)	0.45359	Kilograms
"	16	Ounces(avdp.)
Pounds of water	0.01602	Cu. feet of water
"	0.1198	Gallons(U.S., liq.)
Pounds/acre	0.0005	Short ton/acre
Pounds/sq. inch	0.06805	Atmospheres
"	5.1715	Cm. of mercury(O°C)
"	27.6807	In. of water(39.2°F)
Quarts(U.S., liq.)	0.25	Gallons(U.S., liq.)
"	0.9463	Liters
"	32	Ounces(U.S., liq.)
"	2	Pints(U.S., liq.)
Radians	57.30	Degrees
Sq. centimeters	0.0010764	Sq. feet
"	0.1550	Sq. inches
Sq. feet	2.2957×10^{-5}	Acres
"	0.09290	Sq. meters
Sq. inches	6.4516	Sq. centimeters
"	0.006944	Sq. feet
Sq. kilometers	247.1	Acres
"	1.0764×10^{7}	Sq. feet
"	0.3861	Sq. miles
Sq. meters	10.7639	Sq. feet
"	1.196	Sq. yards
Sq. miles	640	Acres
"	2.788×10^{7}	Sq. feet
"	2.590	Sq. kilometers
Sq. yards	9(or 3×3)	Sq. feet
"	0.83613	Sq. meters
Tons, long	1016	Kilograms
"	2240	Pounds(avdp.)
Tons(metric)	1000	Kilograms
"	2204.6	Pounds(avdp.)
Tons, metric/hectare	0.446	Short ton/acre
Tons(short)	907.2	Kilograms
"	2000	Pounds(avdp.)
Watts	3.4144	Btu./hr.
"	0.05691	Btu./min.
"	14.34	Calories/min.
"	0.001341	Horsepower
"	1	Joule/sec.
Watts/sq. cm.	3172	Btu./sq. foot/hr.
Watt-hours	3.4144	Btu.
"	860.4	Calories
"	0.001341	Horsepower-hours
Yards	3	Feet
"	0.9144	Meters

INDEX

For additional information, see the listings of manufacturers and products and the Bibliographies following individual chapters.

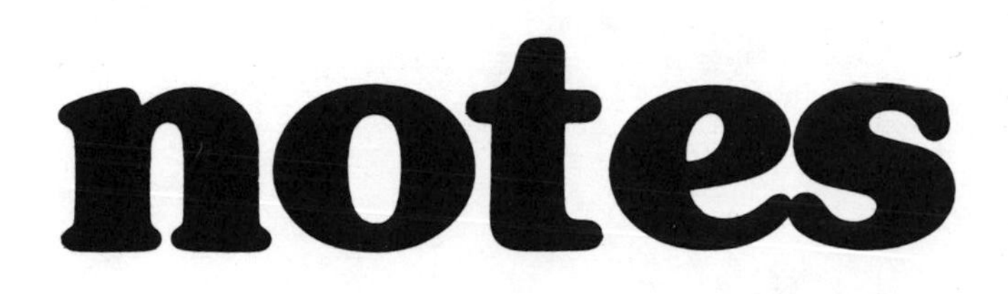
notes